Advanced Climate Simulation and Observation

Advanced Climate Simulation and Observation

Editors

Zengyun Hu
Xuguang Tang
Qinchuan Xin

MDPI • Basel • Beijing • Wuhan • Barcelona • Belgrade • Novi Sad • Cluj • Manchester

Editors

Zengyun Hu
Chinese Academy of Sciences
Urumqi, China

Xuguang Tang
Hangzhou Normal University
Hangzhou, China

Qinchuan Xin
Sun Yat-sen University
Guangzhou, China

Editorial Office
MDPI
St. Alban-Anlage 66
4052 Basel, Switzerland

This is a reprint of articles from the Special Issue published online in the open access journal *Atmosphere* (ISSN 2073-4433) (available at: https://www.mdpi.com/journal/atmosphere/special_issues/Advanced_Climate_SO).

For citation purposes, cite each article independently as indicated on the article page online and as indicated below:

Lastname, A.A.; Lastname, B.B. Article Title. *Journal Name* **Year**, *Volume Number*, Page Range.

ISBN 978-3-0365-9138-4 (Hbk)
ISBN 978-3-0365-9139-1 (PDF)
doi.org/10.3390/books978-3-0365-9139-1

Contents

atmosphere

Editorial

Advanced Climate Simulation and Observation

Zengyun Hu [1,2,3,*], Xuguang Tang [4] and Qinchuan Xin [3,5]

1 Xinjiang Institute of Ecology and Geography, Chinese Academy of Sciences, Urumqi 830011, China
2 Research Center for Ecology and Environment of Central Asia, Chinese Academy of Sciences, Urumqi 830011, China
3 University of Chinese Academy of Sciences, Beijing 101408, China
4 School of Geographical Sciences, Southwest University, Chongqing 400715, China
5 School of Geography and Planning, Sun Yat-sen University, Guangzhou 510275, China
* Correspondence: huzengyun@ms.xjb.ac.cn

Citation: Hu, Z.; Tang, X.; Xin, Q. Advanced Climate Simulation and Observation. *Atmosphere* **2023**, *14*, 364. https://doi.org/10.3390/atmos14020364

Received: 30 December 2022
Accepted: 9 February 2023
Published: 13 February 2023

Global climate changes, particularly extreme weather events, can directly or indirectly affect freshwater availability and food production, and cause disease outbreaks, floods and droughts. Therefore, there is an urgent and necessary need to develop advanced climate simulation and observation approaches and models, especially ones related to extreme climate events. Advanced climate simulations and observations can improve the accuracy of climate change predictions and long-term trends, which can mitigate the impacts of climate events on social and economic development, as well as human lives.

Under these conditions, this Special Issue entitled "Advanced Climate Simulation and Observation" aims to introduce advanced approaches in climate simulation and observation for use in various practical studies related to climate variations.

A total of 22 papers have been published in this Special Issue, with 8 original research articles reporting on climate change.

In their paper, Torsri et al. [1] evaluated the capability of the state-of-the-art atmospheric GCM of the Institute of Atmospheric Physics (IAP-AGCM) in simulating summer rainfall over Thailand by comparing the model's results with ground-truth observation during 1981–2012. It was found that the IAP climate model creditably reproduced the spatial patterns of the first three dominant modes of summer rainfall in Thailand, and that the correlation between the observed rainfall anomalies and the Niño 3.4 index could be reproduced through the use of the IAP model.

In order to study the sensitivity of meteorological factors in the western Tianshan Mountain region in China concerning different parameterization schemes of climate models, Cheng et al. [2] used the regional climate model RegCM4.5 to simulate the meteorological factor occurring in the western Tianshan Mountain region from 2012 to 2016, to investigate the effects of different cumulus convective schemes (Grell, Tiedtke and Emanuel). The results show that different combinations of cumulus convection schemes can improve the simulation performance of meteorological factors.

Based on Beijing's air quality index (AQI) and concentration changes of the six major pollutants from 2019 to 2021, Liu T et al. [3] through descriptive statistics visualized the results, and the air pollution status and influencing factors of Beijing's AQI were analyzed using the ARIMA model and neural network. The results show that PM2.5, PM10 and O3 of the six major pollutants had the greatest impact on the AQI. Meanwhile, the forecast effect of the neural network model was better than that of the ARIMA model.

Chen et al. [4] proposed a weather radar nowcasting method based on the temporal and spatial generative adversarial network (TSGAN), which can obtain accurate forecast results, especially in terms of spatial details, by extracting spatial and temporal features, combining attention mechanisms, and using a dual-scale generator and a multiscale discriminator.

Using the advanced research version of the weather research and forecasting model (ARWv3) and a hydrostatic wind speed change equation, Liu et al. [5] assessed the effects of four CPSs on a 10 m wind speed simulation over mainland China in the summer of 2003. The sensitivity of the wind speed simulation to CPSs was found to be the highest in eastern and southern China, followed by the Tibetan Plateau and then northwest China. In addition, the main physical processes influencing wind speed varied greatly with subregions.

In their paper, Gulakhmadov A et al. [6] evaluated the applicability of three gridded datasets in different combinations against observational data for predicting the hydrology of the Upper Vakhsh River Basin (UVRB) in Central Asia. The water balance components were computed, the results were calibrated with the SUFI-2 approach using the calibration of the soil and water assessment tool model (SWAT–CUP) program and the performance of the model was evaluated. The simulation for the calibration, validation and overall scales showed an acceptable correlation between the observed and simulated monthly streamflow for all combination datasets.

Uwamahoro et al. [7] clarified the precipitation types in two selected catchments by verifying the influence of accumulated and maximum temperatures on snow melting using a separation algorithm of rain and snow that incorporated the temperatures. The novel snow-melting process utilizing the algorithm in the soil and water assessment tool model (SWAT) was also developed by considering the temperatures.

The paper by Al-Helal et al. [8] shows that a 3 m depth was optimal to bury EAHE pipes, where the ground temperature was 32 °C in the summer and 29 °C in the winter. These temperatures would provide a maximum cooling/heating capacity of 1000/890 MJ day^{-1} for each 1 m^3 of humid air exhausted from a greenhouse. If the EAHE pipes were to operate in a closed loop with a greenhouse, the condensation of water vapor in them would be impossible during the cooling process.

There are six papers in this Special Issue that reported on the impact of climate change on society and the economy.

The paper by Li et al. [9] examines the decoupling between carbon emissions per capita and HDI and the welfare output of carbon emissions by using data from 189 countries, from 1990 to 2019; it also decomposes the drivers of the decoupling index and carbon emission performance (CEP) in the example countries. The results show that most countries that achieved strong decoupling had a very high human development, while the worst case was that a few countries with an extremely low human development achieved strong decoupling.

Li et al. [10] used panel data from 236 prefecture-level cities in China from 2001 to 2012 to verify the impact of urban population agglomeration on haze pollution and its mechanism based on a spatial lag model. They found that China's urban haze pollution had a significant positive spatial spillover effect, and presented a spatial distribution state of high–high and low–low agglomeration.

In their paper, Li et al. [11] measured the green innovation efficiency of 30 provinces in China from 2009 to 2019 using the SBM (slack-based measure) of super efficiency based on the undesirable output. The results reveal that the green innovation efficiency of the 30 provinces showed a fluctuating upward trend, but that the differences among provinces were relatively significant.

Based on data of historical floods in 31 provinces and municipalities in China from 2006 to 2018, Chen et al. [12] compared five machine learning methods to predict direct economic losses. Among them, GBR performed the best, with a goodness-of-fit of 90%. The results of the data showed that, in China, provinces heavily reliant on agriculture suffered the most with the proportion of direct economic losses to provincial GDP exceeding 1‰.

With a set of panel data released from Hubei and Hunan provinces in China, Liu et al. [13] adopted the mediating effect model to explore the relationship between rural labor migration and air pollution caused by agricultural activity in China. They found that the increase in labor migration had intensified the comprehensive index of air pollution caused by agricultural activity by changing the supply of labor force in the agricultural sector, the budget line of rural residents, the scale of agricultural production and crop planting structure.

Abdul-Rahim et al. [14] investigated the environmental Kuznets curve (EKC) for haze in 31 cities and provinces across China using the spatial data for a period of 15 years, from 2000 to 2014. The results of the GWR model found the spatial variability of each variable and showed significant spatial heterogeneity in the EKC across regions.

Three papers reported on the impact of climate change on agriculture in this Special Issue.

Miao et al. [15] examined the spatiotemporal pattern of China's drought conditions and cropland exposure to droughts under global warming of 1.5 °C and 2 °C, along with the avoided impacts (as evaluated through the cropland exposure to droughts) when limiting global warming to 1.5 °C instead of 2 °C. The results suggest that drought conditions could be alleviated when the projected rise in mean global temperature is limited to 1.5 °C rather than 2.0 °C. In addition, the total cropland exposure to droughts across China exhibited an increasing trend in response to the 0.5 °C of additional global warming.

The paper by Shen et al. [16] used beans, the food crop with the largest supply and demand gap in China, as the research object, and established a panel spatial error model consisting of multiple indicators of four factors, including the climate environment, economic market, human planting behavior and technical development level of 25 provinces in China, from 2005 to 2019, to explore the impact of climate environmental changes on the yields of beans.

Tao et al. [17] conducted [CO_2] (ambient and enriched up to 500 μmol moL^{-1}) and temperature (ambient and increased by 1.5~2.0 °C)-controlled experiments from 2015 to 2017, as well as in 2020 in two free-air CO_2 enrichment (FACE) sites. They provide evidence that SPAD readings are significantly linearly correlated with the rice leaf chlorophyll a + b content (chl a + b) and N content, while the relationships are profoundly affected by elevated [CO_2] and warming.

The impact of climate change on human health was studied in five articles of this Special Issue.

In their paper, Zhang et al. [18] investigated the impact of air pollutants on the respiratory system and its action mechanism by using information on inpatients with respiratory diseases from two IIIA (highest) hospitals in Wuhan from 2015 to 2019, information on air pollutants and meteorological data, as well as relevant demographic and economic data in China. According to the findings, the economic losses caused by PM2.5, PM10, SO_2, NO_2 and CO exposure totaled USD 454.46 billion, or, approximately, 0.20% of Wuhan's GDP in 2019.

In the study by Hao et al. [19], 26 environmental variables, namely, climatic, geographical and 2 socioeconomic indicators, were collected from regions where MT-ZVL patients were detected during the period from 2019 to 2021, with the aim of creating 10 ecological niche models. They found multiple ensemble ecological niche models based on climatic and environmental variables to be effective at predicting the transmission risk of MT-ZVL in China.

Based on the core collection of the Web of Science and CNKI databases, Gao et al. [20] used CiteSpace software to draw and comment on maps of Chinese and English keywords, publishing times, authors, countries and research institutions concerning the relationship between air pollution and public health. The results point out that the number of studies on the relationship between air pollution and health had increased year by year. Meanwhile, the three areas of sustained pollution exposure, indirect consequences of negative health effects of air pollution and air pollution and climate change may be the future focus of the field.

Du et al. [21] used the 2018 China Health and Retirement Longitudinal Study (CHARLS) project database for their paper. The multivariate linear regression analysis and binomial logistic regression model were applied to detect the impact of the subjective evaluation of air quality on QOL. The results show that there is a significant positive correlation between the subjective evaluation of air quality and the two dimensions of QOL.

In the paper by Gao et al. [22], a penalized distributed lag nonlinear model was applied to explore the influence of meteorological factors on the PTB incidence in Xinjiang from

2004 to 2019. Moreover, they firstly used a comprehensive index (apparent temperature, AT) to access the impact of multiple meteorological factors on the incidence of PTB. Overall, it was indicated that environments with low air temperature, suitable relative humidity and wind speed were more conducive to the transmission of PTB, and low AT was significantly associated with an increased risk of PTB in Xinjiang.

Author Contributions: Conceptualization, Z.H. and X.T.; validation, Z.H., X.T. and Q.X.; writing—original draft preparation, Z.H. and X.T.; writing—review and editing, Z.H., X.T. and Q.X. All authors have read and agreed to the published version of the manuscript.

Funding: This study was supported by the National Natural Science Foundation of P.R. China (Grant 42230708, 41971386), the Alliance of International Science Organizations (Grant ANSO-CR-KP-2021-02), Western Scholars of the Chinese Academy of Sciences (2020-XBQNXZ-010), the National Youth Talent Project (E1190301), the Third Xinjiang Scientific Expedition Program (Grant2021xjkk1300), and the Shenzhen science and technology innovations Committee (JCYJ20210324101406019).

Acknowledgments: We thank Hao Zhang from the Xinjiang Institute of Ecology and Geography, Chinese Academy of Sciences for their assistance during this study.

Conflicts of Interest: The authors declare no conflict of interest.

References

1. Torsri, K.; Lin, Z.; Dike, V.N.; Zhang, H.; Wu, C.; Yu, Y. Simulation of Summer Rainfall in Thailand by IAP-AGCM4.1. *Atmosphere* **2022**, *13*, 805. [CrossRef]
2. Cheng, Q.; Li, F. Performance of RegCM4.5 in Simulating the Regional Climate of Western Tianshan Mountains in Xinjiang, China. *Atmosphere* **2021**, *12*, 1544. [CrossRef]
3. Liu, T.; You, S. Analysis and Forecast of Beijing's Air Quality Index Based on ARIMA Model and Neural Network Model. *Atmosphere* **2022**, *13*, 512. [CrossRef]
4. Chen, X.; Wang, M.; Wang, S.; Chen, Y.; Wang, R.; Zhao, C.; Hu, X. Weather Radar Nowcasting for Extreme Precipitation Prediction Based on the Temporal and Spatial Generative Adversarial Network. *Atmosphere* **2022**, *13*, 1291. [CrossRef]
5. Liu, S.-J.; Wang, M.; Yi, X.; Shao, S.-B.; Zheng, Y.-Q.; Zeng, X.-M. Assessing the Impact of Cumulus Parameterization Schemes on Simulated Summer Wind Speed over Mainland China. *Atmosphere* **2022**, *13*, 617. [CrossRef]
6. Gulakhmadov, A.; Chen, X.; Gulakhmadov, M.; Kobuliev, Z.; Gulahmadov, N.; Peng, J.; Li, Z.; Liu, T. Evaluation of the CRU TS3.1, APHRODITE_V1101, and CFSR Datasets in Assessing Water Balance Components in the Upper Vakhsh River Basin in Central Asia. *Atmosphere* **2021**, *12*, 1334. [CrossRef]
7. Uwamahoro, S.; Liu, T.; Nzabarinda, V.; Habumugisha, J.M.; Habumugisha, T.; Harerimana, B.; Bao, A. Modifications to Snow-Melting and Flooding Processes in the Hydrological Model—A Case Study in Issyk-Kul, Kyrgyzstan. *Atmosphere* **2021**, *12*, 1580. [CrossRef]
8. Al-Helal, I.; Alsadon, A.; Marey, S.; Ibrahim, A.; Shady, M.; Abdel-Ghany, A. Geothermal Energy Potential for Cooling/Heating Greenhouses in Hot Arid Regions. *Atmosphere* **2022**, *13*, 105. [CrossRef]
9. Li, D.; Shen, T.; Wei, X.; Li, J. Decomposition and Decoupling Analysis between HDI and Carbon Emissions. *Atmosphere* **2022**, *13*, 584. [CrossRef]
10. Li, X.; Zhou, M.; Zhang, W.; Yu, K.; Meng, X. Study on the Mechanism of Haze Pollution Affected by Urban Population Agglomeration. *Atmosphere* **2022**, *13*, 278. [CrossRef]
11. Li, S.; You, S.; Song, Z.; Zhang, L.; Liu, Y. Impacts of Climate and Environmental Change on Bean Cultivation in China. *Atmosphere* **2021**, *12*, 1591. [CrossRef]
12. Chen, A.; You, S.; Li, J.; Liu, H. The Economic Loss Prediction of Flooding Based on Machine Learning and the Input-Output Model. *Atmosphere* **2021**, *12*, 1448. [CrossRef]
13. Liu, Y.; You, S.; Li, N.; Fang, J.; Jia, J.; Li, X.; Ren, J. Study on the Agricultural Air Pollution Aggravated by the Rural Labor Migration. *Atmosphere* **2022**, *13*, 174. [CrossRef]
14. Abdul-Rahim, A.S.; Kim, Y.; Yue, L. Investigating Spatial Heterogeneity of the Environmental Kuznets Curve for Haze Pollution in China. *Atmosphere* **2022**, *13*, 806. [CrossRef]
15. Miao, L.; Zhang, J.; Kattel, G.R.; Liu, R. Increased Exposure of China's Cropland to Droughts under 1.5 °C and 2 °C Global Warming. *Atmosphere* **2022**, *13*, 1035. [CrossRef]
16. Shen, T.; Li, D.; Jin, Y.; Li, J. Impact of Environmental Regulation on Efficiency of Green Innovation in China. *Atmosphere* **2022**, *13*, 767. [CrossRef]
17. Tao, Y.; Zhang, J.; Song, L.; Cai, C.; Wang, D.; Wei, W.; Gu, X.; Yang, X.; Zhu, C. Projected Elevated [CO2] and Warming Result in Overestimation of SPAD-Based Rice Leaf Nitrogen Status for Nitrogen Management. *Atmosphere* **2021**, *12*, 1571. [CrossRef]
18. Zhang, H.; You, S.; Zhang, M.; Liu, D.; Wang, X.; Ren, J.; Yu, C. The Impact of Atmospheric Pollutants on Human Health and Economic Loss Assessment. *Atmosphere* **2021**, *12*, 1628. [CrossRef]

19. Hao, Y.; Luo, Z.; Zhao, J.; Gong, Y.; Li, Y.; Zhu, Z.; Tian, T.; Wang, Q.; Zhang, Y.; Zhou, Z.; et al. Transmission Risk Prediction and Evaluation of Mountain-Type Zoonotic Visceral Leishmaniasis in China Based on Climatic and Environmental Variables. *Atmosphere* **2022**, *13*, 964. [CrossRef]
20. Gao, Y.; You, S.; Xu, Y.; Wang, N. The Hot Topics, Frontiers and Trends about Research on the Relationship between Air Pollution and Public Health—Visual Analysis Based on Knowledge Map. *Atmosphere* **2022**, *13*, 892. [CrossRef]
21. Du, Y.; You, S.; Zhang, M.; Song, Z.; Liu, W.; Li, D. Analysis of Correlation between Quality of Life and Subjective Evaluation of Air Quality—Empirical Research Based on CHARLS 2018 Data. *Atmosphere* **2021**, *12*, 1551. [CrossRef]
22. Gao, C.; Wang, Y.; Hu, Z.; Jiao, H.; Wang, L. Study on the Associations between Meteorological Factors and the Incidence of Pulmonary Tuberculosis in Xinjiang, China. *Atmosphere* **2022**, *13*, 533. [CrossRef]

atmosphere

MDPI

Article

Simulation of Summer Rainfall in Thailand by IAP-AGCM4.1

Kritanai Torsri [1,2,3], Zhaohui Lin [1,2,*], Victor Nnamdi Dike [1,4], He Zhang [1], Chenglai Wu [1] and Yue Yu [1,5]

1 International Center for Climate and Environment Sciences, Institute of Atmospheric Physics, Chinese Academy of Sciences, Beijing 100029, China; kritanai@hii.or.th (K.T.); victor@mail.iap.ac.cn (V.N.D.); zhanghe@mail.iap.ac.cn (H.Z.); wuchenglai@mail.iap.ac.cn (C.W.); yuyue@mail.iap.ac.cn (Y.Y.)
2 College of Earth and Planetary Sciences, University of Chinese Academy of Sciences, Beijing 100049, China
3 Hydro-Informatics Institute, Ministry of Higher Education, Science, Research and Innovation, Bangkok 10900, Thailand
4 Energy, Climate, and Environment Science Group, Imo State Polytechnic Umuagwo, Owerri 464119, Nigeria
5 State Key Laboratory of Satellite Ocean Environment Dynamics, Second Institute of Oceanography, Ministry of Natural Resources, Hangzhou 310012, China
* Correspondence: lzh@mail.iap.ac.cn

Abstract: Thailand is located in the Southeast Asian region, where the summer rainfall exhibits strong interannual variability, and the successful simulation of rainfall variation in Thailand by current climate models remains a challenge. Therefore, this paper evaluates the capability of the state-of-the-art Atmospheric GCM of the Institute of Atmospheric Physics (IAP-AGCM) in simulating summer rainfall over Thailand by comparing the model's results with ground-truth observation during 1981–2012. Generally, the model shows a certain skill in reproducing the observed spatial distribution of the summer rainfall climatology and its interannual variability over Thailand, although the model underestimated both rainfall amount and its variability. Using Empirical Orthogonal Function (EOF) analysis, it is found that the IAP climate model reproduced creditably the spatial patterns of the first three dominant modes of summer rainfall in Thailand, whereas it underestimated the explained variance of the observed EOF-1 and overestimated the explained variance of the observed EOF-2 significantly. It was further found that the correlation between the observed rainfall anomalies in Thailand and the Niño3.4 index can be reproduced by the IAP model. However, the observed negative correlation is largely underestimated by the IAP climate model, and this could be the reason for the underestimation of explained variance of the EOF-1 by the IAP model. The evaluation results would be of great importance for further model improvement and thus potential application in seasonal prediction in the region.

Keywords: model evaluation; rainfall simulation; interannual variation; IAP-AGCM; Thailand

Citation: Torsri, K.; Lin, Z.; Dike, V.N.; Zhang, H.; Wu, C.; Yu, Y. Simulation of Summer Rainfall in Thailand by IAP-AGCM4.1. *Atmosphere* **2022**, *13*, 805. https://doi.org/10.3390/atmos13050805

Academic Editors: Zengyun Hu, Xuguang Tang and Qinchuan Xin

Received: 20 April 2022
Accepted: 13 May 2022
Published: 14 May 2022

1. Introduction

With the advances in scientific understanding and improvements in computing capabilities, the current general circulation models (GCMs) have involved many components of the Earth system and can be used for long-term simulations, ranging from seasons to decades, of historical climate and projection of future climate change [1]. These GCMs have been widely applied in climate studies and for seasonal climate forecasts from the global scale to the regional scale, and have been improved not only in spatial and vertical resolution but also in parameterizations, to obtain a better representation of the different processes within climate and earth system [2]. However, before adopting one specific model for the simulation and prediction of climate anomalies or hydro-meteorological disasters in any region, we need to first verify the model's performance in the region [3]. It is also essential to identify the model's systematic biases and the possible reasons for these biases, so that we can know the direction for further model improvement [4], as well as bias correction for better model applications.

Given the need for an enhanced understanding of rainfall variation and its impacts on societies and ecology, many studies have attempted to develop GCMs to provide this important information in advance in different time scales [1,2]. However, applying different climate models for a region could yield different results, as demonstrated by Li et al. [5], when evaluating the performance of GCMs in reproducing rainfall patterns over the Asian–Australian monsoon region. Following their results, it was found that climate models can generally reproduce the observed rainfall patterns over Southeast Asia (SEA) region, but still overestimated its intensity. It is also reported that rainfall climatology over the SEA region is still difficult to be simulated by an individual GCM, but can be improved when a multi-model ensemble approach is employed [6]. Furthermore, the skill of climate models in estimating rainfall over an area is also dependent upon the sea surface temperature (SST) specified during the model integration [7]. Besides, many efforts have been made to improve the model performance by increasing the GCM's resolution [8]. Based on the improved high-resolution GCM outputs (20–50 km) that participated in the Coupled Model Intercomparison Project Phase 6 (CMIP6) [9], it was found that monsoon onset and rainfall climatology over the SEA can be simulated better than that of its original resolution [10]. Moreover, applying a bias correction method to a GCM output can also yield more realistic results than its original output [11]. However, model bias in rainfall simulation exhibits spatial variations, depending on the area and model configuration [7,11–13]; therefore, a systematic model evaluation is important before applying the climate model in the study region.

In Thailand, summer rainfall exhibits strong interannual variation; hence, it is exposed to frequent floods and drought conditions, leading to adverse impacts on many sectors (e.g., agriculture and economy) [14,15]. For instance, the 2011 flood was particularly severe as a result of record-breaking rainfall extremes that caused huge economic losses, estimated at 30 billion USD [16]. In turn, Khadka et al. [17] showed that devastating drought events occur every 6 years in the northeastern part of Thailand, with a notable impact on the agricultural sector. Moreover, summer rainfall in Thailand is largely influenced by the interannual variation in SST anomalies (SSTA) over the tropical Pacific Ocean (ENSO events) [15,18], which serves as a predictor of summer rainfall in the country [19–21]. Nonetheless, strong ENSO events are linked to summer rainfall extremes, as in the case that triggered the 2011 devastating flood event over Thailand [19]. Therefore, monthly and seasonal climate predictions are important for guidance in managing risks in the water resources of the country. During the past decade, the capability of dynamical regional climate models in reproducing rainfall characteristics has also been examined for Thailand [22,23]. Based on the previous studies, the simulation of rainfall characteristics in Thailand has remained a challenge, which could be largely ascribed to the complexity of the topography of the country [22,23], in addition to the inability of the models to reproduce the driving impact of ENSO events on summer rainfall [24–26]. Meanwhile, the model performance in simulating the rainfall characteristics is shown to be spatially and seasonally different [23] and was also dependent on the model's physics parameterizations [22]. Hence, before applying any climate model for the climate simulation and prediction in a region, the behavior of the model in reproducing its historical climate variation must be evaluated first.

The atmospheric component of the Chinese Academy of Sciences–Earth System Model (CAS ESM), the IAP-AGCM (Atmospheric GCM of the Institute of Atmospheric Physics, Chinese Academy of Sciences), has been widely applied for climate simulation and prediction studies over different parts of the world [27–31]. For example, it is found that the model can reproduce the observed relationship between the summer rainfall anomalies in East Asia and the East Asian subtropical western jet [30], and has shown promising applications for extreme event simulation and prediction over mid-latitude regions, particularly in China [27,29]. For the tropical region, it also shows good skills in the simulation of temperature and rainfall variations in West Africa [28]. In an attempt to apply the IAP climate model for a climate simulation and future climate change projection study, it is imperative to understand whether the IAP model can reproduce the observed characteristics of summer rainfall over Thailand. Moreover, for the potential application of the IAP climate model in the seasonal prediction for disaster management in Thailand, it is very important to understand whether the IAP model can reproduce the observed relationship between ENSO and summer rainfall anomalies in the region, as ENSO has already been identified as the key driver and predictor for summer rainfall anomalies in Thailand [19–21,32]. Furthermore, it is also interesting to understand how the model's capability in reproducing the ENSO and summer rainfall relationship can be linked to the model's capability in simulating the observed rainfall variation in the country. The rest of the paper is organized as follows: a brief description of the IAP climate model, study region, data, and analytical method used are explained in Section 2; the results are presented in Section 3; and the discussion and conclusions are summarized in Section 4.

2. Model Description, Study Region, Data, and Methods

2.1. Model Description and Experimental Setup

The model used in this study is IAP-AGCM version 4.1 (IAP-AGCM4.1), with a horizontal resolution of about 1.4° × 1.4° and with 30 vertical levels, and the model top is at 2.2 hPa. Its dynamic core is formulated based on the transformed velocity as the control variable of air motion by the finite-difference method [33]. The model was originally developed based on a two-level atmospheric general climate model [34] and has been continually improved in its dynamic core, parametrizations, and adding a more realistic view of Earth's complexity in later versions [35–38].

The model uses a finite-difference scheme with a terrain-following sigma vertical coordinate [39]. The model grid system is built from a two-dimensional horizontal staggered Arakawa C-grid [40]. Formulation of the governing equations and the finite-difference schemes of the current IAP-AGCM version is based on the baroclinic primitive equations with subtraction of the standard stratification and conserves the total available energy, which is a summation of kinetic energy, the available potential energy, and the available surface potential energy rather than total energy. Compared to its previous version, IAP-AGCM4.1 incorporates the more advanced physics parameterizations from the Community Atmosphere Model version 5 (CAM5) physics packages. A general evaluation of IAP-AGCM4.1 was done by Zhang et al. [33] and the model shows a reasonable performance.

An Atmospheric Model Intercomparison Project (AMIP)-type global simulation with IAP-AGCM4.1 was performed for the 1978–2012 period, which is driven by observed SST and sea-ice from the Hadley Centre Sea Ice and Sea Surface Temperature dataset (HadISST) [41]. The greenhouse gas concentrations, anthropogenic aerosol, and precursor gas emissions from the Coupled Model Intercomparison Project Phase 5 (CMIP5) [1] were applied during the simulation, which varies from year to year. The simulation was initialized from a former 10-year AMIP-like simulation forced by the observed climatology of the SST and sea ice concentration. The last 32 years (1981–2012) of the model's results were used for the analysis, by discarding the first 3 years (1978–1980) as a spin-up period.

2.2. Study Region, Observation Data, and Model Evaluation Methods

Located in the SEA region, Thailand is significantly affected by the Asia monsoon system, with abundant rainfall recorded in the summer season, as well as a strong inter-annual variability of rainfall in the summer season. Meanwhile, the climate pattern over Thailand can be divided into five distinct sub-regions, namely, Central, Eastern, Northeastern, Northern, and Southern (see Figure 1), by which the first four sub-regions (i.e., the North, Northeast, Central, and East) may be aggregately called Upper Thailand [42].

As seen in Figure 1, the geography of Thailand is quite different, in which the Northern part is mostly characterized by hilly and mountainous terrain. Meanwhile, the Northeastern part of Thailand is dominated by a high land plain, generally called the northeast plateau. For the Central part, the sub-region is mostly a low-level large plain area, with a mountainous range extending from the northern sub-region to the western part of the Central sub-region. In the Eastern part, most of the sub-region is plain land and valleys with small hills in its northern, central, and eastern parts, while its southern and southwestern parts are adjacent to the Gulf of Thailand. Whereas the Southern part of Thailand is a peninsula with the Andaman Sea adjacent to its western part, and the South China Sea in its eastern part.

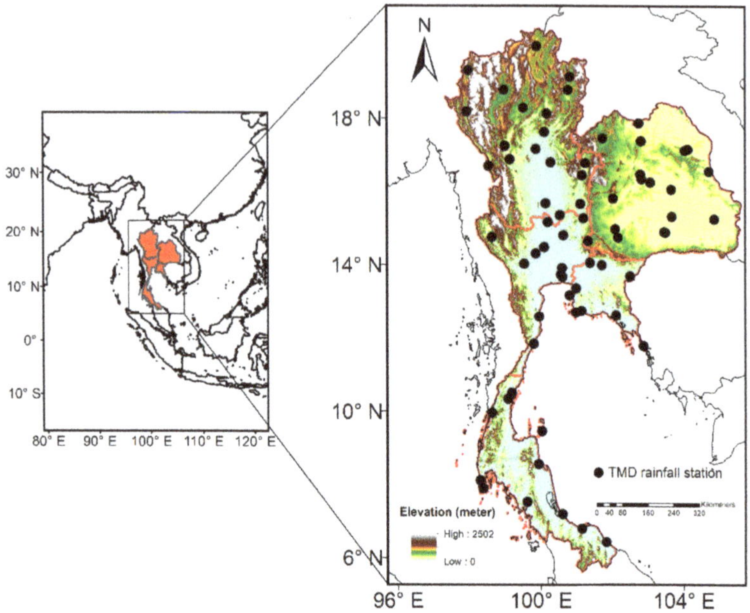

Figure 1. Spatial distribution of the Thai Meteorological Department's rainfall stations (black dots) used in this study. The red line represents the border of Thailand and its five sub-regions (i.e., the North, Northeast, Central, East, and South), superimposed on the terrain elevation obtained from the 30-m Shuttle Radar Topography Mission (SRTM) elevation data.

In this study, rain gauge observations from 69 stations of the Thai Meteorological Department (TMD), covering the period of 1981 to 2012, were used for model evaluation. Originally, we obtained daily rainfall datasets for 120 rain gauge stations from the TMD. However, after general quality checks (e.g., negative rainfall, missing values, and length of data record), only 69 stations' data have a long-historical record from 1981 to 2012. The selected 69 stations contain at least 80% of the observed daily rainfall (i.e., less than 20% missing values). The spatial distribution of the selected TMD stations is depicted in Figure 1. As seen, there is a relatively high density of TMD stations distributed across Thailand, consisting of 19 stations in the North, 18 stations in the Northeast, 12 stations in the Central, 9 stations in the East, and the rest in the South.

As summer is the rainy season in Thailand, the simulated summer (June–July–August; JJA) rainfall by the IAP climate model was examined. Besides the summer mean and the variability, we also investigated the model's capability in reproducing the spatial–temporal variation in the observed rainfall in Thailand using the Empirical Orthogonal Function (EOF) analysis, which is one of the most popular and widely used methods to characterize the important features of the climate variables [43–46]. Meanwhile, the temporal correlation coefficient (TCC), as well as pattern correlation coefficients (PCC) between the observation and the model simulation, were also computed for comparison, following Equations (1) and (2), respectively.

$$\mathrm{TCC} = \frac{1}{\mathrm{N}-1}\sum_{t=1}^{\mathrm{N}} \frac{(\mathrm{TMD}-\overline{\mathrm{TMD}})}{\mathrm{SD}_{\mathrm{TMD}}}\frac{(\mathrm{IAP}-\overline{\mathrm{IAP}})}{\mathrm{SD}_{\mathrm{IAP}}} \quad (1)$$

$$\text{PCC} = \frac{1}{\text{M}-1}\sum_{i=1}^{\text{M}}\frac{(\text{TMD}-\overline{\text{TMD}})}{\text{SD}_{\text{TMD}}}\frac{(\text{IAP}-\overline{\text{IAP}})}{\text{SD}_{\text{IAP}}} \quad (2)$$

where SD is the standard deviation of the observation (TMD) and simulation (IAP) over time (t) space for TCC (N is the number of years) and grid space (i) for PCC (M is the number of grid cells).

To quantify the error statistics between the observation and simulation, the monthly gridded rainfall based on the 69 TMD stations' data was constructed with a 0.5° × 0.5° latitude/longitude resolution, via an iterative objective analysis [47]. Meanwhile, the simulated monthly rainfall was bi-linearly interpolated into a 0.5°-grid resolution to facilitate its comparison with the observation.

Moreover, regional statistics are herein provided for the aforementioned five sub-regions of Thailand (see Figure 1). Note that to obtain a regional daily or monthly time series, daily or monthly values from all stations or the grid cells in a sub-region were firstly aggregated and averaged.

3. Results

3.1. Climatological Distribution of Summer Rainfall

Figure 2a shows the 32-year mean of the observed summer rainfall over Thailand. Essentially, the seasonal rainfall shows a strong spatial variation in the sub-regions. For instance, in Northern Thailand, the average rainfall intensity ranges between 2.0 and 10.0 mm day^{-1} in the summer season over the sub-region. However, a lower rainfall intensity (<4.0 mm day^{-1}) is observed in the narrow part of Central and a small part of Southern Thailand. Meanwhile, the highest intensity of the summer rainfall is found in Upper Thailand and some parts of the Southern sub-region (>10.0 mm day^{-1}). Nonetheless, Figure 2a further suggests that rainfall distribution in the center of the upper part of Thailand is largely homogeneous, with a magnitude of about 6.0 mm day^{-1}.

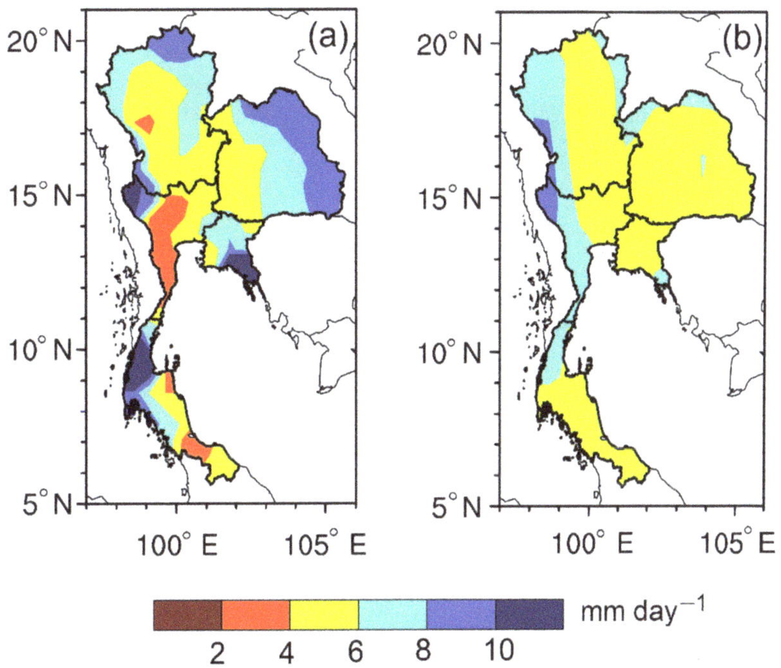

Figure 2. Spatial distribution of summer rainfall (mm day^{-1}) over Thailand averaged over 1981–2012: (**a**) observed; (**b**) simulated by the IAP AGCM4.1.

Furthermore, it is observed that during the summer season, the narrow end of Central Thailand experiences a drier condition relative to other parts of the sub-region, while a wetter condition is observed in the east, northeast, and western part of Central Thailand, and some parts of Southern Thailand. Notably, the foregoing demonstrates that the magnitude of rainfall in the Eastern sub-region and west coast of Southern Thailand are similar, perhaps because of their proximity to the Gulf of Thailand and the Andaman Sea, respectively. Meanwhile, a drier summer is observed on the southeast coast of Southern Thailand, with a rainfall intensity of less than 6.0 mm day^{-1}. Therefore, during the summer season, Southern Thailand is remarkably characterized by dry conditions on its east coast and wet conditions on its west coast area. This is consistent with the findings of Wang et al. [48], which also demonstrated that a high rainfall intensity occurs on the southwest coast during the summer season. The authors further suggest that, in July, heavy rainfall is centered northward around 10° N, which is associated with the seasonal migration of the intertropical convergence zone (ITCZ) [48].

Figure 2b depicts the spatial distribution of the 32-year mean of the simulated summer rainfall as simulated by the IAP model. Interestingly, similar to the observation, the model simulated summer rainfall with an intensity of 10.0 mm day^{-1}. As compared to the observation, the model reproduced largely the spatial patterns of rainfall distribution in the Northern part, Northeast, Central, East, and Southern parts of Thailand with PCC of 0.10, 0.07, 0.34, 0.44, and 0.33, respectively. This indicates that the spatial distribution of the simulated rainfall pattern suggests that the intensity varies depending on the sub-region (Figure 2b). In general, the model shows better performance in reproducing the spatial distribution pattern of observed rainfall intensity in areas where the rainfall intensity is 6.0 mm day^{-1}, especially on the Central and east coast of Southern Thailand. However, in some sub-regions, the model could not reproduce the magnitude of the observed rainfall intensity but the model captured, to a considerable extent, the magnitude and the spatial distribution of rainfall intensity in the drier and moderate sub-regions. In turn, it underestimated the magnitude of the rainfall intensity in the wetter sub-regions (see, Figure 2b).

Furthermore, we show in Figure 3 the overall performance of the model in reproducing the average summer rainfall in each sub-region. It was found that the model seems to underestimate the summer rainfall, with a mean bias of −1.24 mm day^{-1} (Figure 3a). However, the dry bias is seemingly apparent in the Eastern part of Thailand and other areas characterized by mountainous terrain (Figure 3c–f), with bias = −3.65, −1.45, −0.14, and −1.75 mm day^{-1} for the East, Northeast, North, and the South, respectively. Although the IAP model seems to overestimate and underestimate the summer rainfall intensity in the Central and Northern parts of Thailand, respectively, the magnitude of the difference between the model and the observation is relatively small, with a magnitude between 0.49 and −0.14 mm day^{-1}. Therefore, it can be inferred that the models performed better in simulating the magnitude of summer rainfall in the Central and Northern parts of Thailand. Meanwhile, we speculate that the model's gross underestimation of the summer rainfall intensity over the Eastern part of Thailand may be related to its proximity to the Gulf of Thailand, where local impacts induced by the sea breeze process could be an important factor controlling the local and sub-regional rainfall intensity [49–51]. Moreover, it has been noted that global climate models find it difficult to resolve rainfall processes in areas smaller than their grid cell, especially in coastal regions affected by a range of physical processes [52–54]. Interestingly, Feng et al. [55] noted that the considerable systematic underestimation of summer rainfall over Thailand is one of the limitations of GCMs. Despite a significant improvement in CMIP6, a satisfactory accuracy of the rainfall simulations over a region is still a challenge for the GCM models [12].

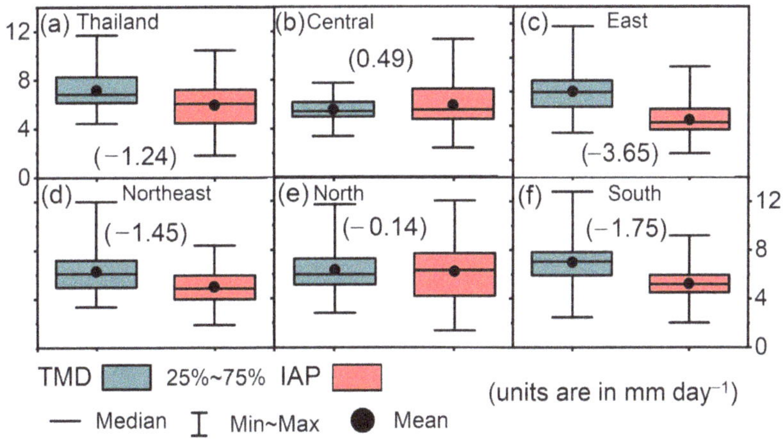

Figure 3. The box–whisker plots show the maximum, minimum, mean, lower quartile, upper quartile, and median of the observed and simulated rainfall over the 32-year period (1981–2012) for (**a**) Thailand, (**b**) Central, (**c**) East, (**d**) Northeast, (**e**) North, and (**f**) South. Numbers in parenthesis present the bias between the mean of the observed and simulated rainfall.

3.2. Distribution of Interannual Variability of Summer Rainfall

Figure 4a shows the observed standard deviation (SD) that is commonly used to represent the interannual variability in rainfall. Notably, it is observed that the summer season rainfall variability is higher than 2.0 mm day^{-1} in most parts of Thailand, except in the Central part where the variability of the summer rainfall is relatively small (SD < 2.0 mm day^{-1}). However, it was found that higher rainfall variability, with SD greater than 3.5 mm day^{-1}, occurs in the upper part of the Northern, the fringes of the Northeast, the narrow tip of the East, and the west coast of the Southern sub-regions. In addition, a high summer rainfall variability is also observed in the western part of Central Thailand close to the mountainous areas, adjacent to Myanmar, whereas on the east coast of Southern Thailand, a lower rainfall variability, with SD less than 2.0 mm day^{-1} is observed.

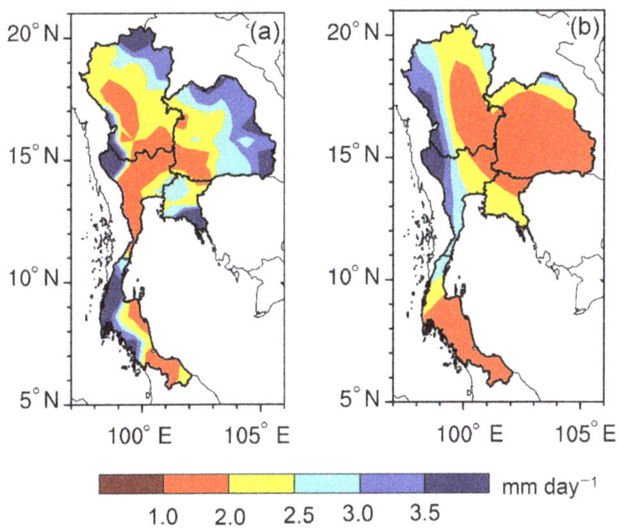

Figure 4. Spatial pattern of the standard deviation of summer rainfall (mm day^{-1}) over Thailand during 1981–2012: (**a**) observed; (**b**) simulated by IAP AGCM4.1.

As compared to the observation, the model's capability in reproducing the spatial pattern of the observed variability is also largely dependent on the sub-region (Figure 4b). Overall, the model shows better performance in reproducing the spatial pattern of the rainfall variability in Central Thailand and followed by the Eastern and Southern parts, with a PCC equal to 0.70, 0.64, and 0.59, respectively. Additionally, it also was found that the IAP model reproduced to a reasonable extent the spatial pattern of the observed rainfall variability in Northeastern Thailand, with the PCC = 0.47, and fails to capture the spatial pattern of the observed rainfall variability over Northern Thailand, with a PCC less than 0.05. It is imperative to note that the IAP model underestimated the magnitude of the summer rainfall variability in some parts of Thailand where the SD is greater than 3.0 mm day^{-1}, except in the western part of Central Thailand, where the simulated SD matches the observation. In fact, the model shows a good agreement with the observation in reproducing rainfall variability in some portions of Central, Northern, Northeast, and Southern Thailand. Moreover, the spatial pattern suggests that the model captured to a certain extent the distribution of the dry and wet zone on the east and west coasts of Southern Thailand.

3.3. Dominant Modes of Summer Rainfall and Their Variations

Next, we present comparatively the spatial–temporal variation of the simulated and observed dominant modes of summer rainfall over Thailand. Figure 5 shows the first three EOF modes of the observed summer rainfall and their corresponding IAP model simulated EOF modes. Essentially, Figure 5a–c indicates that the first EOF (EOF-1) and the second EOF (EOF-2) modes account for 23.1% and 20.7% of the total variance, respectively while 10.2% is accounted for by the third EOF (EOF-3). Moreover, it was found that the observed EOF-1 exhibits a dipolar structure, with a reverse signal in rainfall pattern in Central Thailand and elsewhere. It was also observed that the explained variances of the first two dominant modes are nearly equal, indicating these two dominant modes are equally important for rainfall variation patterns in Thailand. However, the EOF-2 exhibits a uniform pattern in most parts of Thailand. Whereas, the EOF-3 exhibits a dipolar mode with opposite rainfall signs in the northeast and elsewhere in Thailand.

Figure 5d indicates that the simulated summer rainfall EOF-1 mode is found to be relatively uniform and accounts for 76.7% of the total variance. Meanwhile, EOF-2 and EOF-3 exhibit a dipole structure that accounts only for 10.8%, and 3.4%, respectively, of the total variance (Figure 5d–f). As compared to the corresponding EOF mode of the observation, it can be found that the IAP model seems not to reproduce the spatial pattern of its corresponding observed EOF modes for the first two dominant patterns, with PCC = 0.05 for the EOF-1 and PCC = 0.15 for EOF-2. Remarkably, the spatial pattern of the simulated EOF-1 and observed EOF-2 show relatively similar patterns (see Figure 5b,d). This suggests that there is a shift in the simulated EOF models, such that the model reproduced the observed EOF-1 in its EOF-2 mode, as shown in Figure 5a,e, with the PCC = 0.25. Besides, the spatial pattern of the observed EOF-3 is similar to the simulated EOF-3 mode, with the PC = 0.22. Hence, we infer that the model can reproduce the spatial structures of the observed dominant EOF modes of summer rainfall over Thailand, albeit with the overestimation and underestimation of the observed explained variance of EOF-1 and EOF-2 respectively.

Furthermore, we examined the temporal variation in the observed dominant modes of the summer rainfall (Figure 6a–c; red lines). The PC-1 and PC-2 suggest that after the 1990s, the summer rainfall of Thailand exhibited strong interannual variation (Figure 6a,b), while PC-3 shows variation at a longer time scale (Figure 6c). Notably, the variation in SST anomalies (SSTA) in the tropical Pacific Ocean (ENSO event) plays an important role in controlling rainfall variability over Asia [56]. Hence, we further demonstrate the relationship between summer rainfall variation in Thailand and the ENSO index, the so-called Oceanic Niño Index (ONI) [57,58]. Meanwhile, the index is computed based on the variations in the 3-month running means of SST in the east-central tropical Pacific region defined as Niño3.4 [57].

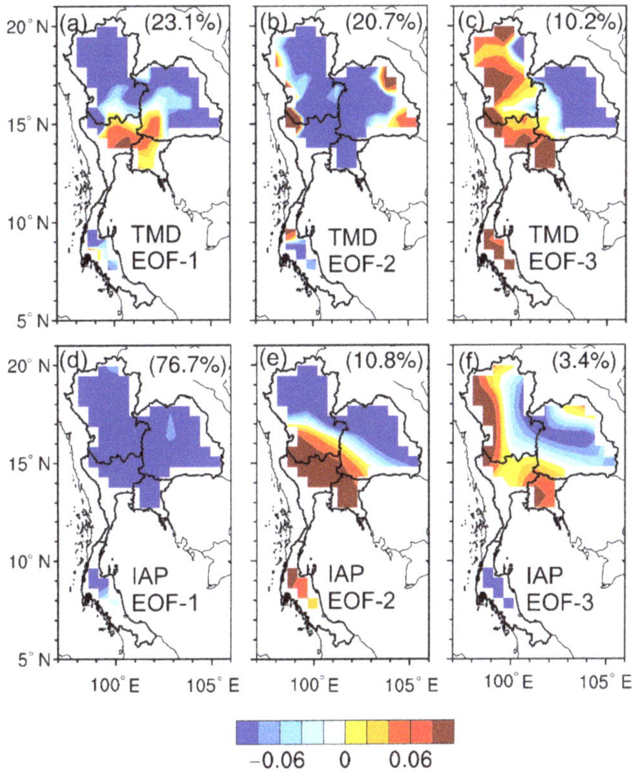

Figure 5. Spatial distribution of the first three dominant EOF modes of summer rainfall over Thailand during 1981–2012: (**a–c**) observed; (**d–f**) IAP-AGCM4.1 simulation. Numbers in parenthesis indicate the explained variance.

The results show that the first two dominant PC time series of the observed rainfall are significantly correlated with ENSO, such that the temporal variation of EOF-1 shows a positive correlation with SSTA in the Niño3.4 region, with the TCC = 0.31 (statistically significant at 90% confidence level), while the observed PC-2 variation shows a negative correlation with the ENSO index (TCC = −0.43, statistically significant at 95% confidence level). This suggests that SSTA over the Niño3.4 region influences summer rainfall variation in Thailand. Moreover, many studies have revealed the inverse relationship between ENSO and Asian summer monsoon rainfall variation, such that drier conditions are experienced in an ENSO warm phase and wet conditions are recorded as a response to the ENSO cold phase [56]. This interannual variation pattern is found between summer rainfall and the ENSO episodes over Thailand, consistent with the findings of [18]. Furthermore, this relationship seems to be more pronounced during extreme rainfall events in the country [18]. The foregoing shows that the first two dominant EOF modes of the summer rainfall and their PC time series are largely influenced by the interannual variation in SSTA in the Niño3.4 region.

As compared to the observation, the simulated PC time series of the first three EOF modes also show strong interannual variation. However, the PC time series did not match well with their corresponding observed PC modes, with the TCC = 0.18, 0.09, and 0.12 for PC-1, PC-2, and PC-3, respectively. As indicated in Figure 5, the observed EOF-1 and the simulated EOF-2 exhibit relatively similar patterns and vice versa for the observed

EOF-2 and the simulated EOF-1. Hence, we compared the TCC between the observed PC-1 and simulated PC-2 and vice versa. The results show that the observed PC-1 and the simulated PC-2 are similar, with the TCC equal to 0.25 (Figure 7a), which is higher than the TCC obtained when compared with their corresponding PC time series directly. A better relationship is obtained for the observed PC-2 and simulated PC-1, with the TCC equal to 0.44 (Figure 7b). This thus demonstrates that the model can reproduce the spatial and temporal pattern of the observed dominant EOF modes of summer rainfall over Thailand.

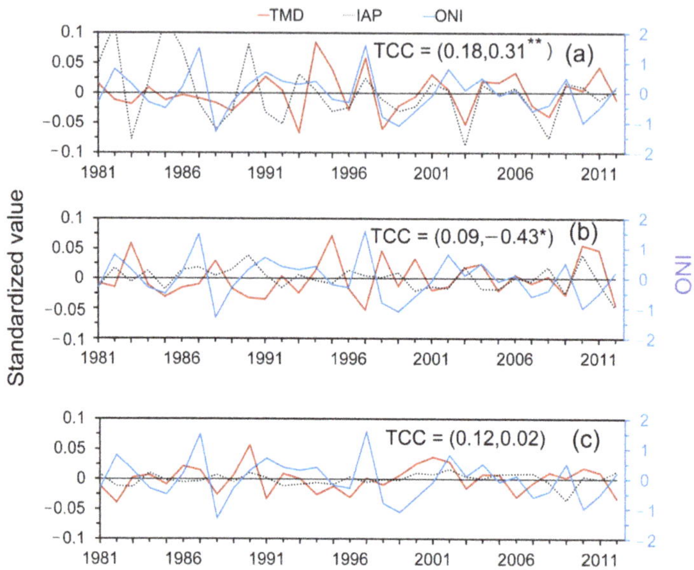

Figure 6. PC time series for the first three dominant modes of observed (solid red line) and IAP-AGCM simulated (Dotted black line) summer rainfall over Thailand during 1981–2012 for (**a**) EOF-1, (**b**) EOF-2, and (**c**) EOF-3, together with the ONI value (solid blue line). The first TCC value in parenthesis indicates the correlation coefficient between TMD and IAP, while the second value is for the coefficient between TMD and ONI. A single asterisk (*) indicates significance at $p < 0.05$, while double asterisks (**) is for significance at $p < 0.10$.

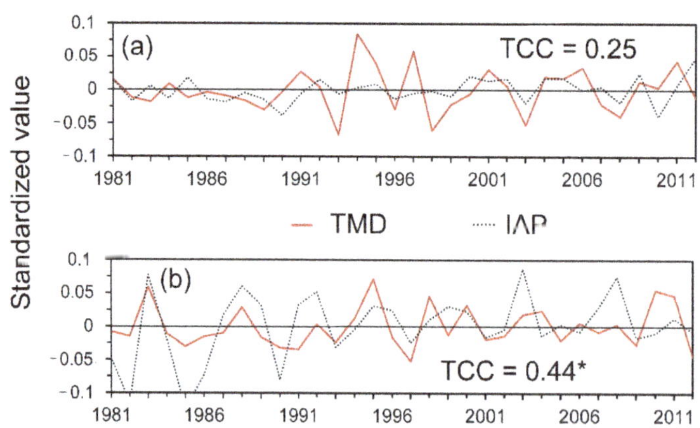

Figure 7. PC time series for shifting modes of the observed (solid red line) and IAP-AGCM simulated (dotted black line) summer rainfall over Thailand during 1981–2012 for (**a**) TMD PC-1 versus IAP PC-2 and (**b**) TMD PC-2 versus IAP PC-2. A single asterisk (*) at the TCC value indicates significance at $p < 0.05$.

3.4. Relationship between ENSO and Summer Rainfall over Thailand

It is suggested that there is a significant negative relationship between summer rainfall over Thailand and the variation in SSTA in the east-central tropical Pacific Ocean [18]. Given that the study used only a few rain-gauge stations located only in the Central part of Thailand, it is, therefore, fair to state that the spatial–temporal variation in summer rainfall over Thailand in response to the variability of ENSO is not well understood. Hence, we re-examined the relationship between ENSO and the variation in summer rainfall in each sub-region of Thailand, using the 69 TMD rainfall stations (see Figure 1). Furthermore, we also investigated whether the model simulation can be ascribed to the model's response to SSTA in the east-central tropical Pacific region. Hence, the Niño3.4 index was computed using observed SST data, which were also used to force the model simulation. The SST data were further used to calculate the relationship between the Niño3.4 index and summer rainfall anomalies in Thailand for both observation and model simulation. Note that the linear trend in the SST anomaly was removed before calculating the TCC between the rainfall time series and the Niño3.4 SSTA.

Figure 8a shows the spatial distribution of TCC between the Niño3.4 SSTA and observed summer rainfall anomalies in Thailand. It was found that indeed the rainfall variation in Thailand is negatively correlated with the ENSO in most parts of Thailand. Moreover, the results indicate that rainfall anomalies in upper Thailand are significantly and inversely correlated with the Niño3.4 index, with a TCC of about −0.4, especially in the Central, East, and Northern parts of Thailand. Besides the negative response of the rainfall to the ENSO signal, a positive correlation (TCC < 0.2) can also be found in the extremities of the Northeast and western part of Central Thailand, adjacent to Myanmar's border.

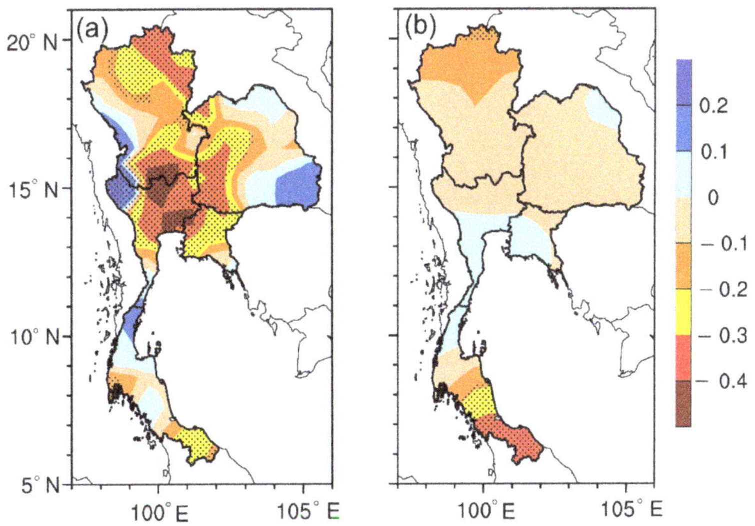

Figure 8. Temporal correlation coefficient (TCC) between (**a**) the detrended anomalies of the observed summer rainfall and sea surface temperature anomalies (SSTA) in the Niño3.4 region during the 1981–2012 period, and (**b**) the same, but for the simulated rainfall and the SSTA in the Niño3.4 region. Dotted regions indicate a significant correlation at the 90% level.

Interestingly, the model can well reproduce the spatial distribution of the observed relationship between the Niño3.4 index and summer rainfall over Thailand (Figure 8b). Although the spatial pattern of the observed relationship between the summer rainfall and Niño3.4 index is significant in most parts of Thailand (at 99% significance levels), the simulated summer rainfall response to the ENSO is relatively weak (at TCC < 0.2) in most parts of Thailand, but the negative response of rainfall anomalies at the tip of the Southern sub-region to the ENSO is well reproduced by the model. The weaker correlation between the summer rainfall anomalies and the ENSO signal in IAP-AGCM4.1 can be ascribed to the fact that atmosphere–ocean coupling process is not considered in the atmospheric general circulation model used in this study [28], and it is further suggested that this underestimation could be the main reason for the underestimation of the explained variance in EOF-1 by the IAP model.

4. Discussion and Conclusions

This study evaluates the performance of the IAP atmospheric general circulation model version 4.1 (IAP-AGCM4.1) in simulating summer rainfall variation over Thailand using the model's AMIP simulation results for the 1981–2012 period. Specifically, in comparison with the observed TMD station-gauge rainfall data, we focused on the model's capability in reproducing the observed summer rainfall distribution, interannual variabilities, and the spatial–temporal variation for the 32-year period. To facilitate comparison, the observed and simulated rainfall datasets were interpolated into the same spatial resolution, and regional statistics were considered for five of Thailand's sub-regions, namely, Central, East, Northeast, North, and Southern Thailand. Moreover, this study evaluated the similarity between the model and the observation using the temporal correlation coefficient (TCC) and pattern correlation coefficient (PCC). The spatial–temporal variation of the simulated rainfall was computed using the EOF analysis and then compared with the observed dominant modes of summer rainfall over Thailand. In addition, this study also examined the capability of IAP-AGCM4.1 in reproducing the observed summer rainfall response to the SSTA in the Niño3.4 region.

Results reveal that the model can reasonably reproduce the observed spatial distribution of summer rainfall over Thailand, with a better performance in the Northern and Central parts of Thailand. However, the model seems to underestimate the observed rainfall in some parts of Thailand, except in Central Thailand where the model overestimated the observed summer rainfall by 0.5 mm day^{-1}. Furthermore, it is also found that the IAP-AGCM4.1 can reproduce the observed spatial distribution of interannual variability in summer rainfall, with better performance in Central, Eastern, and Southern Thailand. However, the low performance of the model in the northern and east coasts of Thailand can also be found, which could be related to the inability of the model in resolving rainfall processes over mountainous terrain because of its horizontal resolution [59,60], especially in the high mountainous areas of the Northern sub-region and upland high plateau in the Northeastern sub-region. A higher resolution IAP model is expected to capture the spatial distribution of summer rainfall in these mountainous areas.

Furthermore, it was found that the observed EOF-1 and EOF-2 account for about 23.1% and 20.7% of the total EOF variance. As such, the EOF-1 and EOF-2 are the most important dominant modes of summer rainfall over Thailand. The spatial pattern of the observed EOF-1 exhibit a dipole mode with an opposite rainfall sign mostly in Central Thailand and elsewhere in Thailand, while the EOF-2 exhibits a uniform pattern, which is found to be the dominant mode that is significantly associated with the ENSO. Interestingly, the simulated EOF-1 exhibits a similar pattern to the observed EOF-2, and vice versa for the simulated EOF-2 and the observed EOF-1. Furthermore, the TCC between the observed PC-1 and simulated PC-2 as well as the observed PC-2 and the simulated PC-1 is higher than the TCC obtained when they are directly compared with their corresponding PC time series. Hence, we infer that the IAP model can reproduce the observed dominant EOF modes of summer rainfall over Thailand, albeit with some level of overestimation of the observed EOF-1 loading.

Based on observation datasets, it was found that the observed summer rainfall over Thailand is negatively associated with SSTA in the Niño3.4 region, with the highest significant correlation of about 0.4 (in absolute terms) found in the Northern, Central, some parts of the Northeastern, and the Eastern sub-regions. The IAP model also captured the spatial pattern of the summer rainfall response to the ENSO signal, but with a certain degree of underestimation when compared with the observation. It is further suggested that the simulated weak response of summer rainfall to the ENSO signal in the model could be the main reason for the underestimation of explained variance of EOF-1 by IAP-AGCM4.1, and later versions of the IAP-coupled climate system model may improve the model's performance over Thailand through a consideration of a robust atmosphere–ocean coupling processes. Therefore, further efforts are required to assess the performance of the IAP atmosphere–ocean coupled model in simulating summer rainfall variations over Thailand, as well as the associated air–sea interaction processes.

It is noteworthy that we only evaluated the relationship between ENSO and summer rainfall in Thailand as simulated in IAP-AGCM in this study. However, previous studies have shown that the SSTA over the Indian Ocean can also influence the summer rainfall anomalies over Thailand to a certain extent [19,32,61], so the model's simulated response of summer rainfall to the Indian Ocean SST anomalies is subject to further verification in future studies, which could be helpful for the further model improvement. Furthermore, besides the model evaluation method and metrics used in this study, many other comprehensive statistical metrics have already been proposed [62,63]. Further climate model assessment with a combination of current and newly developed metrics will be needed, and this will be of great importance for better application of the climate model simulations and thus better seasonal disaster predictions over Thailand and the Southeast Asia region.

Author Contributions: Conceptualization and methodology, Z.L. and K.T.; validation, K.T.; formal analysis, K.T.; data curation, Y.Y.; C.W. and H.Z.; writing—original draft preparation, K.T.; writing—review and editing, Z.L. and V.N.D. All authors have read and agreed to the published version of the manuscript.

Funding: This study was jointly supported by the National Key Research and Development Program of China (2017YFA0604304), and the Thailand Science Research and Innovation (TSRI) under Fundamental Fund No. 48873 and No. 160399. The numerical experiment is conducted at "Earth System Science Numerical Simulator Facility" (EarthLab) which is supported by the National Key Scientific and Technological Infrastructure project.

Institutional Review Board Statement: Not applicable.

Informed Consent Statement: Not applicable.

Data Availability Statement: Not applicable.

Acknowledgments: The authors sincerely thank the Thai Meteorological Department (TMD) for the daily rainfall data for rainfall data. Kritanai Torsri also thanks to the CAS-TWAS President's Fellowship Programme.

Conflicts of Interest: The authors declare no conflict of interest.

References

1. Taylor, K.E.; Stouffer, R.J.; Meehl, G.A. An overview of CMIP5 and the experiment design. *Bull. Am. Meteorol. Soc.* **2012**, *93*, 485–498. [CrossRef]
2. Vitart, F.; Robertson, A.W. The sub-seasonal to seasonal prediction project (S2S) and the prediction of extreme events. *NPJ Clim. Atmos. Sci.* **2018**, *1*, 3. [CrossRef]
3. Dike, V.N.; Lin, Z.; Kece, F.; Langendijk, G.S.; Nath, D. Evaluation and multi-model projection of seasonal precipitation extremes over Central Asia based on CMIP6 simulations. *Int. J. Climatol.* **2022**. [CrossRef]
4. Nahar, J.; Johnson, F.; Sharma, A. Assessing the extent of non-stationary biases in GCMs. *J. Hydrol.* **2017**, *549*, 148–162. [CrossRef]
5. Li, J.; Yang, Y.-M.; Wang, B. Evaluation of NESMv3 and CMIP5 models' performance on simulation of Asian-Australian monsoon. *Atmosphere* **2018**, *9*, 327. [CrossRef]
6. Raghavan, S.V.; Liu, J.; Nguyen, N.S.; Vu, M.T.; Liong, S.-Y. Assessment of CMIP5 historical simulations of rainfall over Southeast Asia. *Theor. Appl. Climatol.* **2017**, *132*, 989–1002. [CrossRef]
7. Tang, W.; Lin, Z.-H.; Luo, L.-F. Assessing the seasonal predictability of summer precipitation over the Huaihe river basin with multiple APCC models. *Atmos. Ocean. Sci. Lett.* **2013**, *6*, 185–190. [CrossRef]
8. Schmidli, J.; Frei, C.M.; Vidale, P.L. Downscaling from GCM precipitation: A benchmark for dynamical and statistical downscaling methods. *Int. J. Climatol.* **2006**, *26*, 679–689. [CrossRef]

9. Haarsma, R.J.; Roberts, M.J.; Vidale, P.L.; Senior, C.A.; Bellucci, A.; Bao, Q.; Chang, P.; Corti, S.; Fučkar, N.S.; Guemas, V.; et al. High Resolution Model Intercomparison Project (HighResMIP v1.0) for CMIP6. *Geosci. Model Dev.* **2016**, *9*, 4185–4208. [CrossRef]
10. Hariadi, M.H.; Schrier, G.v.d.; Steeneveld, G.J.; Sopaheluwakan, A.; Tank, A.K.; Roberts, M.J.; Moine, M.P.; Bellucci, A.; Senan, R.; Tourigny, É.; et al. Evaluation of onset, cessation and seasonal precipitation of the Southeast Asia rainy season in CMIP5 regional climate models and HighResMIP global climate models. *Int. J. Climatol.* **2021**, *42*, 3007–3024. [CrossRef]
11. Xu, Z.; Yang, Z.L. An improved dynamical downscaling method with GCM bias corrections and its validation with 30 years of climate simulations. *J. Clim.* **2012**, *25*, 6271–6286. [CrossRef]
12. Xin, X.; Wu, T.; Zhang, J.; Yao, J.; Fang, Y. Comparison of CMIP6 and CMIP5 simulations of precipitation in China and the East Asian summer monsoon. *Int. J. Climatol.* **2020**, *40*, 6423–6440. [CrossRef]
13. Tan, Y.; Guzman, S.M.; Dong, Z.; Tan, L. Selection of effective GCM bias correction methods and evaluation of hydrological response under future climate scenarios. *Climate* **2020**, *8*, 108. [CrossRef]
14. Komori, D.; Nakamura, S.; Kiguchi, M.; Nishijima, A.; Yamazaki, D.; Suzuki, S.; Kawasaki, A.; Oki, K.; Oki, T. Characteristics of the 2011 Chao Phraya River flood in Central Thailand. *Hydrol. Res. Lett.* **2012**, *6*, 41–46. [CrossRef]
15. Limsakul, A.; Paengkaew, W.; Kummueang, A.; Limjirakan, S.; Suttamanuswong, B. PDSI-based variations of droughts and wet spells in Thailand: 1951–2005. *EnvironmentAsia* **2010**, *4*, 12–20.
16. Gale, E.L.; Saunders, M.A. The 2011 Thailand flood: Climate causes and return periods. *Weather* **2013**, *68*, 233–237. [CrossRef]
17. Khadka, D.; Babel, M.S.; Shrestha, S.; Virdis, S.G.P.; Collins, M. Multivariate and multi-temporal analysis of meteorological drought in the northeast of Thailand. *Weather Clim. Extrem.* **2021**, *34*, 100399. [CrossRef]
18. Singhrattna, N.; Rajagopalan, B.; Kumar, K.K.; Clark, M. Interannual and interdecadal variability of Thailand summer monsoon season. *J. Clim.* **2005**, *18*, 1697–1708. [CrossRef]
19. Imada, Y.; Kanae, S.; Kimoto, M.; Watanabe, M.; Ishii, M. Predictability of Persistent Thailand Rainfall during the Mature Monsoon Season in 2011 Using Statistical Downscaling of CGCM Seasonal Prediction. *Mon. Weather Rev.* **2015**, *143*, 1166–1178. [CrossRef]
20. Singhrattna, N.; Rajagopalan, B.; Clark, M.; Krishna Kumar, K. Seasonal forecasting of Thailand summer monsoon rainfall. *Int. J. Climatol.* **2005**, *25*, 649–664. [CrossRef]
21. Supriyasilp, T.; Pongput, K. El Niño-Southern Oscillation and rainfall characteristics in Northern Thailand: Implications for adaptation in rainfed farms. *Reg. Environ. Chang.* **2021**, *21*, 123. [CrossRef]
22. Chotamonsak, C.; Salathe, E.; Kreasuwan, J.; Chantara, S. Evaluation of precipitation simulations over Thailand using a WRF regional climate model. *Chiang Mai J. Sci.* **2012**, *39*, 623–628.
23. Torsri, K.; Octaviani, M.; Manomaiphiboon, K.; Towprayoon, S. Regional mean and variability characteristics of temperature and precipitation over Thailand in 1961–2000 by a regional climate model and their evaluation. *Theor. Appl. Climatol.* **2012**, *113*, 289–304. [CrossRef]
24. Bellenger, H.; Guilyardi, E.; Leloup, J.; Lengaigne, M.; Vialard, J. ENSO representation in climate models: From CMIP3 to CMIP5. *Clim. Dyn.* **2014**, *42*, 1999–2018. [CrossRef]
25. Dieppois, B.; Rouault, M.; New, M. The impact of ENSO on Southern African rainfall in CMIP5 ocean atmosphere coupled climate models. *Clim. Dyn.* **2015**, *45*, 2425–2442. [CrossRef]
26. Hayashi, M.; Jin, F.-F.; Stuecker, M.F. Dynamics for El Niño-La Niña asymmetry constrain equatorial-Pacific warming pattern. *Nat. Commun.* **2020**, *11*, 4230. [CrossRef]
27. Dong, X.; Xue, F.; Zhang, H.; Zeng, Q.-C. Evaluation of surface air temperature change over China and the globe during the twentieth century in IAP AGCM4.0. *Atmos. Ocean. Sci. Lett.* **2012**, *5*, 435–438. [CrossRef]
28. Adeniyi, M.O.; Lin, Z.; Zhang, H. Evaluation of the performance of IAP-AGCM4.1 in simulating the climate of West Africa. *Theor. Appl. Climatol.* **2018**, *136*, 1419–1434. [CrossRef]
29. Lin, Z.-H.; Yu, Z.; Zhang, H.; Wu, C.-L. Quantifying the attribution of model bias in simulating summer hot days in China with IAP AGCM 4.1. *Atmos. Ocean. Sci.* **2016**, *9*, 436–442. [CrossRef]
30. Yan, Z.-B.; Lin, Z.-H.; Zhang, H. The relationship between the East Asian Subtropical Westerly Jet and summer precipitation over East Asia as simulated by the IAP AGCM4.0. *Atmos. Ocean. Sci. Lett.* **2014**, *7*, 487–492. [CrossRef]
31. Su, T.; Xue, F.; Zhang, H. Simulating the intraseasonal variation of the East Asian summer monsoon by IAP AGCM4.0. *Adv. Atmos. Sci.* **2014**, *31*, 570–580. [CrossRef]
32. Bridhikitti, A. Connections of ENSO/IOD and aerosols with Thai rainfall anomalies and associated implications for local rainfall forecasts. *Int. J. Climatol.* **2013**, *33*, 2836–2845. [CrossRef]
33. Zhang, H.; Zhang, M.; Zeng, Q.-C. Sensitivity of simulated climate to two atmospheric models: Interpretation of differences between dry models and moist models. *Mon. Weather Rev.* **2013**, *141*, 1558–1576. [CrossRef]
34. Zeng, Q.-C.; Zhang, X.-H.; Liang, X.-Z.; Yuan, C.-G.; Chen, S.-F. *Documentation of IAP (Institute of Atmospheric Physics) Two-Level Atmospheric General Circulation Model*; Institute of Atmospheric Physics, Chinese Academy of Sciences: Beijing, China, 1989.
35. Bi, X. IAP 9L AGCM and Climate Simulation. Ph.D. Dissertation, Institute of Atmospheric Physics, Chinese Academy of Sciences, Beijing, China, 1993; 210p. (In Chinese)
36. Liang, X. Description of a nine-level grid point atmospheric general circulation model. *Adv. Atmos. Sci.* **1996**, *13*, 269–298. [CrossRef]
37. Zuo, R. Development of New Generation Grid Point Atmospheric General Circulation Model with High Resolution. Ph.D. Thesis, China People's Liberation Army University of Science and Technology, Changsha, China, 2003; 328p. (In Chinese)

38. Zhang, H. Development of IAP Atmospheric General Circulation Model Version 4.0 and Its Climate Simulations. Ph.D. Thesis, Institute of Atmospheric Physics, Chinese Academy of Sciences, Beijing, China, 2009; 194p. (In Chinese)
39. Phillips, N.A. A coordinate system having some special advantages for numerical forecasting. *J. Meteorol.* **1957**, *14*, 184–185. [CrossRef]
40. Arakawa, A.; Lamb, V.R. Computational design of the basic dynamical processes of the UCLA general circulation model. In *Methods in Computational Physics: Advances in Research and Applications*; Chang, J., Ed.; Elsevier: Amsterdam, The Netherlands, 1977; Volume 17, pp. 173–265.
41. Rayner, N.A.; Parker, D.E.; Horton, E.B.; Folland, C.K.; Alexander, L.V.; Rowell, D.P.; Kent, E.C.; Kaplan, A. Global analyses of sea surface temperature, sea ice, and night marine air temperature since the late nineteenth century. *J. Geophys. Res. Atmos.* **2003**, *108*, 4407. [CrossRef]
42. TMD. Climate of Thailand. Available online: http://www.tmd.go.th/en/archive/thailand_climate.pdf (accessed on 20 March 2022).
43. Navarra, A.; Simoncini, V. Empirical Orthogonal Functions. In *A Guide to Empirical Orthogonal Functions for Climate Data Analysis*; Springer: Amsterdam, The Netherlands, 2010; pp. 39–67.
44. Hu, W.; Newell, R.E.; Wu, Z.-X. Modes of variability of global sea surface temperature, free atmosphere temperature and oceanic surface energy flux. *Clim. Dyn.* **1994**, *10*, 377–393. [CrossRef]
45. Polonskii, A.B.; Basharin, D.V. Large-scale patterns of Eurasian surface meteorological fields influenced by the climate shift of 1976–1977. *Russ. Meteorol. Hydrol.* **2008**, *33*, 280–289. [CrossRef]
46. Rao, A.R.; Hsieh, C.H. Empirical orthogonal function analysis of rainfall and runoff series. *Water Resour. Manag.* **1991**, *4*, 235–250. [CrossRef]
47. Cressman, G.P. An operational objective analysis system. *Mon. Weather Rev.* **1959**, *87*, 367–374. [CrossRef]
48. Wang, Z.; Chang, C.P.; McBride, J.; Liu, C.-H. Annual cycle of Southeast Asia—Maritime continent rainfall and the asymmetric monsoon transition. *J. Clim.* **2005**, *18*, 287–301. [CrossRef]
49. Nounmusig, W. Analysis of rainfall in the eastern Thailand. *Int. J. GEOMATE* **2018**, *14*, 150–155. [CrossRef]
50. Miller, S.T.K.; Keim, B.D.; Talbot, R.W.; Mao, H. Sea breeze: Structure, forecasting, and impacts. *Rev. Geophys.* **2003**, *41*, 1011. [CrossRef]
51. Alomar-Garau, G.; Grimalt-Gelabert, M. Impacts of coastal breezes on the environment and human Life: The case of Mallorca (Western Mediterranean). *Coasts* **2022**, *2*, 2. [CrossRef]
52. Fagundes, M.; Litvin, S.Y.; Micheli, F.; De Leo, G.; Boch, C.A.; Barry, J.P.; Omidvar, S.; Woodson, C.B. Downscaling global ocean climate models improves estimates of exposure regimes in coastal environments. *Sci. Rep.* **2020**, *10*, 14227. [CrossRef]
53. Crosman, E.T.; Horel, J.D. Sea and lake breezes: A review of numerical studies. *Bound.-Layer Meteorol.* **2010**, *137*, 1–29. [CrossRef]
54. Zhong, S.; Takle, E.S. The effects of large-scale winds on the sea land-breeze circulations in an area of complex coastal heating. *J. Appl. Meteorol. Climatol.* **1993**, *32*, 1181–1195. [CrossRef]
55. Feng, J.; Wei, T.; Dong, W.; Wu, Q.; Wang, Y. CMIP5/AMIP GCM simulations of East Asian summer monsoon. *Adv. Atmos. Sci.* **2014**, *31*, 836–850. [CrossRef]
56. Wang, B.; Luo, X.; Liu, J. How robust is the Asian precipitation–ENSO relationship during the industrial warming period (1901–2017). *J. Clim.* **2020**, *33*, 2779–2792. [CrossRef]
57. NOAA. Cold & Warm Episodes by Season. Available online: https://origin.cpc.ncep.noaa.gov/products/analysis_monitoring/ensostuff/ONI_v5.php (accessed on 20 March 2022).
58. Glantz, M.H.; Ramirez, I.J. Reviewing the Oceanic Niño Index (ONI) to enhance societal readiness for El Niño's impacts. *Int. J. Disaster Risk Sci.* **2020**, *11*, 394–403. [CrossRef]
59. Cannon, F.; Carvalho, L.M.V.; Jones, C.; Norris, J.; Bookhagen, B.; Kiladis, G.N. Effects of topographic smoothing on the simulation of winter precipitation in high mountain Asia. *J. Geophys. Res. Atmos.* **2017**, *122*, 1456–1474. [CrossRef]
60. Posada-Marín, J.A.; Rendón, A.M.; Salazar, J.F.; Mejía, J.F.; Villegas, J.C. WRF downscaling improves ERA-Interim representation of precipitation around a tropical Andean valley during El Niño: Implications for GCM-scale simulation of precipitation over complex terrain. *Clim. Dyn.* **2019**, *52*, 3609–3629. [CrossRef]
61. Chansaengkrachang, K.; Luadsong, A.; Aschariyaphotha, N. A Study of the Time Lags of the Indian Ocean Dipole and Rainfall Over Thailand by Using the Cross Wavelet Analysis. *Arab. J. Sci. Eng.* **2015**, *40*, 215–225. [CrossRef]
62. Hu, Z.; Chen, X.; Zhou, Q.; Chen, D.; Li, J. DISO: A rethink of Taylor diagram. *Int. J. Climatol.* **2019**, *39*, 2825–2832. [CrossRef]
63. Zhou, Q.; Chen, D.; Hu, Z.; Chen, X. Decompositions of Taylor diagram and DISO performance criteria. *Int. J. Climatol.* **2021**, *41*, 5726–5732. [CrossRef]

atmosphere

Article

Performance of RegCM4.5 in Simulating the Regional Climate of Western Tianshan Mountains in Xinjiang, China

Quanying Cheng [1,2] **and Fan Li** [3,4,5,*]

[1] State Key Laboratory of Resources and Environmental Information System, Institute of Geographic Sciences and Natural Resources Research, Chinese Academy of Sciences, Beijing 100101, China; chengqy.18b@igsnrr.ac.cn
[2] College of Resources and Environment, University of Chinese Academy of Sciences, Beijing 100049, China
[3] Institute of Forest Resource Information Technique, Chinese Academy of Forestry, Beijing 100091, China
[4] Jiangsu Key Construction Laboratory of IoT Application Technology, School of Internet of Things Engineering, Wuxi Taihu University, Wuxi 214064, China
[5] National Mobile Communications Research Laboratory, School of Information Science and Engineering, Southeast University, Nanjing 211189, China
* Correspondence: lifan@ifrit.ac.cn

Abstract: The western Tianshan Mountains region in China has a complex topography where basins, mountains and glaciers co-exist. It is of great significance to study the sensitivity of meteorological factors in this region to different parameterization schemes of climate models. In this paper, the regional climate model RegCM4.5 is used to simulate the meteorological factor (mean temperature, maximum temperature, minimum temperature, precipitation and wind speed) occurring in the western Tianshan Mountains region from 2012 to 2016, so as to investigate the effects of different cumulus convective schemes (Grell, Tiedtke and Emanuel), including land cumulus convective schemes (LCCs) and ocean convective schemes (OCCs) on annual and seasonal simulations of meteorological factor by using the schemes of RUN1 (Grell for LCC and Tiedtke for OCC), RUN2 (Tiedtke for LCC and Emanuel for OCC), RUN3 (Grell for LCC and Emanuel for OCC) and ENS (the ensemble of RUN1, RUN2 and RUN3). The results show that the simulations of annual and seasonal meteorological factors are not significantly sensitive to the combination of LCCs and OCCs. In the annual simulations, RUN2 scheme has the best simulation performance for the maximum, average and minimum temperatures. However, other schemes of precipitation simulation outperform RUN2 scheme, and there is no difference among the four schemes for wind speed simulation. In the seasonal simulations, RUN2 scheme still performs well in the simulation of the average, maximum and minimum temperatures for four seasons, except for the simulation of the average temperature in spring and summer. For the simulation of the maximum temperature in summer, RUN2 scheme performs the same as ENS. For the simulation of other seasons, different meteorological factors have different performances in four seasons. Overall, the results show that different combinations of cumulus convection schemes can improve the simulation performance of meteorological factors in the western Tianshan Mountains of Xinjiang.

Keywords: regional climate model; RegCM4.5; western Tianshan Mountains; parameterization scheme

Citation: Cheng, Q.; Li, F. Performance of RegCM4.5 in Simulating the Regional Climate of Western Tianshan Mountains in Xinjiang, China. *Atmosphere* **2021**, *12*, 1544. https://doi.org/10.3390/atmos12121544

Academic Editors: Zengyun Hu, Xuguang Tang and Qinchuan Xin

Received: 27 October 2021
Accepted: 20 November 2021
Published: 23 November 2021

1. Introduction

Located in the hinterland of Eurasia, Xinjiang is the largest province with the longest border line and the largest number of neighbors in China, and it has become the most convenient access to Central Asia from the Chinese mainland. It is a very important region in Asia. Xinjiang has a temperate continental climate, characterized by large temperature differences and abundant sunshine, but little precipitation and a dry climate. The presence of Tianshan Mountains divides Xinjiang into two parts: the north and the south. The complex topography of western Tianshan Mountains in Xinjiang, where basins, hills and glaciers co-exist, has an important influence on the formation of climate in Xinjiang, China. Although Xinjiang is generally deep inland and far from the sea, and the climate types are all temperate continental, there are still significant differences in climate between the southern and northern Xinjiang. Accurate modeling and prediction of the climate in the region is important to address climate change [1,2].

At present, an important means of climate research is simulation by climate model, which mainly includes global climate models (GCMs) and regional climate models (RCMs). GCMs are not directly applicable to regional simulations due to the large differences in resolution between the grid spacing and the end-user's needs. Therefore, RCMs are widely used for regional simulation due to their high resolution [3–5]. However, the proper selection of horizontal resolution, land surface models [6,7] and convective parameterization schemes (CPS) [8,9] remains problematic, and their selection mostly relies on empirical study. Many studies have shown different simulation results using different CPS [10,11] for the same region. The comparison of the simulated near-surface temperature with observational data over the European region indicates a cold bias with both Grell scheme configurations, but the bias can be reduced when the Emanuel convective scheme is applied, and the simulated precipitation is not systematically positive in the southern part of Eastern Europe [12]; Tchotchou and Kamga found that monsoonal precipitations are sensitive to the choice of cumulus parameterization and closure schemes in West Africa [13]; Raju's research results reveal that the mixed convective scheme responds better to the simulation of precipitation and temperature over the Indian subcontinent, and the circulation features and the annual cycle of precipitation and temperature are also well simulated with the mixed convection scheme [14]. The authors of ref. [15] found that the Emanuel scheme shows an overall overestimation of precipitation in China, while the other three CPSs overestimate only in northern and northwestern China, and seasonally, the Tiedtke scheme simulates the best annual variability of precipitation in China compared with the others CPSs. In the RegCM4 version, the CPSs are described in more detail in two parts, and it is able to run different convection schemes over land and ocean. Studies have shown that different CPSs have different simulation performance over different regions, especially in land and sea areas. Bhatla's research found that out of six CPSs, Tiedtke and Mix99 (Grell over land and Emanuel over ocean) correctly simulated the onset date, and they also found that the Mix99 CPSs performed well in simulating the synoptic features during the monsoon phases. Existing research shows that these physical processes cannot be accurately simulated over the entire Earth. Choosing an appropriate convective parameterization scheme (CPS) in RCMs is critical for region simulations [16–18], as it allows to simulate the region in a more refined way. At the same time, the annual and seasonal climate simulations are also very important for determining short-term adaptation measures [19,20].

In this paper, we apply the Regional Climate Model version 4.5 (RegCM4.5) to the western Tianshan Mountain area in China. The objective of this study is to evaluate the performance of three mixed CPSs and their ensemble averages, namely, Grell over land and Tiedtke over ocean (RUN1), Tiedtke over land and Emanuel over ocean (RUN2), and Grell over land and Emanuel over ocean (RUN3), with the ensemble being the average of RUN1, RUN2 and RUN3. The remainder of this paper is organized as follows. Section 2 provides a brief description of the model and three convection schemes. Section 3 presents the experimental design and Section 4 describes the results and discusses the sensitivity experiments with different convection schemes. Section 5 discusses the results and summarizes the conclusion.

2. The Convective Parameterization Schemes

2.1. Grell Scheme

Grell's [21] convective scheme is based on the Fritsch-Chappell (FC) assumption [22]. The scheme applies a simple cloud model assuming that the large amount of flux in the upstream and downstream is constant with the height, and that the original level of upstream and sinking air flow are given by the maximum and minimum ambient humid static energy, respectively. The scheme is activated when the parcel finally reaches the wet convection from the original horizontal up-draft. A large amount of flux (m_0) in the downdraft is proportional to the updraft (m_b), and it is calculated by the following relationship.

$$m_0 = \frac{\beta I_1}{I_2} m_b \quad (1)$$

I_1 is the condensed normalized updraft, I_2 is the evaporative normalized downdraft, β is a small part of the re-evaporation of updraft condensate during the descent and depends on the change in the wind direction and varies mainly between 0.3 and 0.5. Hence, the equation is as follows:

$$P^{CU} = I_1 m_b (1 - \beta) \quad (2)$$

where P is precipitation. The meanings of other parameters are shown in Formula (1).

2.2. Tiedtke

Tiedtke is a mass flux and moisture convergence scheme originally designed for global climate models with special attention to the correct display of deep tropical convection. Tiedtke consists of three types of convection. Only one type of convection is allowed per step in each grid, and the collection of clouds appearing in each type of convection is assumed by the updraft and downdraft. The Tiedtke scheme considers only one ascent and descent of the overall airflow, which is equivalent to the ascent and descent of all cumulus monomer airflows in the grid. In the actual simulation, deep or shallow convection occurs when there is an unstable convection on the cloud base with a large-scale increase. Assuming that no deep or shallow convection can occur, then mid-stream convection is applied.

2.3. Emanuel

The Emanuel scheme assumes that interactions in the cloud are random and non-uniform, and increases convective fluxes based on an idealized condition of sub-cloud-scale downdrafts and updrafts, and the convection model is activated when the cloud base level is below the height of the neutral buoyancy layer. Assuming that clouds and the mixture of ambient air are mixed in a uniform spectrum, when the air rises between the cloud base and the zero-buoyancy layer, the condensed water vapor partly forms precipitation and the rest forms clouds. The Emanuel scheme incorporates the automatic conversion formula for clouds and water, and sets the water content of the initial cloud in the formula to a temperature-related function, in order to take into account the microphysical process of cloud ice, which is added to a single downdraft that is hydrostatic and unsaturated and carries water and heat when precipitation is generated.

3. Model and Experimental Design

3.1. Description of the RegCM Model

RegCM is regional climate model developed by Abdus Salam International Center for Theoretical Physics. The first version of RegCM was developed in the late 1980s [23,24], which has since then been upgrades to RegCM2 [25], RegCM3 [26] and RegCM4 [27]. The dynamical core of the RegCM model is similar to the hydrostatic version of the mesoscale model MM5 [28]. The PBL computations are parameterized by using the scheme of Holtslag et al. [29], the land surface model is the Biosphere-Atmosphere Transfer Scheme (BATS) and the Community Land Model (CLM) [30] schemes, and the radiation scheme is the modified NCAR Community Climate Model version 3 (CCM3) [31]. The model used here is RegCM4.5, which has many upgrades in the model physics. One of the main enhancements in this version of the model is the ability to run different convection schemes over land and ocean, also known as 'mixed convection'. Many studies have shown that different schemes have different performance over different regions, especially over land versus ocean areas, and it currently includes three options for representing cumulus convection [32].

The scheme of Grell is the implementation of Giorgi [33]. The other two schemes are Tiedtke [34] and Emanuel scheme [35,36]. In this paper, we apply a combination of the three schemes to simulate the meteorological factors when other parameters are consistent, describe the following Table 1, where ENS is the average of the three schemes RUN1, RUN2 and RUN3, and icup_lnd and icup_ocn are the cumulus parameterization schemes that must be set before the model simulation.

Table 1. Parameterization scheme.

Schemes	icup_lnd	icup_ocn
RUN1	Grell	Tiedtke
RUN2	Tiedtke	Emanuel
RUN3	Grell	Emanuel
ENS	Ensemble_mean	Ensemble_mean

3.2. *Experimental Design and Data Check*

The model domain covers the western Tianshan Mountain region in Xinjiang (36°–50° N, 70°–100° E), as shown in Figure 1. The grid center of the simulated area is located at 43.15° N, 82.88° E. A model grid with a horizontal resolution of 20 km is selected for the simulation experiments. The number of grid cells is 121 × 100 (East-West × North-South). The reanalysis data from the National Centers for Environmental Prediction and National Center for Atmospheric Research (NCEP&NCAR) are selected as the initial field and side boundary values of the model [37,38]. The dynamic module uses MM4 frame, with a step size of 6 h for the model output. The experiment was conducted for a period of four years from 1 January 2012 to 31 December 2016. The results of the model-driven data from January 2012 are used as model spin-up, and therefore are not analyzed. The observations used for comparison with the model were based on the daily terrestrial climate datasets from 66 sites in the western Tianshan Mountains of Xinjiang provided by the China Meteorological Data Network (http://data.cma.cn, accessed on 12 March 2017), which are continuous observation data, and the triangles in Figure 1 show the locations of the 66 observation sites.

3.3. *Verification Method*

3.3.1. Correlation Coefficient

According to the existing analysis methods [39,40], the results are analyzed by calculating the correlation coefficient and root mean square error between the model simulated value and the observed value. The correlation coefficient is defined as follows:

$$\mathrm{CORREL} = \frac{\sum_{i=1}^{N}(x_i - \overline{x})(y_i - \overline{y})}{\sqrt{\sum_{i=1}^{N}(x_i - \overline{x})^2 \sum_{i=1}^{N}(y_i - \overline{y})^2}} \tag{3}$$

In the above formula, x_i and y_i are the simulated values and observed values in the same position, $\overline{x}$ and $\overline{y}$ are their average values, N represents the number of grid points.

3.3.2. Root Mean Square Error

The root mean square error (RMSE) reflects the average deviation between the simulations and observations, which is a trade-off standard of the total error and is one of the ways to reflect the simulation performance. Typically, the greater the root mean square error is, the lower the accuracy, and in turn, the higher the simulation accuracy would be. The root mean square error formula is defined as follows:

$$\mathrm{RMSE} = \sqrt{\frac{\sum_{t=1}^{N}(\mathrm{Sim}_t - \mathrm{Obs}_t)^2}{N}} \tag{4}$$

In the above formula, Sim_t stands for the simulation results of RegCM4.5, Obs_t is the actual observation value, and N represents the number of grid points.

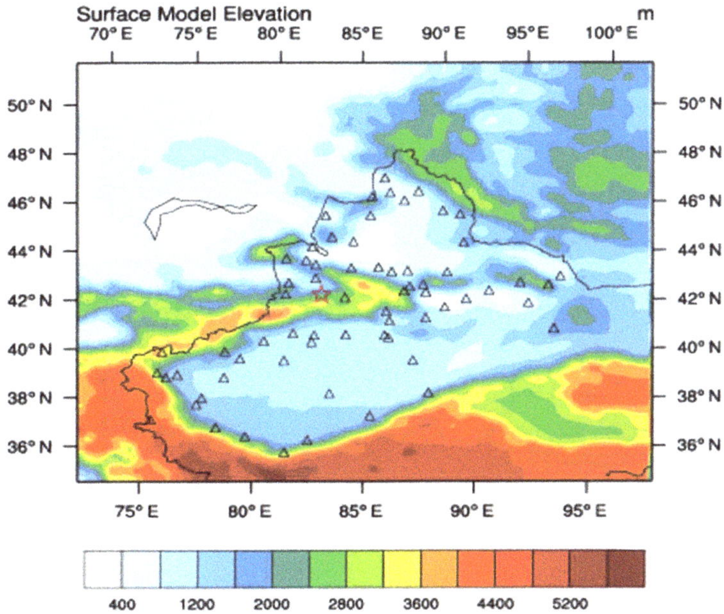

Figure 1. RegCM simulated surface model and the area. The black lines are the simulation area, the pentagram is the center point of the simulation and the triangles are the locations of the 66 observation sites.

4. Results and Discussion

4.1. Annual Simulation and Inspection

4.1.1. The Analysis of Annual Temperature

It can be seen from Figure 2 that the three parameterization schemes and sets of the model have a good simulation effect on the temperature distribution in the western Tianshan Mountains of Xinjiang, and there is no significant difference between the simulated and observed values using different schemes, indicating that the annual average temperature in the western Tianshan Mountains of Xinjiang is not sensitive to different parameterization schemes of the model. The temperature decreases from the center of the region to the border, which is consistent with the climate change in the basin. Meanwhile, a small part of the central and western regions has lower temperatures. Compared with the simulated elevation results in Figure 1, it can be seen that the higher altitude of this small part of the region leads to lower temperatures, and the simulation result is consistent with the actual situation. From the whole simulation area, the temperature in the north is lower than that in the south, which reflects the temperature difference between the north and the south. Comparing the northern and southern parts of the region, it can be seen that the difference between the simulated and observed values in the south is not obvious, and the simulation results are better, while the difference between the simulated values and the observed values in the north is more obvious, and the difference is greater the further north, which indicates that the model simulates the temperature better in the south of Tianshan Mountains in Xinjiang than in the north.

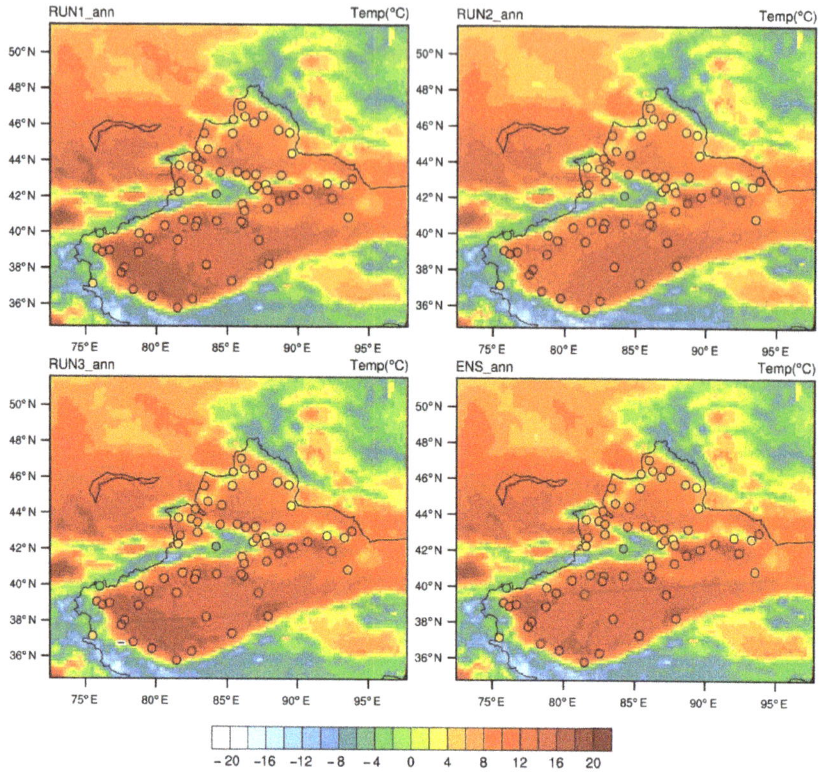

Figure 2. Comparison between simulation and observation of annual average temperature using four cumulus parameterization schemes, where the circles are observed values, the rest are simulated values, and the black lines is the simulation area. The same applies to Figures 3 and 4.

Table 2 shows the calculated correlation coefficients and RMSEs, in which the correlation coefficients are calculated by Formula 3, and RMSEs are calculated by Formula 4. Both the simulated value and the observed value are multi-year averages. Combined with Table 2, it can be found that the model can well simulate the annual average, maximum and minimum temperatures, and the simulation results of the four schemes for each temperature factor do not differ significantly. Moreover, the correlation coefficients between the simulated and observed annual average and minimum temperatures are the same in the four schemes. However, we find that the RUN2 scheme has the lowest RMSE in the annual average and minimum temperature simulations, which means that this scheme has the best performance in simulating temperature. For the simulation of the maximum temperature, the CORREL do not differ greatly. In this case, we still choose the scheme with the lowest RMSE as the optimal scheme.

4.1.2. The Analysis of Average Annual Precipitation

As can be seen from Figure 3, the three parameterization schemes and sets of the model can reflect the distribution of precipitation in the western Tianshan Mountains of Xinjiang, China. The precipitation at the edge of the region is higher than that in the middle of the region, and it is closely related to the terrain. Generally, it is consistent with the law that the precipitation is higher in the mountain area than in the basin, higher in the basin than in the center of the basin, and higher in the windward slope of the mountain area than the leeward slope. On the whole, the precipitation distribution simulated by the model is consistent with the impact of the actual terrain on the precipitation in the western Tianshan Mountains, where the precipitation is scarce and the climate is dry. Combined with Table 2, it is found that the CORREL and RMSE of RUN1, RUN3 and ENS schemes are the same, and all of them are better than the RUN2 schemes.

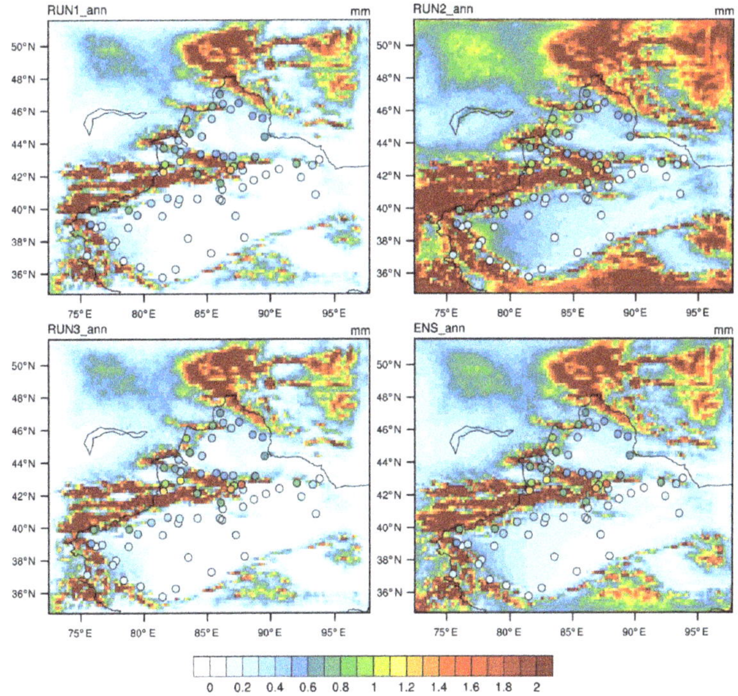

Figure 3. Comparison between simulation and observation of precipitation using four cumulus parameterization schemes.

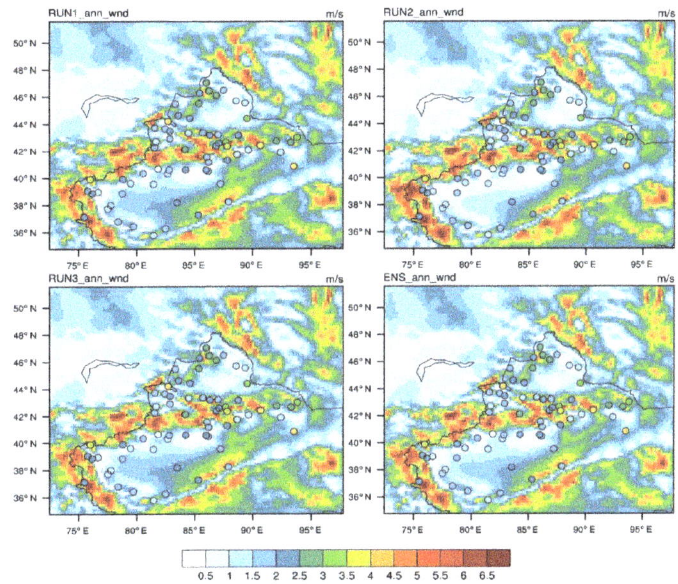

Figure 4. Comparison between simulated and observed wind speeds using four parameterization schemes.

Table 2. CORREL and RMSE of temperature, precipitation and wind speed simulated by different parameterization schemes; the units for the RMSE of temperature, precipitation and wind speed are °C, mm and m/s, respectively.

Meteorological Factors	Parameterization Schemes	Correlation Coefficient	Root Mean Square Error
Average annual temperature	RUN1_ann	0.8	4.8
	RUN2_ann	0.8	4.5
	RUN3_ann	0.8	4.8
	ENS_ann	0.8	4.7
Maximum temperature	RUN1_maxTemp	0.9	5.2
	RUN2_maxTemp	0.8	4.9
	RUN3_maxTemp	0.9	5.2
	ENS_maxTemp	0.8	5.1
Minimum temperature	RUN1_minTemp	0.7	4.4
	RUN2_minTemp	0.7	4.1
	RUN3_minTemp	0.7	4.4
	ENS_minTemp	0.7	4.3
Precipitation	RUN1_ann	0.4	0.7
	RUN2_ann	0.2	0.9
	RUN3_ann	0.4	0.7
	ENS_ann	0.4	0.7
Wind speed	RUN1_ann_wnd	0.4	1.2
	RUN2_ann_wnd	0.4	1.2
	RUN3_ann_wnd	0.4	1.2
	ENS_ann_wnd	0.4	1.2

4.1.3. The Analysis of Average Annual Wind Speed

In Figure 4, we find that there is no difference between simulation and observation among the four schemes. Combined with Table 2, it is found that the CORREL and RMSE of wind speed simulations also have no difference among the four schemes. The main reason for this situation is that the wind speed is highly uncertain.

4.1.4. The Results of Annual Simulation

Through the analysis in Sections 4.1.1–4.1.3, we draw the conclusion in Table 3. In the simulation of western Tianshan Mountains in Xinjiang, China, the RUN2 scheme is selected for temperature, maximum temperature and minimum temperature, which can make the simulation results better and closer to the observed values. The RUN1, RUN3 and ENS scheme is selected for annual precipitation, and all schemes are the same for wind speed simulation

Table 3. The optimal scheme of different meteorological factors in the simulation area.

Simulation Parameter	Optimal Scheme
Average temperature	RUN2
Maximum temperature	RUN2
Minimum temperature	RUN2
Annual precipitation	RUN1/RUN3/ENS
Wind speed	RUN1/RUN2/RUN3/ENS

4.2. RegCM Seasonal Numerical Simulation and Inspection

Four schemes are used to simulate the meteorological factors in four seasons in the western Tianshan Mountains of Xinjiang, China. The simulated and observed CORREL and RMSE are obtained, as shown in Tables 4 and 5, respectively.

Table 4. The CORREL of seasonal simulation and observation.

Meteorological Factors	Parametrization Schemes	Spring	Summer	Autumn	Winter
Mean Temperature	RUN1	0.8	0.8	0.8	0.7
	RUN2	0.8	0.8	0.8	0.6
	RUN3	0.8	0.8	0.8	0.7
	ENS	0.8	0.8	0.8	0.7
Maximum Temperature	RUN1	0.9	0.8	0.9	0.7
	RUN2	0.8	0.8	0.9	0.6
	RUN3	0.9	0.8	0.9	0.7
	ENS	0.9	0.8	0.9	0.7
Minimum Temperature	RUN1	0.8	0.7	0.7	0.6
	RUN2	0.7	0.7	0.7	0.5
	RUN3	0.8	0.7	0.7	0.6
	ENS	0.8	0.7	0.7	0.6
Annual Precipitation	RUN1	0.4	0.4	0.3	0.5
	RUN2	0.2	0.3	0.1	0.2
	RUN3	0.4	0.4	0.3	0.5
	ENS	0.4	0.4	0.2	0.5
Wind Speed	RUN1	0.4	0.4	0.4	0.2
	RUN2	0.4	0.4	0.4	0.2
	RUN3	0.4	0.4	0.4	0.2
	ENS	0.4	0.4	0.4	0.2

Table 5. The RMSE of seasonal simulation and observation. the units for the RMSE of temperature, precipitation and wind speed are °C, mm and m/s, respectively.

Meteorological Factors	Parametrization Schemes	Spring	Summer	Autumn	Winter
Mean Temperature	RUN1	5.4	5.4	4.6	4.7
	RUN2	5.1	5.3	4.4	4.1
	RUN3	5.4	5.4	4.6	4.7
	ENS	5.3	4.4	4.5	4.4
Maximum Temperature	RUN1	5.9	5.5	5.1	5.6
	RUN2	5.6	5.4	5.0	4.8
	RUN3	5.9	5.5	5.1	5.6
	ENS	5.8	5.4	5.1	5.2
Minimum Temperature	RUN1	4.7	5.2	4.2	4.4
	RUN2	4.5	5.0	4.0	3.9
	RUN3	4.7	5.2	4.2	4.4
	ENS	4.6	5.1	4.1	4.1
Annual Precipitation	RUN1	0.7	1.3	0.6	0.3
	RUN2	1.0	1.5	0.8	0.3
	RUN3	0.7	1.3	0.6	0.3
	ENS	0.7	1.2	0.6	0.3
Wind Speed	RUN1	1.4	1.4	1.3	1.5
	RUN2	1.4	1.4	1.3	1.6
	RUN3	1.4	1.4	1.3	1.5
	ENS	1.4	1.4	1.3	1.5

The RegCM4.5 model was used to simulate seasonal meteorological factors in the western Tianshan Mountains from 2012 to 2016. From Table 4, we can see that the correlation coefficients of the four schemes for the simulation of the mean and maximum temperatures are higher in spring, summer and autumn, and the values are greater than 0.8. Here, we choose the scheme with lower RMSE as the optimal one. In the simulation of the minimum temperature in four seasons, the correlation coefficients are lower than those of the mean and maximum temperatures, and their RMSEs are also relatively low. In the simulation of annual precipitation and wind speed of four schemes, the correlation coefficients are lower than those of the temperature simulations. Combined with Tables 4 and 5, the optimal schemes for the four-seasons simulations are shown in Table 6.

Table 6. The optimal solution for different meteorological factors in spring, summer, autumn and winter.

Parameterzation Schemes	Spring	Summer	Autumn	Winter
Temperature	RUN2	ENS	RUN2	RUN2
Maximum Temperature	RUN2	RUN2/ENS	RUN2	RUN2
Minimum Temperature	RUN2	RUN2	RUN2	RUN2
Precipitation	RUN1/RUN3/ENS	ENS	RUN1/RUN3/ENS	ALL
Wind Speed	ALL	ALL	ALL	RUN1/RUN3/ENS

5. Conclusions

In this study, we used the regional climate model RegCM4.5 and selected different cumulus parameterization schemes to simulate annual and seasonal meteorological factors in the western Tianshan Mountains region of Xinjiang from 2012 to 2016, and investigate the applicability of different cumulus parameterization schemes to regional climate simulations. The evaluation of the experimental results shows that: (1) In the annual simulations, the RUN2 scheme performs best in simulating the average, maximum and minimum temperatures, indicating that the RUN2 scheme is more suitable for the simulation of temperature, while the other three schemes perform better in simulating precipitation. In the simulation of wind speed, there is no difference among the four schemes. (2) In the simulation of four seasons, the RUN2 scheme still performs better in the simulation of average, maximum and minimum temperatures in four seasons, except for the simulation of average temperature in spring and summer. In the simulation of the maximum temperature in summer, the RUN2 scheme is the same as that of ENS. In the simulation of precipitation, all other schemes perform better than the RUN2 scheme. In the simulation of wind speed, there is no difference among the four schemes in spring, summer, autumn and winter. In the simulation of winter, other schemes perform better than the RUN2 scheme.

This study provides additional information for the selection of available accurate cumulus parameterization simulation scheme for regional climate simulation in western Tianshan Mountains, Xinjiang, which effectively solves the problem of inaccurate simulation caused by random selection of cumulus parameter schemes. However, there are still some problems, for example, the current simulation time span is relatively short due to the limitation of the data collected from the observation sites, and we will collect more observation data later and lengthen the simulation period in the future.

Author Contributions: Conceptualization, Methodology, Validation, Formal Analysis, Q.C. and F.L.; Software, Q.C. and F.L.; supervision, Q.C. and F.L.; Funding Acquisition, F.L.; Writing—original draft preparation, Q.C. and F.L.; Writing—review and editing, Q.C. and F.L. All authors have read and agreed to the published version of the manuscript.

Funding: This research was funded by National Science and Technology Major Project of China's High Resolution Earth Observation System (grant numbers: 21-Y30B02-9001-19/22).

Data Availability Statement: The data presented in this study are available in Section 3.2.

Acknowledgments: Thanks to NCEP/NCAR for providing reanalysis data, and thanks to China Meteorological Data Network for providing observation data.

Conflicts of Interest: The authors declare no conflict of interest.

References

1. Zhou, X.; Huang, G.; Piwowar, J.; Fan, Y.; Wang, X.; Li, Z.; Cheng, G. Hydrologic Impacts of Ensemble-RCM-Projected Climate Changes in the Athabasca River Basin, Canada. *J. Hydrometeorol.* **2018**, *19*, 1953–1971. [CrossRef]
2. Shi, Y.; Wang, G.; Gao, X. Role of resolution in regional climate change projections over China. *Clim. Dyn.* **2018**, *51*, 2375–2396. [CrossRef]

3. Grassi, B.; Redaelli, G.; Visconti, G. Arctic Sea Ice Reduction and Extreme Climate Events over the Mediterranean Region. *J. Clim.* **2013**, *26*, 10101–10110. [CrossRef]
4. Díez, E.; Orfila, B.; Frías, M.D.; Fernández, J.; Cofiño, A.S.; Gutiérrez, J.M. Downscaling ECMWF seasonal precipitation forecasts in Europe using the RCA model. *Tellus A Dyn. Meteorol. Oceanogr.* **2011**, *63*, 757–762. [CrossRef]
5. Sangelantoni, L.; Ricchi, A.; Ferretti, R.; Redaelli, G. Dynamical Downscaling in Seasonal Climate Forecasts: Comparison between RegCM- and WRF-Based Approaches. *Atmosphere* **2021**, *12*, 757. [CrossRef]
6. Llopart, M.; da Rocha, R.P.; Reboita, M.; Cuadra, S. Sensitivity of simulated South America climate to the land surface schemes in RegCM4. *Clim. Dyn.* **2017**, *49*, 3975–3987. [CrossRef]
7. Chung, J.X.; Juneng, L.; Tangang, F.; Jamaluddin, A.F. Performances of BATS and CLM land-surface schemes in RegCM4 in simulating precipitation over CORDEX Southeast Asia domain. *Int. J. Clim.* **2018**, *38*, 794–810. [CrossRef]
8. Ali, S.; Dan, L.; Congbin, F.U.; Yang, Y. Performance of Convective Parameterization Schemes in Asia Using RegCM: Simu-la-tions in Three Typical Regions for the Period 1998–2002. *Adv. Atmos. Sci.* **2015**, *5*, 141–156. [CrossRef]
9. Koné, B.; Diedhiou, A.; Touré, N.E.; Sylla, M.B.; Kobea, A.T. Sensitivity study of the regional climate model RegCM4 to dif-ferent convective schemes over West Africa. *Earth Syst. Dyn.* **2018**, *9*, 1261–1278. [CrossRef]
10. Bhatla, R.; Mandal, B.P.; Verma, S.; Ghosh, S.; Mall, R.K. Performance of Regional Climate Model in Simulating Monsoon Onset Over Indian Subcontinent. *Pure Appl. Geophys.* **2018**, *176*, 409–420. [CrossRef]
11. Bhatla, R.; Ghosh, S.; Mall, R.K.; Sinha, P.; Sarkar, A. Regional Climate Model Performance in Simulating Intra-seasonal and Interannual Variability of Indian Summer Monsoon. *Pure Appl. Geophys.* **2018**, *175*, 3697–3718. [CrossRef]
12. Zanis, P.; Douvis, C.; Kapsomenakis, I.; Kioutsioukis, I.; Melas, D.; Pal, J.S. A sensitivity study of the Regional Climate Model (RegCM3) to the convective scheme with emphasis in central eastern and southeastern Europe. *Theor. Appl. Clim.* **2009**, *97*, 327–337. [CrossRef]
13. Tchotchou, L.; Kamga, F.M. Sensitivity of the simulated African monsoon of summers 1993 and 1999 to convective parameterization schemes in RegCM3. *Theor. Appl. Climatol.* **2010**, *100*, 207–220. [CrossRef]
14. Raju, P.V.S.; Bhatla, R.; Almazroui, M.; Assiri, M. Performance of convection schemes on the simulation of summer monsoon features over the South Asia CORDEX domain using RegCM-4.3. *Int. J. Clim.* **2015**, *35*, 4695–4706. [CrossRef]
15. Zhang, S.; Lu, S.; Bao, Y.; Ma, D. Sensitivity of precipitation over China to different cumulus parameterization schemes in RegCM4. *J. Meteorol. Res.* **2015**, *29*, 119–131. [CrossRef]
16. Pal, J.S.; Giorgi, F.; Bi, X.; Elguindi, N.; Solmon, F.; Rauscher, S.A.; Steiner, A.L. Regional climate modeling for the develop-ing world: The ICTP RegCM3 and RegCNET. *Bull. Am. Meteorol. Soc.* **2007**, *88*, 1395–1409. [CrossRef]
17. Pieczka, I.; Pongrácz, R.; André, K.S.; Kelemen, F.D.; Bartholy, J. Sensitivity analysis of different parameterization schemes using RegCM4.3 for the Carpathian region. *Theor. Appl. Clim.* **2016**, *130*, 1175–1188. [CrossRef]
18. Gadzhev, G.; Ivanov, V.; Ganev, K.; Chervenkov, H. TVRegCM Numerical Simulations-Preliminary Results. In *International Conference on Large-Scale Scientific Computing*; Springer: Cham, Switzerland, 2018; pp. 266–274.
19. Patari, M.; Brankovi, E. Skill of 2-m Temperature Seasonal Forecasts over Europe in ECMWF and RegCM Models. *Mon. Weather Rev.* **2012**, *140*, 1326–1346. [CrossRef]
20. Sangelantoni, L.; Ferretti, R.; Redaelli, G. Toward a Regional-Scale Seasonal Climate Prediction System over Central Italy based on Dynamical Downscaling. *Climate* **2019**, *7*, 120. [CrossRef]
21. Grell, G.A. Prognostic Evaluation of Assumptions Used by Cumulus Parameterizations. *Mon. Weather Rev.* **1993**, *121*, 764–787. [CrossRef]
22. Fritsch, M.; Chappell, C.F. Numerical prediction of convectively driven mesoscale pressure systems. Part I: Convective parameterization. *Atmos. Sci.* **1980**, *37*, 1722–1733. [CrossRef]
23. Giorgi, F.; Bates, G.T. The climatological skill of a regional model over complex terrain. *Mon. Weather Rev.* **1989**, *117*, 2325–2347. [CrossRef]
24. Dickinson, R.E.; Errico, R.M.; Giorgi, F.; Bates, G.T. A regional climate model for the western United States. *Clim. Chang.* **1989**, *15*, 383–422. [CrossRef]
25. Giorgi, F.; Marinucci, M.R.; Bates, G.T. Development of a second generation regional climate model (RegCM2). Part I. Boundary layer and radiative transfer processes. *Mon. Weather Rev.* **1993**, *121*, 2794–2813. [CrossRef]
26. Pal, J.S.; Small, E.; Eltahir, E.A.B. Simulation of regional-scale water and energy budgets: Representation of subgrid cloud and precipitation processes within RegCM. *J. Geophys. Res.* **2000**, *105*, 29579–29594. [CrossRef]
27. Giorgi, F.; Coppola, E.; Solmon, F.; Mariotti, L.; Sylla, M.B.; Bi, X.; Brankovic, C. RegCM4: Model description and preliminary tests over multiple CORDEX domains. *Clim. Res.* **2012**, *52*, 7–29. [CrossRef]
28. Grell, G.; Dudhia, J.; Stauffer, D. *Description of the Fifth Generation Penn State/NCAR Mesoscale Model (MM5)*; Ncar Technical Note: Boulder, CO, USA, 1994; p. 121.
29. Holtslag, A.; Bruijn, E.; Pan, H.L. A High Resolution Air Mass Transformation Model for Short-Range Weather Forecasting. *Mon. Weather Rev.* **1990**, *188*, 1561–1575. [CrossRef]
30. Oleson, K.W.; Niu, G.-Y.; Yang, Z.-L.; Lawrence, D.; Thornton, P.; Lawrence, P.J.; Stöckli, R.; Dickinson, R.E.; Bonan, G.B.; Levis, S.; et al. Improvements to the Community Land Model and their impact on the hydrological cycle. *J. Geophys. Res.* **2008**, *113*, 1021–1026. [CrossRef]

31. Kiehl, J.T.; Hack, J.J.; Bonan, G.B.; Boville, B.A.; Briegleb, B.P. Description of the NCAR Community Climate Model (CCM3). *Ncar Tech. Note* **1996**, *108*, 55–60.
32. Giorgi, F.; Coppola, E.; Solmon, F.; Mariotti, L.; Sylla, M.B.; Bi, X.; Elguindi, N.; Diro, G.T.; Nair, V.; Giuliani, G.; et al. RegCM4: Model description and preliminary tests over multiple CORDEX domains. *Clim. Res.* **2011**, *936*, 577X.
33. Giorgi, F.; Marinucci, M.R.; Bates, G.T. Development of a 2nd-Generation Regional Climate Model (Regcm2). Part2. Con-vec-tive Processes and Assimilation of Lateral Boundary Conditions. *Mon. Weather Rev.* **1993**, *121*, 2814–2832. [CrossRef]
34. Tiedtke, M. A comprehensive mass-flux scheme for cumulus parameterization in large-scale models. *Mon. Weather Rev.* **1989**, *117*, 1779–1800. [CrossRef]
35. Emanuel, K.A. A Scheme for Representing Cumulus Convection in Large-Scale Models. *J. Atmos. Sci.* **1991**, *48*, 2313–2329. [CrossRef]
36. Emanuel, K.; Marina, R. Development and Evaluation of a Convection Scheme for Use in Climate Models. *J. Atmos. Sci.* **1999**, *56*, 1766–1782. [CrossRef]
37. Kalnay, E.; Kanamitsu, M.; Kistler, R.; Collins, W.; Deaven, D.; Gandin, L.; Iredell, M.; Saha, S.; White, G.; Woollen, J.; et al. The NCEP/NCAR 40-Year Reanalysis Project. *Renew. Energy* **2018**, *74*, 146–194. [CrossRef]
38. Kistler, R.; Kalnay, E.; Collins, W.; Saha, S.; White, G.; Woollen, J.; Chelliah, M.; Ebisuzaki, W.; Kanamitsu, M.; Kousky, V.; et al. The NCEP/NCAR 50-year reanalysis: Monthly means CD-ROM and documentation. *Bull. Am. Meteorol. Soc.* **2001**, *82*, 247–268. [CrossRef]
39. Torma, C.Z. Detailed validation of EURO-CORDEX and Med-CORDEX regional climate model ensembles over the Carpa-thian Region. *Idojaras* **2019**, *123*, 217–240.
40. Kotlarski, S.; Keuler, K.; Christensen, O.B.; Colette, A.; Déqué, M.; Gobiet, A.; Goergen, K.; Jacob, D.; Lüthi, D.; van Meijgaard, E.; et al. Regional climate modeling on European scales: A joint standard evaluation of the EURO-CORDEX RCM ensemble. *Geosci. Model Dev.* **2014**, *7*, 1297–1333. [CrossRef]

atmosphere

MDPI

Article

Analysis and Forecast of Beijing's Air Quality Index Based on ARIMA Model and Neural Network Model

Tingyi Liu and Shibing You *

School of Economics and Management, Wuhan University, Wuhan 430072, China; 2019101050047@whu.edu.cn
* Correspondence: sbyou@whu.edu.cn

Abstract: Based on Beijing's Air Quality Index (AQI) and concentration changes of the six major pollutants from 2019 to 2021, the results are visualized through descriptive statistics, and the air pollution status and influencing factors of Beijing's AQI are analyzed using the ARIMA model and neural network. A forecast system is built and the fitting effects of the two models are compared. The results show that PM2.5, PM10, and O_3 of the six major pollutants have the greatest impact on AQI. Beijing's air quality now shows a trend of improvement in recent years; however, there is obvious seasonal evidence that the summer pollution index has been high. Therefore, special attention should be paid to the treatment of ozone pollution in summer. Both models are useful for the forecast of AQI, but the forecast effect of the neural network model is better than that of the ARIMA model. Moreover, when using the additive seasonal model for the long-term forecast of monthly data, it is found that the Beijing AQI still shows seasonal cyclicality and has a slightly decreasing trend in the next two years. This research provides a basis for the forecast of air quality and policy enlightenment for environmental protection departments to deal with air pollution.

Keywords: AQI; visual analysis; heat map; ARIMA model; neural network model

Citation: Liu, T.; You, S. Analysis and Forecast of Beijing's Air Quality Index Based on ARIMA Model and Neural Network Model. *Atmosphere* **2022**, *13*, 512. https://doi.org/10.3390/atmos13040512

Academic Editors: Zengyun Hu, Xuguang Tang and Qinchuan Xin

Received: 25 February 2022
Accepted: 21 March 2022
Published: 23 March 2022

1. Introduction

In the past few years, air pollution problem in China, especially Beijing, has been so severe that it has received widespread attention from all over the world. Cities are dense areas of economic activities, and therefore, populations, and Beijing is the political and economic center of China. After a stage of radical pursuit of economic growth, improving air quality and overall living environment is the current focus of China's realization of green development. Therefore, it is important to study Beijing's air quality issues to find ways to tackle air pollution problems and provide a reference for other cities.

The correlation between human activities and the atmospheric system in urban ecosystems has been increasing year by year [1]. Domestic research on air quality conditions began in the 1990s, behind abroad [2]. In terms of air quality characteristics, some researchers have studied the temporal and spatial distribution characteristics of China's AQI, finding that the national air quality situation shows a spatial clustering effect. High pollution and low pollution regions show a pattern of north–south differentiation, and the overall air quality of the country shows the distribution characteristics of slightly lighter in the south and lighter in the east [3,4]. The AQI showed a downward trend from 2016 to 2019, showing a "U" shape in the middle of the month [5]. The state has put forward clear pollution control requirements. Many local government departments regard pollution prevention and control as their primary task, and air quality monitoring has become an urgent need. Therefore, an accurate air quality forecast system can reflect the air conditions promptly and provide preparation information for the Ministry of Environmental Protection [6].

The statistical forecast is to analyze data through mathematical modeling, using correlation analysis [7], multiple regression [8,9], principal component analysis [10], gray model [11], fuzzy comprehensive evaluation method [12], harmonic regression [13], and other methods to predict air quality. However, it is difficult to provide air quality data in a timely and rapid manner due to the long forecast period. With the development of data collection and processing technology and the integration of various disciplines, data mining and machine learning methods have been used to analyze environmental information, and obtain timely and accurate air information and provide guiding suggestions [14–18].

Firstly, this paper compares data of Beijing's AQI and the concentration data of six major pollutants from 2019 to 2021, comprehensively evaluates its air quality, and explores factors affecting the air quality. Secondly, it uses time series models and data mining methods to establish predictive models. The ARIMA model is constructed based on the time series data of AQI, and the three-layer neural network model is constructed based on the daily average data and the data of the concentration of six major pollutants. Finally, study shows that the two models are effective for AQI to make short-term forecasts and analyses. Furthermore, this paper analyzes the long-term forecast of Beijing's air quality index based on the seasonal ARIMA model and compares it with the short-term forecast to draw a comprehensive conclusion, which could be helpful to provide references for relevant departments for urban air and environmental governance.

Compared with previous research on air quality, this paper not only uses the combination of visual analysis and time series model, but also considers the delayed effect of air pollution. On the basis of short-term forecast, the long-term forecast of air quality index is added, which makes the results more convincing and representative. Additionally, this paper includes a cluster analysis on the air quality index of Beijing in different periods and a multi-layer perceptron (MLP) neural network model based on the built-in algorithm of data mining technology to classify and evaluate the air quality level of the city. Finally, the classification rules of the six pollutants are used to explore a classification model with high accuracy, and a comprehensive comparison is made with the previous descriptive analysis, which effectively avoids the problems of chance and errors caused by the use of a single method. Research results are time-sensitive and have strong practical significance.

2. Analysis of Beijing's Air Quality

According to the "Technical Regulations on Ambient Air Quality Index" [19], the sub-index of the air quality of each pollutant is calculated based on the monitoring concentration of pollutants. The lower the value of AQI, the better the air quality; on the contrary, the higher the value, the worse the air quality is. This paper crawls the daily and monthly data of Beijing's AQI and six pollutant concentrations from 1 January 2019 to 15 November 2021, including nine effective fields such as date, AQI value, and air quality grade. After data screening and testing, there is no missing information or errors, with a total of 1050 valid data points day by day.

2.1. *Analysis of the Change Characteristics of AQI*

2.1.1. Temporal Characteristics of Beijing's AQI over the Years

Due to the different number of days in February each year and the fact that the data after November 2021 are not used, the blank data are represented by the value 0. The number of good days of air quality is an important index to measure the quality of air quality in a city. It can be seen from Figure 1 that the proportion of excellent and good air quality in the whole year has increased significantly year by year. The proportion of light pollution level decreased compared with the previous two years. As of 31 October, although the moderate pollution in 2021 increased slightly compared with 2020, it still decreased significantly compared with 2019. Only two days of serious pollution occurred in 2021, indicating that the air quality in Beijing fluctuated greatly in 2021, but the air quality was excellent, accounting for 29.93%, and the air quality was good, accounting for 47.37%. Although some extreme weather conditions occurred, the overall trend of air quality throughout the year was good.

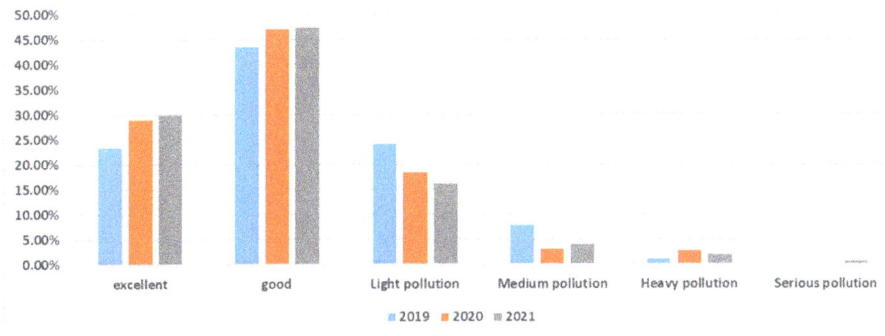

Figure 1. 2019–2021 Beijing's AQI level comparison chart.

2.1.2. The Characteristics of Monthly and Seasonal Changes in Beijing's AQI from 2019 to 2021

According to the four seasons in the northern hemisphere, spring is from March to May, summer is from June to August, autumn is from September to November, and winter is from December to February [20]. We analyzed the monthly data of Beijing's AQI, as shown in Figure 2. The monthly average of AQI reached a peak of 107.33 in June, followed by March, and the monthly average of AQI was 99. The least polluted month of the year is October. Seasonal changes in air quality are further considered and an AQI seasonal index was derived. Compared with the overall average (81.18), the AQI in spring (87.67) and summer (91.56) was above average, whereas the AQI in autumn (66.17) and winter (79.33) was slightly below average. This is consistent with the conclusion of the monthly average of AQI, indicating that there are significant seasonal differences in air quality in Beijing, and the severe air pollution in summer may be related to the severe excess of ozone.

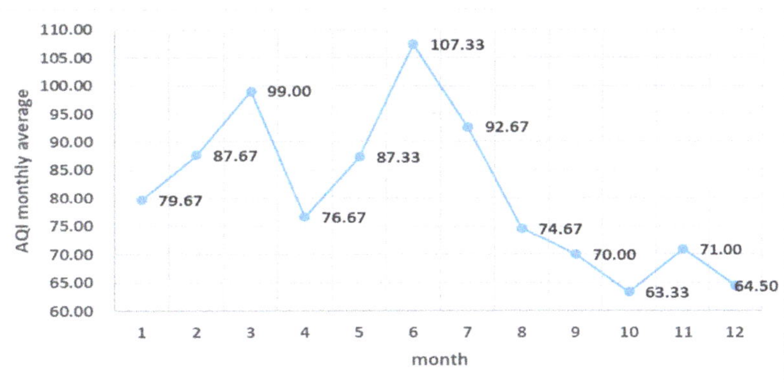

Figure 2. Visualization of the monthly average value of Beijing's AQI 2019–2021.

2.2. Analysis of the Change Characteristics of the Concentration of Six Pollutants

2.2.1. The Time Characteristics of the Concentration Changes of the Six Pollutants

The time trend of the annual average concentration of the six pollutants is shown in Figure 3. It can be seen from the chart that the annual average concentration changes of the six pollutants have the same trend, and they all show a trend of the declining year by year. Although the ozone concentration in 2021 has slightly increased compared with 2020, it still has a downward trend compared with 2019. It shows that Beijing's air control has been somewhat effective in the past three years, but in the process of air pollution control in the next few years, it is necessary to focus on the control of ozone pollution.

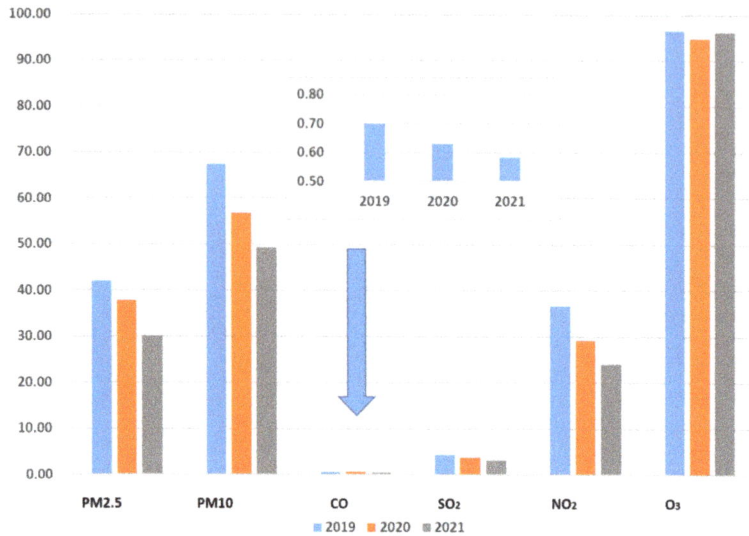

Figure 3. The temporal characteristics of the six pollutants in Beijing from 2019 to 2021.

2.2.2. Monthly and Seasonal Variation Characteristics of the Six Pollutants

Cluster analysis is carried out on the average values of six types of air pollutants in each month, as shown in Figure 4. The color depth in the heat map indicates the expression amount of the index. The greater the expression amount, the darker the color. The tree on the left shows the clustering results of different months. It is found that a year can be divided into two parts by the expression of air pollutants. This corresponds to whether Beijing is in the heating period. During the heating period, a large amount of fuel is consumed and the exhaust emission increases, resulting in a high content of pollutants in the air; In the north, the weather is dry and cold in winter, the probability of people choosing to drive increases, and vehicles will produce a lot of tail gas. At the same time, atmospheric inversion often occurs, which is not conducive to air convection, and pollutants are difficult to diffuse, resulting in poor air quality.

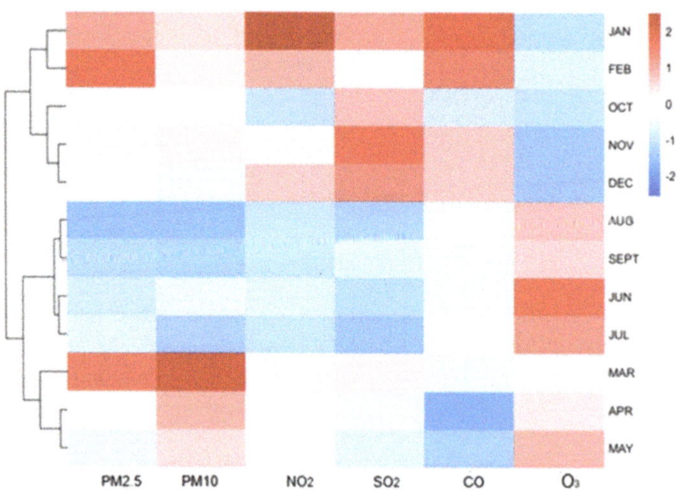

Figure 4. Heat map of six air pollutant indicators.

After the overall description of the six main pollutants in Beijing from 2019 to 2021, the concentration variation characteristics are analyzed monthly. It can be seen from Figure 5 that the particulate pollutant PM2.5. The concentration of PM10 is higher in spring and winter, and tends to be lower in summer. The lowest mass concentration occurs in August. The change trend of the concentration values of pollutants CO, SO_2 and NO_2 is roughly the same, showing a downward trend first and then an upward trend. On the whole, it presents a "U"-shaped structure, in which CO concentration reaches its lowest value in April; the concentration of SO_2 reached its lowest value in July. NO_2 also shows the variation characteristics of low concentration in summer and high concentration in winter. There is a small recovery from February to March, and the concentration reaches its lowest value in July. It can be seen that the concentrations of these five pollutants are lower in summer, lighter in pollution and higher in autumn and winter

As mentioned above, the monthly average index of Beijing's AQI showed the highest trend in June, which may be caused by the increase in ozone concentration. In order to verify our conjecture, a monthly analysis of ozone concentration from 2019 to 2021 was carried out. It can be seen from the monthly change trend of ozone that the annual ozone concentration reaches its peak in June and is lower in December, showing an inverted U-shaped structure different from the other five types of pollutants. Then, a seasonal analysis of the concentrations of six pollutants was conducted, finding that the concentrations of the five major pollutants, as shown in Figure 5a–f PM2.5, PM10, SO_2, CO, and NO_2, are low in summer and high in winter, whereas the concentration of O_3 was the opposite, high both in summer in winter. The concentration is lower, which indicates that the poor air quality in summer is mainly caused by the increase in ozone concentration.

2.3. Correlation between AQI and the Concentration of Six Pollutants

Correlation analysis is carried out on the data of six pollutant indicators every day. The main reference indicator is the Pearson correlation coefficient, which is used to measure the degree of correlation between two variables. In the correlation heat map, the numbers in the grid are the correlation coefficients, the red squares indicate the positive correlation between the indicators, and the blue squares indicate the negative correlation between the indicators. The heavier the color, the stronger the correlation between the indicators. It can be seen from Figure 6 that at a significance level of 5%, the correlation between particulate pollutants PM2.5 and PM10 is the largest, with a correlation coefficient as high as 74%, whereas the content of O_3 is not significant between PM2.5, PM10, and NO_2. The correlation between the amount of ozone and the content of PM2.5, PM10, and NO_2 does not affect each other; the content of PM2.5, PM10, and O_3 has a relatively large and positive correlation with AQI, that is, these three the higher the concentration of these pollutants, the larger the corresponding AQI and the worse the air quality. This is exactly the same as the results of the previous analysis.

In short, Beijing's AQI and the concentration characteristics of the six major pollutants are analyzed together, and the time development trend and the changes in months and seasons are visualized to provide a comprehensive and intuitive understanding. The current situation of air pollution in Beijing in the past three years. Studies have shown that the three indicators that have the greatest impact on AQI are PM2.5, PM10, and O_3. Beijing's air quality changes show obvious seasonal characteristics. In the past three years, Beijing's air quality reached the worst in June. This is due to the significant increase in ozone concentration during summer. However, on the whole, Beijing's AQI and the concentration of six pollutants have shown a trend of the declining year by year. AQI has improved significantly, with more and more days showing good grades, mainly due to the recent years. The government has taken many measures to improve air quality and can provide timely countermeasures when there are significant changes in air quality, especially in the treatment of pollutant emissions during the heating period in winter. However, the continuous high content of ozone is still a thorny issue facing today. Therefore, in the future air pollution control, in addition to continuing to control particulate pollutants in winter, we should also focus on ozone pollution in summer.

Figure 5. Monthly average data of the six major pollutants in Beijing for the same period of the three years from 2019 to 2021. (**a**) Monthly average data of the PM2.5 concentration in Beijing for the same period of the three years from 2019 to 2021. (**b**) Monthly average data of the PM10 concentration in Beijing for the same period of the three years from 2019 to 2021. (**c**) Monthly average data of the CO concentration in Beijing for the same period of the three years from 2019 to 2021. (**d**) Monthly average data of the SO_2 concentration in Beijing for the same period of the three years from 2019 to 2021. (**e**) Monthly average data of the NO_2 concentration in Beijing for the same period of the three years from 2019 to 2021. (**f**) Monthly average data of the O_3 concentration in Beijing for the same period of the three years from 2019 to 2021.

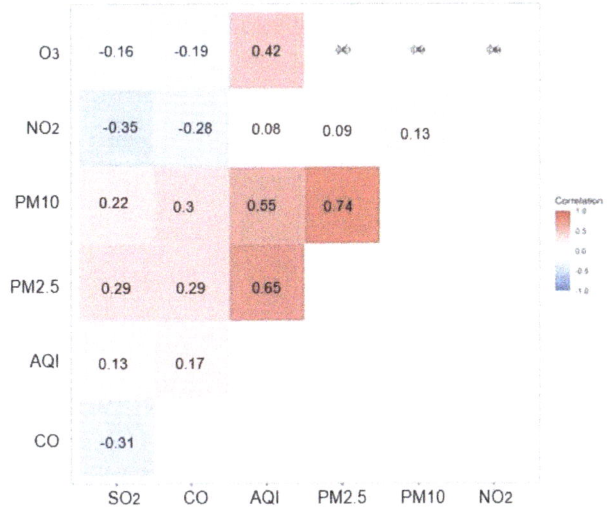

Figure 6. Correlation heat map of six air pollutant indicators.

3. A Forecast of AQI

3.1. Establish an ARIMA Forecast Model

3.1.1. Data Selection and Description

When using the time series analysis method to predict Beijing's AQI, considering the completeness of time and the accuracy of the forecast, we selected the AQI data from 1 November 2020 to 31 October 2021 as a training set to build an ARIMA model and make forecasts, with a total of 365 valid data points; we selected the AQI data from 1 November 2021 to 15 November 2021 as the test set to verify the fitting effect of the model.

3.1.2. Empirical Analysis of the ARIMA Model

(1) The stability test of the original sequence

Time series mapping of AQI of Beijing 1 November 2020–31 October 2021, as shown in Figure 7. From the time series chart, it can be seen that in the past year, Beijing's AQI fluctuated greatly: there were two abnormal peaks, and the AQI value was not always in a constant value near the fluctuation. In order to further verify the stability of Beijing's AQI, we have carried out a graph test of the self-correlation coefficient.

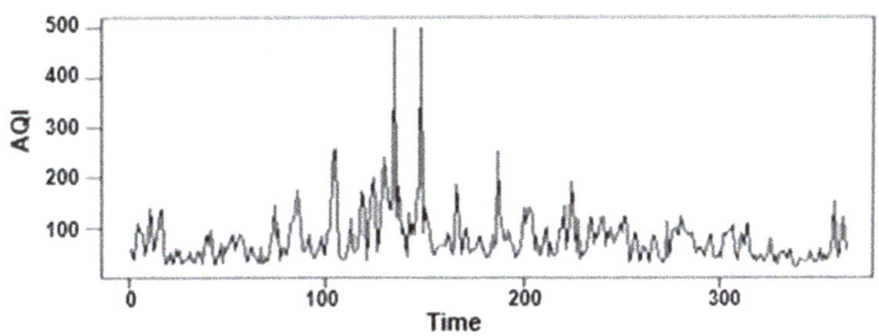

Figure 7. Time series of Beijing AQI original data.

Observing from Figures 8 and 9, the autocorrelation coefficient of the original Beijing's AQI has long-term training, the rate of decay to zero is relatively slow, and the self-correlation coefficient after decay to double the standard deviation has a cyclical trend, which directly indicates that there is a long-term correlation between the original time series data. Since the method of graph test has a certain degree of subjectivity, it is further tested by unit root test, and the results are consistent. Therefore, the sequence can be judged to be non-stable.

(2) Smooth processing of data

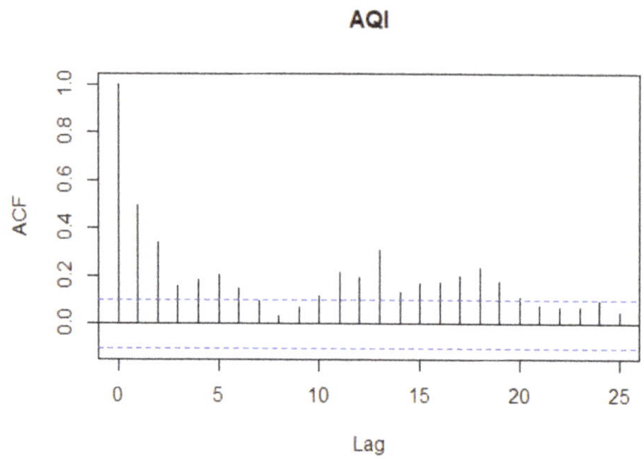

Figure 8. An apocalypse coefficient plot of the original AQI data in Beijing.

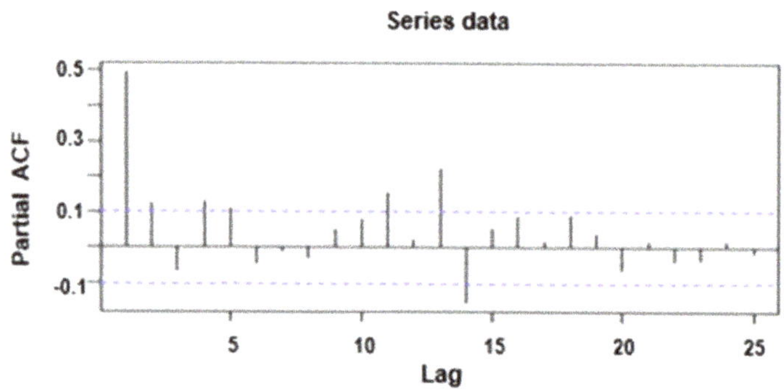

Figure 9. Partial correlation diagram of the original AQI data in Beijing.

The original data is smoothed, and the first-order difference is made on it, and the time series diagram of AQI after the first-order difference is drawn. Verify the stationarity of the sequence after the first-order difference. It can be seen from Figure 10 that after the first-order difference of Beijing's AQI, the time series graph fluctuates around a constant value, essentially showing a steady state.

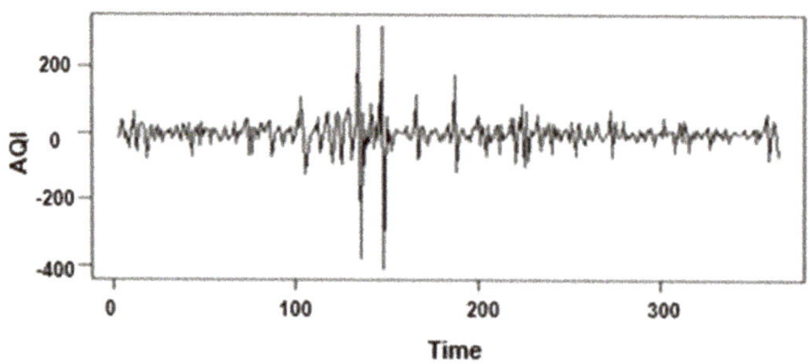

Figure 10. Time series diagram of Beijing's AQI after first-order difference.

In order to further verify the conjecture, the autocorrelation and partial autocorrelation coefficient graphs of AQI after the first-order difference was made. As shown in Figures 11 and 12, the autocorrelation coefficient of the Beijing's AQI series after the first-order difference quickly decays to zero, indicating that the series after the first-order difference has a short-term correlation, and it is preliminarily determined that the series is stable after the difference. The unit root test method is used again to verify as shown in Table 1. The p value is less than 0.01, and the null hypothesis of the unit root is rejected, which is consistent with the conclusion of the graph test; that is, the AQI data is stable after the first order difference.

(3) White noise test of stationary series

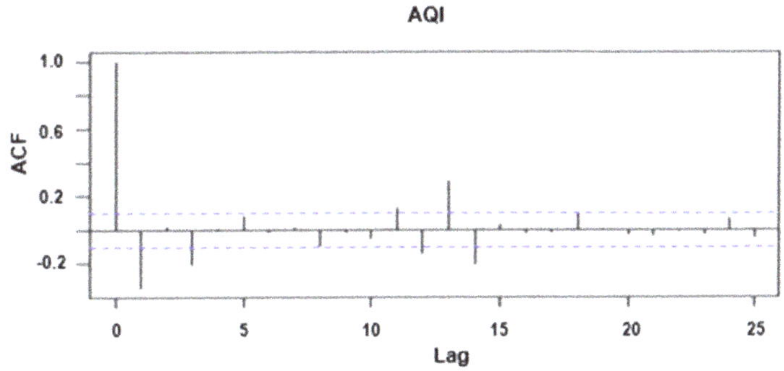

Figure 11. Autocorrelation coefficient after the first-order difference of Beijing AQI.

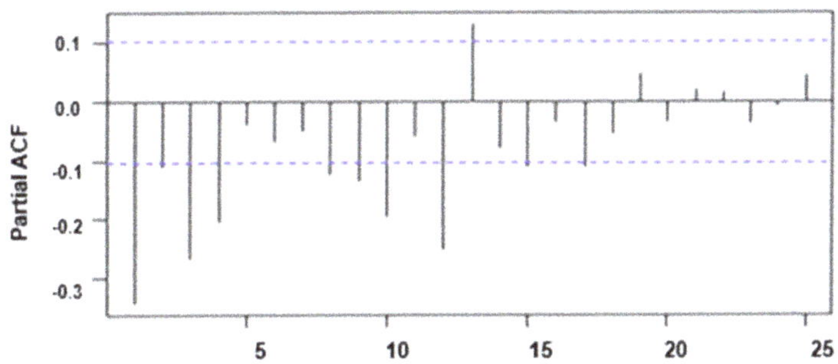

Figure 12. The partial autocorrelation coefficient after the first-order difference of Beijing AQI.

Table 1. Unit root test of the first-order difference sequence.

ADF Statistics	p-Value
−9.8993	<0.01

After the differentiated sequence passes the stationarity test, it needs to be tested for pure randomness to prevent the stationary sequence from being a pure white noise sequence. If it is a white noise sequence, then the sequence does not have any value for further research. Therefore, the R software is used to perform the Ljung–Box method for the pure randomness test of the 6-period and 12-period lags on differential AQI column, and the results are listed in Table 2. The analysis shows that in the first-order difference series with a lag of 6 and 12 periods, the p-value is less than the significance level of 0.05. Therefore, the null hypothesis that the differenced series is a white noise series is rejected, and the result is significant. From this, it can be judged that the time series after the first-order difference is not a purely random white noise series, it can be researched to a certain extent, and the subsequent modeling analysis can be carried out on it.

(4) Identification and order determination of ARIMA model

Table 2. White noise test of difference sequence.

X-Squared	df	*p*-Value
60.504	6	3.555×10^{-11}
78.253	12	8.878×10^{-12}

Since the autocorrelation coefficient and the order of the partial autocorrelation coefficient of the model after the first-order difference are not obvious, the identification and order determination of the ARIMA model has brought some obstacles. Therefore, using the automatic order-setting model of R software and various manual repeated attempts, many more reasonable models are compared and analyzed, and the results are listed below. The results of other attempts are not listed one by one. According to the model's Akaike information criterion, the smaller the model's AIC value, the better the model's fitting effect. After repeated experiments, it can be found that the model with more significant parameters is ARIMA (5,1,4), and the fitting results are shown in Table 3.

(5) ARIMA model fitting effect test

Table 3. Model fitting results.

ARIMA Model	σ^2 Estimated	Log-Likelihood	Aic
ARIMA (5,1,2)	1864	−1888.06	3792.12
ARIMA (5,1,4)	1822	−1885.28	3790.57
ARIMA (4,1,2)	1885	−1890.33	3974.67

Perform a residual white noise test on the fitted ARIMA (5,1,4) model to see whether it extracts the effective information completely, as shown in Table 4 below. After analysis, it can be seen that the *p*-values of the lag 6 and 12 lag white noise tests are both greater than the significance level of 0.05, and the original hypothesis that the residual is white noise cannot be rejected, indicating that the fitting model has basically extracted effective information, and the model is the significant sex.

(6) Model forecast

Table 4. Residual white noise test after model fitting.

X-Squared	df	*p*-Value
7.3767	6	0.2874
11.792	12	0.4625

Use the established ARIMA (5,1,4) to predict AQI for the next 30 periods, and select the data of the first 15 periods in the future as the test set to verify the error of the forecast model. The forecast results are shown in Figure 13. It can be seen from the forecast result graph that in the next 30 days, although the air quality in Beijing will fluctuate to a certain extent, the overall difference is not big, and the air quality is still showing a good trend.

In order to consider the fitting effect of the model, we use 1 November 2021–15 November 2021 as the test set, compare the predicted value obtained by the ARIMA (5,1,4) model with the true value, and combine the error indicators specified in the previous section to compare the results Include under. The comparison of the predicted value and the real value of the model, the histogram of the absolute error and relative error, as shown in Figures 14–16, indicates that the error between the real value and the predicted value in the first six cycles is slightly larger, but after the six periods, the predicted value gradually moves closer to the true value. In the thirteenth issue, the difference between the predicted value and the true value is very small, indicating that the fitting forecast result of the ARIMA model is more accurate, but there is still room for improvement.

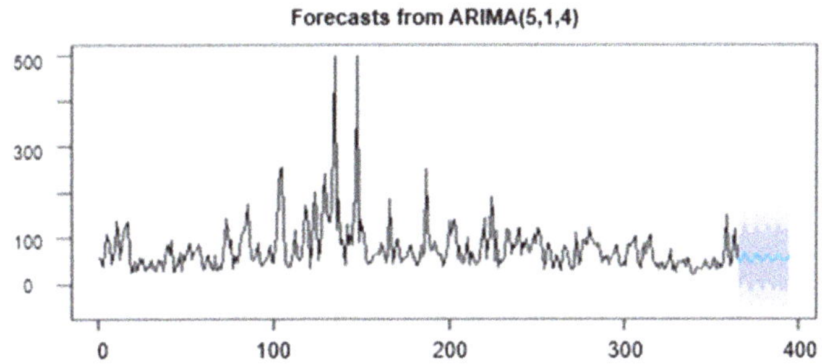

Figure 13. ARIMA model predicts Beijing's AQI map.

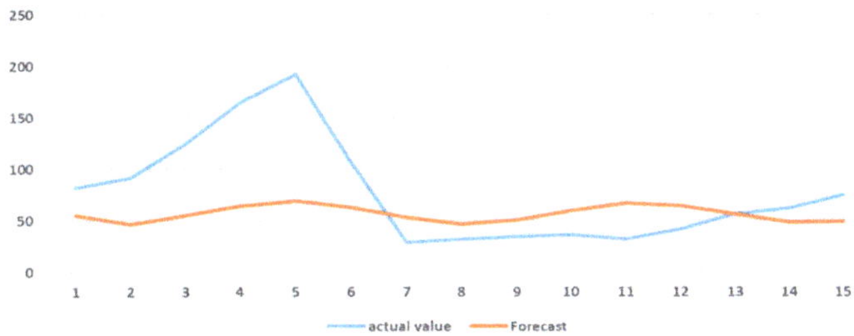

Figure 14. Comparison of predicted value and the true value of ARIMA model.

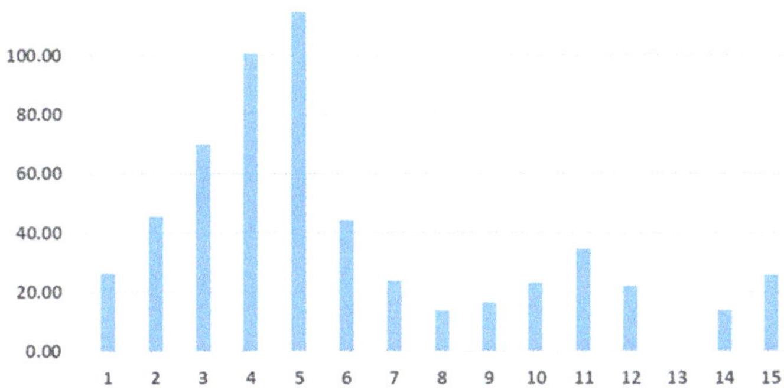

Figure 15. Absolute error histogram of ARIMA forecasting model.

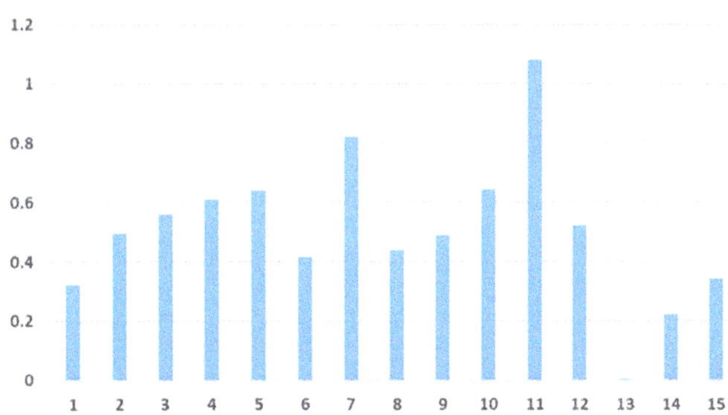

Figure 16. The relative error histogram of the ARIMA forecasting model.

3.2. *Establish a Neural Network Forecast Model*

3.2.1. Data Selection and Description

The data used to build the neural network model are the daily average data of Beijing's AQI from 1 November 2020 to 15 November 2021, and the concentration of six major air pollutants, a total of 380 items, and the most classic three-tier structure to build a model. Through the dimensionless processing of the data, the construction of a training sample set, and a test sample set to build a model, the model learns the changing laws of historical pollutant concentration data and realizes short-term forecast of air quality changes.

3.2.2. Theoretical Overview of Neural Network Models

The neural network is derived from neuroanatomy and neurophysiology. It is a technology that simulates the intelligent processing of the human brain. It is a mathematical model of the structure and function of biological neural networks and has the ability to process multi-dimensional functions. The neural network structure is composed of multiple neurons combined with each other. Each neuron input has a specific weight, and the learning process of the neural network is the process of constant adjustment of the weight in the iterative process.

The neural network consists of three layers: input layer, hidden layer and output layer. The input layer is not responsible for calculation but is mainly responsible for the information of input variables. The number of nodes is the number of influencing factors designed; the hidden layer is between the input layer and the output layer. The middle of the output layer contains unobservable network nodes, which mainly transform the sample variables. Each hidden node is a function of the sum of input weights so that it has corresponding learning rules to train the network; the output layer is mainly responsible for outputting the final forecast. As a result, it transmits information to the outside world, and the number of nodes is the number of predictors required. The sample is usually divided into a training set and a test set. The training set is used to build the model, and the test set is used to test the fitting effect of the model.

Artificial neural networks are increasingly closely related to other subject areas. The neural network field mainly includes multilayer perceptron models, back-propagation neural networks, convolutional neural networks, and so on. The air quality seems disorderly on the surface, but its changing law is affected by many factors such as pollution sources, coal burning, and transportation for a long time. It is a complex non-linear system. Multi-layer perceptron is also called multi-layer feedforward neural network. Information is transmitted in one direction and different layers are fully connected. It has an excellent nonlinear mapping and generalization capabilities and can perform air quality control based on the inherent connection of the data itself.

Based on the built-in algorithm of data mining technology, this paper establishes a neural network model of multi-layer perceptron (MLP) to predict the air quality in Beijing, correlate various dimensions in a large amount of data, train and learn the data, and mine the associated information of the data. In total, 80% of the data is selected as the training set for the learned model by a fixed random seed number, and the remaining 20% of the sample data is used for testing.

3.2.3. Empirical Analysis of Neural Network Model

The fitting results of the true air quality value and the predicted value of the model are shown in Figure 17, and the changes between the two are basically the same. The effect of the model is judged by indicators such as average absolute error (MAE) and average absolute percentage error (MAPE). The smaller the value, the higher the accuracy of the model.

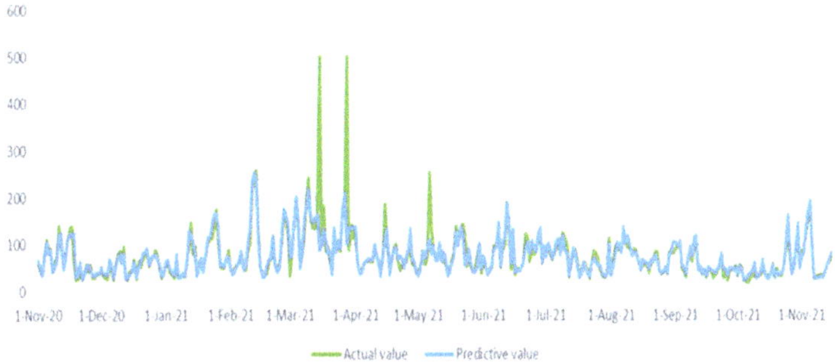

Figure 17. Neural network model forecast diagram.

After obtaining the concentration values of the six major pollutants, the high/low values of the concentration limits close to the concentration of the pollutants, and the corresponding air quality sub-index, the air quality sub-index is calculated, and then the maximum AQI value is selected as AQI. The final selection of values takes a long time. Although accurate AQI values and levels can be obtained, there is a lag, and air quality early warning cannot be effectively provided.

Therefore, this article is also based on the data mining algorithm and uses the daily data of six pollutants to classify the air quality levels layer by layer, through regularization and the use of Dropout and other methods to avoid data overfitting, the air quality in Beijing on the day was finally determined. The pollutants that have the greatest impact on air pollution levels can also be obtained. The accuracy of the model is 89.8%. The classification levels of air quality results are shown in Figure 18.

Actual value \ Predictive value	1	2	3	4	5
1	92.6%	7.4%	0.0%	0.0%	0.0%
2	3.6%	94.2%	2.2%	0.0%	0.0%
3	2.7%	13.5%	83.8%	0.0%	0.0%
4	0.0%	11.1%	33.3%	55.6%	0.0%
5	0.0%	0.0%	16.7%	50.0%	33.3%

Figure 18. Accuracy of the classification model.

The daily air quality level is used as the typed dependent variable, and the six air pollutant indicators are used as independent variables for importance analysis. It can be seen from Figure 19 that PM2.5, PM10, and O_3 have a greater impact on the air quality level, and are the main factors that determine the specific value of AQI and the air quality level. Their importance is 23%, 19%, and 18%, respectively. The result corresponds to the result of the correlation heat map. PM2.5 has the greatest impact on air quality. It stays in the air for a long time and is rich in a lot of harmful substances. It not only affects human health but also affects the global climate. Air governance is not one day's work; thus, it is necessary to accelerate industrial transformation, advocate the use of clean energy by society, and strengthen waste gas treatment procedures.

Figure 19. Analysis of the importance of six air pollutant indicators.

3.3. Model Comparison

In this chapter, the ARIMA model and the multi-layer perceptron neural network model are used to predict Beijing's AQI, and a variety of evaluation indicators such as root mean square error and average absolute percentage error are selected for these two models. Comparing the fitting effect of the model, these indicators take into account the error between the predicted value and the true value, which can be a good analysis of the pros and cons of the model fit. The expressions of these indicators are:

$$MAE = \frac{1}{n}\sum_{i=1}^{n}|y_i - \hat{y}_i| \tag{1}$$

$$RMSE = \sqrt{\frac{1}{n}\sum_{i=1}^{n}(y_i - \hat{y}_i)^2} \tag{2}$$

$$MAPE = \frac{1}{n}\sum_{i=1}^{n}\left|\frac{y_i - \hat{y}_i}{y_i}\right| \times 100\% \tag{3}$$

$$SMAPE = \frac{1}{n}\sum_{i=1}^{n}\frac{|y_i - \hat{y}_i|}{(|y_i| + |\hat{y}_i|)/2} \times 100\% \tag{4}$$

Among them, is the predicted value of the test set data, and is the true value.

Combining the selected test set data, using these four indicators to compare the pros and cons of the two models, the results are shown in Table 5. Combined with error analysis, it is found that the forecast effect of the neural network model is far better than that of the time series ARIMA model. However, the ARIMA model also has certain advantages. It can predict future data based on historical data of endogenous variables without using other variables. The forecast results within a certain period still have a certain reference value. The neural network model has a good self-adaptive ability, and the forecast results are relatively good, but it cannot accurately and specifically describe the mathematical relationship between data and variables. Therefore, in the process of use, the two forecast methods can be combined, or further consider the related influence of other factors, and further optimize the forecast of Beijing's AQI.

In this chapter, the data from 1 November 2019 to 15 November 2021 are selected, and Beijing's AQI is fitted and predicted using the ARIMA model and the neural network model. Among them, the ARIMA model uses the data of the last 15 days as the test set, and the neural network randomly selects the test set using the eight-to-eight method and uses the error-index to comprehensively evaluate the fitting effect of the two models. The results show that both the ARIMA model and the neural network model are significant in predicting AQI, and the established models are reasonable and effective. Through comparison, it is found that the fitting effect of the neural network is better than that of the ARIMA model. The features can be referred to each other and used in combination.

Table 5. Comparison of model errors.

Model	MAE	RMSE	MAPE	SMAPE
ARIMA model	39.06	51.21	50.84%	51.93%
Neural network model	13.66	43.17	12.81%	12.75%

4. Long-Term AQI Forecast Based on Seasonal Model

Based on the results of the previous analysis of Beijing air quality visualization, it can be seen that the AQI of Beijing shows more obvious seasonal characteristics. In the above paper, short-term forecast was made for the daily data of Beijing AQI, and in this chapter, long-term forecast of Beijing AQI is made based on the seasonal model of ARIMA model, so the model was built by selecting 83 monthly air quality data from January 2015 to November 2021. Therefore, 83 monthly air quality data points from January 2015 to November 2021 were selected as the experimental data, and the data from December 2021 to February 2022 were used as the data set to verify the model fitting effect, and the original monthly data points were pre-processed in the following section.

4.1. Data Preprocessing

(1) Smoothness test

The time series plot of the AQI monthly data is drawn using R software, as Figure 20 shown below. From the time series plot of the monthly data, it is known that from January 2015 to November 2021, the overall AQI of Beijing shows a decreasing trend and has a more obvious seasonal effect. Subsequently, the graphical test of autocorrelation and partial autocorrelation coefficients was conducted, as shown in Figures 21 and 22, and its autocorrelation coefficients have a long-term trailing and periodic trend, and the monthly data of Beijing air quality index is initially inferred to be unsteady by the graphical test observation, and in order to further evaluate objectively the steadiness, the series is concluded to be a non-steady time series after unit root test using R software.

(2) Pure randomness test

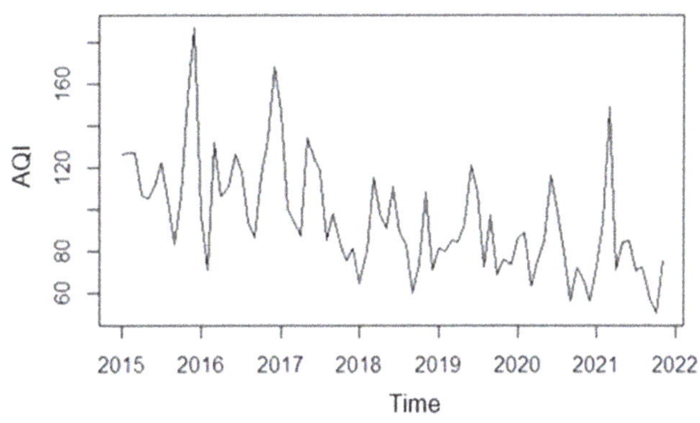

Figure 20. Time series of Beijing AQI monthly data.

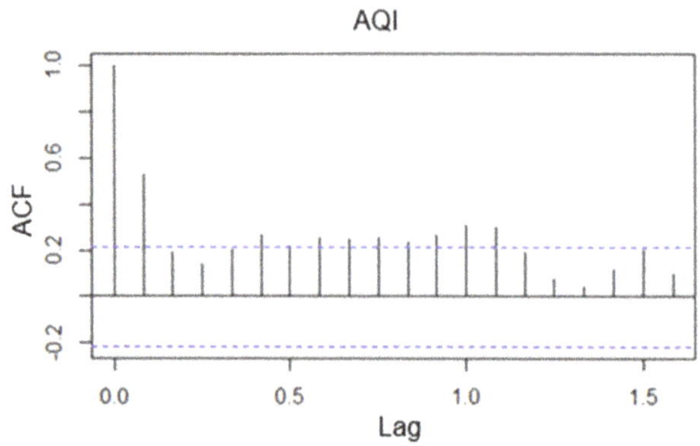

Figure 21. Monthly AQI series autocorrelation chart.

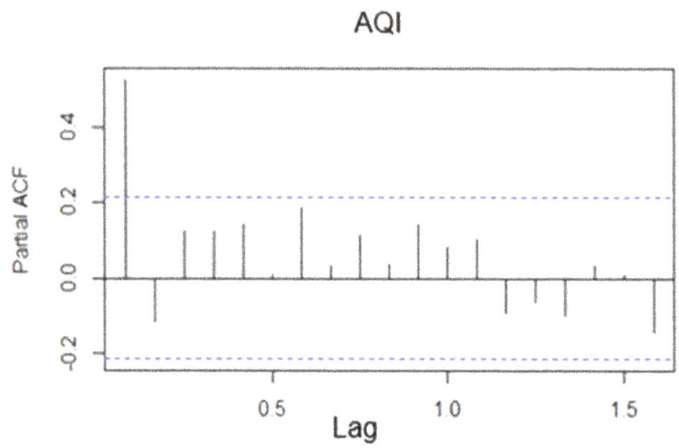

Figure 22. Monthly data AQI bias autocorrelation graph.

Similar to the ARIMA model for short-term forecast of AQI daily data, a pure randomness test is performed on the original series before the modeling analysis in order to investigate whether there is any correlation between the series and whether there is value for further study. The pure randomness test was performed using the Box.test function in the R software, and the results are shown in Table 6. It can be seen that the p-values of delayed 6 periods and delayed 12 periods are significantly less than 0.05; therefore, the original hypothesis is rejected, and the monthly data series of Beijing AQI is not a white noise series, which can be used for subsequent modeling analysis.

Table 6. Results of pure randomness test.

Delayed Orders	p-Value
Delayed by 6 periods	2.754×10^{-7}
Delayed by 12 periods	3.188×10^{-11}

4.2. Construction of Seasonal Model

From the above time series graph, we can see that the original series shows the change of year as the cycle, and the selected air quality data is monthly data, so the cycle length s = 12. To make the original time series smooth, we need to eliminate the linear trend and seasonal periodicity of the series. Therefore, the monthly AQI data of Beijing are first differenced to eliminate the linear trend, and then differenced to eliminate the seasonal periodicity in 12 steps. The series after the first-order twelve-step differencing is denoted as AQI-diff12, and its time series is plotted as shown in Figure 23. The series after trend differencing and seasonal differencing has no obvious upward or downward trend and no obvious periodicity, fluctuating around the zero value, which can be initially judged as a smooth time series after differencing. The autocorrelation coefficients and partial autocorrelation coefficients of the series after differencing are verified by the graph test method, as shown in Figures 24 and 25. The autocorrelation coefficient quickly decays to zero, and the p-value of the pure randomness test of the differenced series is 0.01, which is smaller than the significance level of 0.05. The original hypothesis is rejected, indicating that the series is smooth after eliminating the linear trend and seasonal trend. The p-value of the differenced series after the pure randomness test is 0.019, which is less than the significance level of 0.05. Therefore, the differenced series is still a non-white noise series, and the next modeling analysis is conducted for this series.

(1) Model identification and model ranking

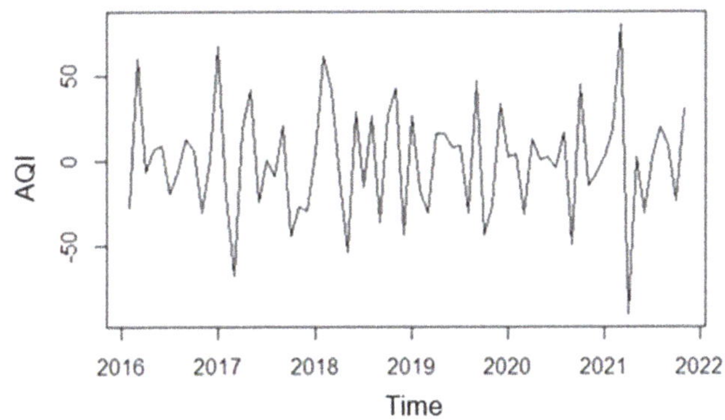

Figure 23. Time series of monthly AQI data after first-order twelve-step differencing.

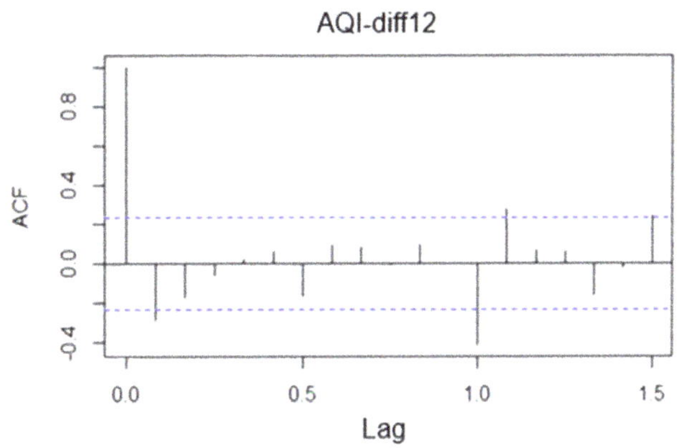

Figure 24. Autocorrelation of monthly AQI data after first-order 12-step differencing.

Based on the above autocorrelation and partial autocorrelation plots after differencing, the first step is to consider the characteristics of the autocorrelation coefficients and partial autocorrelation coefficients within 12 orders of the series after trend differencing and seasonal differencing in order to determine the short-term correlation model. In the autocorrelation and partial autocorrelation plots of the differenced series, the autocorrelation coefficients and partial autocorrelation coefficients up to order 12 are not truncated, so an ARMA(1,1) model is attempted to extract the short-term autocorrelation information of the differenced series.

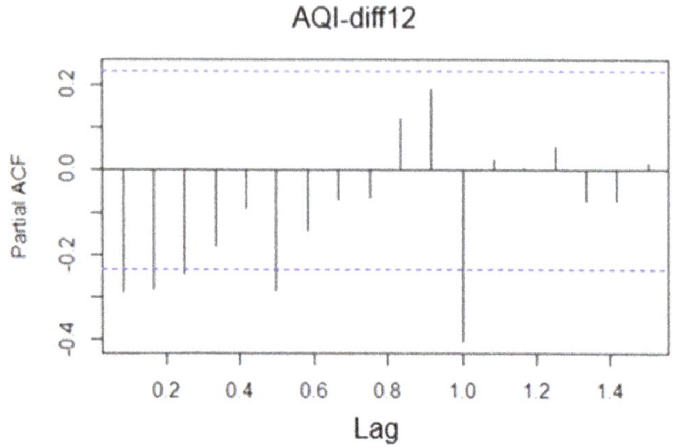

Figure 25. Biased autocorrelation of monthly AQI data after first-order 12-step differencing.

The second step considers the autocorrelation characteristics of the season in question in order to confirm the choice of additive or multiplicative seasonal model. The approach is to consider the characteristics of autocorrelation coefficients and partial autocorrelation coefficients in autocorrelation plots and partial autocorrelation plots with delayed 12th order, 24th order, etc. with the length of the period as the unit. According to the autocorrelation and partial autocorrelation plots, the autocorrelation coefficients and partial autocorrelation coefficients of the delayed 12th and 24th orders fall within the range of 2 times the standard deviation, and the corresponding values of the delayed 24th order are smaller, which shows that there is no significant seasonal effect in the differenced series, so we initially consider a simple seasonal model, i.e., an additive seasonal model. At this point, the seasonal differencing order D = 1, $p = 0$, and Q = 0.

Combined with the previous first-order twelve-step differencing information, the additive seasonal model fitting ARIMA (1,(1,12),1) was finally determined, and its model structure is as follows.

$$\nabla_1 \nabla^1 x_t = \frac{(1-\theta_1 B)}{(1-\phi_1 B)} \varepsilon_t \tag{5}$$

(2) Parameter estimation of the model

The final fitted model has been determined in the previous step of the analysis, and the next step is to determine the caliber of this model based on the observed values of the series, which means that the values of the unknown parameters in the fitted model need to be estimated. Using R software, the parameters of the fitted additive seasonal model were estimated according to the maximum likelihood estimation method, and the following results were obtained, as shown in Table 7.

Table 7. Estimated values of parameters.

Parameter	p-Value	Standard Deviation
$\hat{\phi}_1$	0.2895	0.1163
$\hat{\theta}_1$	−1.0000	0.0501

Based on the above results, the caliber of the fitted additive seasonal model can be seen as

$$\nabla_1 \nabla^1 x_t = \frac{(1+B)}{(1-0.2895B)} \varepsilon_t \tag{6}$$

where B is the delay operator and ε_t is the white noise sequence, i.e., $\varepsilon_t \sim WN(0, \sigma^2)$.

(3) Model testing

A white noise test of the residuals was performed on the established additive seasonal model in order to determine the significance of the model. Next, the Box.test function in the R software was used to test whether the residual series is a white noise series, and the test results are shown in Table 8. According to the results of the white noise test of the residuals, the p-value corresponding to the LB statistic at each order of delay is significantly greater than the significance level of 0.05; therefore, it can be considered that the residual series of the fitted additive seasonal model is a white noise series, which means that the established model is significantly valid.

Table 8. White noise test.

X-Squared	df	p-Value
2.0741	6	0.9128
18.393	12	0.1043

4.3. Forecast Analysis of the Additive Seasonal Model

Based on the established additive seasonal model, the Beijing air quality index from December 2021 to February 2022 was selected as the test set to verify whether the model had a more accurate fit. Using the same short-term correlation criteria as above, the results are shown in Table 9. As can be seen from the graphs, the differences between the predicted and true values are small and the error values are within acceptable limits, indicating that the additive seasonal model is appropriate and valid for extrapolating the future long-term Beijing AQI, with high forecast accuracy and reasonable and credible results.

Table 9. Model goodness of fit.

Model	MAE	RMSE	MAPE	SMAPE
ARIMA (1,(1,12),1)	15.55	24.87	34.57%	23.89%

The predicted results of Beijing AQI for the next 24 periods are shown in Figure 26. It is observed that the AQI index still shows seasonal cycles and still has a slightly decreasing trend in the next two years.

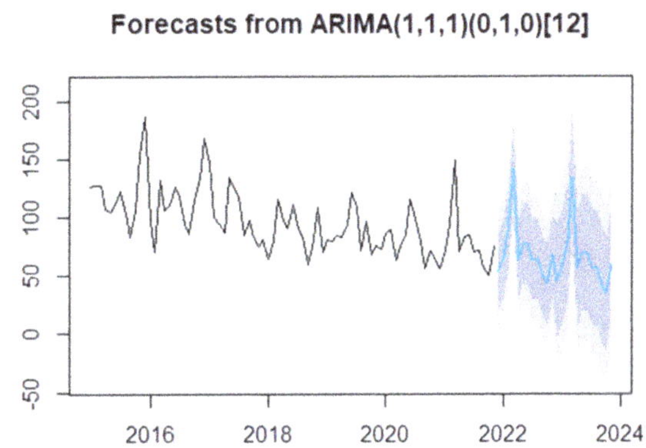

Figure 26. ARIMA model long-term forecast graph.

4.4. Section Subsection

In this section, the long-term forecast of AQI in Beijing is based on the seasonal model of ARIMA model, which shows an overall decreasing trend of AQI in Beijing from January 2015 to November 2021 with a more obvious seasonal effect. The parameters of the fitted additive seasonal model are estimated according to the maximum likelihood estimation method, and the AQI of Beijing from December 2021 to February 2022 is predicted according to the additive model, and the results show that the AQI still shows a seasonal cycle and still has a slightly decreasing trend in the next two years.

5. Summary and Outlook

Based on Beijing's AQI from January 2019 to November 2021 and the daily average and monthly data of six major air pollutants, this article uses descriptive statistical analysis, correlation analysis, and cluster analysis to visualize air quality development trends; Using time series analysis and data mining algorithms to build models and make short-term forecasts of Beijing's air quality, the following conclusions are obtained:

Using statistical methods to analyze the air quality level, AQI and the distribution of the six types of pollution concentration changes, the daily analysis results show that with the continuous deepening of air pollution prevention and control work, the air quality in Beijing continues to improve, AQI has improved significantly, and the level is excellent. The proportion of days has increased year by year. The monthly analysis results show that in the past three years, the air pollution level was the most serious in June, which was mainly related to the serious excess of ozone content. The changes in air quality in Beijing show obvious seasonal characteristics. The five main pollutants PM2.5, PM10, SO_2, CO, and NO_2 have low concentrations in summer and high concentrations in winter; only O_3 is the opposite of other pollutants. Because of the high concentration in summer and low concentration in winter, the persistently high content of ozone is still a thorny issue facing today, and the air quality varies greatly between the heating period and non-heating period.

The short-term forecast of Beijing air quality index using time series model and neural network model overcomes the lag of the current air quality monitoring system, and the AQI index high and low is determined by the co-construction of six air pollutants. The results show that both ARIMA model and neural network model are significant for the forecast of air quality index, and the established models are reasonable and effective, and it is found by comparison, the fitting effect of the neural network is better than that of the ARIMA model, but both models have their own characteristics. It was also found that PM2.5, PM10, and O_3 have a greater influence on the air quality class, and are the main factors to determine the specific value of AQI and air quality class. When using the additive seasonal model for long-term forecast of monthly data, it was found that the Beijing AQI still shows seasonal cyclicality and still has a slightly decreasing trend in the next two years. In summary, based on the conclusions of the article, we can propose measures to improve air quality from the three perspectives of the government, society, and individuals.

The government must increase implementation of environmental protection policies and investment in environmental protection technology. Environmental protection departments should strengthen environmental management, earnestly implement national and local laws and regulations, comprehensively use technical means and administrative measures, and manage air quality through legislation, monitoring, and protection.

The analysis shows that PM2.5, PM10, and O_3 have a greater impact on air quality levels. Therefore, environmental protection management agencies have been established at all levels from the central to the local level to use monitoring technology tools to publish monitoring data promptly, inspect and dispose of pollution sources, and control building dust, Pollution behaviors such as burning coal for heating and burning straw. Increase investment in the field of environmental protection technology, develop reasonable treatment equipment, reduce waste of resources, and improve sewage treatment technology. Optimize the industrial structure, lower pollution standards, and increase pollution punishment. Resource control policies such as pollutant discharge fees have a significant impact on pollution control costs. The development of a washing energy industry with high energy utilization and low pollution, and making good use of renewable resources such as solar and wind energy. Air pollution has fluidity and regional characteristics, and its changes are synchronized. Pollution between regions affects each other. Pollution prevention and control is not just an administrative region's problem. It is necessary to establish a regional cooperation system, regional joint prevention and control, to solve cross-regional air pollution problems, for example, the Beijing-Tianjin-Hebei simultaneous implementation of the "Regulations on the Prevention and Control of Emission Pollution from Motor Vehicles and Non-road Mobile Machinery", and so on. Improve urban green coverage, borrow the characteristics of plants to absorb dust and purify the air, provide zoning control strategies for the in-depth fight against pollution, and continue to promote precise, scientific, and legal pollution control.

The society must vigorously promote environmental protection knowledge, raise awareness of protecting the atmospheric environment, and advocate low-carbon life. Prevention and control work increasingly requires scientific and refined management. The city should adhere to project emission reductions and management emission reductions according to changes in air quality in months and seasons, and promote the formation of a spatial pattern, industrial structure and lifestyle that conserves resources and protects the environment. The aims are to deepen the "one microgram" action, focus on the coordinated governance of PM2.5, PM10 and O_3, and achieve green transformation of the industrial structure, green and low-carbon energy structure, green optimization of vehicle structure, and green and clean urban appearance.

Another aim is to establish an action pattern led by the government and public participation. With the expansion of the scale of cities and the improvement of the level of economic activities, the number of motor vehicles has increased, and cars emit a large amount of NO_2 and inhalable particulate matter, which will seriously damage the environment and affect people's health. Therefore, it is necessary to consciously eliminate old motor vehicles and improve awareness of the purification and treatment of polluting vehicle exhausts, supporting the development and use of new energy vehicles. The general public should actively participate in environmental protection activities and environmental protection supervision, consciously practice a simple and moderate, green and low-carbon lifestyle, and offer advice and suggestions for a more beautiful Beijing.

Author Contributions: T.L. conducted the analysis and drafted the manuscript; S.Y. initiated the study and conducted the data interpretation. All authors contributed to checking and proofreading. All authors have read and agreed to the published version of the manuscript.

Funding: This research received no external funding.

Institutional Review Board Statement: Not applicable.

Informed Consent Statement: Not applicable.

Data Availability Statement: The data presented in this study are openly available in the platform of China Air Quality Online Monitoring and Analysis at: https://www.aqistudy.cn/historydata/ (accessed on 24 February 2022).

Acknowledgments: The authors would like to thank the editor and anonymous reviewers for their valuable comments and suggestions to this paper.

Conflicts of Interest: The authors declare no conflict of interest.

References

1. Wu, J.; Yang, Z.H.; Hu, L. Study on the benefits of urban eco-spatial air quality improvement based on spatial dependence—Beijing as an example. *China Environ. Sci.* **2021**, 1–10. [CrossRef]
2. Ma, T. *Analysis and Prediction of AQI in Jinan*; Shandong University: Jinan, China, 2021. [CrossRef]
3. Zhang, X.; Luo, S.; Li, X.; Li, Z.; Fan, Y.; Sun, J. Temporal and Spatial Variation Characteristics of Air Quality in China. *Sci. Geogr. Sin.* **2020**, *40*, 190–199.
4. Xiao, Y.; Tian, Y.; Xu, W.; Liu, J.; Wan, Z.; Zhang, X.; Liu, X. Characteristics of Urban Air Pollution and its Socio-economic Impact in China. *Ecol. Environ. Sci.* **2018**, *27*, 518–526.
5. Guo, W.; Chen, Y.; Liu, G.; Song, K.; Tao, B. Analysis on the Characteristics and Influencing Factors of Air Quality of Urban Agglomeration in the Middle Reaches of the Yangtze River in 2016 to 2019. *Ecol. Environ. Sci.* **2020**, *29*, 2034–2044.
6. Liu, Y.; Zhang, Y.; Zhu, C.; Hao, J.; Liu, Q. Intelligent Forecasting and Monitoring of Air Index Based on Big Data and Internet of Things. *J. Commun.* **2017**, *38*, 129–138.

7. Zhou, M.; Yang, Y.; Sun, Y.; Zhang, F.; Li, Y. Analysis of air quality spatial and temporal distribution characteristics and influencing factors in Shandong Province from 2016 to 2020. *Environ. Sci.* **2022**, 1–12. [CrossRef]
8. Ziheng, J.; Xizhang, G.; Baolin, L.; Dechao, Z.; Jie, X.; Fei, L. Analysis on the temporal and spatial distribution pattern and influencing factors of air quality in Sichuan and Chongqing areas. *J. Ecol.* **2022**, *11*, 1–10.
9. Jędruszkiewicz, J.; Czernecki, B.; Marosz, M. The Variability of PM10 and PM2.5 Concentrations in Selected Polish Agglomerations: The Role of Meteorological Conditions, 2006–2016. *Int. J. Environ. Health Res.* **2017**, *27*, 441–462. [CrossRef] [PubMed]
10. Lu, B. Application of Statistical Analysis and Machine Learning Method to Air Quality Prediction in Yinchuan City. In *Engineering Science and Technology*; CNKI: Beijing, China, 2021; Volume 1. [CrossRef]
11. Wang, W.; Zheng, Z.; Liang, X.; Zhou, H. Application of Grey System Model in Forecasting Ambient Environment Quality in Jinan City. *J. Green Sci. Technol.* **2013**, *3*, 123–126.
12. Dong, S.; Su, T.; Huang, L.; Lv, Z. Fuzzy Comprehensive Evaluation and Curve Fitting Prediction Models for Air Quality. *Math. Pract. Theory* **2018**, *48*, 102–108.
13. Akdi, Y.; Gölveren, E.; Ünlü, K.D.; Yücel, M.E. Modeling and forecasting of monthly PM2.5 emission of Paris by periodogram-based time series methodology. *Environ. Monit. Assess.* **2021**, *193*, 622. [CrossRef] [PubMed]
14. Wu, Q.; Lin, H. Daily Urban Air Quality Index Forecasting Based on Variational Mode Decomposition, Sample Entropy and LSTM Neural Network. *Sustain. Cities Soc.* **2019**, *50*, 101657. [CrossRef]
15. Rahimi, A. Short-term Forecast of NO_2 and NOx Concentrations Using Multilayer Perception Neural Network: A Case Study of Tabriz, Iran. *Ecol. Processes* **2017**, *6*, 4.
16. Zhao, Y.; Zhang, X.; Chen, M.; Gao, S.; Li, R. Regional Variation of Urban Air Quality in China and Its Dominant Factors. *Acta Geogr. Sin.* **2021**, *76*, 2814–2829.
17. Yang, S.; Zhao, L. Application of Random Forest Algorithm in Urban Air Quality Forecast. *Stat. Decis.* **2017**, *20*, 83–86. [CrossRef]
18. Akbal, Y.; Ünlü, K.D. A deep learning approach to model daily particular matter of Ankara: Key features and forecasting. *Int. J. Environ. Sci. Technol.* **2021**, 1–17. [CrossRef]
19. *HJ633-2012*; Technical Regulation on Ambient Air Quality Index (ON Trial). Ministry of Environmental Protection of the People's Republic of China: Beijing, China, 2012.
20. Yin, C.; Li, H.; Yu, L.; Wang, A. Characteristics of Air Quality in Jinan. *J. Shandong Meteorol.* **2012**, 32, 24–26.

atmosphere

Article

Weather Radar Nowcasting for Extreme Precipitation Prediction Based on the Temporal and Spatial Generative Adversarial Network

Xunlai Chen [1,2], Mingjie Wang [1,2], Shuxin Wang [1,2], Yuanzhao Chen [1,2,*], Rui Wang [1,2], Chunyang Zhao [1,2] and Xiao Hu [1,2]

1 Shenzhen Meteorological Bureau, Qixiang Road, Futian, Shenzhen 518040, China
2 Shenzhen Key Laboratory of Severe Weather in South China, Shenzhen 518040, China
* Correspondence: chenyuanzhao@weather.sz.gov.cn; Tel.: +86-8251-1633

Abstract: Since strong convective weather is closely related to heavy precipitation, the nowcasting of convective weather, especially the nowcasting based on weather radar data, plays an essential role in meteorological operations for disaster prevention and mitigation. The traditional optical flow method and cross-correlation method have a low forecast accuracy and a short forecast leading time, while deep learning methods show remarkable advantages in nowcasting. However, most of the current forecasting methods based on deep learning suffer from the drawback that the forecast results become increasingly blurred as the forecast time increases. In this study, a weather radar nowcasting method based on the Temporal and Spatial Generative Adversarial Network (TSGAN) is proposed, which can obtain accurate forecast results, especially in terms of spatial details, by extracting spatial-temporal features, combining attention mechanisms and using a dual-scale generator and a multi-scale discriminator. The case studies on the forecast results of strong convective weather demonstrate that the GAN method performs well in terms of forecast accuracy and spatial detail representation compared with traditional optical flow methods and popular deep learning methods. Therefore, the GAN method proposed in this study can provide strong decision support for forecasting heavy precipitation processes. At present, the proposed method has been successfully applied to the actual weather forecasting business system.

Keywords: weather radar nowcasting; generative adversarial network (GAN); Temporal and Spatial GAN (TSGAN); heavy precipitation

Citation: Chen, X.; Wang, M.; Wang, S.; Chen, Y.; Wang, R.; Zhao, C.; Hu, X. Weather Radar Nowcasting for Extreme Precipitation Prediction Based on the Temporal and Spatial Generative Adversarial Network. *Atmosphere* **2022**, *13*, 1291. https://doi.org/10.3390/atmos13081291

Academic Editors: Zengyun Hu, Xuguang Tang, Qinchuan Xin and Gianni Bellocchi

Received: 5 July 2022
Accepted: 11 August 2022
Published: 14 August 2022

1. Introduction

Precipitation is an important weather phenomenon and an important part of the water cycle, which has a profound impact on all aspects of people's lives. Extreme precipitation is one of the important factors that cause natural disasters. The accurate and timely prediction of upcoming extreme precipitation can avoid economic losses and help protect the safety of people's lives and property. Using algorithms such as the Z–R relationship in business [1], weather radars can effectively observe precipitation. The precipitation derived from a series of radar echoes could be used to forecast precipitation in the next 1 to 2 h and provide information on the development and change of precipitation, which is helpful to making the right decisions about the possible effects of precipitation. However, the spatiotemporal characteristics of the precipitation development process have great uncertainty, resulting in difficulties in predicting its change and movement trends. Therefore, accurately predicting the future changes of radar echoes is the key to improving the accuracy of precipitation prediction.

Convective weather nowcasting refers to the forecast of the occurrence, development, evolution and extinction of thunderstorms and other disastrous convective weather in the next few hours, which is crucial for meteorological disaster prevention and mitigation.

Weather radars are the primary tool for convective weather nowcasting in 0–2 h. Currently, the operational nowcasting methods mainly include the identification and tracking of thunderstorms and the automatic extrapolation forecasting technology based on radar data, such as the single centroid method [2], cross-correlation method [3] and optical flow method [4]. The traditional extrapolation methods based on radar echoes only use the shallow-level feature information of the radar images, and their application is limited to a single unit with strong radar echoes and a small range. Therefore, these methods are unreliable for predicting large-scale precipitation. The TREC (Tracking Radar Echoes by Cross-correlation) technique and its improved algorithm usually treat the echo variation as linear. However, the evolution of the intensity and shape of radar echoes is relatively complex during the generation and extinction of a convective process in the actual atmosphere. Moreover, these traditional methods have a low utilization of historical radar observations. Therefore, the forecast leading time is usually less than one hour, and the forecast accuracy can no longer meet the needs of high-precision prediction.

In recent years, artificial intelligence technology represented by deep learning has analyzed, associated, memorized, learned and inferred uncertain problems, whose applications have made significant progress in image recognition, nowcasting, disease prediction, environment changes and other fields [5–10]. As an advanced nonlinear mathematical model, deep learning technology contains multiple layers of neurons and has an excellent feature learning capability, which can automatically learn from massive data to extract the intrinsic characteristics and physical laws of the data and is widely used to build complex nonlinear models. Convective weather nowcasting is a sequence of forecast problems based on time and space. Some scholars have applied deep learning technology to weather nowcasting and have achieved satisfactory results [11,12].

Spatiotemporal forecasts based on deep learning involve two essential aspects, namely, spatial correlation and temporal dynamics, and the performance of a forecast system depends on its ability to memorize relevant structural information. Currently, there are two main types of neural network models for spatiotemporal sequence forecasting, i.e., image sequence generation methods based on a convolutional neural network (CNN) and image sequence forecast methods based on a recurrent neural network (RNN).

The CNN-based method converts the input image sequence into an image sequence of one or more frames on a certain channel, and many scholars have proposed implementation schemes based on this method [13,14]. For example, Kalchbrenner et al. [15] proposed a probabilistic video model called Video Pixel Network (VPN). Xu et al. [16] proposed a PredCNN network, which stacks multiple extended causal convolutional layers. Ayzel et al. [14] proposed a CNN model named DozhdyaNet. Compared with traditional radar echo extrapolation methods, the CNN-based methods can use a large amount of historical radar echo observations during training and learn their variation patterns, including the enhancing and weakening processes of rainfall intensity. However, the unchanged position of the convolution structure makes the radar images show the same rainfall field transformation. Thus, the CNN-based methods have certain limitations and are not widely used in radar echo extrapolation.

The long short-term memory RNN (LSTM-RNN) [17] with convolutional LSTM units has dramatically improved the forecast accuracy of precipitation with an intensity of more than 0.5 mm per hour. Predictive RNN with spatiotemporal LSTM units has achieved significant performance gains in practical applications. The LSTM units with a spatiotemporal memory unit have a certain ability to predict the intensity variation of the radar reflectivity factor [18]. Shi et al. designed the Convolutional LSTM (ConvLSTM) model based on previous research, which can capture spatiotemporal motion features by replacing Hadamard multipliers with convolution operations in the internal transformation of the LSTM [19]. A well-known variant of ConvLSTM is the Convolutional Gate Recurrent Unit (ConvGRU). However, the spatial position is unchanged due to the introduction of convolution kernels, which is a disadvantage for weather patterns with rotation and deformation. Wang et al. [20,21] proposed a spatial-temporal LSTM (ST-LSTM) with a zigzag connection structure model, which can transfer memory states horizontally across states and vertically transfer memory states among different layers. Shape deformation and motion trajectories can be effectively modeled by introducing spatiotemporal memory units. However, the spatial-temporal LSTM also faces the problem of vanishing gradients. For this, several scholars proposed a PredRNN++ model [22] and Memory in Memory (MIM) method [23], which can capture long-term memory dependencies by introducing a gradient highway unit module. Shi et al. [24] developed the TrajGRU model to overcome the problem of spatial consistency by generating a neighborhood set with parameterized learning subnetworks for each location. Eidetic 3D LSTM(E3D-LSTM) [25] utilizes the self-attention [26] module to preserve the long-term spatiotemporal correlation. Jing et al. [27] designed the Hierarchical Prediction RNN for long-term radar echo extrapolation, which can meet the needs of long-term extrapolation in actual precipitation predictions. This model employs a hierarchical forecasting strategy and a coarse-to-fine round-robin mechanism to alleviate the accumulation of forecast errors over time and therefore facilitate long-term extrapolation.

However, the extrapolation results of all existing deep learning methods inevitably suffer from blur, i.e., as the forecast leading time increases, the diffusion of forecast echoes becomes more and more serious, resulting in blur. Therefore, how to reduce the blur of the predicted echo and improve the forecast accuracy at the same time is an urgent issue to be solved in the current forecast operational applications. In this study, we propose a radar echo prediction method based on the Temporal and Spatial Generative Adversarial Network (TSGAN), which can extract the spatiotemporal features of input radar images through the three-dimensional convolution and attention mechanism module and can use a dual-scale generator and a multi-scale discriminator to restore the detailed information of the predicted echoes. Therefore, the main advantage of the proposed method is that it obviously improves the forecasts of the echo details while ensuring the accuracy of the forecast results and effectively reducing the blur of the predicted echoes.

The remainder of this manuscript is organized as follows. Section 2 describes the basic principle of the Generative Adversarial Network. The proposed methodology for weather radar nowcasting, including the dual-scale generator, multi-scale discriminator and loss function, is introduced in Section 3, followed by the experiments and results of two typical strong convective weather nowcasting, i.e., the squall line and typhoon. Further conclusions are offered in Section 5, and a brief summary of this work is also given.

2. Generative Adversarial Network

Inspired by the zero-sum game, the training process of the model in the GAN is designed as a confrontation and game between the two networks: the generative network G and the discriminant network D. The schematic diagram of the overall GAN model structure for radar echo extrapolation is as follows (Figure 1).

In the generative network G, the random noise vector z, obeying the standard normal distribution $N(0, 1)$, is taken as the input, and the generated image G(z) is the output. The generative network tries to generate images that make the discriminative network indistinguishable during training. The generative network G is responsible for the data generation task in the generative adversarial network. For the random distribution of the input samples, the "generated" samples are as similar as possible to the "real" samples. For the generative network G in the GAN network, a new data distribution is generated through complex nonlinear network transformation. In order to make the generated data distribution approach the real data distribution, it is necessary to minimize the difference between the generated data distribution and the real data distribution.

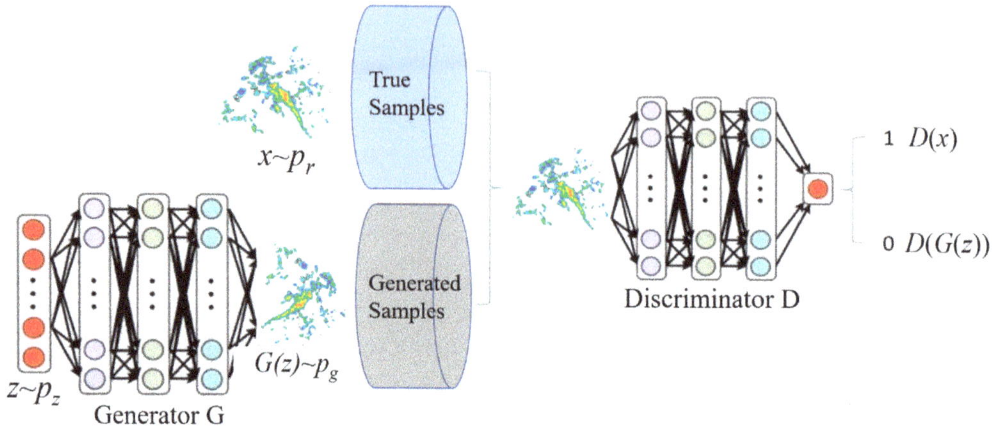

Figure 1. The basic structure of the Generative Adversarial Network.

The role of the discriminant network D in the training process is equivalent to that of a binary classification network, which is used to distinguish the actual training image x from the generated image G(z). In this way, the two networks continue to conduct adversarial training, and both are optimized in the process of mutual confrontation. After optimization, the two networks continue to confront each other, and the generated images obtained from the generative network G become closer and closer to the actual images.

Under normal circumstances, the discriminant network assigns the label "1" to the actual image and the label "0" to the generated image. The generative network tries to make the discriminative network "misjudge" the generated image as "1". Suppose *Pr* represents the data distribution of the real image *x*, *Pg* denotes the data distribution of the generated image G(z) and *Pz* indicates the prior distribution *N(0, 1)* of the random noise vector *z*. G and D denote the generation network and the discriminant network, respectively. By using the cross-entropy loss function, the optimization objective of the GAN can be expressed as the following equation (Equation (1)).

$$\min_{G}\max_{D} V(G, D) = E_{x \sim Pr}[\log D(x)] + E_{z \sim Pz}[\log(1 - D(G(z)))] \tag{1}$$

where *E* is the mathematical expectation. The former term $E_{x \sim Pr}[\log D(x)]$ represents the probability that the discriminant model judges the real original data, and the latter term $E_{z \sim Pz}[\log(1 - D(G(z)))]$ represents the probability that the generated data is judged to be false. The GAN optimizes G and D alternately through a Max-Min game until they reach the Nash equilibrium point. Simultaneously, as the alternate optimization proceeds, D will gradually approach the optimal discriminator. When this proximity reaches a certain level, the optimization objective of the GAN is approximately equivalent to minimizing the Jensen–Shannon Divergence between the data distribution of the actual image (*Pr*) and the data distribution of the generated image (*Pg*). In other words, the principle of the GAN is based on the zero-sum game in game theory, which is equivalent to the optimization of the distribution distance between the actual and generated data.

For the training process of the GAN model only, D is equivalent to a binary classifier. Each update to D enhances its ability to distinguish between the actual and generated images, i.e., correctly assigning two kinds of labels to the two kinds of data and dividing the correct decision boundary between the two kinds of data. The update of G tries to classify generated images as actual images. Thus, the newly generated images are closer to the decision boundary and the actual images. As the alternate iterations continue, the generated images will continue to approach the actual images, eventually making D indistinguishable. Therefore, G can highly and realistically fit the actual data.

3. Temporal and Spatial Generative Adversarial Network (TSGAN)

The GANs are theoretically feasible through mathematical derivation, but they face many problems in the actual training process, the most important of which are gradient disappearance and mode collapse. The reason for the disappearance of the gradient is that the probability of non-negligible overlap between the real distribution and the generated distribution is very small. Therefore, the discriminant network can easily divide the generated data and the real data. The generative network can hardly obtain gradient updates, so it is difficult to optimize the network iteratively. Furthermore, the reason for the mode collapse is that the optimization of the distance between the generated data and the real data distribution is very difficult to control, resulting in the degradation of the generative model and the inability to capture all the changes in the real data distribution.

For the above reasons, inspired by Pix2PixHD [28], the TSGAN proposed in this study consists of two parts, namely, a dual-scale generator and a multi-scale discriminator. The dual-scale generator uses two radar echo sequences with different resolutions to extract spatiotemporal features. Then, the UNet structure and attention mechanism are used to generate predicted echo sequences. The multi-scale discriminator distinguishes the generated predicted echo sequences at multiple scales. Subsequently, the dual-scale generator is guided to generate higher-quality predicted echo sequences.

3.1. Dual-Scale Generator

The task of the generator is to use an input radar echo sequence to generate the subsequent 20 frames of the radar echo sequence while retaining as much detailed information of the echoes as possible. Therefore, the spatiotemporal features of the radar echoes need to be considered during this process. A deeper network structure can generate better sequences, but it also faces the problems of overfitting and training difficulties. Therefore, this study is conducted on two scales to take into account the generation effect and network scale. We use three-dimensional convolution to extract the spatial-temporal features of radar echo sequences and employ the UNet structure to restore the spatial details of the generated echo sequences.

The basic structure of the dual-scale generator is shown in Figure 2:

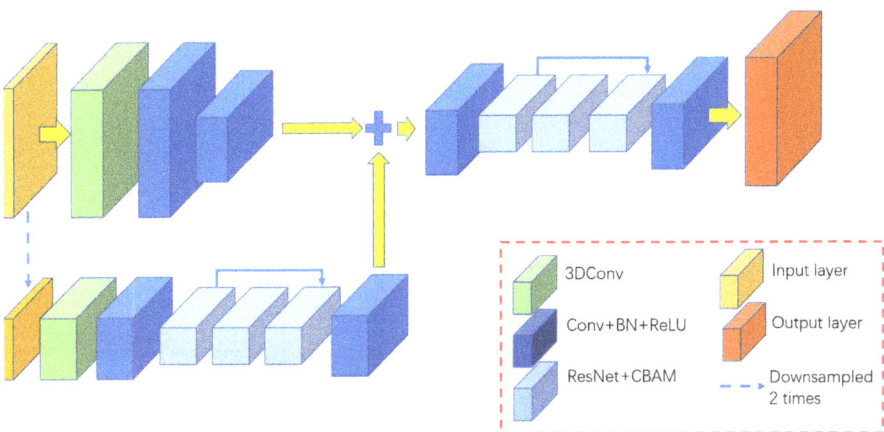

Figure 2. The structure of the dual-scale generator.

The input of the generator is the radar echo data of 1 h before the current time. Since the time resolution is 6 min, the input is the radar echo data of 10 consecutive moments, with a size of 896 × 896 × 10 pixels. In terms of the dual-scale generator, the second scale is half of the original scale. At the original scale, the input radar echo sequence (896 × 896 × 10 pixels) passes through several three-dimensional convolutional layers and ordinary convolutional pooling layers to obtain a series of feature maps with a size of 448 × 448 pixels. At the second scale, the input radar echo sequence is down-sampled by a factor of 2, and then the size is changed to 448 × 448 × 10 pixels. The down-sampled data also pass through the three-dimensional convolutional layer and the ordinary convolutional pooling layer. Then, this sequence proceeds through a ResUNet-structured module consisting of the ResNet module [29] and the CBAM attention mechanism [30]. The UNet-structured module is composed of modules that resize the output feature maps to the size of the original feature maps, i.e., 448 × 448 pixels. ResUNet replaces the convolution layers in the conventional UNet model with the ResNet module, whose role is to preserve the spatial details of different feature maps as much as possible. The CBAM attention mechanism consists of spatial attention and channel attention, and its role is to preserve the more important information on the space and channel as much as possible. After adding the feature maps of the two scales, the output is restored to the size of the original input radar echoes (896 × 896 pixels) through the convolution pooling layer and the UNet structure module of another ResNet + CBAM layer. Therefore, the final 2 h predicted radar echo sequence is obtained with a size of 896 × 896 × 20.

Two scales are used in the generator. The spatial resolution of the original scale is consistent with that of the input radar echo sequence, which is conducive to retaining the spatial details of the predicted echoes. Because the original scale has the highest spatial resolution, in the process of radar echo time series prediction, almost all of the algorithms will face the problem that the predicted echo becomes more and more blurred as the forecast time increases. The main reason for this is that the spatial detail information is gradually weakened in the process of gradual extrapolation. Using the original scale data, we hope that the spatial details are preserved as much as possible in the network. Meanwhile, the spatial resolution of the second scale is half that of the input radar echo sequence, which facilitates a more thorough control of the orientation of the generator network. The reduction in spatial resolution is equivalent to increasing the receptive field of each convolution kernel, which is beneficial to the network obtaining more global information, thereby controlling the generator network to better fit the trend of future echoes. The balance between the generation effect and the network training can be achieved through the joint action of the spatiotemporal features extracted by three-dimensional convolution and the two scales, obtaining the extrapolation results that not only conform to the development law of radar echoes but also maintain the spatial details.

3.2. *Multi-Scale Discriminator*

The generated images have high spatial resolution and rich spatial details. Therefore, the discriminator generally needs a deeper network or a larger convolution kernel to ensure that the discriminant network has a larger receptive domain. However, the discriminator may lead to overfitting due to the excessive network capacity and requires more GPU memory for network training.

Therefore, a multi-scale discriminant network is adopted in this study to identify the generated images from different scales, i.e., three discriminators are utilized. The three discriminators all have the same network structure but operate on images of different sizes. Specifically, we down-sample the real and generated images by factors of 2 and 4, respectively, to create image pyramids at three scales. Three discriminators are trained by using different real and generated images of the three sizes. Although the structures of the discriminators are the same, the discriminator with four times down-sampling has the largest receptive field, which ensures that it has more global perspective information and can guide the generator to generate overall consistent images. Additionally, the discriminator at the original scale favors the generator to generate finer details, which also makes the training of the generator easier.

The structure of the discriminant network is shown in Figure 3, consisting of a series of convolutional layers, pooling layers and fully connected layers. The input size of the original scale discriminator is 896 × 896, and for the second and third scales, it is 448 × 448 and 224 × 224, respectively.

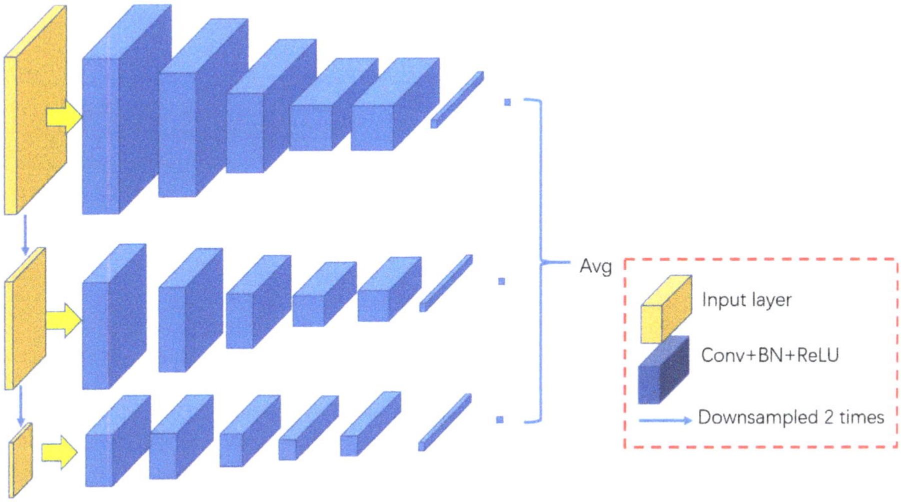

Figure 3. The structure of the multi-scale discriminator.

3.3. Loss Function

The loss function consists of three parts, namely, adversarial loss, multi-scale feature loss and overall content loss. Assuming that the discriminator of the network has three scales, "*in*" represents the input radar echo sequence, "*tar*" represents the future real radar echo sequence, "G" represents the output of the generator network, and D_k represents the output of the discriminator at the *k*-th scale (*k* = 1, 2, 3). The total loss function is expressed as Equation (2).

$$Loss = \min_G \{ [\max_{D_1, D_2, D_3} \sum_{k=1,2,3} l_{GAN}(G, D_k)] + \alpha \sum_{k=1,2,3} l_{feature}(G, D_k) + \lambda l_{content}(G, tar) \} \quad (2)$$

where α and λ are the weights of the multi-feature loss $l_{feature}$ and the overall content loss $l_{content}$. The adversarial loss l_{GAN} in the above formula can be expressed as Equation (3)

$$l_{GAN}(G, D_k) = \mathrm{E}_{(in,tar)}[\log D_k(in, tar)] + \mathrm{E}_{in}[\log(1 - D_k(in, G(in))] \quad (3)$$

The multi-feature loss $l_{feature}$ can be represented as Equation (4).

$$l_{feature}(G, D_k) = \sum_i \left| D_k^i(in, tar) - D_k^i(in, G(tar)) \right| \quad (4)$$

where *k* (*k* = 1, 2, 3) denotes the number of discriminators, and *i* represents the *i*-th layer of the discriminant network.

The overall content loss $l_{content}$ can be obtained according to Equation (5).

$$l_{content}(G, tar) = L_1(G, tar) \quad (5)$$

where L_1 is the L1 loss, i.e., the MAE loss. The adversarial loss is mainly used to recover the detailed information of the predicted echoes. The multi-scale feature loss and the overall content loss characterize the difference in content between the predicted and observed echoes in the aspects of deep features and pixels. The joint effect of the three loss functions guides the results from the generator to gradually approach the actual radar echoes.

4. Experiments and analysis

The study area is Guangdong Province in southern China. The whole area is located at a latitude of 20°13′ to 25°31′ north and a longitude of 109°39′ to 117°19′. It belongs to the subtropical monsoon climate region. The land spans the northern tropics, the southern subtropics and the central subtropics from south to north. The airflow in this area is particularly strong, and the hot and cold flows frequently meet and collide in this area, resulting in frequent strong convective weather and abundant precipitation. There are many meteorological disasters in the region. The main disasters are: low temperature and rain, strong convection (hail, tornado, strong thunderstorm and strong wind), rainstorm and flood, typhoon, drought, cold dew wind, cold wave, etc. Among them, tropical cyclones and rainstorms have a high frequency and high intensity, ranking first in the country. Meteorological disasters have caused heavy losses to the national economy. For example, Typhoon No. 9615 caused losses of nearly CNY 17 billion to western Guangdong. Therefore, it is extremely important to improve the nowcasting technology.

In this research, the reflectivity factor mosaic data during 2015–2021 from 11 new-generation S-band Doppler radars in Guangdong are used for the experiments. The data in 2015–2019 are selected as the training dataset, the data in 2020 are selected as the validation dataset and the data in 2021 are selected as the test dataset. These original data have horizontal grid points of 700 × 900, with spatiotemporal resolutions of 1 km × 1 km and 6 min. Each sample contains the radar echo input sequence of 10 moments in 1 h and the radar echo target sequence of 20 moments in the next 2 h.

In order to verify the forecast performance of the TSGAN method on extreme convective rainfall, over 80,000 cases in 2021 every 6 min were analyzed. For the page limitation, we only select the squall line process on 4 May 2021 and the typhoon process on 8 October 2021 as study cases for radar echo extrapolation forecasts visualization. Moreover, for comparing the forecasting effectiveness of various methods, the results from the TSGAN method are compared with those from the optical flow [4], ConvGRU [19], PredRNN [21] and PredRNN V2 [21] methods, which are widely used in the existing operations. The optical flow method employs the Lucas–Kanade algorithm to calculate the optical flow and performs the extrapolation by using the semi-Lagrangian method. The ConvGRU, PredRNN and PredRNN V2 are trained by using the official codes. All of the employed methods should be evaluated by many aspects and multi-dimensions [31,32]. By comparing the observed radar echo images, we perform a grid-by-grid test for the prediction accuracy in this study. Additionally, the prediction ability at different radar reflectivity levels is investigated according to the radar reflectivity factors of different intensities. Finally, the critical success index (CSI) is used to evaluate the forecast results quantitatively.

The expression of the CSI is shown in Equation (6).

$$CSI_k = \frac{NA_k}{NA_k + NB_k + NC_k} \tag{6}$$

where NA_k denotes the number of the correct grid points, NB_k denotes the number of false grid points, NC_k denotes the number of missing grid points and k (k = 20 dBz, 30 dBz, 40 dBz and 50 dBz) denotes the threshold value of the different intensities of radar reflectivity. The validation method is adopted according to the forecast leading time and the threshold value. The calculation is performed grid-by-grid, i.e., the predicted and observed values at the same grid point are selected for testing and comparison (Table 1).

Table 1. The validation of the radar echo predictions at different reflectivity levels.

Obs \ Pred	$\geq$k	<k
$\geq$ k	NA_k	NC_k
$< k$	NB_k	ND_k

As mentioned earlier, the results obtained by most extrapolation methods suffer from blur, i.e., as the forecast leading time increases, the predicted echoes become more and more blurred, and more details are lost. However, the TSGAN method proposed in this study can recover the detailed information of the radar echoes to a certain extent. To enrich spatial details that are characterized, two indicators, definition and spatial frequency, are introduced in this study. The expression of the definition is as follows (Equation (7)).

$$Definition = \frac{1}{(M-1)(N-1)}\sum_{i=1}^{M}\sum_{j=1}^{N}|I(i,j)-I(i,j-1)|+|I(i,j)-I(i-1,j)| \tag{7}$$

The spatial frequency is defined by the frequency in both vertical and horizontal directions. The frequency in the vertical direction is defined as follows (Equation (8)).

$$RF = \sqrt{\frac{1}{MN}\sum_{i=0}^{M}\sum_{j=1}^{N-1}[I(i,j)-I(i,j-1)]^2} \tag{8}$$

The frequency in the horizontal direction is defined as follows (Equation (9)).

$$CF = \sqrt{\frac{1}{MN}\sum_{i=1}^{M-1}\sum_{j=0}^{N}[I(i,j)-I(i-1,j)]^2} \tag{9}$$

Therefore, the overall spatial frequency can be expressed as Equation (10).

$$SF = \sqrt{RF^2 + CF^2} \tag{10}$$

4.1. Squall Line Process on 4 May 2021

On 4 May 2021, a squall line process swept across Guangdong Province, resulting in extreme heavy rainfall in several areas. In this research, the initial forecast time is 16:00 China Standard Time (CST, same as below) on 4 May 2021, and the echoes for the next 2 h are predicted. Figures 4–7 show the forecast results of this squall line process for the next 0.5 h, 1 h, 1.5 h and 2 h by using each method.

In terms of the overall trend, the difference between the forecast results of the optical flow method and the observations is the largest, where the echo intensity and shape are basically the same, while the difference in the spatial position is the largest among all methods. The other four methods can better grasp the evolution trend of the echoes within 2 h and can also predict the position of the strong echoes. However, except for the TSGAN method, the overall intensity predicted by all the methods decreases rapidly with increasing forecast time. For the detail retained, the forecasts of both the optical flow method and the TSGAN method can present richer detailed information, while those of the other methods become more and more blurred as the forecast time increases. The details predicted by the PredRNN V2 method are slightly better than those predicted by the PredRNN method, and the results of the ConvGRU method are the most blurred. The TSGAN method can retain richer detailed information, and its results do not become more blurred with increasing forecast time.

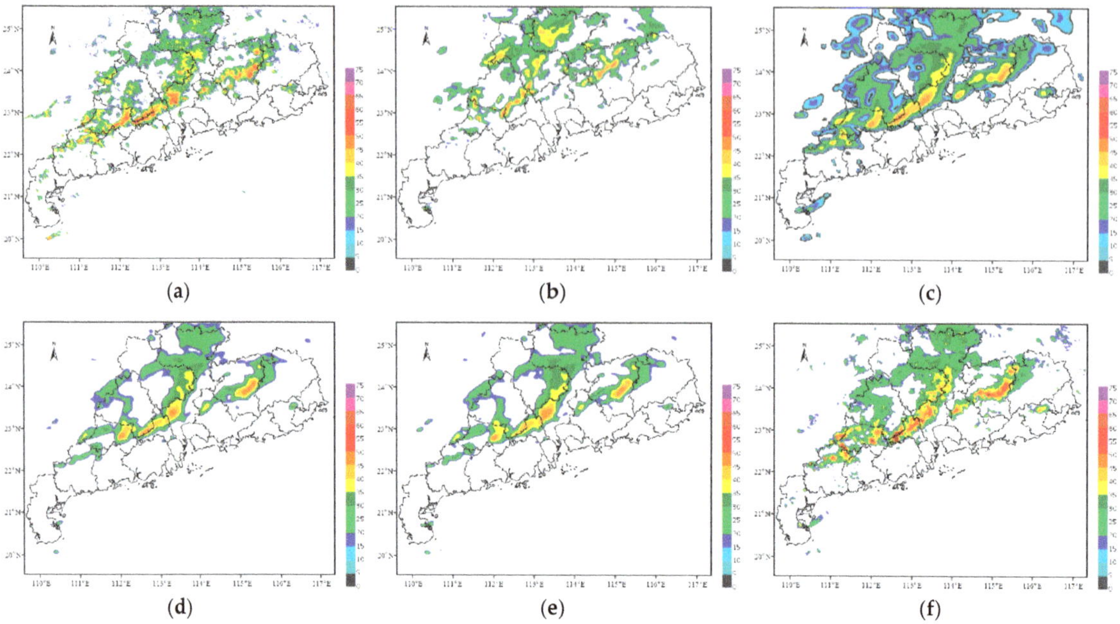

Figure 4. Nowcasting results of the squall line process for the next 0.5 h. (**a**) Observation; (**b**) Opticalflow; (**c**) ConvGRU; (**d**) PredRNN; (**e**) PredRNN V2; (**f**) TSGAN.

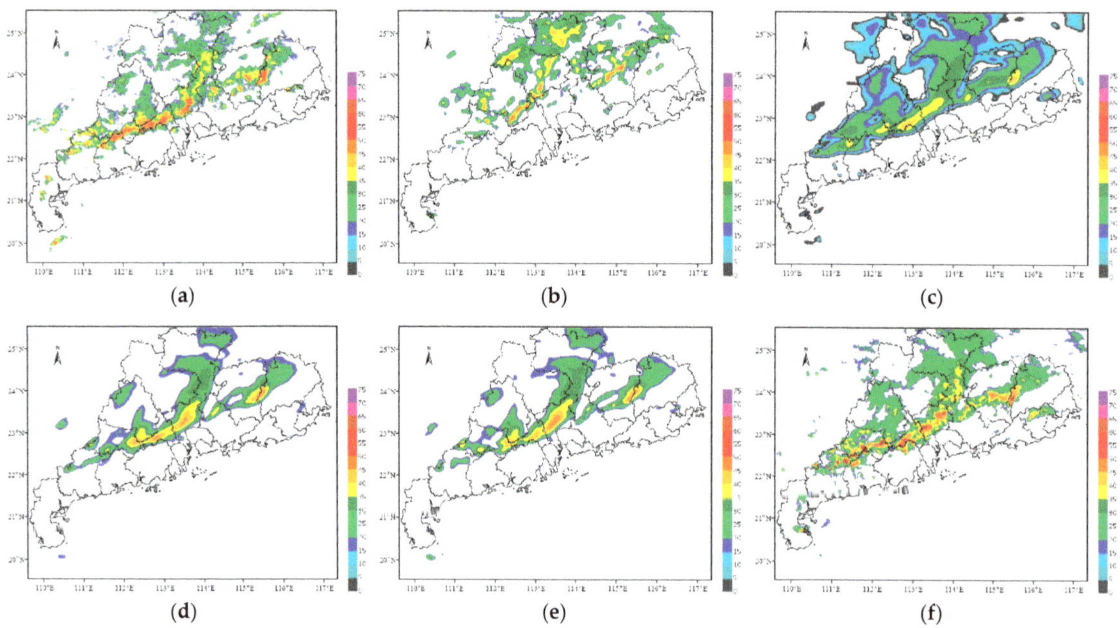

Figure 5. Nowcasting results of the squall line process for the next 1 h. (**a**) Observation; (**b**) Opticalflow; (**c**) ConvGRU; (**d**) PredRNN; (**e**) PredRNN V2; (**f**) TSGAN.

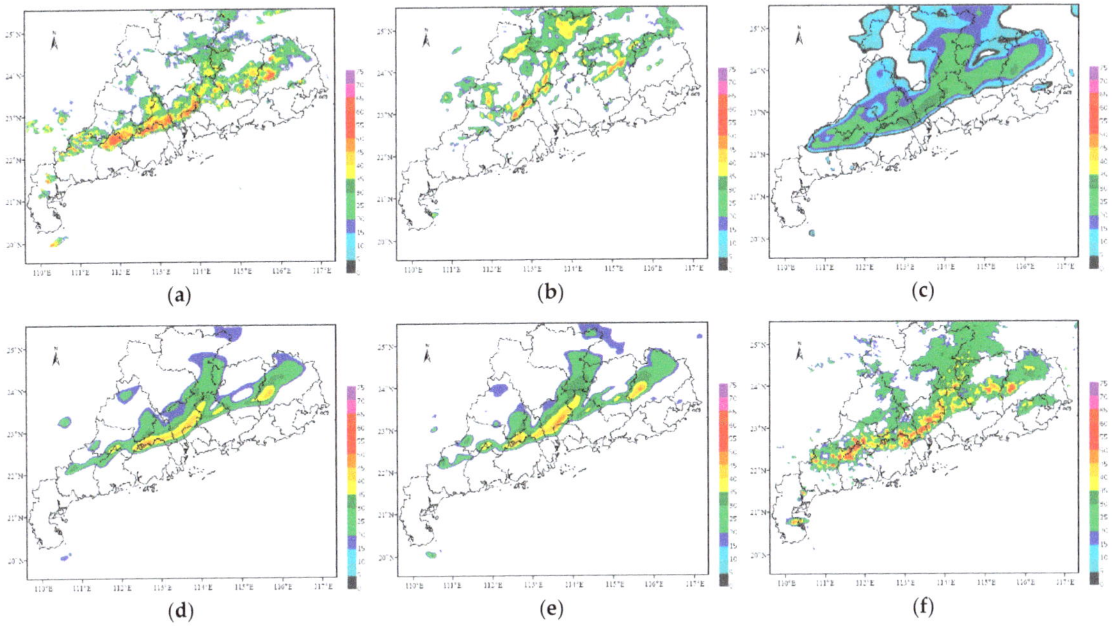

Figure 6. Nowcasting results of the squall line process for the next 1.5 h. (**a**) Observation; (**b**) Opticalflow; (**c**) ConvGRU; (**d**) PredRNN; (**e**) PredRNN V2; (**f**) TSGAN.

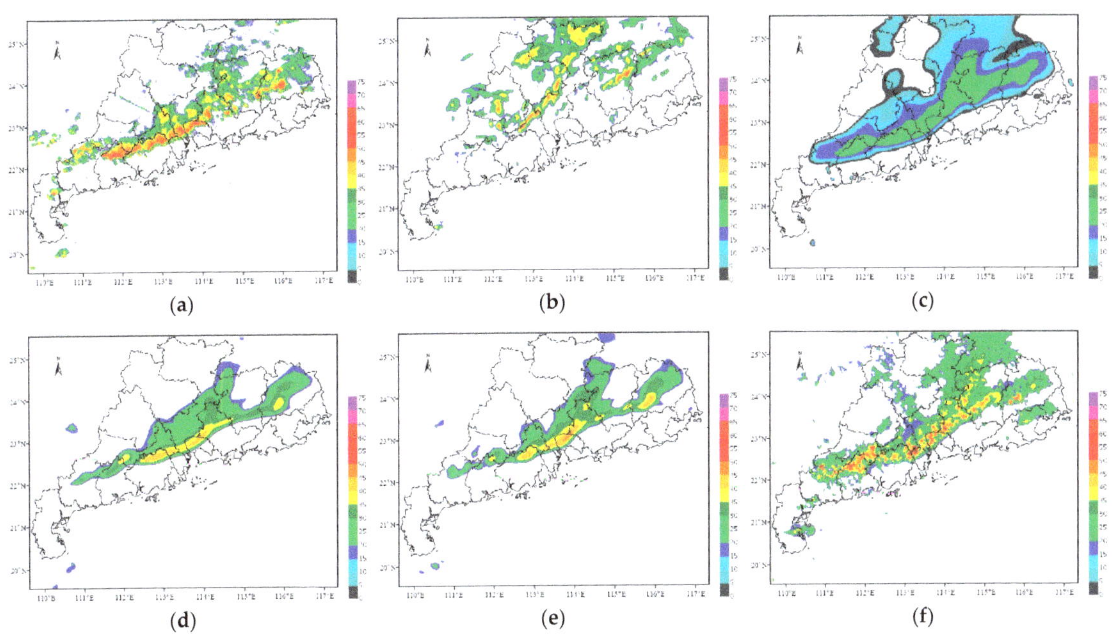

Figure 7. Nowcasting results of the squall line process for the next 2 h. (**a**) Observation; (**b**) Opticalflow; (**c**) ConvGRU; (**d**) PredRNN; (**e**) PredRNN V2; (**f**) TSGAN.

Figures 8 and 9 present the objective assessment scoring results for each method every 6 min over the 2 h period, and the labels of the horizontal axis are the prediction leading times.

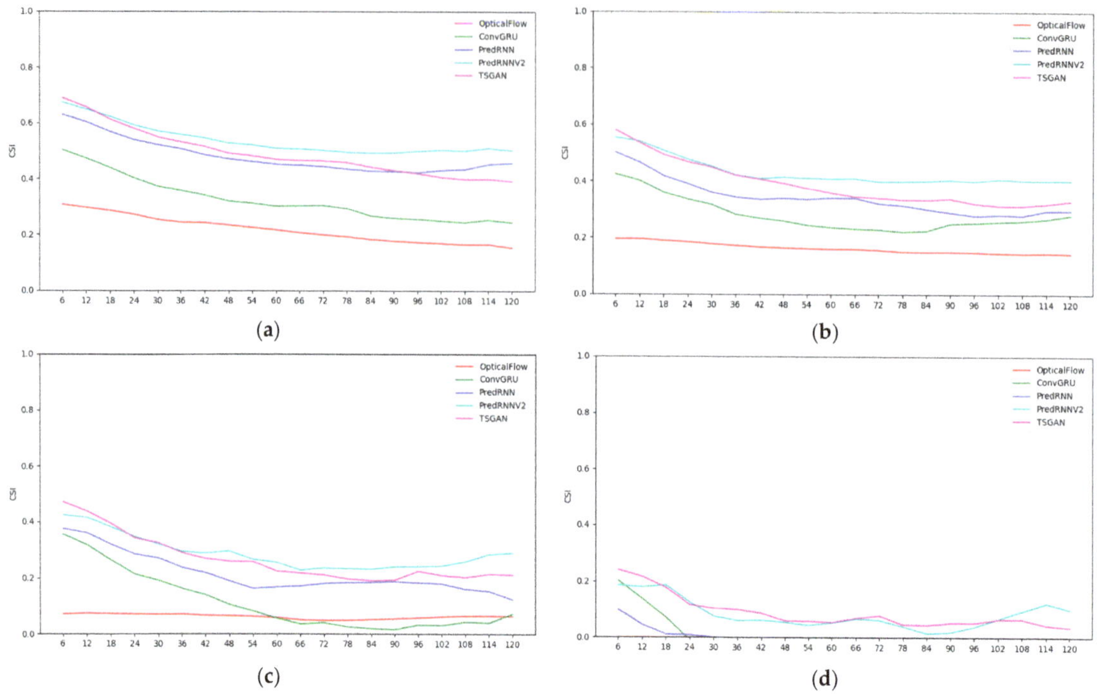

Figure 8. The scores of the critical success index for the forecasts of the squall line process. (**a**) 20 dBZ; (**b**) 30 dBZ; (**c**) 40 dBZ; (**d**) 50 dBZ.

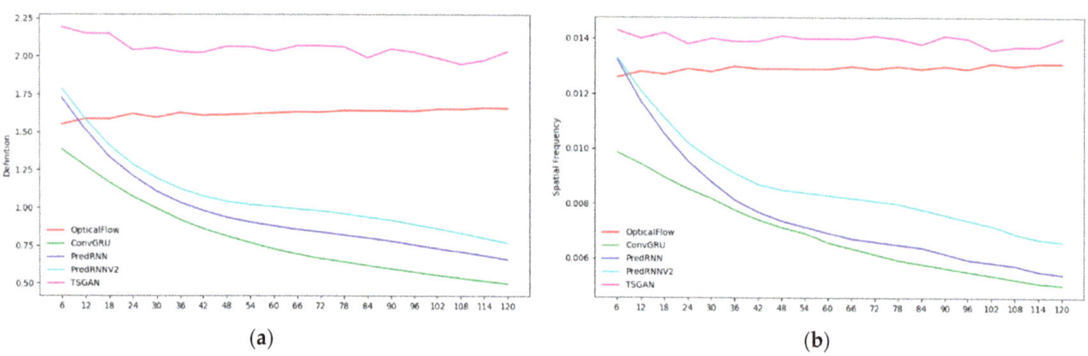

Figure 9. The scores of the spatial details for the forecasts of the squall line process. (**a**) Definition; (**b**) Spatial frequency.

The CSI scores suggest that the CSI values of all methods decrease with the increase in the forecast time, indicating that the longer the forecast time is, the lower the forecast accuracy is. The higher the radar reflectivity is, the more dramatic the CSI value of each method decay is, which means that the longer the forecast time is, the more difficult it is to predict strong echoes. Overall, the PredRNN V2 algorithm performs the best in all reflectivity levels. The PredRNN and TSGAN methods have a little difference between each other, followed by the ConvGRU method, and the optical flow method has the lowest CSI value due to the large deviation in the predicted echo position. In terms of the high-intensity echoes at the 50 dBZ level, the CSI values of the TSGAN and PredRNN V2 differ slightly. The definition and spatial frequency indicators of the ConvGRU, PredRNN and PredRNN

V2 methods all show a continuous decreasing trend with increasing forecast time, which is consistent with the fact that their results become more and more blurred. However, for the optical flow method and the TSGAN, the definition and spatial frequency indicators have no obvious decreasing trend, and the definition and spatial frequency values of the TSGAN are higher than those of the optical flow method. This finding indicates that the TSGAN method has obvious advantages in retaining spatial details. Therefore, the comprehensive analysis of the CSI and the spatial information indexes indicates that the TSGAN method performs the best in predicting the squall line process.

4.2. Typhoon Lion Rock on 8 October 2021

On 8 October 2021, Typhoon Lion Rock was generated, and strong wind and rainfall occurred in the east of Hainan Island and in the south of Guangdong. The precipitation within 6 h in Shenzhen and Shanwei exceeded 80 mm. Furthermore, Shanwei experienced short-term heavy rainfall from 10:00 to 11:00, and the hourly rain intensity reached 34.3 mm. In addition, gusts of 17 m s^{-1} and above occurred in Qiongshan of Haikou City and Mulantou of Wenchang City in Hainan Province and in Pinghu of Shenzhen City and Jiuzhou Port of Zhuhai City in Guangdong Province. Figures 10–13 present the forecast results of Typhoon Lion Rock. The initial forecast time is 06:00 on 8 October 2021, and the leading time is 2 h, with an interval of 6 min.

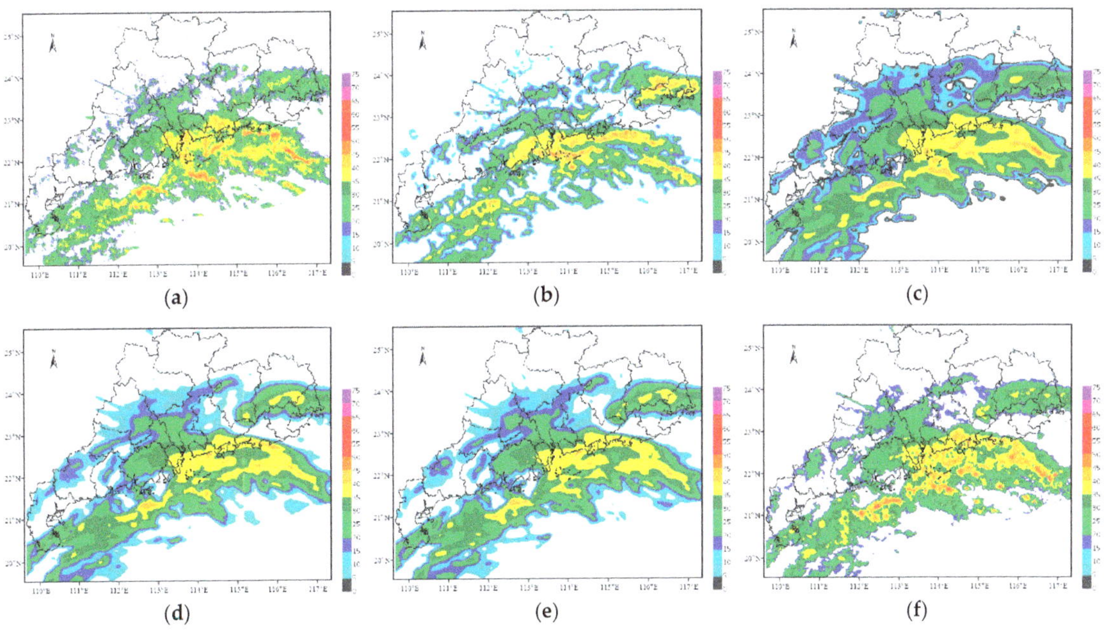

Figure 10. Nowcasting results of the typhoon process for the next 0.5 h. (**a**) Observation; (**b**) Opticalflow; (**c**) ConvGRU; (**d**) PredRNN; (**e**) PredRNN V2; (**f**) TSGAN.

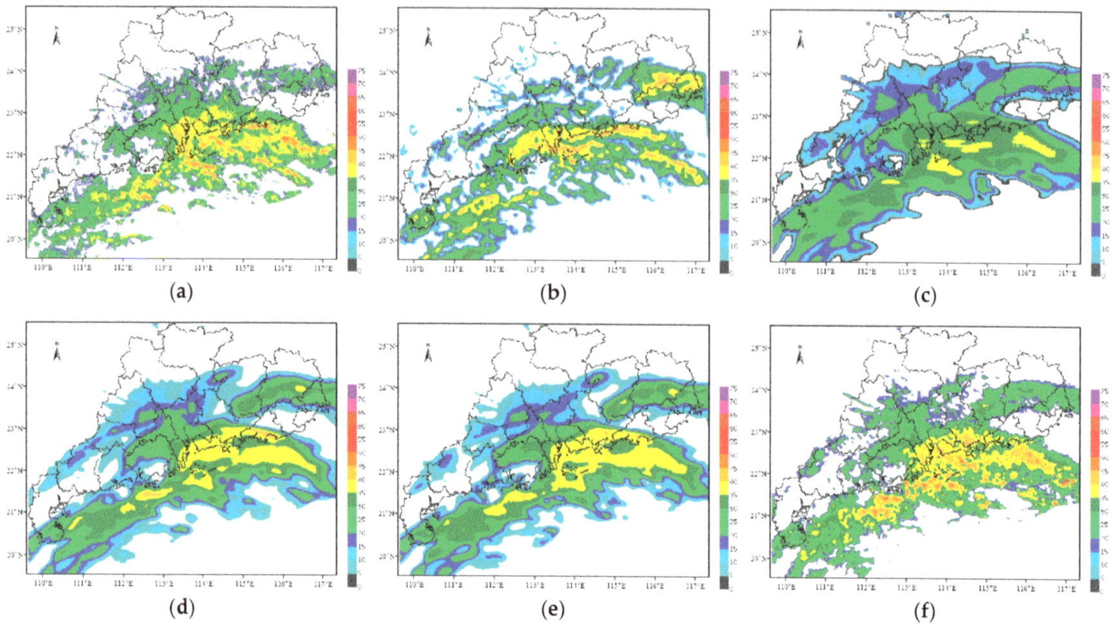

Figure 11. Nowcasting results of the typhoon process for the next 1 h. (**a**) Observation; (**b**) Opticalflow; (**c**) ConvGRU; (**d**) PredRNN; (**e**) PredRNN V2; (**f**) TSGAN.

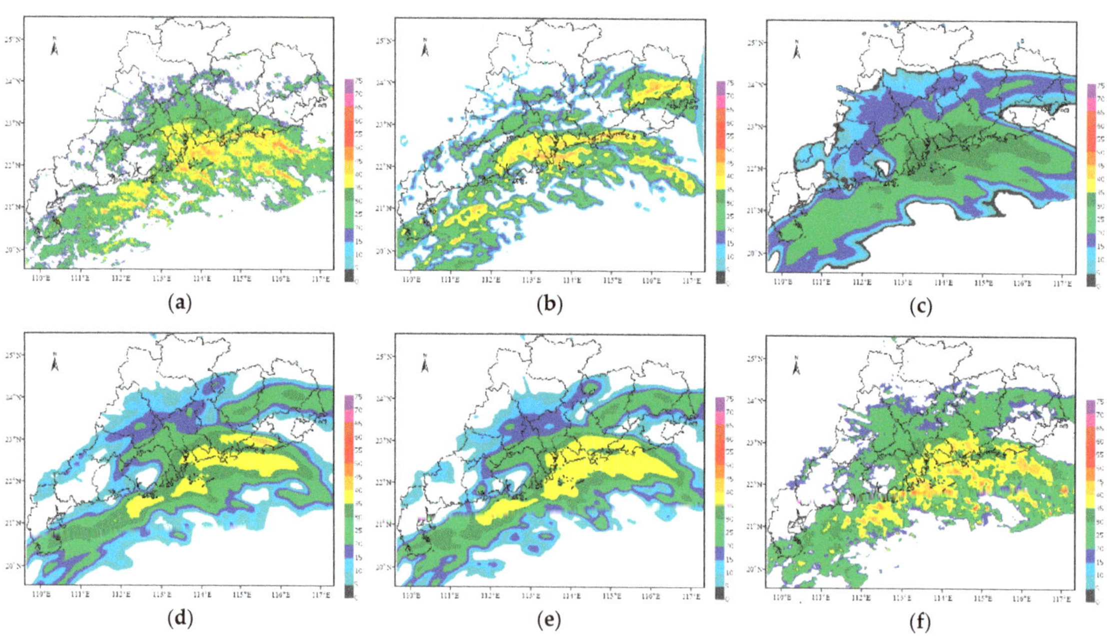

Figure 12. Nowcasting results of the typhoon process for the next 1.5 h. (**a**) Observation; (**b**) Opticalflow; (**c**) ConvGRU; (**d**) PredRNN; (**e**) PredRNN V2; (**f**) TSGAN.

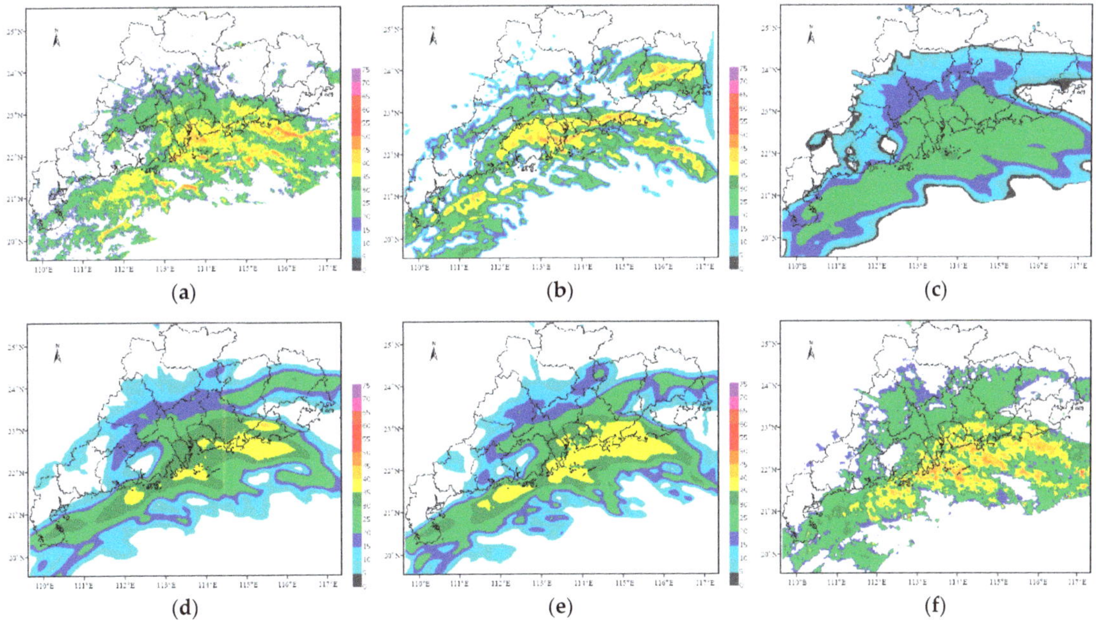

Figure 13. Nowcasting results of the typhoon process for the next 2 h. (a) Observation; (b) Opticalflow; (c) ConvGRU; (d) PredRNN; (e) PredRNN V2; (f) TSGAN.

Overall, the forecast results of this typhoon case from each method can better display the development trend of typhoon echoes, and the predicted position is similar to the actual observation. Similar to the previous case on 4 May 2021, the ConvGRU, PredRNN and PredRNN V2 methods still have the problem that as the forecast time increases, the forecast results become more and more blurred, and the predicted radar intensity also weakens considerably. The PredRNN V2 method improves the results of the PredRNN method in detail but still has the problem of blur prediction results. Although the position predicted by the optical flow method changes somewhat within 2 h, the predicted intensity remains basically constant, resulting in strong echoes appearing in the east of Guangdong Province, which is determined by the principle of the optical flow method itself. The forecast results of the TSGAN method retain rich spatial details and are consistent with the observations in intensity and spatial position.

The objective assessment results of the forecasts from each method are presented in Figures 14 and 15. The labels of the horizontal axis are the prediction leading times of the future 2 h every 6 min.

The results of the objective assessment indicators have a certain similarity with those of the squall line process. Due to the accurate predicted location and shape of radar echoes, the PredRNN V2, TSGAN and PredRNN methods show apparent advantages in the CSI scores. The spatial location of the optical flow method is not satisfactory in terms of accuracy. Thus, the CSI scores of the optical flow method are lower than those of the ConvGRU method. The PredRNN V2 method has a noticeable improvement in detail compared with the PredRNN method but still suffers from blur. In terms of definition and spatial frequency indicators, similar to the case on 4 May 2021, the TSGAN and optical flow methods can maintain stable spatial detail forecasts in each forecast time, while the other three methods become more and more blurred with increasing forecast time, resulting in more detail loss. In summary, the objective assessment suggests that the TSGAN method has certain advantages in forecasting this typhoon process while retaining more spatial details.

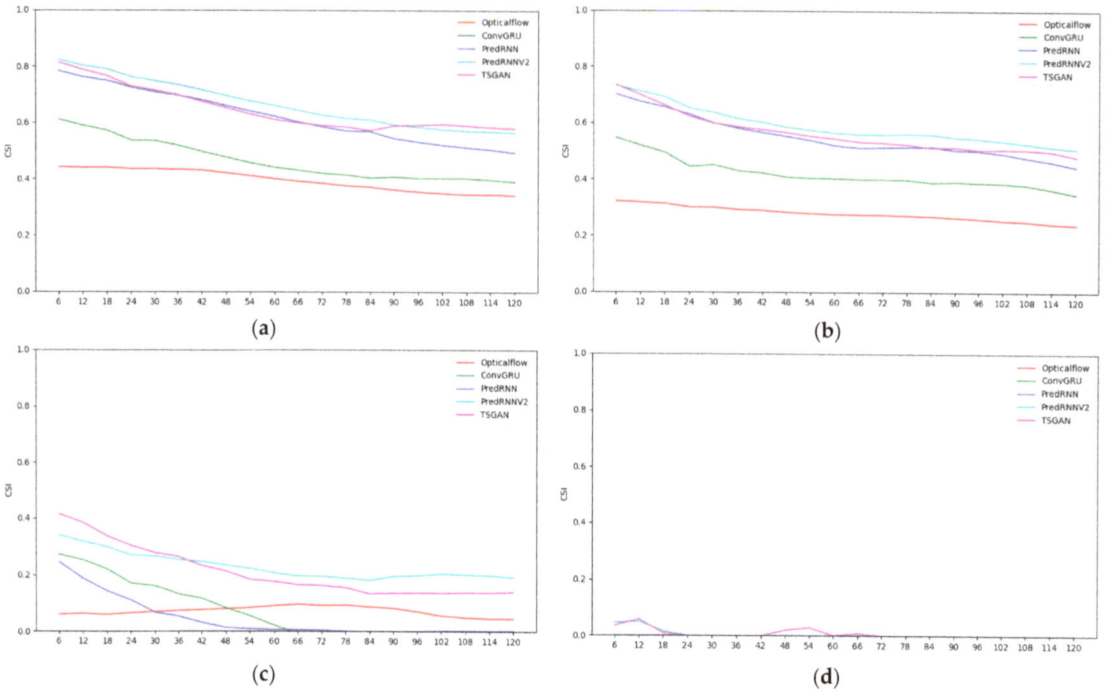

Figure 14. The scores of the critical success index for the forecasts of the typhoon process. (**a**) 20 dBZ; (**b**) 30 dBZ; (**c**) 40 dBZ; (**d**) 50 dBZ.

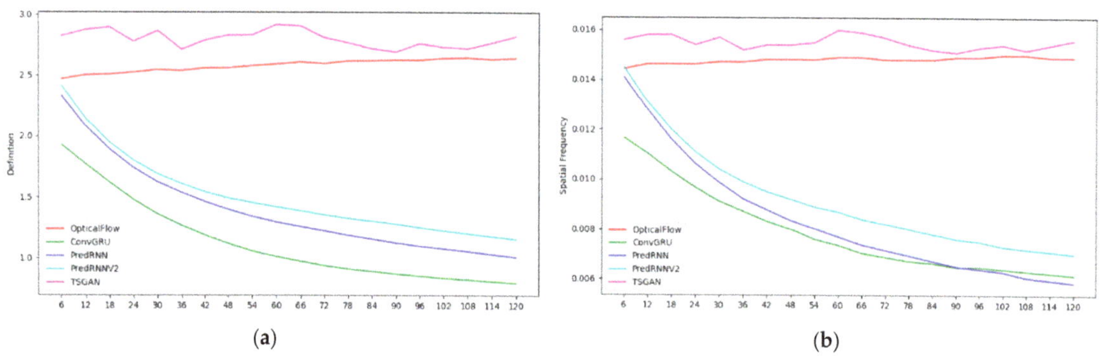

Figure 15. The scores of the spatial details for the forecasts of the typhoon process. (**a**) Definition; (**b**) Spatial frequency.

5. Conclusions

Currently, artificial intelligence forecasting methods based on weather radar data generally suffer from the problem that, with the increase in forecast time, the forecast results become increasingly blurred. In order to address this problem, an artificial intelligent forecasting method based on the GAN is proposed in this study. The spatiotemporal features of the radar echo sequence are extracted by three-dimensional convolution, and the local receptive domain is enlarged by the dual-scale generator and multi-scale discriminant network. Then, combining the attention mechanism and the training method of generative confrontation, we proposed a TSGAN method that can effectively mitigate the common problem suffered by artificial intelligence methods. The testing results of the two cases demonstrate that the TSGAN method can better predict the position and shape of radar echoes while retaining rich spatial details. Although the TSGAN method shows distinct advantages in predicting spatial details, the increase in spatial details does not necessarily lead to an increase in the CSI score due to the comprehensiveness and grid-to-grid calculation method of the CSI. Therefore, in future studies, more types of weather processes will be selected as test cases, and the idea of the spatial neighborhood will be introduced to optimize the algorithm further, improving the operational application of the algorithm.

Author Contributions: Conceptualization, X.C. and Y.C.; methodology, Y.C. and X.C.; validation, M.W.; investigation, Y.C. and C.Z.; data curation, S.W. and R.W.; writing—original draft preparation, X.C., M.W. and Y.C.; writing—review and editing, C.Z., S.W., R.W. and X.H.; visualization, X.H.; supervision, X.C.; project administration, X.C.; funding acquisition, X.C. All authors have read and agreed to the published version of the manuscript.

Funding: This research was jointly funded by the National Key Research and Development Program of China for Intergovernmental Cooperation (2019YFE0110100), the Science and Technology Innovation Team Project of the Guangdong Meteorological Bureau (GRMCTD202104), the Innovation and Development Project of the China Meteorological Administration (CXFZ2022J002, CXFZ2021Z012) and the Shenzhen Hong Kong Macao Science and Technology Plan Project (SGDX20210823103537035).

Conflicts of Interest: The authors declare no conflict of interest.

References

1. Licznar, P.; Krajewski, W.F. Precipitation Type Specific Radar Reflectivity-rain Rate Relationships for Warsaw, Poland. *Acta Geophys.* **2016**, *64*, 1840–1857. [CrossRef]
2. Dixon, M.; Wiener, G. Titan: Thunderstorm identification, tracking, analysis, and nowcasting—A radar-based methodology. *J. Atmos. Ocean. Technol.* **1993**, *10*, 785–797. [CrossRef]
3. Woo, W.; Wong, W. Operational application of optical flow techniques to radar-based rainfall nowcasting. *Atmosphere* **2017**, *8*, 48. [CrossRef]
4. Marrocu, M.; Massidda, L. Performance Comparison between Deep Learning and Optical Flow-Based Techniques for Nowcast Precipitation from Radar Images. *Forecasting* **2020**, *2*, 194–210. [CrossRef]
5. Yao, H.; Tang, X.; Wei, H.; Zheng, G.; Li, Z. Revisiting spatial-temporal similarity: A deep learning framework for traffic prediction. In Proceedings of the AAAI Conference on Artificial Intelligence, Honolulu, HI, USA, 27 January–1 February 2019; pp. 5668–5675.
6. Donahue, J.; Anne Hendricks, L.; Guadarrama, S.; Rohrbach, M.; Venugopalan, S.; Saenko, K.; Darrell, T. Long-term recurrent convolutional networks for visual recognition and description. In Proceedings of the IEEE Conference on Computer Vision and Pattern Recognition (CVPR), Boston, MA, USA, 11–12 June 2015; pp. 2625–2634.
7. Kipf, T.; Fetaya, E.; Wang, K.C.; Welling, M.; Zemel, R. Neural relational inference for interacting systems. In Proceedings of the International Conference on Machine Learning (ICML), Stockholm, Sweden, 10–15 July 2018; pp. 2688–2697.
8. Sun, C.; Chao, L.; Li, H.; Hu, Z.; Zheng, H.; Li, Q. Modeling and Preliminary Analysis of the Impact of Meteorological Conditions on the COVID-19 Epidemic. *Int. J. Environ. Res. Public Health* **2022**, *19*, 6125. [CrossRef] [PubMed]
9. Gao, C.; Wang, Y.; Hu, Z.; Jiao, H.; Wang, L. Study on the Associations between Meteorological Factors and the Incidence of Pulmonary Tuberculosis in Xinjiang, China. *Atmosphere* **2022**, *13*, 533. [CrossRef]
10. Hao, Y.; Luo, Z.; Zhao, J.; Gong, Y.; Li, Y.; Zhu, Z.; Tian, T.; Wang, Q.; Zhang, Y.; Zhou, Z.; et al. Transmission Risk Prediction and Evaluation of Mountain-Type Zoonotic Visceral Leishmaniasis in China Based on Climatic and Environmental Variables. *Atmosphere* **2022**, *13*, 964. [CrossRef]
11. Yipeng, H.; Wanbiao, L.; Yuchun, Z.; Lanqiang, B. A Review of Radar-and Satellite-based Observational Studies and Nowcasting Techniques on Convection Initiation. *Adv. Earth Sci.* **2019**, *34*, 1273–1287.
12. Sokol, Z.; Szturc, J.; Orellana-Alvear, J.; Popová, J.; Jurczyk, A.; Célleri, R. The Role of Weather Radar in Rainfall Estimation and Its Application in Meteorological and Hydrological Modelling-A Review. *Remote Sens.* **2021**, *13*, 351. [CrossRef]
13. Shi, E.; Li, Q.; Gu, D.; Zhao, Z. A method of weather radar echo extrapolation based on convolutional neural networks. In Proceedings of the International Conference on Multimedia Modeling, Bangkok, Thailand, 5–7 February 2018; pp. 16–28.
14. Ayzel, G.; Heistermann, M.; Sorokin, A.; Nikitin, O.; Lukyanova, O. All convolutional neural networks for radar-based precipitation nowcasting. *Procedia Comput. Sci.* **2019**, *150*, 186–192. [CrossRef]
15. Kalchbrenner, N.; Oord, A.; Simonyan, K.; Danihelka, I.; Vinyals, O.; Graves, A.; Kavukcuoglu, K. Video pixel networks. In Proceedings of the 34th International Conference on Machine Learning (ICML), Sydney, Australia, 11 August 2017; pp. 1771–1779.
16. Xu, Z.; Wang, Y.; Long, M.; Wang, J. Predcnn: Predictive learning with cascade convolutions. In Proceedings of the Twenty-Seventh International Joint Conference on Artificial Intelligence (IJCAI), Stockholm, Sweden, 13–19 July 2018; pp. 2940–2947.
17. Hochreiter, S.; Schmidhuber, J. Long short-term memory. *Neural Comput.* **1997**, *9*, 1735–1780. [CrossRef] [PubMed]

18. Tran, Q.K.; Song, S.K. Multi-channel weather radar echo extrapolation with convolutional recurrent neural networks. *Remote Sens.* **2019**, *11*, 2303. [CrossRef]
19. Shi, X.; Chen, Z.; Wang, H.; Yeung, D.Y.; Wong, W.K.; Woo, W.C. Convolutional LSTM network: A machine learning approach for precipitation nowcasting. In Proceedings of the Advances in Neural Information Processing Systems (NeurIPS), Montreal, QC, Canada, 7–12 December 2015; pp. 802–810.
20. Wang, Y.; Long, M.; Wang, J.; Gao, Z.; Yu, P.S. Predrnn: Recurrent neural networks for predictive learning using spatiotemporal lstms. In Proceedings of the Advances in Neural Information Processing Systems (NeurIPS), Long Beach, CA, USA, 4–9 December 2017; pp. 879–888.
21. Wang, Y.; Wu, H.; Zhang, J.; Gao, Z.; Wang, J.; Yu, P.; Long, M. PredRNN: A Recurrent Neural Network for Spatiotemporal Predictive Learning. *IEEE Trans. Pattern Anal. Mach. Intell.* **2022**. [CrossRef] [PubMed]
22. Wang, Y.; Gao, Z.; Long, M.; Wang, J.; Philip, S.Y. Predrnn++: Towards a resolution of the deep-in-time dilemma in spatiotemporal predictive learning. In Proceedings of the International Conference on Machine Learning (ICML), Stockholm, Sweden, 10–15 July 2018; pp. 5123–5132.
23. Wang, Y.; Zhang, J.; Zhu, H.; Long, M.; Wang, J.; Yu, P.S. Memory in memory: A predictive neural network for learning higher-order non-stationarity from spatiotemporal dynamics. In Proceedings of the IEEE Conference on Computer Vision and Pattern Recognition, Long Beach, CA, USA, 16–20 June 2019; pp. 9154–9162.
24. Shi, X.; Gao, Z.; Lausen, L.; Wang, H.; Yeung, D.Y.; Wong, W.K.; Woo, W.C. Deep learning for precipitation nowcasting: A benchmark and A new model. In Proceedings of the Advances in Neural Information Processing Systems (NeurIPS), Long Beach, CA, USA, 4–9 December 2017; pp. 5622–5632.
25. Wang, Y.; Jiang, L.; Yang, M.H.; Li, L.J.; Long, M.; Fei-Fei, L. Eidetic 3D LSTM: A Model for Video Prediction and Beyond. In Proceedings of the International Conference on Learning Representations, New Orleans, LA, USA, 6–9 May 2019; 2019; pp. 1–14.
26. Vaswani, A.; Shazeer, N.; Parmar, N.; Uszkoreit, J.; Jones, L.; Gomez, A.N.; Kaiser, Ł.; Polosukhin, I. Attention is all you need. In Proceedings of the Advances in Neural Information Processing Systems, Long Beach, CA, USA, 4–9 December 2017; pp. 5998–6008.
27. Jing, J.; Li, Q.; Peng, X.; Ma, Q.; Tang, S. Hprnn: A hierarchical sequence prediction model for long-term weather radar echo extrapolation. In Proceedings of the IEEE International Conference on Acoustics, Speech and Signal Processing (ICASSP), Barcelona, Spain, 4–8 May 2020; pp. 4142–4146.
28. Wang, T.C.; Liu, M.Y.; Zhu, J.Y.; Tao, A.; Kautz, J.; Catanzaro, B. High-Resolution Image Synthesis and Semantic Manipulation with Conditional GANs. In Proceedings of the IEEE Conference on Computer Vision and Pattern Recognition (CVPR), Salt Lake City, UT, USA, 18–22 June 2018.
29. Qian, H.; Wang, X.; Chen, X.; Yang, Z. Research on Noise Suppression Technology of Marine Optical Fiber Towed Streamer Seismic Data Based on ResUNet. *Energies* **2022**, *15*, 3362. [CrossRef]
30. Woo, S.; Park, J.; Lee, J.Y.; Kweon, I.S. CBAM: Convolutional Block Attention Module. In Proceedings of the European Conference on Computer Vision (ECCV), Munich, Germany, 8–14 September 2018.
31. Zhou, Q.; Chen, D.; Hu, Z.; Chen, X. Decompositions of Taylor diagram and DISO performance criteria. *Int. J. Climatol.* **2021**, *41*, 5726–5732. [CrossRef]
32. Hu, Z.; Chen, X.; Zhou, Q.; Chen, D.; Li, J. DISO: A rethink of Taylor diagram. *Int. J. Climatol.* **2019**, *39*, 2825–2832. [CrossRef]

atmosphere

Article

Assessing the Impact of Cumulus Parameterization Schemes on Simulated Summer Wind Speed over Mainland China

Si-Jie Liu [1,2], Ming Wang [3], Xiang Yi [4], Shuai-Bing Shao [2], Yi-Qun Zheng [1,*] and Xin-Min Zeng [1,2,*]

1 College of Hydrology and Water Resources, Hohai University, Nanjing 210098, China; sijieliu214@gmail.com
2 CMA-HHU Joint Laboratory for Hydrometeorological Studies, Hohai University, Nanjing 210098, China; shaosbing2015@gmail.com
3 Air Traffic Control and Navigation College, Air Force Engineering University, Xi'an 710000, China; backdream817@163.com
4 Unit No. 75839 of PLA, Guangzhou 510510, China; yxiang2022@gmail.com
* Correspondence: zhengyiqun2021@gmail.com (Y.-Q.Z.); xinmin.zeng@hhu.edu.cn (X.-M.Z.)

Abstract: Wind speed is an important meteorological parameter, whose simulation is influenced by various physical process parameterizations. However, the impact of cumulus parameterization schemes (CPSs) on wind speed simulation at the climate scale has not been sufficiently investigated in previous studies. Using the Advanced Research version of the Weather Research and Forecasting model (ARWv3) and hydrostatic wind speed change equation, we assessed the effects of four CPSs on a 10 m wind speed simulation over mainland China in the summer of 2003. In general, different CPSs can reproduce the wind speed distribution. Meanwhile, the sensitivity of wind speed simulation to CPSs was found to be the highest in East and southern China, followed by the Tibetan Plateau, and then Northwest China. We found that the main physical processes influencing wind speed (i.e., the pressure gradient (PRE), diffusion (DFN), and convection (CON) terms) vary greatly with sub-regions. CPSs mainly affect the secondary CON that regulates the balance between the dominant terms PRE and DFN, and also has a significant effect on PRE. For example, for CON, the difference index (DIF) between the Kain–Fritsch (KF) and previous KF (pKF) CPSs is larger than 20%, corresponding to a PRE DIF of about 14%. The term of local wind speed change (V_t) is significantly more sensitive to the CPSs than the other terms with a DIF of 283% over the Tibetan Plateau, suggesting high CPS sensitivity of the simulated wind speed. In addition, we explained the causes of the CPS-induced sensitivities. This work helps understand the Weather Research and Forecasting model (WRF) performance and emphasizes the importance of the CPS choice in simulating/forecasting wind speed.

Keywords: 10 m wind speed; cumulus parameterization schemes; sensitivity of physical processes; WRF; mainland China

Citation: Liu, S.-J.; Wang, M.; Yi, X.; Shao, S.-B.; Zheng, Y.-Q.; Zeng, X.-M. Assessing the Impact of Cumulus Parameterization Schemes on Simulated Summer Wind Speed over Mainland China. *Atmosphere* **2022**, *13*, 617. https://doi.org/10.3390/atmos13040617

Academic Editors: Zengyun Hu, Xuguang Tang and Qinchuan Xin

Received: 24 February 2022
Accepted: 8 April 2022
Published: 12 April 2022

1. Introduction

As one of the key variables of meteorological fields, the simulation and prediction of wind has been a goal of intensive research in various academic and industrial fields [1]. On the one hand, changes in wind direction are important, especially in coastal areas where they affect hydrodynamic factors such as waves and storm surges [2]. On the other hand, the change and distribution of wind speed have a significant influence on the change in the thermal structure of the boundary layer [3], surface fluxes [4], heat transfer, and mass transport [5], and even the estimation of pollutants [6]. Therefore, the studies of the processes affecting wind speed change and the sensitivity of wind speed simulation to different physical schemes are of importance for reference to further understand atmospheric motion. Meanwhile, from the perspective of society and economics, global warming leads to the frequent occurrence of extreme weather (e.g., storms and typhoons), so the forecast of wind speed is also extremely important to reduce economic losses and personal injury [7].

To explore the factors affecting wind speed simulation, it is necessary to understand the effects of atmospheric and land processes on wind speed. Previous studies on wind speed have been mainly based on the direct momentum balance method, i.e., the surface wind vector is determined by pressure gradient force, Coriolis force, gravity, and friction. For example, using this method in a weather case, Van den Broeke and Van Lipzig [8] suggested that large-scale pressure gradient forces controlled the momentum budget on the near-surface and contributed the most to wind speed, while the Coriolis force and gravity wave drag also affected wind speed. Horvath et al. [9] showed through a case study in winter that local and regional thermally driven circulation could influence the air mass through pressure gradient force within the pressure system, which was an important part of wind speed formation. In addition, the effects of land surface properties and their changes on wind speed cannot be ignored. For example, Wen et al. [10] studied the heat circulation of an oasis in the Gobi desert, and their results showed that the changes in land use types and land surface parameters could also change the land surface processes and affect the lower atmosphere and boundary layer with the change in atmospheric circulation situation, humidity, and temperature. Carvalho et al. [1] found that due to the limitation of topographic data for the area with complex terrain, a worse simulation could be produced by the WRF model. Lin et al. [11] studied the wind speed on the Qinghai–Tibet Plateau and pointed out that global warming and cooling caused by atmospheric thermal adaptation would greatly change the surface wind speed at the regional scale, and the surface wind speed on the Qinghai–Tibet Plateau changed more than other regions in mainland China. In addition, Zhang et al. [12] indicated that the pressure gradient force decreased when the wind speed dropped over the Qinghai–Tibet Plateau in summer, mainly due to the adjustment of atmospheric circulation. Zeng et al. [13] derived an equation affecting wind speed change, and the simulation results showed that the Coriolis force had no effect on wind speed, while the main physical processes affecting wind speed change were pressure gradient force, convection and turbulence, and these physical processes showed large differences with regional land perturbations, i.e., the simulation of these processes could be very sensitive to the choice of land surface schemes.

As can be seen from previous simulations of wind speed, there have been many influencing factors involved, among which the relationship between cumulus development and the wind field cannot be ignored. Previous studies of cumulus parameterization schemes (CPSs) were basically related to the study of precipitation, e.g., how cumulus development can affect heavy rainfalls [14], while the influence of CPSs on wind speed simulation has been rarely investigated [13,15]. Srinivas et al. [16] employed a mesoscale model of ARWv3.2, and found that compared to cloud microphysics and boundary layer schemes, the selection of cumulus convection parameterization schemes was more sensitive to tropical cyclone intensity and path. Therefore, different CPSs had influences on the simulation of surface wind speed to varying degrees. In fact, in addition to calculating cloud water content, CPSs are also intended to consider the effects of unresolved deep and shallow convective clouds, in which one of the main purposes is to calculate vertical fluxes caused by subgrid updrafts and downdrafts, as well as corresponding horizontal motions to compensate the change in vertical velocity, i.e., CPSs directly affect grid-scale wind speeds in the model. For example, Asai [17] found that vertical wind shear tended to inhibit the development of convection in the vertical plane parallel to the wind. Sui and Yanai [18] estimated the influence of cumulus clouds on the rotating part of the large-scale momentum field by using the vorticity balance residual, and found that cumulus clouds could slow down the mean airflow, and in the lower troposphere, such deceleration tended to reduce the vertical wind shear of the environment. The relationship between cumulus clouds and wind fields is more complicated in summer when convection is frequent. Das et al. [19] used different CPSs to simulate the Indian summer monsoon, showing that the simulation of typical characteristics of the Indian summer monsoon was quietly different with different schemes, among which the velocity potential and divergent winds obtained by Kuo-type CPS simulation were weak. Using the obliquely rotated principal component analysis,

Rao et al. [20] found that the summer topographic convection in China was controlled not only by topographic thermal conditions but also by the dynamic force of increasing wind speed in the mountainous area of the northeast Pearl River Delta. However, these studies have seldom focused on CPS-induced effects on simulated wind speed on the climate scales.

Therefore, the objective of this work is to assess the impact of CPSs on simulated wind speed over mainland China on a seasonal scale. The summer of 2003 was a very special season, during which extreme high temperature and rainstorm events occurred in southern/eastern China [21,22]. In this paper, the sensitivity of simulated summer wind speed to CPSs over typical underlying surfaces of different subregions of China (i.e., Northwest China, NW; East China, EC; and the Tibetan Plateau, TP) was evaluated, with the summer of 2003 taken as the study period. The following section describes the model, designed experiments, data, and relevant methods used in the study. Section 3 presents simulation results, and a summary and discussion are given in Section 4.

2. Methodology and Data

2.1. Model and Experimental Design

This work is a continuation of Zeng et al. [13], who assessed the sensitivity of 10 m wind speed to land surface schemes and the processes affecting wind speed in China during the summer of 2003 using the third version of the Advanced Research WRF (ARWv3) mesoscale model. In the summer of 2003, extremely heavy precipitation [21] and continuous high-temperature weather [22] occurred in eastern China, and there was an overall slightly positive anomaly of 10 m wind speed in mainland China (Figure 1b,c). The model selected in this study is the ARWv3 [23]. The physical schemes used in all experiments were the rapid radiative transfer model (RRTM) for long-wave radiation scheme, Dudhia short-wave radiation scheme, the WRF single moment 5-class microphysics scheme (WSM5), Monin-Obukhov near-surface scheme, Yonsei University (YSU) planetary boundary scheme, and Noah land surface scheme [13].

It should be noted that as a continuation of the work on the impact of land surface schemes (LSSs) on simulated wind speed by Zeng et al. (2018), the present study employs the same model and almost the same suite of model configurations, except for the CPSs, and intends to compare the impact induced by LSSs with that by CPSs. Because Zeng et al. (2018) had used four LSSs, correspondingly, for the sake of comparability, this study uses the four widely-used CPSs as follows: (1) The Kain–Fritsch (new Eta) scheme (KF hereafter; [24]) is an adjustment of the old Kain–Fritsch scheme in the Eta model. Its closure hypothesis is consistent with that of the old KF scheme, and the simple cloud model with water vapor rising and sinking is used to consider the role of entrainments and detrainments with relatively rough microphysical processes. In addition, the scheme inhibits large-scale convections in both marginal unstable and dry environments [25]. (2) The Betts–Miller–Janjic scheme (BMJ hereafter; [26]) is an adjusted and improved Betts–Miller scheme; the thermodynamic profile is relaxed at a given time, and the convective mass flux can consume a certain amount of effective buoyancy. Furthermore, the scheme considers the role of both deep and shallow convection processes. (3) The Grell–Devenyi ensemble scheme (Grell hereafter; [27]) is characterized by a parameterization framework of a simple scheme based on a previous convective parameterization, and this simple scheme was expanded to allow for a series of different assumptions that are commonly used in convective parameterizations with large sensitivity in model simulations, in which values for the assumed parameters are perturbed to obtain the ensemble scheme. (4) The previous Kain–Fritsch (pKF hereafter; [25]) is a one-dimensional entraining/detraining plume model for cumulus clouds, which is characterized by its representation of environmental entrainment and updraft detrainment rates, e.g., the mass exchange between clouds and their environment is regulated at model levels by a buoyancy sorting mechanism.

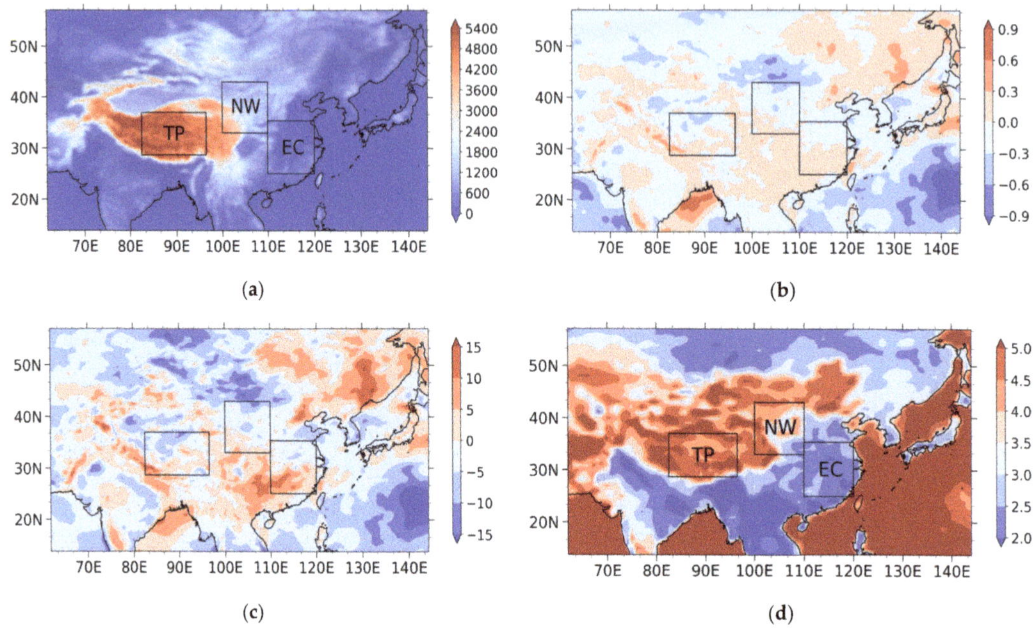

Figure 1. The studied domain and its 10 m wind speed distributions. (**a**) The terrain of the simulation domain (unit: m) and distribution of the studied sub-regions (i.e., Northwest China, NW; East China, EC; and the Tibetan Plateau, TP); (**b**) the distribution of the wind speed anomaly during the summer of 2003 (units: m s^{-1}); (**c**) the distribution of the wind speed anomaly in percentage in the summer (unit: %); (**d**) the seasonal mean wind speed of FNL analysis [13].

As in Zeng et al. [13], the simulation domain of in this study is centered at (37° N, 103° E) for mainland China. The total number of grids in the study area is 144 × 116 with horizontal and vertical resolutions of 40 km and 28 levels, and the top pressure of the model is 50 hPa. This area includes three sub-regions (Figure 1a) with typical and contrastive land surface features: NW (arid and semi-arid areas with a large topographic variability), EC (subtropical humid and temperate monsoon climate with high vegetation coverage and flat terrain), and TP (cold climate with high altitudes). The simulation period of all experiments was from 1 May 2003 to 1 September 2003, and the integral time step was 180 s. In order to make the simulation results more robust, here we conducted 12-ensemble experiments, with a small disturbance (i.e., at a 6-h interval) of the starting time of each experiment, i.e., 0000 UTC on 1 May 2003 to 1800 UTC on 3 May 2003 as the initial time of each ensemble. In the consecutive four-month tests, we took the first month as the model spin-up time, and used the simulation results of the ensemble simulation tests in June, July and August to conduct seasonal scale analysis. It should be noted here that we will use the names of CPSs (i.e., KF, BMJ, Grell, and pKF) to represent the ensemble averages of the simulations of the corresponding CPSs. Due to all experiments adopting the same physical parameterization schemes and the same simulation area and time period (i.e., June–August), the differences in simulation results were caused by the differences among the CPSs.

2.2. The Data

The ARW modeling system includes a number of datasets for default options or for users' selection, e.g., the present study employs a background albedo dataset and a dataset for vegetation indices based on the Moderate Resolution Imaging Spectroradiometer (MODIS) data, and a United States Geological Survey 10 min resolution topography dataset. Following Zeng et al. [13], the FINAL (FNL) 1° × 1° analysis data with a 6 hourly interval provided by the National Centers for Environmental Prediction (NCEP) were used for the initial fields, boundary conditions and validation data for simulation results. The NCEP Climate Forecast System Reanalysis (CFSR) monthly average data with a resolution of 0.5° × 0.5° were used in the anomaly distribution of summer wind speed in 2003 relative to the summer average over the last 30 years (Figure 1b,c).

2.3. Wind Speed Change Equation

Zeng et al. [13] derived the equation for full wind speed (other than wind components), which can be expressed as

$$V_t = \text{ADV} + \text{PRE} + \text{CON} + \text{DFN}, \tag{1}$$

where the 4 terms ADV, PRE, CON and DFN indicate the influences on local full wind speed change V_t, by advection, pressure gradient force, convection, and turbulent diffusion, respectively, where

$$\text{ADV} = -\int_t (u\frac{\partial V}{\partial x} + v\frac{\partial V}{\partial y})dt, \tag{2}$$

$$\text{PRE} = -\int_t \frac{1}{V}\left(\alpha u\frac{\partial p}{\partial x} + \alpha v\frac{\partial p}{\partial y} + u\frac{\partial \Phi}{\partial x} + v\frac{\partial \Phi}{\partial y}\right)dt, \tag{3}$$

$$\text{CON} = -\int \dot{\sigma}\frac{\partial V}{\partial \sigma}dt, \tag{4}$$

$$\text{DFN} = \int_t \frac{\boldsymbol{V}\cdot\boldsymbol{F}}{V}dt, \tag{5}$$

where u and v are zonal and meridional wind speed components, respectively. $\boldsymbol{V}$ and V are the wind vector and full wind speed ($V = \sqrt{u^2 + v^2}$), respectively, p is pressure, $\dot{\sigma}$ is the vertical velocity of the coordinate system, α is the specific volume, Φ is the geopotential, and $\boldsymbol{F}$ is the friction. For the above formulas, V_t, ADV, PRE, and CON can be calculated directly through the model outputs, then DFN can be calculated using Equation (1), and therefore the relative contribution of each factor/term to the wind speed change can be investigated.

Compared with the traditional momentum balance equation, the advantage of Equation (1) is that it is a variant of the prognostic equation for full wind speed, and can give exact quantitative results of the physical processes that affect the full wind speed, in which the influence of the Coriolis force on full wind speed is reasonably excluded although it must be considered for wind speed components [13].

By selecting different CPSs, the terms in Equations (1)–(5) can be affected in the following two ways. First, the CPSs can affect the thermodynamic structure of the atmosphere. The CPS-related processes of clouds and precipitation can directly heat the air mass, which changes the air density and also leads to modifications of geostrophic winds and hydrostatic stability in the horizontal and vertical directions, respectively, i.e., the terms of ADV, PRE, and CON, which are, respectively, associated with the wind, temperature (or air density), and stability, can be altered. Furthermore, when precipitation is received at the land surface, the surface energy balance is affected, which also results in a change in hydrostatic stability at the land surface. All of the stability changes would lead to changes in convective activities. Second, the CPSs can affect the dynamic structure of the atmosphere. Affected by cumulus entraining/detraining processes at the subgrid scale, the environmental wind field would be changed for compensation, which further causes the grid-cell averaged change of wind at different heights and can further result in wind shear-related turbulence, i.e., DFN can be altered.

2.4. Measures for Assessment

In order to quantitatively compare the consistency between simulation results and NCEP analysis data and investigate the differences among the CPSs, following Zeng et al. [13], several measures were used for the assessment, i.e., the seasonal mean bias of simulations and reference data (BIAS), the standard deviation (STD; STD_M and STD_O are for simulations and reference data, respectively), the spatial correlation coefficient (CR_{MO}), and the difference index (DIF) of simulated processes affecting the wind speed in different CPSs, which are computed as follows:

$$\text{BIAS} = \overline{M} - \overline{O}, \tag{6}$$

$$\text{STD}_\text{M} = \sqrt{\frac{1}{N}\sum_{i=1}^{N}\left(M_i - \overline{M}\right)^2}, \tag{7}$$

$$\text{STD}_\text{O} = \sqrt{\frac{1}{N}\sum_{i=1}^{N}\left(O_i - \overline{O}\right)^2}, \tag{8}$$

$$\text{CR}_\text{MO} = \frac{\sum\limits_{i=1}^{N}(M_i - \overline{M})(O_i - \overline{O})}{\sqrt{\sum\limits_{i=1}^{N}(M_i - \overline{M})^2}\sqrt{\sum\limits_{i=1}^{N}(O_i - \overline{O})^2}}, \tag{9}$$

$$\text{DIF} = \frac{\overline{T_{k_1}} - \overline{T_{k_2}}}{\frac{1}{k_m}\sum\limits_{k=1}^{k_m}\left|\overline{T_k}\right|} \times 100\%, k_1 \neq k_2, k_1 < k_m, k_2 < k_m, \tag{10}$$

where N is the number of grid points involved in the assessment within a region, M and O represent the simulated and observed values of a quantity, respectively, k_m is the number of the CPSs (4 in this case) for the difference index DIF, $\overline{T_k}$ is the regional mean value of a term for a certain physical process in Equation (1) from the CPS as labeled by k. DIF indicates the relative differences in the simulations between the different schemes.

3. Simulated Results

Because the same initial and boundary conditions and almost the same physical options were used in this study, in this section we focus on the CPS-induced differences of the 10 m full wind speed simulations in terms of the 12-member ensemble summer mean results.

3.1. The 10 m Wind Speed

3.1.1. Spatial Distributions

Figure 2 presents the spatial distributions of the seasonal average 10 m surface wind speed simulated by different CPSs. It can be found that except for southeastern mainland China, the CPSs generally well produced the wind speed distributions over almost the entire study domain as compared with the reference FNL data (e.g., Figure 1d vs. Figure 2a). Specifically, with a quite high simulation–reference correlation over the land portion of the study domain (Table 1), the general characteristics of wind speed were successfully simulated by the WRF model, e.g., the low values over the area north of 50° N and over the Indochina peninsula, and high values over North China, Northwest China, the Tibetan Plateau, and the surrounding oceans, showing a larger part of the study domain with simulated wind consistent with the reference data. Relatively, the area with values quite different from/higher than the reference data is small (i.e., over southern China and part of North China).

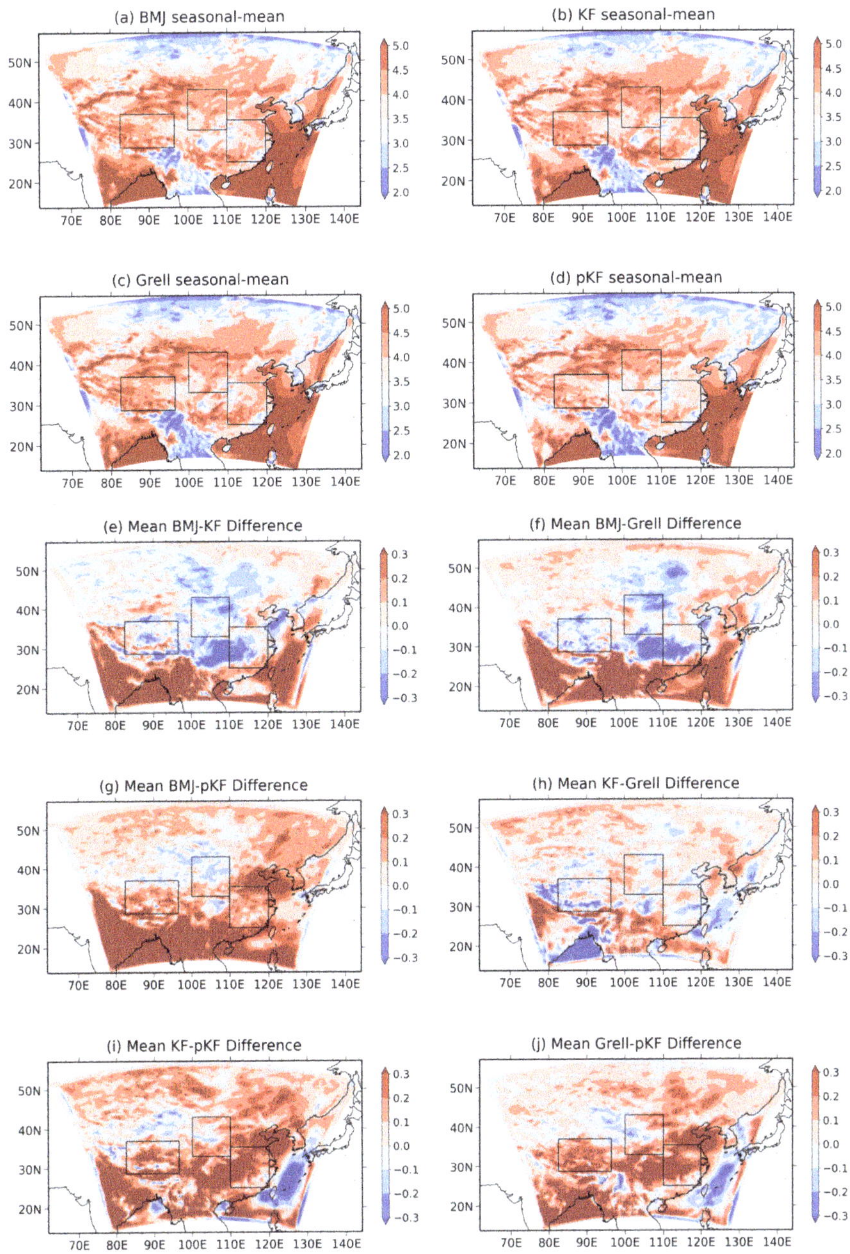

Figure 2. Ensemble mean seasonal spatial distributions of the 10 m wind speed in the CPS simulations (units: m s^{-1}), where grid cells are marked with grey dots for significant differences at the 0.10 significance level.

Table 1. Ensemble seasonal mean correlation coefficients (CR_{MO}) for Northwest China (NW), East China (EC), the Tibetan Plateau (TP), and the land portion of the model domain (ALL).

Scheme	NW	EC	TP	ALL
BMJ	0.31	0.80	0.31	0.69
KF	0.31	0.75	0.28	0.71
Grell	0.24	0.74	0.42	0.72
pKF	0.33	0.77	0.39	0.76

The wind field distributions by the CPSs appear to be close to each other due to the same initial and boundary conditions, and almost the same physical options as well. Overall, the CPSs tend to moderate the wind speed, i.e., the simulated wind speed is shown to be higher in the reference low wind speed zone and lower in the reference high wind speed zone (Figure 1d, Figure 2a–d). For the three subregions, the simulated wind speeds of the four CPSs decrease from west to east in TP, are larger in the north of NW than in the south, and decrease from southwest to northeast in EC.

However, a closer comparison would lead to clearer and larger differences among the CPSs. It can be seen more clearly from the CPS-induced different fields (Figure 2e–j) that the differences in simulated wind speed between different CPSs are relatively smaller in northern China and larger in southern China. For example, large KF–BMJ differences with absolute values of more than 0.3 m s^{-1} (Figure 2e) mainly exist in South China, and also a similar amplitude of the BMJ-pKF differences can be observed in the TP, South China, and the EC area around the Shandong Peninsula. This is mainly due to the fact that in the summer, the area of South/East China is a wind convergence region of the East Asian monsoon zone, where prevailing northwesterly wind from Siberia is relatively cold and dry in contrast with the warm and moist southwesterly wind from the Indian Ocean [28]. These two types of contrastive winds converge with approximately the same strength, and therefore fronts are formed which lead to the updrafts in the front zones and favor the release of the convective available potential energy; as a result, a large number of convective activities occur in South/East China. For the TP, the summer climate is controlled by the thermal low, which further brings the biggest portion of convective activities and hence precipitation of the year. All of these suggest that the above areas can be most affected by the CPSs. In addition, the CPS differences can enlarge the differences in simulated wind speed. For example, the pKF scheme is a mass flux parameterization using the Lagrangian parcel method to estimate whether instability exists, whether existing instability can become available for cloud growth, and what the properties of convective clouds could be. In contrast, the KF CPS considers a set of parameterizations including the convective trigger function, the mass flux formulation, and the closure assumptions [24], which makes it very different from pKF in simulating wind speed, e.g., KF presents significantly different wind speed compared with pKF in EC and the TP, with differences as large as about 0.3 m s^{-1} (Figure 2i). In addition, the Grell CPS makes use of a large variety of assumptions previously introduced in earlier formulations, with the assumptions generating a large spread in the solution for the ensemble scheme [27], which makes the CPS very unique as compared to a single scheme such as the other CPSs in the present study. As a result, the Grell–BMJ difference is apparently large in southern China (Figure 2f).

Correspondingly, the Student's *t*-test results at the 0.10 significance level [29] show that there is a quite large total area with significant differences, mostly in the southeast of mainland China. This further confirms that wind speed simulation is sensitive to the CPSs in areas with high convective precipitation. At the same time, there are some differences in the results of these CPS-produced simulations, e.g., the BMJ-pKF, KF-pKF and Grell-pKF differences are mainly distributed in the southeast of mainland China, while the total area with significant differences between KF and Grell is small, which is consistent with the above-mentioned results.

3.1.2. Assessment Results

Figure 3a shows the BIAS of the average seasonal wind speed simulated by different CPSs in the total area and each subregion. It can be seen that the wind speed values of the total area and subregions simulated by the four CPSs are similar to those of different land surface schemes [13]. For example, for simulated wind speeds of the total area, NW and EC are higher than the reference, while the TP value is lower. This result is closely associated with the distributions of simulated surface air temperatures. Generally, the model produced higher surface air temperatures over different parts of mainland China except for the TP (not shown). This means that the CPSs produced low-level stratifications that were more stable than they should have been over the TP, which does not favor the downward momentum transport and results in a lower wind speed in the TP. Similarly, simulated wind speed for subregions other than the TP is higher.

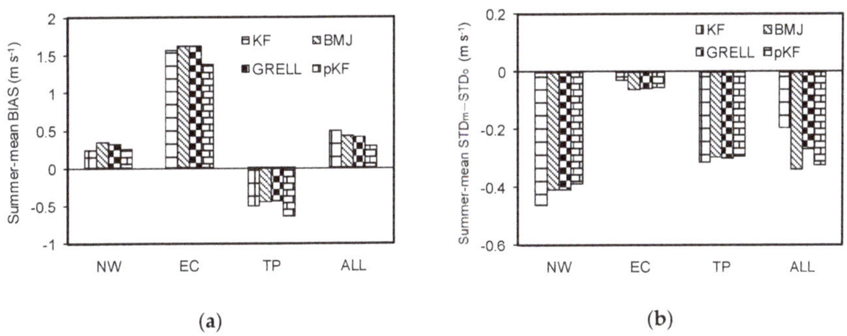

Figure 3. Ensemble seasonal mean wind speed results of BIAS (**a**) and STD_M minus the STDo (**b**) for each sub-region, where the regional means of wind speed from reference FNL data for NW, EC, TP, and ALL (land portion of the model domain) are 3.83, 2.42, 4.61, 3.58 m s^{-1}, respectively [13]. The FNL data are available at: https://rda.ucar.edu/datasets/ds083.2 (accessed on 24 February 2022).

In terms of BIAS, the CPSs present an overall higher wind speed for the total area (with the BIASs less than 0.50 m s^{-1}) while sub-regions might show different results, e.g., the BIAS of East China is much larger, and there exists a negative BIAS in the TP. In addition, wind speed in NW (EC) is the best (worst) simulated among the sub-regions, which is consistent with the above-mentioned results of wind speed distributions, e.g., there are much higher wind speed values in EC compared with the reference data (Figure 2 vs. Figure 1d).

The sensitivities of simulated wind speed to CPSs vary in different sub-regions. For example, although CPS-induced BIAS differences are relatively small for EC, while for NW, TP, and the total area, the differences can be much larger between the least and the largest, e.g., compared with the Grell BIAS of about −0.40 m s^{-1}, the amplitude of the pKF BIAS is approximately 50% larger for the TP, and for the total area, the BIAS differences also show the largest amplitude change of 70% as compared to the pKF BIAS of 0.30 m s^{-1} with the BMJ BIAS of 0.50 m s^{-1}.

Figure 3b shows the CPS-induced difference between STD_M and STD_O for the seasonal wind speed in the total area and sub-regions. Obviously, among the study sub-regions, the difference amplitude of the four schemes in EC is the smallest (less than 0.05 m s^{-1}), while the difference amplitude in NW is the largest (around −0.4 m s^{-1}), showing that although for different sub-regions simulated spatial variability is quite different with small CPS-induced changes, the CPS-induced changes for the total area are very large, e.g., the STD_M -STD_O difference amplitude by BMJ is 60% larger than that by KF.

It is noteworthy that in Figure 3b, the fluctuations of wind speeds are all smaller than the reference data. Possible reasons for this result are as follows. (1) Some influences of land surface heterogeneity are probably missing from the simulation. The subgrid heterogeneity could induce modifications of grid-scale changes in surface variables such as wind speed. For example, microscopic topography could dynamically change the atmospheric flow [30], subgrid variations in the surface moisture could thermally induce mesoscale circulations [31], and all of these could lead to grid-scale changes [32]. (2) The model has some deficiencies in simulating the wind speed, e.g., the simulated temperature fields were not well simulated, which would affect vertical momentum transport and then wind speed (as stated above). (3) Models are generally difficult to reproduce extremes of observations, suggesting that they tend to give moderate results compared to observations.

In addition, Table 1 lists the correlation (CR_{MO}) values by the CPSs for the sub-regions. Overall, the CPSs present wind speed patterns quite consistent with the reference (i.e., the CR_{MO} values are approximately 0.7). However, the values differ from sub-regions greatly, and moreover, large CPS-induced CR_{MO} difference can be seen, e.g., the KF–Grell difference amounts to up to 0.14 for the TP.

In general, the assessment measures show that the simulated wind speed is sensitive to the CPSs with different sub-regional characteristics, suggesting that the CPS choice or improvement is important for seasonal wind speed simulations or forecasts. Meanwhile, it is worth noting that the CPS-induced sensitivity is less than that induced by land surface schemes [13].

3.2. *Processes Affecting Wind Speed Change*

According to the summer mean integral results of V_t, ADV, PRE, CON, and DFN (Table 2), the main processes affecting the wind speed changes are PRE, DFN, and CON, and the effects of PRE and DFN are much larger than that of CON, having a positive and a negative contribution to wind speed, respectively, which is consistent with Zeng et al. [13]. Meanwhile, due to the influence of climate characteristics, there are certain differences in the factors affecting wind speed variation in different subregions. For subregion NW, PRE and DFN of KF have the greatest impacts among those of the CPSs, with both mean integral values of five and four times as large as that of CON, respectively, while PRE and DFN of the other schemes are also 3–5 times higher than CON. For EC, the absolute values of the mean integral PRE and DFN are the largest by BMJ, which are about three and four times larger than that of CON, respectively, but the signs of PRE and DFN are opposite (i.e., they have opposite effects on wind speed change). For the TP, a striking feature of CON appears: the CON contribution to the wind speed change is negative (i.e., the weak updraft formed by the thermal low pressure in summer restrained downward momentum transfer), which is the opposite of the facts observed for EC and NW. In addition, the integral means of PRE and DFN by Grell are the largest for the TP, reaching 12 and 11 times that of CON, respectively.

Table 2. Different CPS-ensemble area-averaged summer mean integral results of the five terms in Equation (1) (units: m s^{-1}).

	V_t			ADV			PRE			CON			DFN		
	NW	EC	TP	NW	EC	TP	NW	EC	TP	NW	EC	TP	NW	EC	TP
KF	−0.99	0.13	−0.26	57.37	31.52	17.91	4.27×10^3	2.77×10^3	1.72×10^4	1.01×10^3	8.47×10^2	-1.43×10^3	-5.34×10^3	-3.65×10^3	-1.58×10^4
BMJ	−1.52	−0.09	−0.53	64.27	24.17	8.07	4.30×10^3	2.62×10^3	1.88×10^4	1.04×10^3	8.83×10^2	-2.45×10^3	-5.40×10^3	-3.53×10^3	-1.64×10^4
Grell	−1.33	0.18	−0.21	56.96	30.91	13.46	4.19×10^3	2.54×10^3	1.89×10^4	1.06×10^3	8.94×10^2	-1.26×10^3	-5.31×10^3	-3.46×10^3	-1.77×10^4
pKF	−1.51	−0.35	−0.65	53.69	36.94	11.05	3.72×10^3	2.43×10^3	1.68×10^4	1.27×10^3	9.22×10^2	-9.91×10^2	-5.04×10^3	-3.39×10^3	-1.59×10^4

As can be seen from the difference index DIF of the terms in Equation (1) simulated by different CPSs (Table 3), the least term V_t is much more sensitive to CPSs than the other terms. Among them, the V_t DIF values in NW are less than 40%, while the DIF by pKF and Grell in EC and TP can reach 283% and 106%, respectively. Except V_t, the sensitivities of the other terms to CPSs are complicated. For example, CON is most sensitive to CPSs in NW, with a maximum DIF of 24%, while DFN is the least sensitive. For EC, the sensitivity of each term to CPSs is the lowest among the three sub-regions, with absolute DIF values within 13%. For the TP, CON has the largest effect (corresponding to a maximum DIF amplitude of −95%), followed by ADV, while the DIF maxima of PRE and DFN are lower than 12%. Compared with the sub-regions of EC and NW that show moderate CPS-induced sensitivities in the three major terms (i.e., PRE, DFN, and CON), the TP displays an apparently higher sensitivity, e.g., the DIF between KF and pKF can reach 95% for CON, suggesting that the influence of CPSs on wind speed processes varies greatly with sub-regions. For the overall wind speed change, it suggests that the CPS-induced influence on V_t is great in EC, probably due to frequent convective activities in the summer monsoon; for an even higher CPS-induced influence on V_t in the TP, the high topographic elevations that lead to a thin troposphere in this area could be a main cause.

Table 3. The difference index (DIF) in different CPS-ensemble area-averaged summer mean integral results of the five terms in Equation (1) for the sub-regions, corresponding to the value of the vertical CPS minus the one of the horizontal CPS (unit: %).

		V_t			ADV			PRE			CON			DFN		
		BMJ	**Grell**	**pKF**	**BMJ**	**Grell**	**pKF**	**BMJ**	**Grell**	**pKF**	**BMJ**	**Grell**	**pKF**	**BMJ**	**Grell**	**pKF**
NW	KF	40	25	39	−12	1	6	−1	2	14	−3	−5	−24	1	−1	−6
	BMJ	-	−14	−1	-	13	18	-	3	14	-	−2	−21	-	−2	−7
	Grell	-	-	14	-	-	6	-	-	12	-	-	−19	-	-	−5
EC	KF	−25	−168	115	6	0	−4	6	9	13	−4	−5	−8	−3	−5	−8
	BMJ	-	−143	140	-	−5	−10	-	3	8	-	−1	−4	-	−2	−4
	Grell	-	-	283	-	-	−5	-	-	4	-	-	−3	-	-	−2
TP	KF	66	−11	95	78	35	54	−9	−10	2	67	−11	−28	3	12	0
	BMJ	-	−77	29	-	−43	−24	-	−1	11	-	−77	−95	-	8	−3
	Grell	-	-	106	-	-	19	-	-	12	-	-	−18	-	-	−11

3.3. Associated Boundary-Layer Parameters

3.3.1. Near-Surface Fluxes

The change in convective activities can affect the change in surface energy fluxes, while the change in surface sensible and latent heat fluxes will affect the convective processes and wind speed through energy transfer and release in the vertical direction with land surface disturbances. Figure 4 shows the average sensible and latent heat fluxes simulated by different CPSs. It can be seen that although CPS-induced sensible and latent heat flux differences are relatively small over most of mainland China, the largest differences still exist in EC (e.g., with an amplitude of over 20 w m^{-2}; Figure 4c,g), which is consistent with the above-mentioned result that over this area, simulated convective activities are affected by the CPSs to quite an extent.

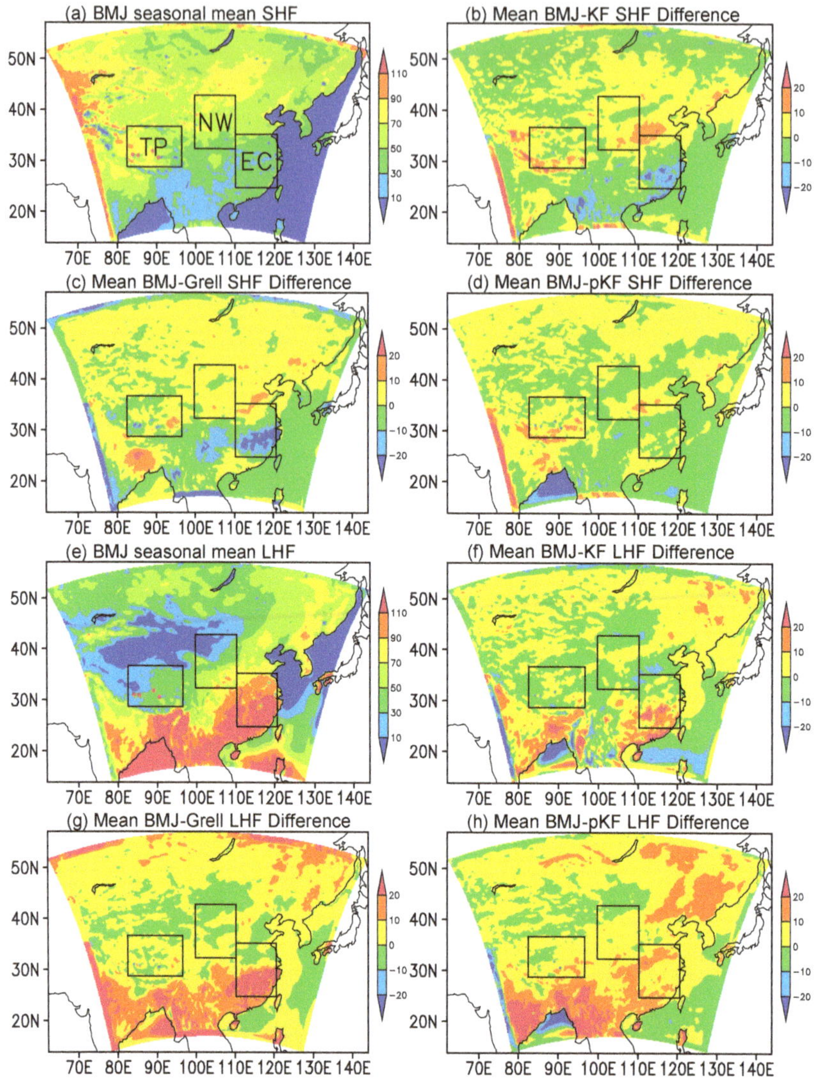

Figure 4. Ensemble seasonal mean distributions of sensible heat flux (SHF), latent heat flux (LHF), and their differences (units: W m^{-2}).

By comparing the difference field of wind speed with that of sensible/latent heat flux, there appears no clear correlation. This suggests that the CPS-induced sensitivities are complicated. Although the land surface plays an important role in modifying 10 m wind speed [13], the changes in surface fluxes are probably less important in CPS-induced changes in wind speed compared with the changes in the interior atmosphere.

3.3.2. Atmospheric Boundary Layer Stability

On the weather scale, different CPSs can influence simulated momentum exchange by influencing the stability of the simulated atmospheric boundary layer, i.e., the stability of the atmospheric boundary layer can be affected by the energy release in the convective process, which will lead to a difference in the momentum exchange between the upper and lower levels and thus affect the wind speed [1]. Figure 5 shows the CPS-produced seasonally average Richardson number (R_i), which can be used as a measure for atmospheric boundary layer stability. On the whole, similar to the simulation by different land surface schemes that presented the vertical distributions of atmospheric stability in the three sub-regions [13], it can be also concluded that the CPS-produced instability layer over the TP is the thickest, followed by NW and EC (e.g., Figure 5d). Consistent with the above-mentioned results for the sub-regions, the R_i for NW is not sensitive to the CPSs, while the TP shows the highest sensitivity, e.g., there is a quite large R_i difference between Grell and pKF (Figure 5c), corresponding to the highest sensitivity of CON to the CPSs, as addressed in Section 3.2. In addition, Grell produces a more unstable boundary layer than pKF does, which favors the downward momentum transfer, further strengthens the 10 m wind speed, and results in a higher Grell wind speed compared to pKF (Figure 2j). Note that the simulated wind speed sensitivity for the TP is not the highest of the CPSs., i.e., the DIF of V_t for the TP is not the largest among the sub-regions (Table 3). The reason for this complexity is that vertical momentum transfer is a result of convection (i.e., CON), while there are other important processes responsible for wind speed change (e.g., PRE).

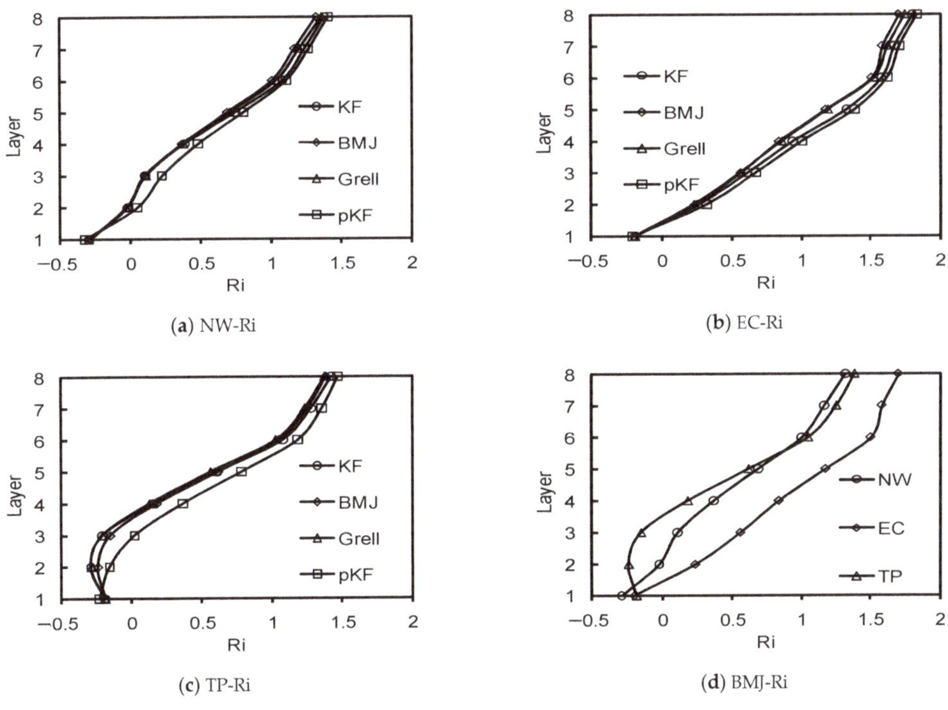

Figure 5. Ensemble seasonal mean Richardson number (R_i) in the selected atmospheric boundary layer in the three sub-regions, where Layers 1–8 are approximately 10, 30, 100, 200, 320, 470, 600, and 900 m above ground level, respectively, with $R_i = \frac{g\Delta\theta\Delta z}{\overline{\theta}[(\Delta\overline{U})^2+(\Delta\overline{V})^2]}$, where $\Delta\theta$ represents the potential temperature difference between the top and bottom of the atmospheric layer, Δz the thickness of the layer, $\overline{\theta}$ the mean potential temperature, and $\Delta\overline{U}$ ($\Delta\overline{V}$) the zonal (meridional) wind speed difference between the top and bottom of the atmospheric layer [33]. (**a**) The four CPS ensembles for NW; (**b**) same as (**a**) but for EC; (**c**) same as (**a**) but for the TP; (**d**) the BMJ ensemble for the three sub-regions.

4. Summary and Discussion

In order to investigate the sensitivity of simulated wind speed and its influencing processes to the CPS choice, we selected four CPSs (KF, BMJ, Grell, and pKF) as employed in the mesoscale model WRFv3 to separately conduct ensemble simulations for the summer of 2003 in mainland China. Using the wind speed change equation under the hydrostatic condition, we calculated the four physical terms responsible for the wind speed change (V_t), i.e., the terms induced by advection (ADV), pressure gradient force (PRE), convection (CON), and turbulent diffusion (DFN), and quantified the relative importance of influencing processes and analyzed their sensitivities to the CPSs. The main results and conclusions are as follows:

(1) By and large, different CPSs can reproduce 10 m wind speed over mainland China, which is also indicated by the simulation–reference correlation efficiency of approximately 0.70. Previous studies of CPSs were basically associated with precipitation [14], while this result indicates an overall performance of simulating wind speed by the CPSs.

(2) In comparison to the CPS ensembles, the largest simulated difference is generally found between Grell and pKF. Although the CPS choice does not greatly modify the simulated wind speed, sub-regions of mainland China show quite a large CPS-induced impact on wind speed. It can be seen that northern China is relatively unaffected by the CPSs, but southern China, East China, and the Tibetan Plateau are affected to quite large extents, as is confirmed by Student's *t*-tests. These high sensitivities are associated with the frequent convective activities in the summer monsoon (e.g., over East China) and a relatively thin troposphere (i.e., over the Tibetan Plateau). Because the influence of CPSs on wind speed simulation has been rarely investigated on a climate scale [13,15], these results clearly indicate where the simulated wind speed is greatly affected in mainland China on the summer scale, and the mechanisms have been revealed.

(3) Among the terms of influencing processes, CON is most affected by the CPSs, followed by PRE and DFN, corresponding to CPS-induced DIF values of 95%, 14%, and 12% for the sub-regions, respectively. ADV is a secondary term for contribution to V_t, with the latter having a large DIF value of 283% for East China. Previous works seldom showed the CPS-induced impacts on complete influencing processes [14]; this study presents the impacts of the turbulence effect (DFN) and revealed that they cannot be conventionally quantified.

(4) The results of the related boundary layer parameters can demonstrate the CPS-induced impact on simulated wind speed, in which surface fluxes do not show clear correlations with wind change while the Richardson number does. This suggests that compared with the CPS-induced changes in wind speed in the interior atmosphere, the CPS-induced changes of surface fluxes are less important. This work makes an incremental advance in wind speed study based on the LSS-induced impact [13], emphasizing the importance of the atmospheric process rather than land surface processes.

Because CPSs are closely associated with precipitation simulations, it is expected that there is a correlation between simulated precipitation and wind speed as induced by the CPSs. We found that simulated precipitation is sensitive to the CPSs over the selected subregions, especially over EC (not shown). However, the precipitation–wind speed correlation is complicated. On the one hand, higher precipitation generally means more frequent convective activities that would strengthen surface wind speed by stronger downward momentum transport. On the other hand, precipitation-related surface cooling would strengthen the low-level stable stratification and then restrain downward momentum transport. All of these are likely to lead to opposite results (not shown), i.e., that no definite correlation between CPS-induced precipitation and wind speed can be validated.

It is noteworthy that this work is a case study focusing on the quantitative assessment of physical processes affecting summer wind speeds and their sensitivities to CPSs with a regional climate model. Due to the complexity of the CPSs, the differences in the physical parametrizations of the specific CPSs have not been discussed here. As addressed in previous studies [1,13], the coupled-model simulation reflects not only the performance of a single scheme that is coupled to the model, but also the overall performance of the modeling system, which suggests that this CPS evaluation is of significance to our understanding of the CPS performance and improving the model to simulate the summer wind speed.

Author Contributions: Conceptualization, X.-M.Z. and Y.-Q.Z.; methodology, S.-J.L., M.W., S.-B.S. and X.Y.; software, M.W.; validation, S.-J.L., S.-B.S. and M.W.; formal analysis, S.-J.L., X.Y. and M.W.; investigation, S.-J.L. and M.W.; resources, S.-J.L. and M.W.; data curation, M.W.; writing—original draft preparation, S.-J.L., X.Y. and X.-M.Z.; supervision, X.-M.Z. and Y.-Q.Z.; project administration, X.-M.Z.; funding acquisition, X.-M.Z. All authors have read and agreed to the published version of the manuscript.

Funding: This research was financially supported by the National Natural Science Foundation of China (Grant Nos. U2240248, 41775087 and 41675007), and the National Key Research and Development Program of China (Grant Nos. 2018YFA0606004 and 2018YFC1508200).

Informed Consent Statement: Not applicable.

Data Availability Statement: The FNL Data Archive is managed by the Data Engineering and Curation Section of the Computational and Information Systems Laboratory at the National Center for Atmospheric Research in Boulder, Colorado. The National Center for Atmospheric Research is sponsored by the National Science Foundation.

Acknowledgments: The authors would like to thank three anonymous reviewers for their comments and suggestions which helped in improving the manuscript.

Conflicts of Interest: The authors declare no conflict of interest.

References

1. Carvalho, D.; Rocha, A.; Gómez-Gesteira, M.; Santos, C. A sensitivity study of the WRF model in wind simulation for an area of high wind energy. *Environ. Model. Softw.* **2012**, *33*, 23–34. [CrossRef]
2. McInnes, K.L.; Erwin, T.A.; Bathols, J.M. Global climate model projected changes in 10 m wind speed and direction due to anthropogenic climate change. *Atmos. Sci. Lett.* **2011**, *12*, 325–333. [CrossRef]
3. He, Y.; Monahan, A.H.; McFarlane, N.A. Diurnal variations of land surface wind speed probability distributions under clear-sky and low-cloud conditions. *Geophys. Res. Lett.* **2013**, *40*, 3308–3314. [CrossRef]
4. Monahan, A.H. The probability distribution of sea surface wind speeds. Part I: Theory and sea winds observations. *J. Clim.* **2006**, *19*, 497–520. [CrossRef]
5. Cakmur, R.; Miller, R.; Torres, O. Incorporating the effect of small-scale circulations upon dust emission in an atmospheric general circulation model. *J. Geophys. Res.* **2004**, *109*, D07201. [CrossRef]
6. Hu, X.M.; Klein, P.M.; Xue, M. Evaluation of the updated YSU planetary boundary layer scheme within WRF for wind resource and air quality assessments. *J. Geophys. Res. Atmos.* **2013**, *118*, 10490–10505. [CrossRef]
7. Schwierz, C.; Köllner-Heck, P.; Mutter, E.Z.; Bresch, D.N.; Vidale, P.; Wild, M.; Schär, C. Modelling European winter wind storm losses in current and future climate. *Clim. Chang.* **2010**, *101*, 485–514. [CrossRef]
8. Van den Broeke, M.R.; van Lipzig, N.P.M. Factors controlling the near-surface wind field in Antarctica. *Mon. Weather Rev.* **2003**, *131*, 733–743. [CrossRef]
9. Horvath, K.; Koracin, D.; Vellore, R.; Jiang, J.; Belu, R. Sub-kilometer dynamical downscaling of near-surface winds in complex terrain using WRF and MM5 mesoscale models. *J. Geophys. Res.* **2012**, *117*, D11111. [CrossRef]
10. Wen, X.H.; Lu, S.H.; Jin, J.M. Integrating remote sensing data with WRF for improved simulations of oasis effects on local weather processes over an arid region in Northwestern China. *J. Hydrometeorol.* **2012**, *13*, 573–587. [CrossRef]
11. Lin, C.G.; Yang, K.; Qin, J.; Fu, R. Observed coherent trends of surface and upper-air wind speed over China since 1960. *J. Clim.* **2013**, *26*, 2891–2903. [CrossRef]
12. Zhang, H.L.; Pu, Z.X.; Zhang, X.B. Examination of Errors in Near-Surface Temperature and Wind from WRF Numerical Simulations in Regions of Complex Terrain. *Weather Forecast.* **2013**, *28*, 893–914. [CrossRef]
13. Zeng, X.M.; Wang, M.; Wang, N.; Yi, X.; Chen, C.H.; Zhou, Z.G.; Wang, G.L.; Zheng, Y.Q. Assessing simulated summer 10-m wind speed over China: Influencing processes and sensitivities to land surface schemes. *Clim. Dyn.* **2018**, *50*, 4189–4209. [CrossRef]
14. Kotroni, V.; Lagouvardos, K. Precipitation forecast skill of different convective parameterization and microphysical schemes: Application for the cold season over Greece. *Geophys. Res. Lett.* **2001**, *28*, 1977–1980. [CrossRef]
15. Cohen, C. A comparison of cumulus parameterizations in idealized sea-breeze simulations. *Mon. Weather Rev.* **2002**, *130*, 2554–2571. [CrossRef]
16. Srinivas, C.V.; Bhaskar, R.D.V.; Yesubabu, V.; Baskaran, R.; Venkatraman, B. Tropical cyclone predictions over the bay of bengal using the high-resolution advanced research weather research and forecasting (ARW) model. *Q. J. R. Meteor. Soc.* **2013**, *139*, 1810–1825. [CrossRef]

17. Asai, T. Cumulus Convection in the Atmosphere with Vertical Wind Shear: Numerical Experiment. *J. Meteorol. Soc. Jpn.* **1964**, *42*, 245–259. [CrossRef]
18. Sui, C.H.; Yanai, M. Cumulus Ensemble Effects on the Large-Scale Vorticity and Momentum Fields of GATE. Part I: Observational Evidence. *J. Atmos. Sci.* **1986**, *43*, 1618–1642.
19. Das, S.; Mitra, A.K.; Iyengar, G.R.; Mohandas, S. Comprehensive test of different cumulus parameterization schemes for the simulation of the Indian summer monsoon. *Arch. Meteorol. Geophys. Bioclimatol. Ser. B* **2001**, *78*, 227–244. [CrossRef]
20. Rao, X.N.; Zhao, K.; Chen, X.C.; Huang, A.N.; Xue, M.; Zhang, Q.H.; Wang, M.J. Influence of Synoptic Pattern and Low-Level Wind Speed on Intensity and Diurnal Variations of Orographic Convection in Summer over Pearl River Delta, South China. *J. Geophys. Res. Atmos.* **2019**, *124*, 6157–6179. [CrossRef]
21. Zeng, X.M.; Wu, Z.H.; Song, S.; Xiong, S.Y.; Zheng, Y.Q.; Zhou, Z.G.; Liu, H.Q. Effects of land surface schemes on the simulation of a heavy rainfall event by WRF. *Chin. J. Geophys.* **2012**, *55*, 16–28. [CrossRef]
22. Zeng, X.M.; Wang, B.; Zhang, Y.; Zheng, Y.; Wang, N.; Wang, M.; Yi, X.; Chen, C.; Zhou, Z.; Liu, H. Effects of land surface schemes on WRF-simulated geopotential heights over China in summer 2003. *J. Hydrometeorol.* **2016**, *17*, 829–851. [CrossRef]
23. Skamarock, W.C.; Klemp, J.B.; Dudhia, J. *A Description of the Advanced Research WRF Version 3*; NCAR/TN-475 + STR.; National Center for Atmospheric Research: Boulder, CO, USA, 2008.
24. Kain, J.S. The Kain-Fritsch convective parameterization, an update. *J. Appl. Met.* **2004**, *43*, 170–181. [CrossRef]
25. Kain, J.S.; Fritsch, J.M. A one-dimensional entraining/detraining plume model and its application in convective parameterization. *J. Atmos. Sci.* **1990**, *47*, 2784–2802. [CrossRef]
26. Betts, A.K.; Miller, M.J. A new convective adjustment scheme Part II: Single column tests using GATE wave, BOMEX, and arctic air-mass datasets. *Q. J. R. Meteor. Soc.* **1986**, *112*, 693–709.
27. Grell, G.A.; Devenyi, D.A. Generalized approach to parameterizing convection combining ensemble and data assimilation techniques. *Geophys. Res. Lett.* **2002**, *29*, 1693. [CrossRef]
28. Chen, H.M.; Zhou, T.J.; Yu, R.C.; Bao, Q. The East Asian Summer Monsoon Simulated by Coupled Model FGOALS_s. *Chin. J. Atmos. Sci.* **2009**, *33*, 155–167.
29. Student. Review of Statistical Methods for Research Workers, by R. A. Fisher. *Eugen. Rev.* **1926**, *18*, 148–150.
30. Zeng, X.; Pielke, R.A. Landscape-induced atmospheric flow and its parameterization in large-scale numerical models. *J. Clim.* **1995**, *8*, 1156–1177. [CrossRef]
31. Yan, H.; Anthes, R.A. The effect of variations in the surface moisture on mesoscale circulations. *Mon. Weather Rev.* **1988**, *116*, 192–208. [CrossRef]
32. Zeng, X.-M.; Zhao, M.; Su, B.-K.; Tang, J.-P.; Zheng, Y.-Q.; Zhang, Y.-J.; Chen, J. Effects of the land-surface heterogeneities in temperature and moisture from the "combined approach" on regional climate: A sensitivity study. *Glob. Planet. Chang.* **2003**, *37*, 247–263. [CrossRef]
33. Stull, R.B. *An Introduction to Boundary Layer Meteorology*; Kluwer Academic Publishers: Dordrecht, The Netherlands, 1987; p. 177.

atmosphere

MDPI

Article

Evaluation of the CRU TS3.1, APHRODITE_V1101, and CFSR Datasets in Assessing Water Balance Components in the Upper Vakhsh River Basin in Central Asia

Aminjon Gulakhmadov [1,2,3,4], Xi Chen [1,2,5,*], Manuchekhr Gulakhmadov [1,2,5,6], Zainalobudin Kobuliev [4], Nekruz Gulahmadov [2,4,5], Jiabin Peng [1,2,7], Zhengyang Li [1,2] and Tie Liu [1,2,7]

1 Research Center of Ecology and Environment in Central Asia, Xinjiang Institute of Ecology and Geography, Chinese Academy of Sciences, Urumqi 830011, China; aminjon@ms.xjb.ac.cn (A.G.); gmanuchekhr@mail.ru (M.G.); pengjiabin17@mails.ucas.edu.cn (J.P.); lizhengyang19@mails.ucas.ac.cn (Z.L.); liutie@ms.xjb.ac.cn (T.L.)

2 State Key Laboratory of Desert and Oasis Ecology, Xinjiang Institute of Ecology and Geography, Chinese Academy of Sciences, Urumqi 830011, China; nekruz.abdujabborovich@mails.ucas.ac.cn

3 Ministry of Energy and Water Resources of the Republic of Tajikistan, Dushanbe 734064, Tajikistan

4 Institute of Water Problems, Hydropower and Ecology of the National Academy of Sciences of Tajikistan, Dushanbe 734042, Tajikistan; kobuliev@mail.ru

5 University of Chinese Academy of Sciences, Beijing 100049, China

6 Committee for Environmental Protection under the Government of the Republic of Tajikistan, Dushanbe 734034, Tajikistan

7 Department of Geography, Ghent University, 9000 Ghent, Belgium

* Correspondence: chenxi@ms.xjb.ac.cn; Tel.: +86-136-0992-3012

Abstract: In this study, the applicability of three gridded datasets was evaluated (Climatic Research Unit (CRU) Time Series (TS) 3.1, "Asian Precipitation—Highly Resolved Observational Data Integration Toward the Evaluation of Water Resources" (APHRODITE)_V1101, and the climate forecast system reanalysis dataset (CFSR)) in different combinations against observational data for predicting the hydrology of the Upper Vakhsh River Basin (UVRB) in Central Asia. Water balance components were computed, the results calibrated with the SUFI-2 approach using the calibration of soil and water assessment tool models (SWAT–CUP) program, and the performance of the model was evaluated. Streamflow simulation using the SWAT model in the UVRB was more sensitive to five parameters (ALPHA_BF, SOL_BD, CN2, CH_K2, and RCHRG_DP). The simulation for calibration, validation, and overall scales showed an acceptable correlation between the observed and simulated monthly streamflow for all combination datasets. The coefficient of determination (R^2) and Nash–Sutcliffe efficiency (NSE) showed "excellent" and "good" values for all datasets. Based on the R^2 and NSE from the "excellent" down to "good" datasets, the values were 0.91 and 0.92 using the observational datasets, CRU TS3.1 (0.90 and 0.90), APHRODITE_V1101+CRU TS3.1 (0.74 and 0.76), APHRODITE_V1101+CFSR (0.72 and 0.78), and CFSR (0.67 and 0.74) for the overall scale (1982–2006). The mean annual evapotranspiration values from the UVRB were about 9.93% (APHRODITE_V1101+CFSR), 25.52% (APHRODITE_V1101+CRU TS3.1), 2.9% (CFSR), 21.08% (CRU TS3.1), and 27.28% (observational datasets) of annual precipitation (186.3 mm, 315.7 mm, 72.1 mm, 256.4 mm, and 299.7 mm, out of 1875.9 mm, 1236.9 mm, 2479 mm, 1215.9 mm, and 1098.5 mm). The contributions of the snowmelt to annual runoff were about 81.06% (APHRODITE_V1101+CFSR), 63.12% (APHRODITE_V1101+CRU TS3.1), 82.79% (CFSR), 81.66% (CRU TS3.1), and 67.67% (observational datasets), and the contributions of rain to the annual flow were about 18.94%, 36.88%, 17.21%, 18.34%, and 32.33%, respectively, for the overall scale. We found that gridded climate datasets can be used as an alternative source for hydrological modeling in the Upper Vakhsh River Basin in Central Asia, especially in scarce-observation regions. Water balance components, simulated by the SWAT model, provided a baseline understanding of the hydrological processes through which water management issues can be dealt with in the basin.

Citation: Gulakhmadov, A.; Chen, X.; Gulakhmadov, M.; Kobuliev, Z.; Gulahmadov, N.; Peng, J.; Li, Z.; Liu, T. Evaluation of the CRU TS3.1, APHRODITE_V1101, and CFSR Datasets in Assessing Water Balance Components in the Upper Vakhsh River Basin in Central Asia. *Atmosphere* **2021**, *12*, 1334. https://doi.org/10.3390/atmos12101334

Academic Editor: Luis Gimeno

Received: 6 September 2021
Accepted: 9 October 2021
Published: 12 October 2021
Corrected: 8 December 2021

Keywords: hydrological modeling; gridded datasets; sensitivity analysis; water balance; snowmelt; SWAT; Upper Vakhsh River Basin

1. Introduction

Watershed-based hydrological models provide a practical approach to evaluating the water cycle's components, particularly snowmelt's contribution to river flow [1,2]. One of the challenges in mountainous regions when modeling watershed hydrology and evaluating water balance components is obtaining weather input data, which are generally among the most essential drivers of watershed models [3]. Unfortunately, observational climate stations are often sparsely located and thus cannot characterize the climate conditions throughout a catchment, particularly if large hydroclimatic gradients exist. Additionally, climate station measurements often do not cover the proposed modeling period, and there may be gaps in the records. In order to solve this issue, the investigation of alternative climate data is essential in mountainous areas.

The applicability of the climate forecast system reanalysis (CFSR), "Asian Precipitation—Highly Resolved Observational Data Integration Toward the Evaluation of Water Resources" (APHRODITE), and Climatic Research Unit (CRU) datasets for hydrological models in water balance components analysis has not been investigated thus far in the UVRB. Similarly, previous studies on the applicability of models to estimate hydrological components in the highlands of Tajikistan (UVRB) in Central Asia have not been conducted. Various hydrological models at the watershed scale have been used for the estimation of water cycle components, including the Hydrologic Engineering Center hydrologic modeling system [4], MIKE SHE [5], the soil and water assessment tool [6], the hydrologic simulation program Fortran [7], and the snowmelt runoff model [2]. The SWAT model is internationally recognized as a robust hydrological model and is widely used, including in several basins that have snowmelt-dominated streamflow [8–14].

Previous research indicated that the SWAT model is a common tool to assess the water balance components of watersheds. Combinations of CFSR datasets with the SWAT model and observational datasets with the SWAT model were applied to different watersheds in the Blue Nile Basin in Ethiopia to assess water-balance components, particularly actual evapotranspiration [15]. In most cases, CFSR weather simulations gave similar or lower evaluations than those obtained when using in situ observations in model inputs. Independent observation datasets and CFSR were used in the SWAT model to estimate water-balance components in the Melka Kuntur watershed in Ethiopia [16]. Analysis of the mean annual water balance demonstrated that higher values of water-balance components were acquired when applying the CFSR datasets to the Melka Kuntur watershed. This may be associated with the relatively high total precipitation in the CFSR dataset for the Melka Kuntur watershed [16]. Adeogun et al. noted that the SWAT model could be a promising tool for predicting water balance and water output for sustainable water management in Nigeria [17]. Gupta et al. noted that SWAT is a powerful tool that very effectively evaluated the hydrological components in a study of water balance and river flow in the Sabarmati River Basin in India [18]. Goswami et al. used the SWAT model and CFSR datasets from 1984 to 2013 in the Narmada River Basin in India and suggested that the SWAT model was able to simulate the water balance components at the basin and sub-basin scales [19]. Himanshu et al. [20] concluded that the SWAT model can accurately simulate the hydrology and water balance components of the Ken River Basin in India. Nasiri et al. [21] applied the SWAT model to the Samalqan Basin in Iran to assess water-balance components. Actual evapotranspiration contributed to the largest water loss from the basin, which was approximately 86%. Nasiri et al. pointed out that the high evapotranspiration rate that was simulated may be related to the vegetation types in the region [21]. The applicability of the SWAT model for the simulation of water-balance components, particularly surface runoff, has been assessed in the Heihe mountain river basin in northwest China [22]. The components of the water balance tended to increase, and the total runoff increased by 30.5% between 1964 and 2013. Rising surface runoff accounted for 42.7% of the total increasing runoff [22]. Pritchard [23] used a combination of CFSR temperature and APHRODITE rainfall datasets in the SWAT model to simulate water-balance components, in particular the actual evapotranspiration in five Asian river basins, including the Aral, Indus, Ganges, Brahmaputra, and Tarim, and the lakes of Issyk-Kul and Balkhash. Regarding the Aral Sea Basin in Central Asia, Pritchard reported that summer evaporation is approximately equal to summer precipitation [23].

The snowmelt runoff model (SRM) and SWAT model with conventional weather data were used to carry out a water balance study of the Karnali River Basin in Nepal and to simulate the contribution of snowmelt to river runoff [2]. Dhami et al. reported that after comparing the results obtained from the SWAT model and the SRM model, it is recommended to use the results obtained from the SWAT model, which is able to control the volume of melting snow compared to the SRM model [2]. Siderius et al. [24] calculated the contribution of snowmelt to river runoff in the Ganges River in the Himalayan arc, using APHRODITE data with the SWAT model. The simulation results showed that approximately 1% and 5% could be considered indicative of the actual total annual contribution of snowmelt to total runoff [24]. Chiphang et al. [1] used the SWAT model in the mountainous Mago River basin, located in the Eastern Himalayan region of India, from 2006 to 2009 to compute the contribution of snowmelt to streamflow and evapotranspiration changes in the basin. The results showed that the contribution of snowmelt runoff to the annual streamflow of the basin was about 8% [1]. Another study was conducted to simulate snowmelt using the SWAT model in the Tizinafu River Basin (TRB) in Xinjiang, in Central Asia, from 2013 to 2014 using observational climate data [25]. Duan et al. found that about 44.7% of the total runoff comes from snowmelt runoff in the TRB [25].

Climate data are regarded as among the most important data for setting up the SWAT model. Therefore, assessment of the reliability of the most commonly used gridded climate data in SWAT modeling and water-balance analysis has become a popular theme in recent times, particularly in developing and less developed countries [26–28]. Malsy et al. [29] examined the performance of hydrologic modeling using four datasets, including the Global Precipitation Climatology Center (GPCC) Reanalysis product v6, APHRODITE, WATCH forcing data (WFD), and CRU in a hydrological model named "Water Global Assessment and Prognosis 3" (WaterGAP 3). According to Malsy et al., the GPCC and APHRODITE datasets, coupled with the WaterGAP 3 hydrological model, showed better hydrological results than CRU and WFD datasets at the Tuul River Basin and Khovd River Basin in Mongolia in East Asia. Due to the lack of data on the Upper Helmand Basin in Afghanistan, which is a neighboring country to Tajikistan, the SWAT model and the global CRU dataset were applied to create long-term hydrological conditions [30]. The results showed the good performance of the SWAT model using CRU data for the study area; therefore, the NSE was 0.84 for the calibration period and 0.82 for the validation period [30]. It is not known if the same results can be generated with a different hydrological model. For instance, Luo et al. [31] used the SWAT and the MIKE SHE hydrological models to assess their performance in the Hotan River Basin in southwestern Xinjiang, in Central Asia. The results demonstrated that the SWAT model performs better than the MIKE SHE model for the same climate input. Liu et al. used the SWAT model with climate data from the China meteorological assimilation driving datasets (CMADS V1.0) and CFSR in the Yellow River Source Basin, Qinghai–Tibet Plateau [32]. The APHRODITE dataset with a SWAT model, in the Yarlung Tsangpo–Brahmaputra River Basin (YTBRS) in Southeast Asia, was used for hydrological modeling. The results showed the validity of APHRODITE estimates in driving the hydrological model in the YTBRB [33]. Tan et al. [34] assessed the capabilities of the APHRODITE, CFSR, and PERSIANN datasets to model river flow using the SWAT model for the Kelantan River Basin and the Johor River Basin in Malaysia, in Southeast Asia. The combination of APHRODITE precipitation data and CFSR temperature data resulted in the accurate simulation of river flow. Tan et al. recommended the use of APHRODITE precipitation and CFSR temperature data in the modeling of water resources in Malaysia [34]. Xu et al. [35] applied a SWAT model with WFD and APHRODITE datasets to the Xiangjiang River Basin (XRB) in China, to simulate river flow. In XRB, APHRODITE data performed better than WFD data, during both calibration and validation periods [35]. The Tropical Rainfall Measuring Mission (TRMM), National Center for Environmental Prediction (NCEP), Global Precipitation Climatology Project (GPCP), CFSR, and APHRODITE datasets were used to assess the performance of SWAT in the Wunna Basin in India. In the Wunna Basin, APHRODITE datasets can be an alternative source for hydrological modeling as APHRODITE simulations perform much better than TRMM, NCEP, GPCP and CFSR [36]. Shen et al. used gridded products, including CFSR, APHRODITE, CRU, TRMM, ERA-Interim and MERRA-2, with the J2000 model to analyze the spatiotemporal patterns of water balance and the distribution of runoff components in the glacierized Kaidu Basin in Central Asia. The results showed that APHRODITE and CRU represented annual and seasonal precipitation dynamics similar to the observational results at most climate points [37]. However, it should be noted that these results are region- and model-dependent. Many studies show that the accuracy of gridded data results varies by region [38,39]. Meanwhile, a hydrological model with a different concept and representation of the streamflow procedure may lead to different conclusions.

The present work focuses on modeling mountainous terrain with insufficient observational climate data. The major goal of this study is to investigate alternative climate data sources for improving the performance of distributed hydrological models, to explore options that could substitute existing observational data in data-scarce areas. The second objective was to investigate the performances of grid-based data combinations of precipitation and temperature data from multiple sources in order to understand the status of water resources by simulating water balance components in general in the UVRB in Central Asia.

2. Materials and Methods

2.1. Study Area

The study presented in this paper was conducted in the Upper Vakhsh River Basin (UVRB) in Central Asia. The watershed area of the UVRB, including the river network and the location of the measured hydro-climatic stations, as well as the CRU, APHRODITE and CFSR, are shown in Figure 1. The Vakhsh River is the second-largest northwestern tributary of the Amu Darya River in the Aral Sea Basin in Central Asia. The UVRB is located in the north-central part of Tajikistan and the south-west part of Kyrgyzstan (latitude 38.52° to 39.48° N, and longitude 69.78 to 73.70° E). Vakhsh is a very seasonal river, with a discharge maximum in July and a minimum in February, as can be seen in Figure 2. River flow is mainly influenced by snowmelt, since a major part of the annual precipitation falls during the winter months, in higher areas, as snow. A seasonal and annual temperature and precipitation trend analysis of the flat and mountainous areas of Tajikistan can be found in our previous study [40]. In the upstream reaches of the Vakhsh River, due to limitations in the availability of suitable land, irrigation is rather limited. Furthermore, water for this small-scale irrigation setup is taken from tributaries of the Vakhsh River and not drawn directly from the river itself. Therefore, the water used for this purpose is not evidenced by measuring the flow of the river.

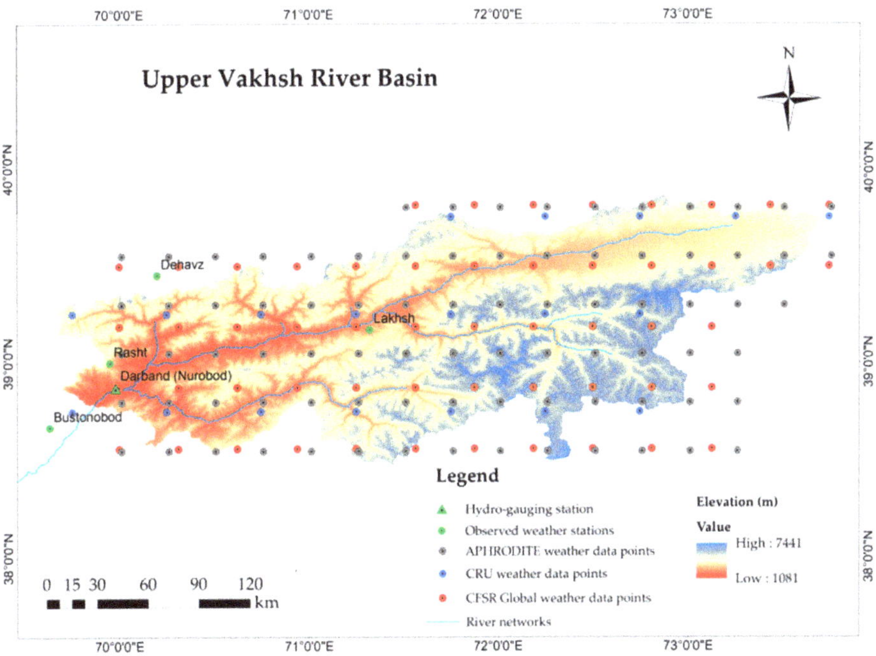

Figure 1. The digital elevation model of the Upper Vakhsh River Basin in Central Asia with the locations of the observed weather and hydro-gauging stations, Climatic Research Unit (CRU), Asian Precipitation Highly Resolved Observational Data Integration Towards the Evaluation of Water Resources (APHRODITE), and Climate Forecast System Reanalysis dataset (CFSR), as well as the global weather data points and streamflow.

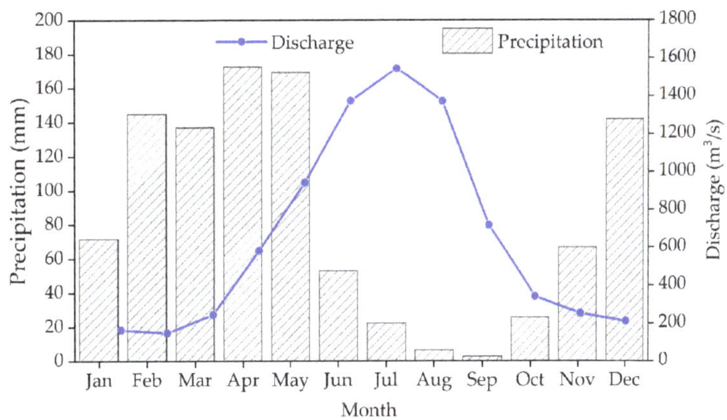

Figure 2. The monthly dynamics of the streamflow and precipitation in the Upper Vakhsh River Basin in Central Asia (Darband station).

2.2. Data

Precipitation is the main factor in hydrological processes, as well as in hydrological modeling, while mountainous regions suffer from a lack of observational climate stations. In order to overcome this issue, most researchers are looking for an alternative option to obtain hydro-climatic data, in order to build hydrological models in mountainous watersheds and to evaluate water-balance components. Comparisons of different datasets with observational data and the combination of different datasets are appropriate objectives that have been considered here. In this study, the water-balance components were derived from the SWAT model results by applying multiple combinations of weather data products to the observational hydro-meteorological data.

Furthermore, in this study, we used the CRU Time Series (TS) version 3.1 data in our hydrological modeling of the UVRB. This data product was produced by the Climatic Research Unit at the University of East Anglia. The CRU TS 3.1 daily maximum and minimum temperatures, as well as precipitation data, were obtained from the website https://www.2w2e.com/home/CRU (accessed on 20 April 2019) for the period of 1979–2006. The reason we derived the data of CRU TS 3.1 from this site is that the historical (1970–2006) reanalysis data of precipitation and maximum and minimum temperatures from CRU TS3.1 are reformatted from NetCDF data into TXT files, which are required by SWAT. The database is updated daily, has a resolution of 0.5° and covers 67,420 files across the world's land areas. The CRU TS 3.1 data have been used in an analysis of the historical (1970–2005) climate variability and extreme weather conditions in the state of California in the United States [41]. Touseef et al. [42] applied the CRU TS 3.1 data to validate the historical daily precipitation measurement-based data in the Xijiang River Basin in China.

The "Asian Precipitation—Highly Resolved Observational Data Integration Towards Evaluation of Water Resources Version 1101" (APHRODITE_V1101) project contains daily gridded precipitation datasets [43]. The Research Institute of Humanity and Nature and the Meteorological Research Institute of Japan's Meteorological Agency created the APHRODITE_V1101 project by combining precipitation station data recorded from thousands of stations throughout Asia, including Japan, the Middle East, Russia, and the Asian monsoon region, to a spatial extent of 15° S–55° N, 60° E–150° E [44]. The APHRODITE_V1101 dataset is available at http://www.chikyu.ac.jp/precip/ (accessed on 6 February 2019). In the SWAT model, precipitation data alone cannot be used to build a hydrological model.

The climate forecast system reanalysis dataset (CFSR) is developed by the National Center for Environmental Prediction (NCEP) and is derived from the Global Forecast System [45]. The CFSR product is widely used in hydrological modeling, considering its high spatial resolution, robustness, and long time series. Publicly available data from January 1979 to July 2014 can be found on the official SWAT website (http://globalweather.tamu, accessed on 15 June 2019) for an almost 36-year period, in the format required by the SWAT model, for a given location. For this study, we obtained all variables of the CFSR data for 54 locations (Figure 1). Previously, many studies have been conducted to compare CFSR climate data with observational datasets to assess the reliability of gridded climate data by applying hydrological models [46,47].

The monthly discharge data for the Darband gauging station during the period of 1979–2006 in the UVRB were derived from the Department of Water Resources of the Ministry of Energy and Water Resources of the Republic of Tajikistan. The measurement-based climate data, including daily maximum and minimum temperatures and daily precipitation, were obtained from the Agency of the Hydrometeorology Committee on Environmental Protection under the Government of the Republic of Tajikistan. From the existing climate stations in Tajikistan within and outside of the UVRB, we found four climate stations, two of them within the basin—Lakhsh station in the central part of the basin, and Rasht station in the eastern part—and two more climate stations were selected from outside of the basin, Dehavz station, near to the northeastern part and Bustonobod, near to the southeastern part of the basin's boundary.

To delineate the watershed boundary and river network of the basin, the digital elevation model (DEM) Shuttle Radar Topographic Mission (SRTM) with a 90 m (Figure 1) spatial resolution was employed, from the Consultative Group for International Agricultural Research (CGIAR) (https://www2.jpl.nasa.gov/srtm/, accessed on 16 December 2018) [48]. Soil data were obtained from the Harmonized World Soil Database (HWSD) version 1.2, with the 1:5,000,000 scale FAO/UNESCO (Food and Agriculture Organization/The United Nations Educational, Scientific and Cultural Organization) Soil Map of the World (http://www.fao.org/soils-portal/data-hub/soil-maps-and-databases/harmonized-world-soil-database-v12/en/, accessed on 29 September 2018) [49]. The area and percentage of the soil type, the latter of which is prominent in the UVRB, are shown in Table 1.

The land-use map was obtained from the Envisat Medium Resolution Imaging Spectrometer (MERIS) with a 300 × 300 m grid-scale. Based on the data from Envisat MERIS, the GlobCover initiative of the European Space Agency (ESA) developed and presented a service for the creation of land cover maps worldwide (https://ladsweb.modaps.eosdis.nasa.gov/missions-and-measurements/meris/, accessed on 12 September 2019) [50]. The land-use map, area, and percentage of land-use types in the UVRB are shown in Table 1. We provided five different ranges of slope classes (0–10%, 10–20%, 20–30%, 30–40%, and >40%) for the hydrologic response unit (HRU) resolution. Slope (in percent) is measured by computing the difference in the height distance (meters), divided by the lateral distance (meters), multiplied by 100. The SWAT model allowed a maximum of five ranges of slope classes. More detailed information regarding the significance of slope in hydrological modeling can be found in the studies of Yacoub et al., where the relative importance of slope discretization, compared with other discretization criteria, was assessed in the streamflow results of the SWAT model in a mountainous basin [51]. Figure 3 shows the area occupied by HRUs in the UVRB, calculated by ArcSWAT, the geographic information system (GIS) interface for SWAT.

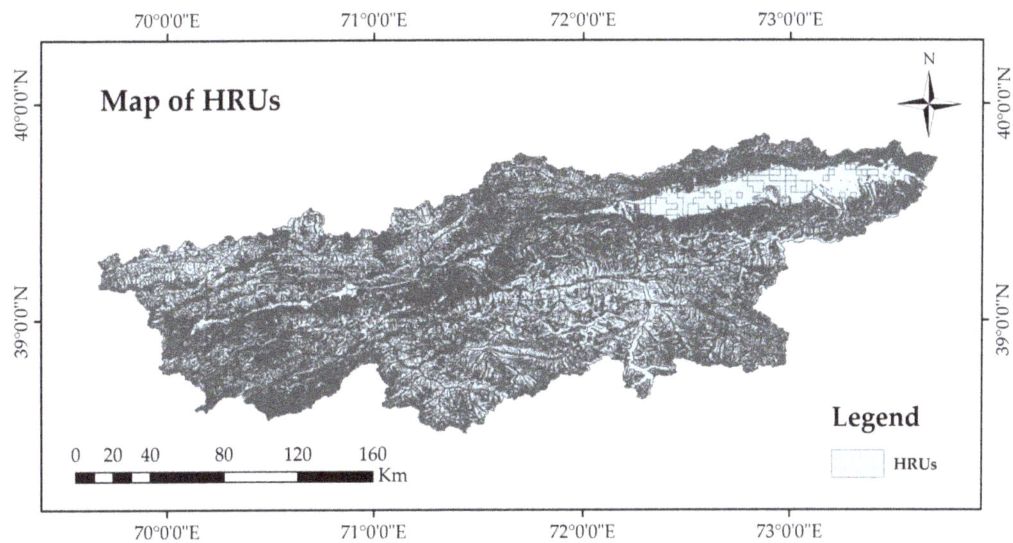

Figure 3. The map of HRUs of the Upper Vakhsh River Basin in Central Asia. HRUs: hydrologic response units.

Distributed models include a large number of parameters and dealing with all these parameters at the calibration stage is not feasible. So, to ensure efficient calibration, a sensitivity analysis was conducted to filter out less influential parameters using a built-in SWAT sensitivity analysis tool. During the model calibration, the monthly observed discharge of 1979–1999 recorded at the Darband discharge station was used, which is located at the outlet of the UVRB.

Table 1. Soil and land-use data of the Upper Vakhsh River Basin, based on area and percentage.

Land Cover Types	Area (% of Basin)	Area (km² of Basin)	FAO Soil Name	Area (% of Basin)	Area (km² of Basin)
Pasture	6.57	1935.31	Acrisols	22.06	6495.10
Agriculture	6.58	1938.49	Gleysols	25.47	7499.95
Forest	1.16	340.38	Leptosols	9.17	2701.29
Grassland	48.63	14,318.06	Phaeozems	7.13	2099.55
Shrubland	4.30	1267.29	Rock outcrops	20.46	6024.15
Urban	0.03	8.07	Eutric cambisols	0.68	199.90
Bare land	16.84	4957.09	Gelic gleysols	0.02	7.33
Water body	0.14	42.08	Glaciers	15.00	4417.40
Ice and snow	15.75	4637.89			
Total	100	29,444.66		100	29,444.66

3. Methodology

In this study, a physical-based, watershed-scale, continuous-time, semi-distributed hydrological model using a SWAT (soil water and assessment tool) was implemented for the evaluation of water availability in various components of the hydrological cycle in the UVRB. The United States Department of Agriculture's Agricultural Research Service (USDA-ARS) developed the SWAT model; a detailed description of this model can be found in the theoretical documentation [52]. The SWAT model has been widely used to support water-resource managers and worldwide research dealing with water quality analyses, hydrological assessment, climate and land-use changes, water supplies, non-point-source pollution, soil erosion/sediment transport, and watershed management impact studies in small- to large-scale river basins [53]. The model does not have any limitations in terms of the river basin areas of study and is compatible with ArcGIS, QGIS, and MapWindow software, as well as providing reliable and useful theoretical documentation that is readily available. Using the ArcGIS version 10.3 interface of SWAT, named ArcSWAT, the UVRB was divided into sub-basins, based on a digital elevation model. Each sub-basin is connected through a stream channel and the model operates by dividing sub-basins into many HRUs (Figure 3), according to a unique homogenous combination of land cover, soil properties, and terrain features. The model performs a modification of the soil conservation service curve number (SCS-CN) method, which identifies the surface runoff from daily precipitation, land use, the area of the hydrological group and the antecedent moisture content for each HRU [54,55].

The UVRB is a mountainous catchment; hence, the observational climate stations in the UVRB are located at lower altitudes. For instance, the Rasht station is located at an elevation of 1316 m, Bustonobod at an elevation of 1964 m, Lakhsh at an elevation of 1998 m, and Dehavz at an elevation of 2561 m. The orographic features of the UVRB mountainous catchment, in terms of temperature and precipitation, led to the splitting of the UVRB into different elevation bands in the SWAT model. In order to simulate the snowmelt in this study, we used a temperature index algorithm employing the elevation band approach [56,57]. We weighted the temperature and precipitation elevation band between the climatic station band and the other elevation band (ED) by using the following mathematical equations:

$$R_{band} = R_{day} + (EL_{band} - EL_{c.s}) \times \frac{plaps}{days_{pcp,yr} \times 1000}, \; R_{day} > 0.01, \quad (1)$$

$$T_{max,band} = T_{max} + (EL_{band} - EL_{c.s}) \times \frac{tlaps}{1000}, \quad (2)$$

$$T_{min,band} = T_{min} + (EL_{band} - EL_{c.s}) \times \frac{tlaps}{1000}, \quad (3)$$

$$T_{mean,band} = T_{mean} + (EL_{band} - EL_{c.s}) \times \frac{tlaps}{1000}, \quad (4)$$

where 1000 serves as the conversion element from meters to kilometers; R_{band} is the precipitation in an EB (mm); R_{day} is the precipitation recorded at the measurement gauge (mm); $T_{max,band}$ shows the daily maximum temperature of the EB (C); $T_{min,band}$ indicates the daily minimum temperature of the EB (C); $T_{mean,band}$ is the daily mean temperature of the (C); T_{max} shows the daily maximum temperature recorded at the measurement gauge (C); T_{min} indicates the daily minimum temperature recorded at the measurement gauge (C); T_{mean} shows the daily average temperature recorded at the measurement gauge (C); tlaps is the lapse rate of temperature (C/km); EL_{band} shows the mean elevation in the EB (m); $EL_{c.s}$ indicates the elevation at the measurement gauge (m); plaps is the precipitation lapse rate (mm/km) and $days_{pcp,yr}$ represents the average annual value of the days when precipitation occurred. The EB approach to the SWAT model has been employed in various mountainous catchments across the globe [58–60]. As in our previous study, Gulakhmadov et al. presented the hydrological model calibration results they obtained with the SWAT–CUP tool before and after the EB approach. The application of EB had a positive impact on the modeling of river flow in a mountain watershed [61].

The model was auto-calibrated for sensitive parameters, such as runoff curve number (CN), Manning's n, and groundwater (GW) parameters (Soil K, Ch_K, Alpha BF, REVAP, ESCO, soil AWC, GW delay, Recharge_DP, Soil Z), based on their rankings. A multiple regression equation was used to identify the sensitive parameters, as follows:

$$g = \alpha + \sum_{i=1}^{m} \beta_i b_i, \tag{5}$$

where g shows the value of the objective function; b_i indicates the parameter of the calibration; α and β_i represent the regression coefficients; and m indicates the selected parameter number [62].

The simulation of the hydrological processes by SWAT is carried out on the basis of the water balance equation:

$$SW_t = SW_0 + \sum_{i=1}^{t} \left(P_{day} - Q_{surf} - E_a - W_{seep} - Q_{gw} \right), \tag{6}$$

where SW_0 shows the initial soil water content on day i (mm H_2O); i is time in days; P_{day} shows the amount of precipitation on day i (mm H_2O); Q_{surf} is the amount of surface runoff on day i (mm H_2O); E_a is the amount of evapotranspiration (ET) on day i (mm H_2O); W_{seep} is the amount of water entering the vadose zone from the soil profile on day i (mm H_2O); Q_{gw} is the amount of return flow on day i (mm H_2O); and SW_t shows the final soil water content (mm H_2O).

On the basis of the average daily air temperature, the SWAT model divides the precipitation into rain or snow. The user of the model will give a threshold temperature in order to categorize precipitation as rain or snow. The precipitation, as snow, will be modeled and the equivalent water will be supplemented to the snowpack if the average air temperature is lower than the temperature threshold. The precipitation will be modeled in the form of liquid rain if the average daily temperature is higher than the temperature threshold. If additional snow falls, the snowpack will be raised and if snowmelt or sublimation occur, the snowpack will be reduced, and the water accumulation in the snowpack will be given as the snow water component.

The SWAT model calculates the snowmelt as a linear function of the divergence between the mean maximum temperature of the snowpack and the snowmelt threshold temperature or base. The snowmelt on a given day is calculated based on the following equation:

$$SNO_{mlt} = b_{mlt} \times SNO_{cov} \times \left[\frac{T_{snow} + T_{mx}}{2} - T_{mlt,\,sno} \right], \tag{7}$$

where b_{mlt} represents the melt factor for the day (mm H_2O/day-°C); the fraction of the HRU area covered by snow is SNO_{cov}; the temperature of the snowpack is T_{snow} (°C); the maximum air temperature is T_{mx}; the base temperature above which snowmelt is allowed is $T_{mlt,\,sno}$(°C); and SNO_{mlt} indicates the amount of snowmelt (mm). Seasonal differences are allowed by the melt factor, with maximum and minimum indices, taking place towards winter and summer solstices:

$$b_{mlt} = \frac{(b_{mlt6} + b_{mlt12})}{2} + \frac{(b_{mlt6} - b_{mlt12})}{2} \times \sin\left(\frac{2\pi}{365} \times (d_n - 81)\right), \quad (8)$$

where b_{mlt6} represents the melt factor for 21 June (mm H_2O/day-°C); the melt factor for 21 December is b_{mlt12} (mm H_2O/day-°C); d_n is the day of the year, and the resulting value (b_{mlt}) shows the melt factor for the day (mm H_2O/day-°C).

The evaluation of evapotranspiration (ET) is essential for water-resource management and hydrological research. The studies of previous researchers suggested that it is acceptable to apply PET (potential evapotranspiration) in models and water allocations [63,64]. In order to estimate PET, there are three methods given in the SWAT model, including the radiation-based Priestley and Taylor method [65], the temperature-based Hargreaves method [66], and the combined Penman–Monteith method [67,68]. For the present study, the Hargreaves method depends on inputted climate data, which were selected to determine the potential evapotranspiration in a mountainous catchment. The Hargreaves approach is the most commonly used method; it is based on temperature and is recommended by the FAO. Li et al. compared the results of the Hargreaves and Penman–Monteith methods in the Ganjiang River Basin in Southern China by using two different datasets [69]. The results of the analysis showed that there is no significant discrepancy between the Hargreaves and Penman-Monteith methods in terms of streamflow simulations with the same spatial scale. The ET was computed as a function of the corrected potential evapotranspiration, soil depth, soil cover, and plants' water uptake [52]. Based on each hydrological response unit, the water balance components were simulated, including precipitation partitioning, precipitation interception, evapotranspiration, snowmelt water, the redistribution of soil water content, return flow from shallow aquifers and lateral subsurface flow from the soil profile.

Model Assessment

The model evaluation was carried out based on the Nash–Sutcliffe efficiency (NSE) measure, the coefficient of determination (R^2), and the percentage bias (PBIAS). Model assessment statistics were evaluated using the NSE, R^2, and Kling–Gupta efficiency (KGE) calculations [70]. In watershed modeling, the NSE, R^2 and KGE are standard regression statistics [71]. NSE ranges from $-\infty$ to 1, with 1 being the best performance. The degree of the linear relationship between measured data and model output is R^2 and it ranges between 0 and 1. KGE is the goodness-of-fit measure initiated by Gupta et al. [70], which gives a decomposition of mean squared error and NSE. In hydrological modeling, the KGE statistic value contributes to the analysis of the relative significance of the correlation, variation, and bias [72]. The model result is more accurate if the KGE output value is closer to 1, and it ranges from $-\infty$ to 1. Moreover, in the model performance, we used the root mean square error (RMSE) observed standard deviation ratio (RSR) and an error-index statistic. The values of NSE > 0.50 and R^2 > 0.60 are considered satisfactory for river discharge on a monthly scale [71]. The values of KGE > 0.5 and RSR < 0.60 are also considered satisfactory levels [70,71]. To assess the strength of the model calibration and uncertainty, two important factors were computed based on the calibration of soil and water assessment tool model (SWAT–CUP) performance, the P-factor and R-factor [73,74]. According to Abbaspour et al. [74], the P-factor describes the percentage of observational data that is covered with a 95% prediction uncertainty (95PPU). It quantifies the model's capability in terms of catching uncertainties, and its magnitude ranges between 0 and 1, where 1 demonstrates that 100% of the station-recorded variability is captured by 95PPU.

The thickness of 95PPU is the value of the R-factor, which presents the ratio of the mean width of the 95PPU band and the standard deviation of measured variability. The model performance is superior if the R-factor value is low. For discharge modeling, in order to compute prediction uncertainty, the studies of Abbaspour et al. [74] recommended that the value of the P-factor be > 0.7 and the value of the R-factor be < 1.5.

The NSE, the R^2, the PBIAS, the RSR, KGE, and MSE are frequently applied measures in hydrological modeling studies [71], which are calculated as:

$$\text{PBIAS} = \frac{\sum_{i=1}^{n}\left(Q_i^{obs} - Q_i^{sim}\right) \times 100}{\sum_{i=1}^{n}(Q_i^{obs})}, \quad (9)$$

$$R^2 = \frac{\left[\sum_{i=1}^{n}\left(Q_i^{obs} - \overline{Q^{obs}}\right)\left(Q_i^{sim} - \overline{Q^{sim}}\right)\right]^2}{\sum_{i=1}^{n}\left(Q_i^{obs} - \overline{Q^{obs}}\right)^2 \sum_{i=1}^{n}\left(Q_i^{sim} - \overline{Q^{sim}}\right)^2}, \quad (10)$$

$$\text{KGE} = 1- \sqrt{(r-1)^2 + (\alpha - 1)^2 + (\beta - 1)^2}, \quad (11)$$

$$\text{RSR} = \frac{\sqrt{\sum_{i=1}^{n}\left(Q_i^{obs} - Q_i^{sim}\right)^2}}{\sqrt{\sum_{i=1}^{n}\left(Q_i^{obs} - \overline{Q^{obs}}\right)^2}}, \quad (12)$$

$$\text{NSE} = 1 - \frac{\sum_{i=1}^{n}\left(Q_i^{sim} - Q_i^{obs}\right)^2}{\sum_{i=1}^{n}\left(Q_i^{obs} - \overline{Q^{obs}}\right)^2}, \quad (13)$$

where n is the whole number of sample couples; Q_i^{obs} is the station-recorded discharge variable; $\overline{Q^{obs}}$ is the mean of the station-recorded discharge parameters; Q_i^{sim} is the simulated discharge parameter; $\overline{Q^{sim}}$ is the mean of the simulated discharge parameters; and i is the ith station-recorded data or simulated data. Moreover, $\alpha = \frac{\sigma_s}{\sigma_m}$ and $\beta = \frac{\mu_s}{\mu_m}$, while r is the linear regression coefficient of the simulated value against station-recorded value, μ_s and μ_m are the averages of the simulated value against the station-recorded value, and σ_s and σ_m are the standard deviations of the simulated value against the station-recorded value [70].

4. Results

In hydrological process analysis, and for reliable hydrological modeling, precipitation data are considered to be the main factor. To apply the precipitation data initially in hydrological modeling, we carried out a correlation analysis to examine the data's suitability for watershed modeling. Figure 4 presents a Taylor diagram with the performances of the three precipitation data sources, the CRU TS3.1, CFSR and APHRODITE_V1101, against measurement-based precipitation data on a monthly scale. Taylor diagrams are capable of providing performance insights by comparing precipitation satellite datasets and measurement-based data sets, in terms of their standard deviation, root mean square error and correlation coefficient. In the Taylor diagram, the radial blue dotted lines show the standard deviation and the red semicircles present the root mean square error. Hence, the black dotted lines describe the correlation coefficient. These three statistic indices are shown solely for the Lakhsh precipitation gauge point in the central part of the catchment on a monthly scale, for the purposes of demonstration (Figure 4). The precipitation correlation showed the highest value (0.86) between the APHRODITE_V1101 datasets and measurement-based datasets. Meanwhile, the precipitation correlation between the CRU TS3.1 and measurement-based data is 0.76, and the correlation between the CFSR and measurement-based datasets is 0.59. The correlation coefficients between all three different combinations, the CRU TS3.1 and measurement-based datasets, CFSR and measurement-based datasets, as well as APHRODITE_ V1101 and measurement-based datasets, showed a good performance, meaning that they could be used for hydrological modeling in the UVRB. In many studies, Taylor diagrams have been applied to evaluate the performance of satellite products against observational datasets [75,76].

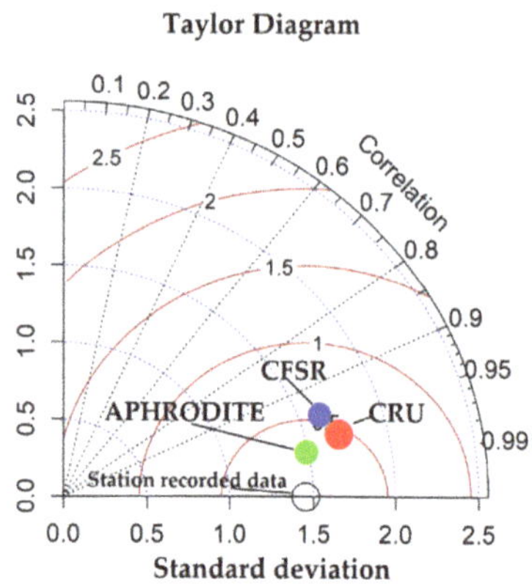

Figure 4. Taylor diagram indicating the performances of CRU, CFSR and APHRODITE precipitation data on a monthly scale at the Lakhsh climate station in the Upper Vakhsh River Basin in Central Asia.

4.1. Parameter Sensitivity Analysis

The hydrological model was calibrated and validated by employing a software application for the SWAT–CUP, SUFI-2 (sequential uncertainty fitting, version 2). The SWAT–CUP software is a semi-automatic calibration and uncertainty analysis tool that was developed by the EAWAG Swiss Federal Institute of Aquatic Science and Technology for the SWAT model [3]. The SUFI-2 algorithm utilizes an inversion modeling technique that determines a wide range of parameters and then carries out several iterations that contain a number of simulations. After running the iterations, the result of each iteration was compared with the result of other iterations and, in this way, the most suitable ranges of the model's parameters were identified [74]. This iterative procedure takes into account the uncertainty of parameters from all types of sources, including model structure, model parameters, weather, etc. By using the global sensitivity approach in the SUFI-2 algorithm, detailed uncertainty and optimization examinations are possible [77]. In order to obtain satisfactory watershed characteristics, the calibration and validation of the hydrological model are essential. Following the outcome of the final modeled simulation, a sensitivity ranking was presented for the appropriate parameters by analyzing the values of the "*t*-stat" and *p*-value statistics. The SWAT–CUP contains multiple parameters that could impact the simulation of the water cycle. The selection of suitable parameters plays an important role in identifying the effectiveness of model calibration.

In this study, the sensitivity analysis was executed using the Latin hypercube global sensitivity approach, which is included in the SWAT–CUP (version 2019) package. Characteristically, sensitivity analysis is required prior to calibration due to the recognition of sensitive parameters and model elements. The global sensitivity approach leads to the attainment of a set calibration with optimal parameters and allows us to find the parameters using the degree of sensitivity of their performance characteristics in the model.

A gridded dataset comparison was performed to evaluate how well gridded datasets, such as CRU TS3.1, APHRODITE_V1101 and CFSR, correlated with the observational data. The analysis years were determined to include similar years from all datasets since CRU daily data availability ended in 2006. The first three years of the total simulation period (January 1999–December 2002) were used as a warm-up to allow the model to reach hydrological equilibrium and were excluded from the analysis. For each of the datasets, the semi-automated calibration process was conducted with an identical range of parameter values and calibration/validation periods for comparison purposes. Semi-automated calibration ensures the consistency of the process for all models, minimizing the model bias due to the modeler in calibration exercises conducted for different precipitation and maximum/minimum temperature sources. Initial parameter ranges were selected based on the professional judgment of the authors and the literature. Each model executed 1000 simulations for each iteration of the semi-automated calibration. An initial 300–500 simulations are recommended for studying model performance and for regionalizing parameters [73]. At the end of the iteration with 1000 simulations, parameter sensitivities were determined through a global sensitivity analysis. Only one iteration was used to avoid re-calibration, using a different range of parameter values for each model in the subsequent calibration. The Nash–Sutcliffe efficiency (NSE) measure was used to estimate model performance during calibration since it is a commonly used statistical measure in SWAT studies [71].

While determining the parameters' distribution and sensitivity, the baseflow alpha factor (ALPHA_BF), moist bulk density (SOL_BD), SCS runoff curve number for moisture condition II (CN2), effective hydraulic conductivity in main channel alluvium (CH_K2), and deep aquifer percolation fraction (RCHRG_DP) are computed as the most sensitive parameters. The results of sensitivity parameters and analyses of statistical indices, such as P-factor and R-factor, in both calibration and validation parts indicated that all climate datasets utilized in this study have acceptable prediction uncertainty and reasonable parameter adjustment. These results indicate the potential of applying gridded datasets for hydrological modeling. It should be noted that gridded datasets are advantageous because they give continuous data at spatial and temporal scales throughout the catchment area and for an extensive duration.

4.2. Calibration and Validation

Figure 5 shows the calibration results of the SUFI-2 algorithm of the SWAT–CUP model, utilizing monthly discharge data at the Darband gauging station for the period of 1982–1999, where a combination of four different datasets was used (Figure 5a–d), including an initial warm-up period of three years (1979–1981). Figure 6 indicates the validation results of the fit between the monthly measurement-based flow and the flow simulated by SWAT. In addition, to demonstrate the flow peaks over a long period of time, in Figure 7, we present the overall calibration hydrographs via the application of five different dataset combinations.

Table 2 shows the ability of gridded datasets to derive the long-term average annual flow from the simulated flow at the Darband gauging site in UVRB in Central Asia. The observation and simulated flow over a 25-year period demonstrated that the average annual flow between the observation and the simulated flow does not differ much, with the exception of a few years. The results shown as Simulated-1 in Table 2 demonstrate that, in most years, the average annual flow was simulated to be less, compared to the other three simulated flows. In Simulated-1, the largest negative ratio value between the observed and simulated average annual flows was found in 1996 (−123.87%) and 1988 (−124.79%), while the largest positive value of the ratio was observed in 1983 (13.58%). The results of Simulated-2 presented the lowest negative ratio value of the average annual flow in 2006 (−59.79%) and the highest positive value in 1989 (35.69%). The maximum negative and positive rates of the average annual flow for Simulated-3 were detected in 1984 (27.66%) and 1998 (−74.62%), while for Simulated-4, the biggest downward and upward rate values compared to observational flow were obtained in 1997 (−42.27%) and 2003 (26.53%). Our results show that, compared with the observational average annual flow, the average annual flow of Simulated-1 indicated the largest negative rate, and the Simulated-4 results showed the lowest negative and positive rates (Table 2). In this study, the four different quantity comparisons of the average annual flow were derived after the application of a well-calibrated hydrological model.

Table 2. Average annual simulated flow of four different simulation results from SWAT–CUP and their rates, compared to observational flow at the Darband hydrological station over the period of 1982–2006 in the Upper Vakhsh River Basin in Central Asia. Simulated-1: results of the combination of the APHRODITE_V1101 precipitation datasets and CFSR maximum/minimum temperature datasets; Simulated-2: results of the combination of the APHRODITE_V1101 precipitation datasets and CRU TS3.1 maximum/minimum temperatures datasets; Simulated-3: results of the CFSR as maximum/minimum temperatures, precipitation, average solar radiation, average wind speed and relative humidity datasets; Simulated-4: results of the CRU ST3.1 precipitation and maximum/minimum temperature datasets.

Year	Observation	Simulated-1		Simulated-2		Simulated-3		Simulated-4	
	Flow (m^3/s)	Flow (m^3/s)	Rate (%)	Flow (m^3/s)	Rate (%)	Flow (m^3/s)	Rate (%)	Flow (m^3/s)	Rate (%)
1982	519.42	518.25	−0.23	731.39	28.98	608.03	14.57	541.45	4.07
1983	573.58	663.75	13.58	727.21	21.13	760.64	24.59	550.86	−4.13
1984	644.50	707.10	8.85	734.60	12.27	890.92	27.66	577.44	−11.61
1985	595.75	595.67	−0.01	765.99	22.23	657.31	9.37	603.96	1.36
1986	492.25	492.34	0.02	633.18	22.26	515.41	4.49	492.89	0.13
1987	683.42	562.95	−21.40	949.26	28.01	545.32	−25.32	658.34	−3.81
1988	711.75	685.50	−3.83	924.93	23.05	764.24	6.87	666.43	−6.80
1989	443.08	440.85	−0.51	688.96	35.69	568.37	22.04	454.13	2.43
1990	624.67	684.65	8.76	791.25	21.05	805.38	22.44	620.63	−0.65
1991	565.17	586.13	3.58	796.42	29.04	659.22	14.27	633.34	10.76
1992	650.42	507.05	−28.27	773.53	15.92	597.96	−8.77	661.16	1.62
1993	694.67	449.89	−54.41	744.46	6.69	551.77	−25.90	678.11	−2.44
1994	631.81	638.18	1.00	811.60	22.15	745.32	15.23	641.08	1.45
1995	568.18	377.26	−50.61	592.66	4.13	491.28	−15.65	606.12	6.26
1996	638.31	285.13	−123.87	595.88	−7.12	382.11	−67.05	634.59	−0.59
1997	600.54	472.78	−27.02	439.89	−36.52	623.95	3.75	422.12	−42.27
1998	828.35	368.49	−124.79	710.06	−16.66	474.37	−74.62	773.38	−7.11
1999	673.37	459.95	−46.40	702.52	4.15	533.08	−26.32	667.83	−0.83
2000	572.29	472.56	−21.10	567.44	−0.85	563.35	−1.59	517.43	−10.60
2001	564.96	610.16	7.41	568.66	0.65	677.75	16.64	500.73	−12.83
2002	726.54	617.11	−17.73	693.48	−4.77	558.38	−30.12	698.52	−4.01
2003	656.57	587.33	−11.79	695.84	5.64	573.49	−14.49	893.66	26.53
2004	649.40	568.10	−14.31	704.03	7.76	555.56	−16.89	779.52	16.69
2005	695.60	472.60	−47.19	587.79	−18.34	593.18	−17.27	660.14	−5.37
2006	647.45	477.57	−35.57	405.19	−59.79	646.72	−0.11	668.40	3.13

APHRODITE_V1101: Asian Precipitation—Highly Resolved Observational Data Integration Toward Evaluation of Water Resources Version 1101; CFSR: Climate Forecast System Reanalysis; CRU ST3.1: Climatic Research Unit Time Series Version 3.1.

The observed and simulated monthly streamflow values for the calibration (1982–1999), validation (2000–2006), and overall values (1982–2006) are shown in Figures 5–7, respectively. Figures 5–7 present the simulated flow hydrographs and peak flows, which are in good agreement with the timing of the observational flow hydrographs and flow peaks comprising the outcomes of the APHRODITE_V1101 and CFSR combination, the APHRODITE_V1101 and CRU TS3.1 combination, and the CRU TS3.1, CFSR, and measurement-based combination. Calibration using monthly river flow data for long-term simulations demonstrates a better performance than short-term simulations. The base flow as well as most of the peak flows are well simulated. The hydrographic results show that the observed and modeled discharges are nearly the same for most of the study period (25 years), except for a few years when high-flow events occurred. For example, the peak flow in the hydrograph for the combination of APHRODITE_V1101 and CFSR in 1987, 1993, 1996, and 1998 is underestimated by the model, while in the years 1984 and 1994 it is overestimated (Figure 7a). The peak flow rate for the combination of APHRODITE_V1101 and CRU TS3.1 in 1989, 1991 and 1997 is insignificantly underestimated (Figure 7b). The model insignificantly underestimated the peak flow for the CFSR dataset in the years 1987, 1989, 1996 and 1998, while in the years 1983, 1984, 1994 and 2001, it slightly overestimated the flow (Figure 7c). For the CRU TS3.1 hydrograph, a slight underestimation of the peak flow is observed in the years 1983, 1987, 2000 and 2001, whereas in the year 2003, the flow is slightly overestimated (Figure 7d). Lastly, the model insignificantly underestimated the peak discharges in 2008 and 2011 for the observational datasets (Figure 7e). Our results showed that better simulation flows were obtained from APHRODITE_V1101 and CRU TS3.1 climate datasets compared to the CFSR, which demonstrates the advantage of using the CRU TS3.1 and APHRODITE_V1101 products in SWAT modeling.

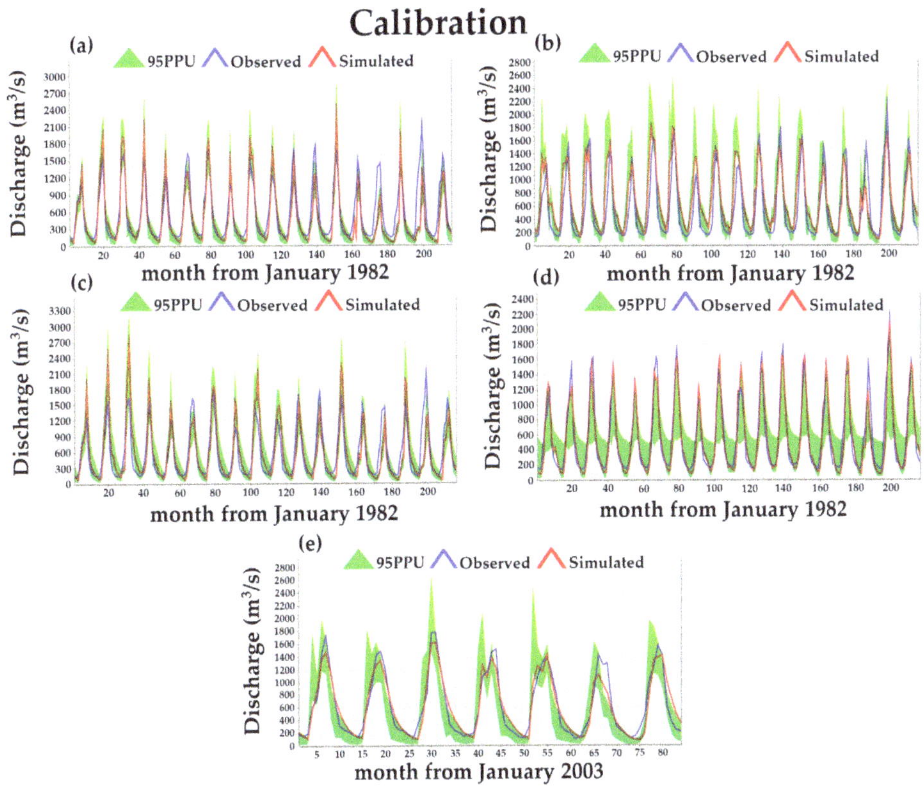

Figure 5. Hydrographic calibration between monthly observed and simulated streamflow, when applying the SWAT–CUP tool, at Darband gauging station in the Upper Vakhsh River Basin in Central Asia. (**a**) Results of the daily precipitation of the APHRODITE_V1101 and daily maximum/minimum temperatures from the CFSR datasets; (**b**) results of the daily precipitation of the APHRODITE_V1101 and daily maximum/minimum temperatures of the CRU TS3.1 product; (**c**) results of the daily maximum/minimum temperatures, precipitation, average solar radiation, average wind speed and relative humidity from the CFSR product; (**d**) results of the daily maximum/minimum temperatures, and precipitation from the CRU TS3.1 product; (**e**) results of the daily maximum/minimum temperatures and precipitation from the observational climate datasets.

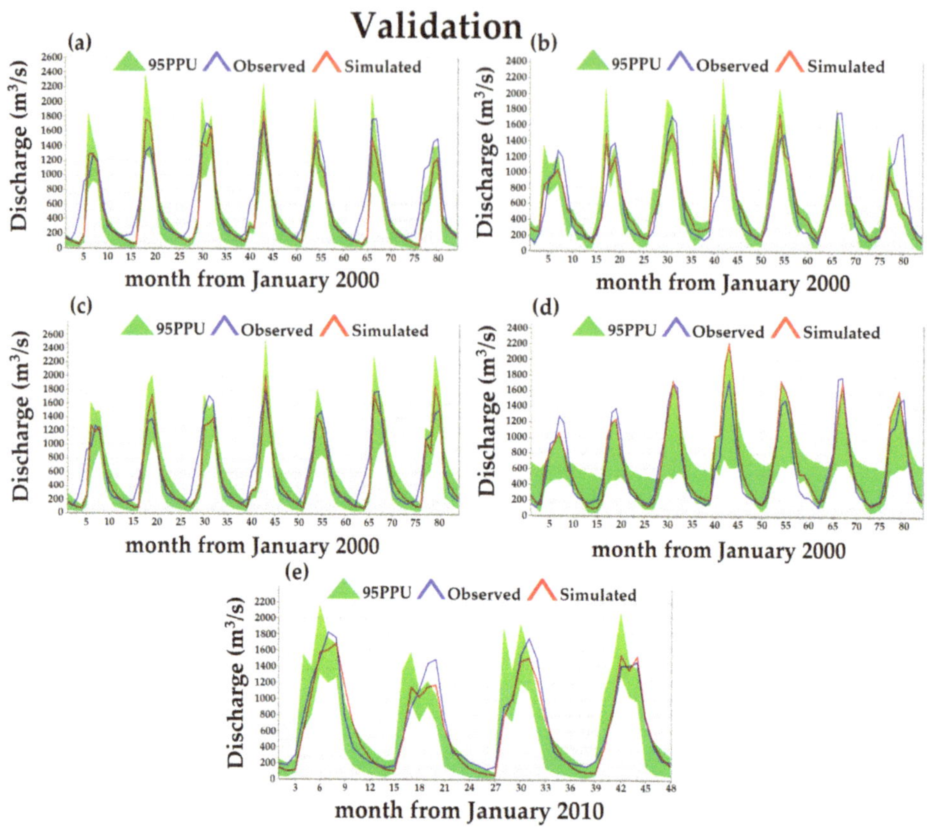

Figure 6. Hydrographic validation between monthly observed and simulated streamflow, when applying the SWAT–CUP tool, at Darband gauging station in the Upper Vakhsh River Basin in Central Asia. (**a**) Results of the daily precipitation of the APHRODITE_V1101 and daily maximum/minimum temperatures from the CFSR datasets; (**b**) results of the daily precipitation of the APHRODITE_V1101 and daily maximum/minimum temperatures of the CRU TS3.1 product; (**c**) results of the daily maximum/minimum temperatures, precipitation, average solar radiation, average wind speed and relative humidity from the CFSR product; (**d**) results of the daily maximum/minimum temperatures, precipitation from the CRU TS3.1 product; (**e**) results of the daily maximum/minimum temperatures and precipitation from the observational climate datasets.

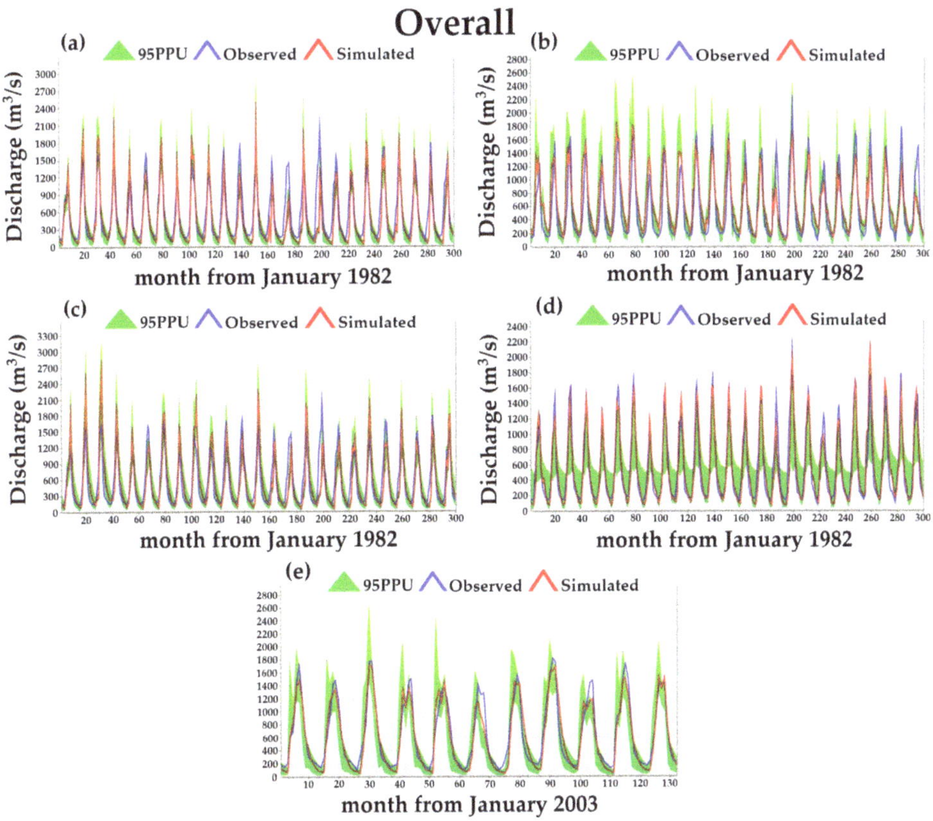

Figure 7. Hydrograph of the overall calibration and validation period between monthly observed and simulated streamflow, when applying the SWAT–CUP tool, at Darband gauging station in the Upper Vakhsh River Basin in Central Asia. (**a**) Results of the daily precipitation of the APHRODITE_V1101 and daily maximum/minimum temperatures from the CFSR datasets; (**b**) results of the daily precipitation of the APHRODITE_V1101 and daily maximum/minimum temperatures of the CRU TS3.1 product; (**c**) results of the daily maximum/minimum temperatures, precipitation, average solar radiation, average wind speed and relative humidity from the CFSR product; (**d**) results of the daily maximum/minimum temperatures, precipitation from the CRU TS3.1 product; (**e**) results of the daily maximum/minimum temperatures and precipitation from the observational climate datasets.

4.3. Performance of the Hydrological Model

Table 3, which presents the statistical values for the calibration, validation and overall periods of the Darband discharge station, confirms the essentially "excellent", "very good" and "good" performance of the model. Firstly, in the SUFI-2 algorithm, we adopted the Nash–Sutcliffe efficiency (NSE) calculation as an objective function for the optimization process. It was used as a goodness-of-fit metric for calibration, in order to set up the adjustment during the calibration period and for the performance examination of the validation and overall periods. The NSE working system has been recognized for its ability to concentrate on the good simulation of peak flows. Its selection and related influence should be considered in view of the results obtained in this study. A more general view, which included all modeling on an overall scale for all datasets, showed that most of the years provided accurate modeling in all angles of the hydrograph.

In general, eight kinds of evaluation indices (R^2, NSE, PBIAS, RSR, MSE, KGE, P-factor and R-factor) were used to evaluate the acquisition accuracy of the hydrological modeling. The statistical evaluation of the model performance based on monthly streamflow is described in Table 3. The streamflow is slightly overestimated by 14.8%, 1.32% and 0.69% for the APHRODITE_V1101+CFSR, CRU TS3.1 and measurement-based data while being slightly underestimated by −17.70% and −0.31% for APHRODITE_V1101+CRU TS3.1 and CFSR during the calibration period. During the validation period, the model results for the employed APHRODITE_V1101+CFSR, APHRODITE_V1101+CRU TS3.1, CFSR and observational datasets indicated an overestimation of the monthly streamflow by 17.22%, 1.3%, 9.58% and 5.51%, respectively, while the model results showed an underestimation by −6.09% for the CRU TS.3.1. The model results of the overall test performance of the calibration indicated an overestimation of the peak flow for the APHRODITE_V1101+CFSR, CFSR, CRU TS3.1, and observational data by 15%, 2%, 0.29% and 5.21% respectively, whereas, for the APHRODITE_V1101+CRU TS3.1, the results showed an underestimation of peak flow by −10.80% (Table 2). However, the combination of the precipitation data from APHRODITE_V101 and the maximum/minimum temperatures data from CFSR exhibited a slightly overestimated streamflow, which can probably be explained by the large amount of precipitation generated by APHRODITE_V1101. As we realized that precipitation is a major factor in hydrological processes, and in an effort to demonstrate the difference in precipitation between the four climate datasets, before implementing the data into the model, we correlated the precipitation between the observational data and APHRODITE_V1101, CFSR, and CRU TS3.1, as presented in Figure 4. In general, for calibration and validation periods, the hydrographs of all utilized datasets are nearly in line with measurement-based data. According to the recommendation of Moriasi et al. [71], the performance of the model is "very good" (PBIAS $\leq$ 10) in the study area, based on four different combinations of datasets.

In the case of monthly calibration, validation and overall scales, the P-factor ranges from 0.66 to 0.82 for all employed datasets. By using the APHRODITE_V1101+CFSR, APHRODITE+CRU TS3.1, CFSR, CRU TS3.1 and observational values, the bracketed values of the 95PPU band for the monthly streamflow data were 66%, 66%, 75%, 82% and 69% during the calibration period, 67%, 70%, 79%, 80% and 81% during the validation period and 66%, 68%, 75%, 81% and 73% on an overall scale, respectively. The R-factor is the average thickness of the 95PPU band, and, for the monthly calibration, validation and overall scales, the R-factor values were 0.66, 0.79, 0.95, 1.01, and 0.80 (calibration), 0.66, 0.66, 0.89, 1.08, and 0.76 (validation), and 0.67, 0.75, 0.95, 1.02, and 0.78 (overall) when coupling the SWAT model with the respective datasets. The wider 95PPU indicates more parameter uncertainties [62]. According to the recommendations of Schuol et al. [78], the perfect simulation is the one that has an R-factor equal to zero; however, values around 1.0 are considered quite reasonable. In this study, the values obtained for the width of the uncertainty band were quite reasonable for the monthly simulation (Table 3).

The model results for the calibration, validation and overall periods were found to produce a reliable assessment of monthly observed and simulated streamflow. The monthly calibration results for streamflow were "very good", with R^2 values of 0.92, 0.90, 0.79, and 0.76 for the CRU TS.3.1, measurement-based, APHRODITE_V1101+CRU TS3.1, and APHRODITE_V1101+CFSR data, whereas for CFSR, the data were found to possess a "good" performance, with an R^2 value of 0.71. The NSE values were "very good" at 0.91 and 0.90 for CRU TS3.1 and observational datasets, "good", at NSE = 0.74 and NSE = 0.70, for APHRODITE_V1101+CRU TS.3.1 and APHRODITE_V1101+CFSR, and "satisfactory" at NSE = 0.64, for CFSR during the calibration period. The R^2 and NSE coefficient values during the validation period were "very good" (R^2 equal to 0.94, 0.89, 0.85, 0.83, and 0.79 and NSE equal to 0.93, 0.88, 0.83, 0.77, and 0.78) for the observational datasets, CRU TS3.1, CFSR, APHRODITE_V1101+CFSR and APHRODITE_V1101+CRU TS3.1, respectively. This is likely due to the availability of data and the mountainous location of the precipitation station in this region, since the distribution of precipitation strongly influences the flow formation and therefore the NSE. The overall R^2 coefficient values were "very good" (0.92, 0.90, 0.78, and 0.76 for the observational datasets, CRU TS3.1, APHRODITE_V1101+CFSR, and APHRODITE_V1101), whereas R^2 was "good" (0.74 for the CFSR (Table 3)). The overall NSE coefficient values were "very good" (0.91 and 0.90 for the observational datasets and CRU TS3.1), while the NSE values were "good" (0.74, 0.72, and 0.68 for the APHRODITE_V1101+CRU TS.3.1, APHRODITE_V1101+CFSR, and CFSR, respectively). According to the model evaluation criteria, the simulation of the observational data performed better than the APHRODITE_V1101+CFSR, APHRODITE_V1101+CRU TS.3.1, CFSR and CRU TS3.1 simulations on an overall scale. The model using observational dataset calibration, validation, and overall scales demonstrated an excellent performance (Table 3). The less accurate results that were obtained overall, when using gridded datasets in mountainous regions, are most probably associated with the fact that there are fewer weather stations that can be used for product development.

Table 3. Summary statistical indices of monthly streamflow periods with different climate datasets, based on the model's performance. NSE: Nash–Sutcliffe efficiency; R^2: coefficient of determination; PBIAS: percentage bias; MSE: mean square error; RSR: root mean square error standard deviation ratio; KGE: Kling–Gupta efficiency; P-factor; R-factor.

Data Source\Statistical Indices	R^2	NSE	PBIAS (%)	RSR	MSE (%)	KGE	P-Factor	R-Factor
					Calibration			
Combination of the APHRODITE_V1101 and CFSR data (1982–1999)	0.76	0.70	14.80	0.55	7.78	0.80	0.66	0.66
Combination of the APHRODITE_V1101 and CRU TS3.1 data (1982–1999)	0.79	0.74	−17.70	0.77	6.72	0.77	0.66	0.79
CFSR data (1982–1999)	0.71	0.64	−0.31	0.60	9.12	0.81	0.75	0.95
CRU TS3.1 data (1982–1999)	0.92	0.91	1.32	0.29	2.23	0.95	0.82	1.01
Observational data (2003–2009)	0.90	0.90	0.69	0.31	2.54	0.90	0.69	0.80
Data source					Validation			
Combination of the APHRODITE_V1101 and CFSR data (2000–2006)	0.83	0.77	17.22	0.48	6.01	0.81	0.67	0.66
Combination of the APHRODITE_V1101 and CRU TS3.1 data (2000–2006)	0.79	0.78	1.30	0.46	5.54	0.80	0.70	0.66
CFSR data (2000–2006)	0.85	0.83	9.58	0.42	4.48	0.87	0.79	0.89
CRU TS3.1 data (2000–2006)	0.89	0.88	−6.09	0.35	3.18	0.91	0.80	1.08
Observational data (2010–2013)	0.94	0.93	5.51	0.26	2.04	0.93	0.81	0.76
Data source					Overall			
Combination of the APHRODITE_V1101 and CFSR data (1982–2006)	0.78	0.72	15.00	0.53	7.20	0.81	0.66	0.67
Combination of the APHRODITE_V1101 and CRU TS3.1 data (1982–2006)	0.76	0.74	−10.80	0.51	6.70	0.78	0.68	0.75
CFSR data (1982–2006)	0.74	0.68	2.00	0.56	8.10	0.83	0.75	0.95
CRU TS3.1 data (1982–2006)	0.90	0.90	0.29	0.31	2.51	0.95	0.81	1.02
Observational data (2003–2013)	0.92	0.91	5.21	0.29	2.34	0.93	0.73	0.78

4.4. Water Balance of the Upper Vakhsh River Basin in Central Asia

Obviously, in order to skillfully tackle water management issues, it is necessary to quantify and study the various hydrological components within the basin. Regardless of the issues explored by the SWAT model, water balance is a critical component in the SWAT model as it includes all processes in the basin [79,80]. A water-balance study was conducted using a simulation of the overall scale, as well as the appropriate SWAT output tables. The mean annual water-balance components of the UVRB are presented in Table 4.

Table 4. The mean annual water balance components of the Upper Vakhsh River Basin in Central Asia during the calibration (1982–1999), validation (2000–2006) and overall (1982–2006) periods generated from the SWAT model. (**a**) Combination of the APHRODITE_V1101 and CFSR datasets; (**b**) combination of the APHRODITE_V1101 and CRU TS3.1 datasets; (**c**) CFSR data source; (**d**) CRU TS3.1 datasets; and (**e**) observational datasets.

Components	Calibration (1982–1999)					Validation (2000–2006)					Overall (1982–2006)				
	a	b	c	d	e	a	b	c	d	e	a	b	c	d	e
Precipitation	1940.5	1287.8	2608.7	1188.8	1077.2	1709.6	1106.1	2145.4	1285.5	1104.1	1875.9	1236.9	2479	1215.9	1098.5
Snowfall	18[illegible]0.24	926.38	2498.15	1027.98	827.7	1579.83	728.25	2075.87	1114.06	856.11	1760.11	870.85	2379.86	1051.98	836.65
Snowmelt	525.24	664.73	612.49	700.66	570.48	518.31	523.78	586.88	759.7	672.69	525.54	625.65	619.19	718.21	574.41
Sublimation	105.07	140.26	43.41	165.83	150.08	124.33	121.3	63.36	171.03	59.73	110.83	134.95	49.95	167.57	156
Surface runoff Q	46.1	295.28	94.32	27.69	244.04	173.16	226.6	76.31	37.27	284.41	46.54	276.79	90.56	30.73	243.22
Lateral soil Q	408.14	387.13	331.26	384.37	294.81	278.43	331.83	309.39	418.27	369.63	413.34	371.49	331.92	394.21	307.92
Ground water (deep AQ) Q	70.93	128.18	180.68	120.7	96.12	82.46	107.48	161.2	131.13	132.42	70.57	122.34	179.25	123.78	103.83
Revap or shallow aquifer recharges	41.96	53.23	32.06	124.19	51.08	59.61	57.35	44.06	131.79	49.44	39.95	48.98	35.11	124.45	46.47
Deep aquifer recharge	70.98	128.16	180.74	121.12	96.37	82.47	105.12	161.68	130.19	131.16	70.64	121.64	179.4	123.83	103.57
Total aquifer recharge	106.04	173.12	274.57	360.35	130.19	111.4	142.01	245.61	387.34	177.18	105.53	164.33	272.53	368.4	139.91
Total water yield	529.52	811.05	669.31	651.75	634.98	534.05	665.94	591.31	721.27	786.48	534.72	771.24	661.98	672.11	654.97
Percolation out of soil	106.05	173.12	274.59	360.64	130.22	111.39	141.87	245.58	386.49	177.24	105.54	164.28	272.53	368.4	139.9
Actual evapotranspiration	179.3	310.6	64.7	254.2	296.3	207.4	328.8	88.6	261.5	142.9	186.3	315.7	72.1	256.4	299.7
Potential evapotranspiration	390.1	672.7	160.3	673.6	668.2	436.7	692.5	220.3	693.1	297	403.1	678.3	177.1	679.1	668.6

The results of the SWAT model for the overall period of the simulation (1982–2006) are presented; the annual precipitation values for the basin are 1875.9 mm, 1236.9 mm, 2479 mm, 1215.9 mm, and 1098.5 mm, out of which about 93.82% (1760.11 mm), 70.41% (870.85 mm), 96% (2379.86 mm), 86.52% (1051.98 mm), and 76.16% (836.65 mm) of precipitation falls as snow, according to the respective datasets employed (Table 4). The CFSR simulations demonstrated a higher amount of precipitation than other utilized, which means that the CFSR overestimated the precipitation in the UVRB. Similarly, Hu et al. [81] reported an overestimation in the results of the CFSR precipitation datasets used in a mountainous region of Central Asia. Mean annual evapotranspiration from the whole catchment is about 9.93%, 25.52%, 2.9%, 21.08%, and 27.28% of the annual precipitation (186.3 mm, 315.7 mm, 72.1 mm, 256.4 mm, and 299.7 mm out of 1875.9 mm, 1236.9 mm, 2479 mm, 1215.9 mm, and 1098.5 mm) when using the respective datasets. Water yield is as streamflow, which is obtainable at the catchment outlet and is determined from surface runoff, lateral flow and baseflow or return flow. Based on the respective datasets on an overall scale, the annual water yields at the catchment outlet are 534.72 mm, 771.24 mm, 661.98 mm, 672.11 mm, and 654.97 mm, from which surface runoff or overland flow can be obtained; these take place across a sloping surface at about 46.54 mm, 276.79 mm, 90.56 mm, 30.73 mm, and 243.22 mm (including channel losses). The lateral subsurface flow or interflow, which originates below the surface but above the rock saturation zone, contribute 413.34 mm, 371.49 mm, 331.92 mm, 394.21 mm, and 307.92 mm (about 77.30%, 48.17%, 50.14%, 58.65%, and 47.01% of the total water yield) to all aforementioned combinations of datasets. The remaining flow is contributed by the base flow, which originates from groundwater (shallow aquifer). The annual mean streamflow during the period of 1979–2006 at the Darband discharge station at the outlet point of the UVRB is 626.08 m^3/s. The threshold depth of water in the shallow aquifer (REVAP) takes into account the volume of water transported from a shallow aquifer to overlying unsaturated terrain during the dry season.

The mean monthly values of precipitation, water yield, and actual evapotranspiration for the UVRB are also estimated by coupling the combination of the respective datasets, as demonstrated in Figure 8. The low precipitation period occurs in July, August and September, while May–September is the high flow period and flow is less affected by precipitation events during this time. Similarly, evapotranspiration is higher from May to September and the large runoff over this period is mainly due to the melting of the snowpack and permanent glaciers. Moderately high and high precipitation occurs from October to June. In this study, the maximum precipitation values of 227.21 mm, 154.87 mm, 279.76 mm, 153.29 mm, and 135.54 mm per month occurred in May, according to the respective datasets. Based on the CRU TS3.1 and observational datasets, July and August are the only months where actual evapotranspiration is higher than total precipitation during the dry period. This may have happened because evapotranspiration is a sustained process that takes place during the day and night. The simulation of all employed datasets revealed that the maximum evapotranspiration occurred in July. The higher mean monthly actual evapotranspiration generated the simulation of the observational datasets, while the CFSR simulation produced lower than average actual evapotranspiration in all months compared to other simulated datasets for the entire catchment. The CRU TS3.1 dataset simulations and the APHRODITE_V1101+CRU TS3.1 simulations of the mean monthly actual evapotranspiration showed a "very good" correlation with the observational dataset simulations. For example, about 70% of the evapotranspiration occurred from May to August and, during this period, the APHRODITE_V1101+CRU TS3.1 datasets simulated about 39.86 mm, 55.01 mm, 62.70 mm, and 49.27 mm, the CRU S3.1 datasets simulated about 36.36 mm, 50.63 mm, 55.02 mm, and 33.93 mm, and the observational datasets simulated about 42.95 mm, 56.69 mm, 60.19 mm, and 48.05 mm of evapotranspiration.

The mean annual values of precipitation for the UVRB are also estimated using the SWAT model by coupling the combination of the respective datasets, as demonstrated in Table 4. Due to the combination of the different dimensions of the gridded datasets, the

SWAT model results showed mean annual precipitation differently. In particular, for scenarios (a) and (c), as shown in Table 4, the average annual precipitation in these two scenarios is higher than in other scenarios because the CFSR datasets were combined in scenario (a) with APHRODITE_V1101, and in scenario (c) the CFSR results were demonstrated independently. The CFSR is a global coupled atmosphere-ocean–land surface–sea ice assimilation system, developed by NCEP at a resolution of 38 km (T382) and APHRODITE_V1101 for monsoons in Asia, used at a resolution of 0.25° × 0.25°.

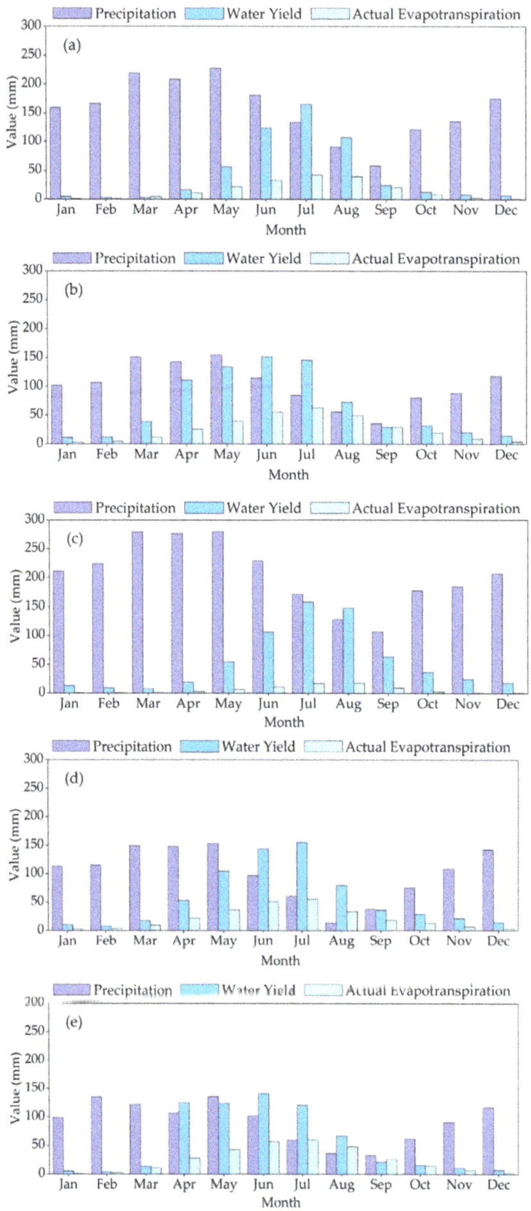

Figure 8. Mean monthly basin values of precipitation, water yield and actual evapotranspiration in the Upper Vakhsh River Basin in Central Asia, using (**a**) APHRODITE_V1101+CFSR, (**b**) APHRODITE_V1101+CRU TS3.1, (**c**) CFSR, (**d**) CRU TS3.1, and (**e**) observational datasets.

The deviation between the mean monthly actual evapotranspiration levels obtained from the simulated CRU TS3.1 datasets and the simulated observational datasets was less than 7 mm in all months except August, when it reached 14 mm (Figure 8d,e). The deviation between the mean monthly actual evapotranspiration levels, produced by the APHRODITE_V1101+CRU TS3.1, and observational dataset simulations for the entire basin was less than 4 mm in all months except October, when it reached 5.28 mm (Figure 8b,e). These results for the simulated actual evapotranspiration can be explained by the fact that in the UVRB, most of the land is covered with seasonal grassland (48.63%) and, after grassland, the prevailing land cover type is bare land (16.84%) and snow/ice cover (15.75%). The amount of evapotranspiration depends mainly on the type of land cover. Evapotranspiration also depends on soil moisture, i.e., the water held in the spaces between soil particles. Nevertheless, the hydrological model of SWAT is a continuous-time model and considers the variation in the moisture content of the soil. It also takes into account the soil moisture from the previous day. Consequently, evapotranspiration occurs on a dry day, and on such days, the soil moisture decreases. Accordingly, during the dry season, evapotranspiration may exceed precipitation. In general, the total precipitation is greater than the annual evapotranspiration.

4.5. Snowmelt Contribution to the Streamflow of the Upper Vakhsh River Basin, Using the SWAT Model

Melting snow in the Pamir-Alay is the main source of groundwater recharge and streamflow in the dry se ason for all perennial rivers in Tajikistan, which supply fresh water for drinking and irrigation to Uzbekistan and Turkmenistan in Central Asia. In addition, snowmelt flow facilitates hydropower production in Tajikistan, which accounts for over 95% of total electricity production. For that reason, it is important to assess the contribution of snowmelt in the UVRB in order to successfully develop, plan, distribute and maximize the efficient and beneficial use of water resources. In this study, the contribution of snowmelt to the streamflow of the Upper Vakhsh River is computed by applying the SWAT model to the snowfall–snowmelt mode.

The hydrology of snowmelt is important for SWAT applications in catchments where the river flows in spring and summer are mainly associated with snowmelt. The SWAT model's snowmelt module uses a linear function based on air temperature, snowpack temperature and melting rate, and measures the amount of snowmelt based on the areal coverage of snow and the snowmelt factor method [8]. In alpine basins with cold weather conditions and rare precipitation, the snowmelt streamflow is influenced along with the air temperature by the slope gradient, aspect, climatic variations, and solar radiation. In the SWAT model, which depends on the temperature index, the melting rate changes only with elevation, due to the air temperature gradient. The SWAT model divided the watershed into ten altitude ranges and, for each band, simulated snow cover and snowmelt separately. In this study, five established elevation bands were incorporated into the SWAT model to account for the spatial variation in the snowmelt parameters across the entire watershed, based on its topographic controls. The output tables for the overall simulation periods of the SWAT model, along with various climatic products, were used to compute the snowmelt streamflow in the UVRB. The results are shown in Tables 5–7.

Table 5. Average monthly snowmelt contribution in the Upper Vakhsh River flow by coupling the SWAT model with the APHRODITE_V1101+CFSR and APHRODITE_V1101+CRU TS3.1 during the overall period of the SWAT model simulation.

Month	Rainfall (mm) (a)	Snowmelt (mm) (b)	Net Rainfall Input (a + b)	Water Yield (mm)	Snowmelt Contribution (mm)	Monthly Snowmelt Contribution to the Streamflow (%)
			Combination of the APHRODITE_V1101 and CFSR datasets			
January	0.08	0.10	0.18	4.79	2.64	55.19
February	0.10	0.33	0.43	3.27	2.51	76.78
March	2.27	4.85	7.12	3.35	2.28	68.10
April	6.23	26.13	32.36	16.32	13.18	80.75
May	18.86	63.04	81.90	56.58	43.55	76.97
June	26.07	131.06	157.13	123.71	103.18	83.41
July	32.56	172.81	205.37	165.00	138.84	84.15
August	21.28	101.46	122.74	106.71	88.21	82.66
September	4.37	16.68	21.05	24.44	19.36	79.24
Octpber	3.22	4.86	8.08	12.69	7.63	60.13
November	0.40	0.94	1.34	8.39	5.88	70.07
December	0.35	0.48	0.83	6.45	3.72	57.66
Total	115.79	522.72	638.51	531.70	430.99	
			Combination of the APHRODITE_V1101 and CRU TS3.1 datasets			
January	2.35	9.22	11.57	11.56	9.22	79.70
February	2.88	13.56	16.44	11.98	9.88	82.49
March	15.51	40.87	56.38	38.55	27.94	72.49
April	37.49	104.65	142.14	110.93	81.68	73.63
May	68.70	113.25	181.95	134.35	83.62	62.24
June	70.65	122.18	192.83	152.15	96.41	63.36
July	65.80	116.29	182.09	145.97	93.22	63.86
August	43.56	45.86	89.42	72.72	37.30	51.29
September	20.86	13.01	33.87	29.57	11.36	38.42
Octpber	23.10	23.74	46.84	31.93	16.18	50.68
November	10.21	15.01	25.22	20.50	12.20	59.52
December	4.95	10.97	15.92	14.33	9.88	68.92
Total	366.06	628.63	994.69	774.54	488.88	

Table 6. Average monthly snowmelt contribution in the Upper Vakhsh River flow, by coupling the SWAT model with the CFSR and CRU TS3.1, during the overall period of the SWAT model simulation.

Month	Rainfall (mm) (a)	Snowmelt (mm) (b)	Net Rainfall Input (a + b)	Water Yield (mm)	Snowmelt Contribution (mm)	Monthly Snowmelt Contribution to the Streamflow (%)
			CFSR datasets			
January	0.04	0.06	0.10	13.26	8.20	61.81
February	0.19	0.22	0.41	9.04	4.88	54.01
March	1.00	3.46	4.46	8.09	6.28	77.57
April	3.64	23.25	26.89	18.93	16.37	86.47
May	13.12	63.62	76.74	53.86	44.65	82.90
June	24.06	135.85	159.91	106.03	90.08	84.95
July	29.53	197.33	226.86	158.66	138.01	86.98
August	19.84	156.09	175.93	148.17	131.46	88.72
September	5.22	30.26	35.48	63.33	54.01	85.29
Octpber	1.95	4.68	6.63	36.37	25.68	70.60
November	0.29	0.80	1.09	24.32	17.84	73.38
December	0.24	0.17	0.41	18.16	7.48	41.18
Total	99.12	615.80	714.92	658.22	544.93	

Table 6. Cont.

Month	Rainfall (mm) (a)	Snowmelt (mm) (b)	Net Rainfall Input (a + b)	Water Yield (mm)	Snowmelt Contribution (mm)	Monthly Snowmelt Contribution to the Streamflow (%)
			CRU TS3.1 datasets			
January	0.69	5.77	6.46	10.35	9.25	89.32
February	0.57	10.05	10.62	8.20	7.76	94.63
March	4.04	32.84	36.88	17.68	15.74	89.05
April	17.96	88.73	106.69	52.67	43.80	83.17
May	37.72	136.36	174.08	104.58	81.92	78.33
June	40.37	160.78	201.15	143.77	114.91	79.93
July	31.73	159.57	191.30	155.30	129.54	83.41
August	5.07	56.39	61.46	79.87	73.28	91.75
September	8.41	20.02	28.43	36.66	25.81	70.42
Octpber	9.28	26.40	35.68	29.17	21.58	73.99
November	5.85	16.09	21.94	21.73	15.93	73.33
December	2.21	7.66	9.87	14.63	11.35	77.60
Total	163.90	720.67	884.57	674.61	550.89	

Table 7. Average monthly snowmelt contribution in the Upper Vakhsh River flow, by coupling the SWAT model with the observational datasets, during the overall period of the SWAT model simulation.

Month	Rainfall (mm) (a)	Snowmelt (mm) (b)	Net Rainfall Input (a + b)	Water Yield (mm)	Snowmelt Contribution (mm)	Monthly Snowmelt Contribution to the Streamflow (%)
			Observational datasets			
January	0.09	0.32	0.41	5.47	4.28	78.20
February	0.53	1.14	1.67	3.96	2.71	68.35
March	5.23	25.77	31.00	13.80	11.47	83.13
April	25.72	152.84	178.56	124.85	106.87	85.60
May	57.90	120.70	178.60	123.98	83.79	67.58
June	63.90	115.64	179.54	140.94	90.78	64.41
July	44.26	100.51	144.77	120.89	83.93	69.43
August	29.86	45.71	75.57	67.34	40.73	60.49
September	18.11	4.68	22.79	21.00	4.31	20.55
Octpber	12.64	3.81	16.45	15.77	3.65	23.14
November	3.32	4.08	7.40	10.95	6.04	55.16
December	0.32	0.97	1.29	7.61	5.72	75.18
Total	261.88	576.19	838.07	656.57	444.28	

The basin-wide monthly snowmelt simulation showed that gridded datasets and observational datasets generated more or less similar outputs, as shown in Tables 5–7. The results acquired from the overall SWAT (1982–2006) simulation revealed that about 81.06% of the annual runoff (out of 531.70 mm of the annual runoff, 430.99 mm is snowmelt runoff) is supplied by snowmelt runoff when using a combination of the APHRODITE_V1101 and CFSR. For the combination of APHRODITE_V1101 and CRU TS3.1, the overall SWAT model results showed that about 63.12% of the annual runoff (out of 774.54 mm of annual runoff, 488.88 mm is the snowmelt runoff) is provided by snowmelt runoff. The simulation of the CFSR in the SWAT model reveals that about 82.79% of the annual runoff (out of 658.22 mm of annual runoff, 544.93 mm is the snowmelt runoff) is supplied by snowmelt runoff. By coupling CRU TS3.1 and the SWAT model, we found that in the annual runoff, about 81.66% is contributed by snowmelt runoff (out of 674.61 mm of the annual runoff, 550.89 mm is the snowmelt runoff). The overall simulation of the SWAT model in the application of the observational revealed that from the annual runoff, about 67.67% is supplied by snowmelt runoff (out of 656.57 mm of the annual runoff, 444.28 mm is the snowmelt runoff). We used five combinations of datasets in the SWAT model and the simulation results of the model showed that, in general, during winter (December–February), the monthly snowmelt and rainfall was estimated to be less than 5 mm for all types of datasets. The minimum rainfall and snowmelt are simulated in winter, including the fact that in winter, the precipitation mostly falls in a solid form rather than a liquid form (rain) in the UVRB in Central Asia. According to the simulated observational datasets, September and October are the only periods in which rainfall in the catchment is dominant.

In addition, the results of all combination datasets show that during the spring and summer (March–August) most of the runoff is provided by the runoff from snowmelt (Tables 5–7). For example, according to the simulated SWAT model based on the APHRODITE_V1101+CFSR, during the period of 1982–2006, about 68.10%, 80.75%, 76.97%, 83.41%, 84.15%, and 82.66% of the river flow is provided by the snowmelt runoff from March to August. The simulation results of the APHRODITE_V1101+CRU TS3.1 showed that the contribution of snowmelt runoff to annual runoff is about 72.49%, 73.63%, 62.24%, 63.36%, 63.86%, and 51.29% (March–August). The overall (1982–2006) simulation of the CFSR datasets showed that between March and August, about 77.57%, 86.47%, 82.90%, 84.95%, 86.98%, and 88.72% of the annual runoff is supplied by snowmelt runoff. The SWAT simulation on the CRU TS3.1 showed that, in spring and summer (March–August), of the total runoff, about 89.05%, 83.17%, 78.33%, 79.93%, 83.41%, and 91.75% originates as snowmelt runoff. Using the observational datasets in the SWAT model, the simulation results revealed that from March to August, about 83.13%, 85.60%, 67.58%, 64.41%, 69.43%, and 60.49% of the annual runoff is contributed by runoff formed due to melting snow. As a result of using different datasets in the SWAT hydrological model, the model results showed that an increase in the contribution of melt runoff to total runoff begins in March and continues with a fairly good contribution until September, while the maximum peak of snowmelt runoff is observed in June and July (Tables 5–7). In winter, the contribution of snowmelt and total runoff is low due to limited rainfall, and snowmelt is constrained by low temperatures in mountainous areas. Our results also showed that the hydrology of the Vakhsh River Basin is dominated by snowmelt. In this study, the amount of simulated snowmelt, based on applied datasets, ranges from 115.64 mm to 160.78 mm in June and from 100.51 mm to 197.33 mm in July. However, the contribution of snowmelt to the total runoff during June and July is considered to be a peak period of contribution, and simulations of the observation datasets presented the lowest amount of snowmelt contribution to the total runoff, compared to other simulated datasets. The reason for these minimal values might be the number of climate stations that were used in the hydrological modeling in this mountain catchment. Similarly, previous studies have shown that mountainous catchments in most regions of the world have very few climate stations. In this study, the situation is the same; we used only four observational climate stations, Lakhsh, Dekhavz, Rasht, and Bustonobod because there are no other operating climatic stations in the catchment. Regarding the maximum contribution of snowmelt to total runoff during peak periods (June–July), in June, simulations of snowmelt using the CRU TS3.1 showed the maximum contribution of snowmelt to total runoff compared to other utilized datasets, while for snowmelt in July, the simulation of snowmelt using the CFSR compared to other datasets showed the maximum contribution of snowmelt to total runoff.

Figure 9 shows the mean monthly snowmelt and rainfall contribution in the Upper Vakhsh River flow, using the SWAT model. The cumulative curves of the runoff input components, including snowmelt and rainfall, are shown in Figure 10. Figures 9 and 10 demonstrate that the various gridded datasets in the hydrological model perform slightly differently in snowmelt and rain simulations. One possible explanation could be the fact that the resolution of the climate inputs varies, i.e., the numbers of gridded points are different. However, the performance of the employed objective functions in hydrological models was almost equal under different gridded climate inputs and this is to be expected since, each time, the model adjusts its parameters during calibration so as to maximize the utility of the input dataset. The results obtained from the SWAT model, using the respective datasets, indicated that an average of about 81.06%, 63.12%, 82.79%, 81.66%, and 67.67% of the annual flow of the Upper Vakhsh River is contributed by the snowmelt runoff.

Snowmelt is the dominant hydrological process in the VRB and flow is less influenced by precipitation events. The contribution of the rain to the annual flow was estimated to be about 18.94%, 36.88%, 17.21%, 18.34%, and 32.33%, according to the simulations of the respective datasets.

Our study revealed that the gridded climate data from APHRODITE, CRU and CFSR can be used as alternative data, especially in areas with fewer climate stations, including the UVRB. These results represent the potential of the use of gridded datasets for hydrological modeling. If complete observational data are available for the reference period, it is more appropriate to use these data. In some cases, because of the provision of continuous data at spatial and temporal scales over a longer period, gridded datasets are preferred over observational datasets. The use of the most recently available gridded datasets in a consistent format and with improved technology has made them easier to use for hydrological modeling. Meanwhile, the combination of different gridded datasets can lead to improved hydrological modeling [82]. Similarly, we used the reference gridded datasets in combination, by combining their corresponding features with the daily measurement-based climate data to achieve evaluation and better model simulations. The snowmelt and rainfall contributions to annual river flow, according to the simulation results of the respective datasets, were different (Figures 9 and 10). The differences in the contributions of the snowmelt and rainfall to annual river flow from the simulations of all combinations of datasets are due to the differences in the climate data. Such a difference may have occurred because the climate data came from different independent sources, with different approaches toward acquisition, processing data, and resolution. However, our results demonstrated that the gridded datasets performed well in capturing peak flows and base flows. The overall modeling results indicate that the SWAT model is potentially useful for studying hydrology and assessing the water yield of catchments.

Spring and summer runoff from the Vakhsh River Basin supplies several Central Asian countries, including Kyrgyzstan, Tajikistan, Uzbekistan and Turkmenistan, with water for domestic and agricultural needs, hydropower production and recreation [83]. Modeling the spring and summer snowmelt runoff is critical to understanding seasonal runoff variability. Water management and flood protection strategies in Central Asia are based on capturing and storing runoff for delivery in the fall months. Water managers track spring and summer runoff as a key factor when planning to meet Central Asia's water supply needs [84]. It is also important to use spring and summer runoff data to predict the water supply and flooding in the basin.

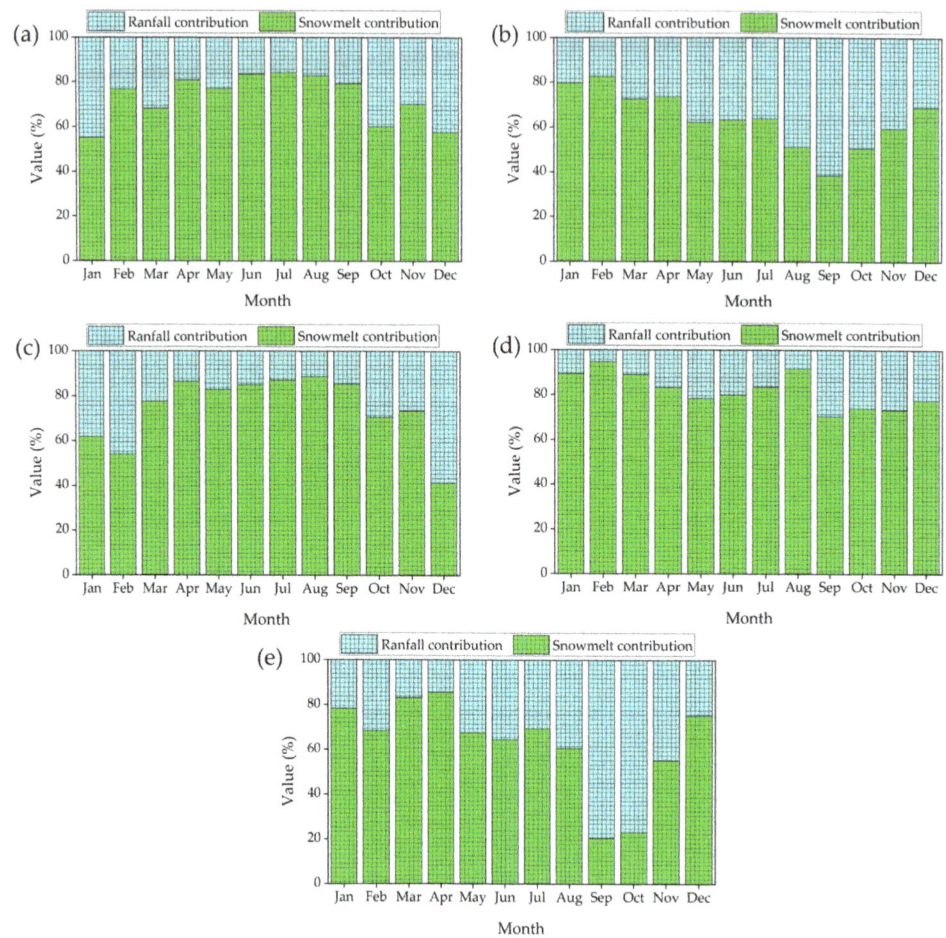

Figure 9. Mean monthly rainfall and snowmelt contribution in the Upper Vakhsh River flow, using the SWAT model with (**a**) APHRODITE_V1101+CFSR, (**b**) APHRODITE_V1101+CRU TS3.1, (**c**) CFSR, (**d**) CRU TS3.1, and (**e**) observational datasets.

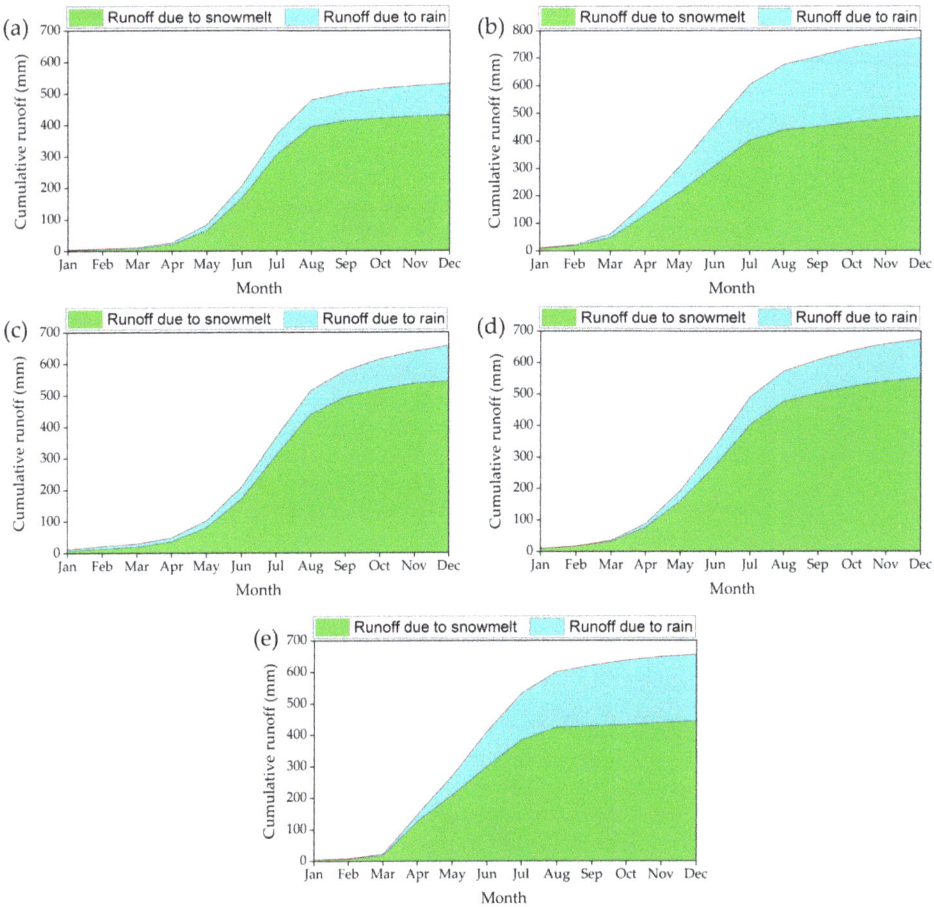

Figure 10. Cumulative curves of the runoff input components for the Upper Vakhsh River, using the SWAT model with (**a**) APHRODITE_V1101+CFSR, (**b**) APHRODITE_V1101+CRU TS3.1, (**c**) CFSR, (**d**) CRU TS3.1, and (**e**) observational datasets.

5. Discussion

To the best of our knowledge, the simulation of the water balance component, particularly snowmelt runoff and its contribution to total runoff in the Vakhsh River Basin, has not been conducted previously. Similarly, an evaluation of the datasets' ability to simulate the hydrological behavior of the UVRB from 1982 to 2006 has not previously been performed, especially not when using a combination of the APHRODITE_V1101+CFSR, AHPRODITE_V1101+CRU TS3.1, CFSR, and CRU TS3.1 datasets in a hydrological SWAT model. Water-balance elements in a catchment are influenced by climate and the physical characteristics of the basin, such as topography, land use, soil properties, glaciers, and human economic activities. For any analyses related to sustainable water resource management, understanding all the hydrological components is important. Snowmelt from the Vakhsh River is the main source of groundwater recharge and runoff in the dry seasons of many perennial rivers in Tajikistan, supplying fresh water for drinking, irrigation and hydropower generation. For this reason, it is very important to assess the contribution of snowmelt to the total runoff of UVRB for effective water resources management. Moreover, when modeling mountainous watershed hydrology, the most important determinant is the provision of accurate and alternative climate inputs in modeling. The lack of data has a large impact on modeling in mountainous regions. Ordinary climate stations are often scattered sparsely and cannot fully reflect the climatic conditions in the basin, especially in mountainous areas. In addition, the records of climate stations often do not cover the proposed modeling period or contain gaps. To address/mitigate this issue in UVRB CA, we examined the performance of gridded precipitation and temperature data combinations from various sources for simulating river flow, using the SWAT model.

In this study, the simulation of the observational-based datasets performed better than the gridded products in the UVRB for modeling river flow. Our results revealed that among the applied gridded datasets, the use of the CRU TS3.1 datasets for the overall scale (1982–2006) showed higher accuracy in river flow simulations, followed by the APHRODITE_V1101 and CRU TS3.1 combination, the APHRODITE_V1101 and CFSR combination, and the CFSR in the UVRB in Central Asia. Our results are consistent with those of Hajihosseini et al., who used the global CRU and SWAT models in the Upper Helmand Basin (UHB) in Afghanistan, which is a neighboring country to Tajikistan, to simulate the long-term hydrological conditions of the basin. Hajihosseini et al. used the CRU as an alternative data source, due to the lack of observational data at the UHB [30]. The results revealed the good performance of the SWAT model, using the CRU dataset for the UHB from 1969 to 1979, while NSE was 0.84 for the calibration period and 0.82 for the validation period [30]. As is similar to the results of our study, CRU datasets were used in the SWAT model to simulate river flows in the African continent over the period of 1968–1995, and the results showed a "good" performance (NSE > 0.6) [85]. However, our results for the CRU TS3.1 simulation contradicted the results of Malsy et al. [29], who examined the performance of hydrological modeling using four datasets, the Global Precipitation Climatology Centre (GPCC) Reanalysis product v6, APHRODITE, WATCH forcing data (WFD), and CRU, in a hydrological model (Water Global Assessment and Prognosis 3, WaterGAP 3) for the period from 1976 to 1999. According to Malsy et al., the GPCC and APHRODITE datasets with the WaterGAP 3 hydrological model showed better hydrological results than the CRU and WFD at the Tuul River Basin and Khovd River Basin in Mongolia in East Asia [29]. This contradiction may be associated with the use of a different hydrological model, a different selected period and a different study area location. It is not known if the same results could be generated in our study area with a different hydrological model. For instance, in the Hotan River Basin in southwestern Xinjiang in Central Asia, from 2004 to 2008, Luo et al. [31] assessed the performance of the SWAT and MIKE SHE hydrological models. The results showed that the SWAT model performs better (NSE = 0.77) than the MIKE SHE model (NSE = 0.66) for the same climate input. Besides hydrological modeling, the CRU dataset was evaluated for climatological studies in Central Asia by other researchers and their results showed that the CRU dataset is applicable and satisfactory for climatological studies in Central Asia [86,87].

Furthermore, our results indicated that a second alternative source for the hydrological simulation of the UVRB with the SWAT model could be the combination of precipitation with APHRODITE_V1101 and maximum/minimum temperature with CRU TS3.1, followed by the combination of APHRODITE_V1101 and CFSR, and CFSR. These results are in agreement with the findings of Tan et al. [34], who assessed the capabilities of the APHRODITE, CFSR and the "Precipitation Estimation from Remotely Sensed Information using Artificial Neural Networks" (PERSIANN) datasets to model river flow using the SWAT model for the Kelantan River Basin (KRB) and the Johor River Basin (JRB) in Malaysia in Southeast Asia, from 1985 to 1999. The combination of APHRODITE precipitation data and CFSR temperatures resulted in an accurate simulation of the river flow of the KRB and JRB. Tan et al. recommend the use of APHRODITE precipitation and CFSR temperature data when modeling the water resources of Malaysia [34]. Similarly, the APHRODITE dataset and the SWAT model were used for hydrological modeling from 1980 to 1989 in the Brahmaputra River Basin (BRB) and Yarlung Tsangpo River Basin (YTRB), which is the main international river flowing through China, Bhutan, India, and Bangladesh. The results showed the validity of APHRODITE estimates in driving the hydrological model in the YTRB and BRB [33]. Comparable results were reported by Xu et al. [35], who applied the SWAT model with APHRODITE and WFD in the Xiangjiang River Basin (XRB) in China to simulate river flow, particularly high flow and low flow. The APHRODITE simulation (NSE = 0.79, NSE = 0.82) performed better than the WFD dataset (NSE 0.69, NSE = 0.71), both during calibration (1991–2005) and validation (2001–2005) periods. WFD modeling leads to more errors in simulating flood events than APHRODITE [35]. The TRMM, NCEP, GPCP, CFSR and APHRODITE were used to assess the performance (NSE) of SWAT in the Wunna Basin in India [36]. APHRODITE dataset simulations performed much better (NSE = 0.68) than TRMM, NCEP, GPCP and CFSR, meaning that APHRODITE can be seen as an alternative source for hydrological modeling in the Wunna Basin [36]. Our results indicated the superiority of the combination of the APHRODITE_V1101 and CRU TS3.1 in the streamflow simulation in the UVRB from 1981 to 2006. However, these results are somewhat inconsistent with the findings of Eini et al. [88], who modeled hydrology systems and evaluated the performance of the SWAT model in the Maharlu Lake Basin in Iran by comparing CRU, NCEP CFSR, and APHRODITE, as well as reanalyzing Asfezari rainfall data using conventional data from 1983 to 2010. In this Iranian catchment, a simulation that was achieved through a combination of APHRODITE and CFSR showed superior performance (NSE = 0.91) compared to the other dataset combinations [88]. However, it should be noted that these results are region- and model-dependent. Many studies show that the accuracy of gridded data results varies by region [38,39]. Meanwhile, a hydrological model with a different concept and representation of the streamflow procedure may lead to different conclusions.

Our results demonstrated that for the overall scale (1982–2006), the CFSR simulated the hydrology of the UVRB with lower accuracy than APHRODITE_V1101+CFSR, APHRODITE_V1101+CRU TS3.1, and CRU TS3.1. Correspondingly, the simulation of river flow using a SWAT model based on the characteristics (NSE) of weather products, with reference to the CFSR, showed that despite the small number of observational climate stations in the basin, the modeling of observational datasets was more accurate for representing climate relationships in the basin than the CFSR. These results are very similar to those of Dile et al. [15] in the Gilgel Abay River and Gumera River in the Lake Tana Basin, and the upper part of the Upper Blue Nile Basin, where SWAT was also set up to assess the performance of CFSR datasets compared with conventional datasets for hydrological predictions from 1994 to 2008. Liu et al. [32] also used the SWAT model with climate data from the "China Meteorological Assimilation Driving Datasets" (CMADS V1.0) and CFSR in the Yellow River Source Basin, Qinghai–Tibet Plateau, from 2009 to 2013. Liu et al. found that the performance of the hydrological model for the monthly scale of CMADS (NSE = 0.78) was higher than that of the CFSR (NSE = 0.69) [32]. However, our results, as found in the UVRB, conflict with the findings of Tolere et al. [16], Fuka et al. [45], Cuceloglu and Ozturk [46], and Grusson et al. [89]. For instance, Tolere et al. [16] reported more successful SWAT streamflow simulation results using the CFSR than conventional datasets from 1990 to 1995 for the Keleta watershed in Ethiopia, where conventional data are scarce [16]. Fuka et al. [45] reported that the SWAT model presented better simulation results with CFSR datasets, compared to using traditional weather-gauging stations in the Catskill Mountains, NY, USA, and the Gumera Watershed in the Blue Nile River in Ethiopia, from 1996 to 2010. Similarly, Cuceloglu and Ozturk [46] evaluated CFSR using the SWAT model as a hydrological simulator in the Black Sea catchment from 2000 to 2012. The results showed that the CFSR gives quite reasonable agreement between simulated and observed river flow, compared to the observational dataset. Another study was conducted by Grusson et al. [89] in the Garonne River Watershed in France, employing CFSR and conventional datasets in hydrological simulations using the SWAT model from 2000 to 2010. The results revealed that the CFSR provided better hydrological simulations than conventional datasets [89]. Differences in climate and geographic conditions are the most likely explanation for such differences between the findings of Tolere et al. [16], Fuka et al. [45], Cuceloglu and Ozturk [46], Grusson et al. [89], and the results presented in this study.

The applicability of the SWAT model for the simulation of water balance components was assessed in the Heihe mountain river basin in northwest China, from 1961 to 1988 [22]. Water-balance components over the Narmada River Basin in India were assessed by Goswami et al. using the SWAT model and CFSR, from 1984 to 2013. The results suggested that the SWAT model was able to simulate the water-balance components at the basin and sub-basin scales [19]. Pathak et al. [90] applied the SWAT model in nine watersheds in India to validate the annual water yield obtained from diverse water-balance models. Moreover, Pathak et al. assessed the applicability of the Lumped Zhang model and InVEST model, together with the SWAT model, to compute water yield in scenarios before and after climate change for 1980, 1990, 2001, and 2015 [90]. Gupta et al. [18] estimated the water-balance components in the Sabarmati River Basin (SRB) in India using the SWAT model, from 1999 to 2005. Gupta et al. noted that SWAT is a powerful tool that very effectively evaluated hydrological components in the study of the water balance and river flow of the SRB [18]. In Nepal, Thapa et al. used the HBV and BTOPMC models, along with the SWAT model, to assess the components of water balance from 2001 to 2010. The results of the three models were similar [91]. A predictive study using three models also offered a reasonable range for runoff and evapotranspiration estimates [91]. Water balance and water yield were predicted by the SWAT model in a basin in the north-central part of Nigeria, from 1985 to 2010 [17]. Adeogun et al. noted that the SWAT model can be a promising tool for predicting water balance, in terms of sustainable water management in Nigeria [17]. The SWAT model was applied by Leta et al. to model water balance components in the Heeia watershed in Hawai'i, an island in the Pacific Ocean, from 2006 to 2013 [92]. This study demonstrated the applicability of SWAT to small island watersheds with large topographic, precipitation, and land-use gradients. Our seasonal and annual precipitation results in the UVRB in Central Asia, using the APHRODITE_V1101 and CRU TS3.1, demonstrated much closer results to the observational dataset. These findings are in agreement with the results of a study by Shen et al. in the Kaidu Basin in Central Asia. Shen et al. [37] used gridded products, including CFSR, APHRODITE, CRU, TRMM, ERA-Interim and MERRA-2, with the J2000 model to analyze the spatiotemporal patterns of water balance and the distribution of runoff components in the glacierized Kaidu Basin in Central Asia. The results showed that APHRODITE and CRU represented annual and seasonal precipitation dynamics that were similar to the observational dataset at most climate points [37]. Similarly, a water balance study with the application of the SWAT model and observational datasets was conducted in the Indian Ken River Basin in South Asia, from 1986 to 2005 [20]. Himanshu et al. concluded that the SWAT model can accurately simulate the hydrology of the Ken River Basin in India. The water balance study of the basin showed that evapotranspiration is more predominant, accounting for about 44.6% of the average annual precipitation [20].

In this study, considering the more accurate performance of the CRU TS3.1 and observational datasets than other studied datasets, the simulations of the CRU TS3.1 and observational datasets showed that the actual evapotranspiration in July is almost equal to the July catchment precipitation values. These results are in accordance with those of Pritchard [23], who used a combination of CFSR temperature and APHRODITE precipitation datasets in the SWAT model to simulate water-balance components, especially the actual evapotranspiration in five Asian river basins, including the Aral, Indus, Ganges, Brahmaputra, Tarim, and the lakes of Issyk-Kul and Balkhash. For the Aral Sea Basin in Central Asia, Pritchard reported that summer evaporation is approximately equal to summer precipitation [23]. In this study, less actual evapotranspiration occurs in December, January, and February for all datasets studied, including CFSR. However, this result contradicts the findings of Goswami et al. [19], who found that the simulations of actual evapotranspiration showed minimal values in May, using CFSR and the SWAT model, in the Narmada River Basin of India in South Asia between 1984 and 2013.

The results of our simulation showed that the average annual actual evapotranspiration is about 2.9% (CFSR), 9.93% (APHRODITE_V1101+CFSR), 21.1% (CRU TS3.1), 25.52% (APHRODITE_V1101+CRU TS3.1), and 27.28% (observational datasets) of the average annual precipitation in the UVRB from 1982 to 2006. The same methodology was applied by Budhathoki et al. [93] in the West Seti River Basin (WSRB) in South Asia, to simulate the mean annual water balance components, including precipitation and evapotranspiration, using the SWAT model and a combination of conventional weather data and APHRODITE, for the period of 1986–2005. The mean annual total evapotranspiration matched about 36% of the mean annual precipitation in the WSRB [93]. In another study, Nasiri et al. [21] applied the SWAT model to the Samalqan Basin in Iran in Western Asia, from 2004 to 2014, to compute actual evapotranspiration using observational weather data. Actual evapotranspiration contributed to the largest water loss from the basin, at approximately 86%. Nasiri et al. pointed out that the high evapotranspiration rate that was simulated may be related to the vegetation types in the region [21].

Our results indicated that the simulation of the CFSR provided lower actual evapotranspiration values than traditional weather data for the UVRB from 1981 to 2006. Similar results were obtained by Dile et al. [16] in the Gumera, Rib and Megech River Basins in Ethiopia in the Horn of Africa, where independent observation datasets and CFSR were used in the SWAT model in 1990–1995 [16]. However, Dile et al. stated that the results in the Melka Kuntur Basin showed higher values for the water-balance components; this may be due to the relatively high total precipitation in the CFSR dataset in the Melka Kuntur Basin [16]. In general, in our study, CFSR also simulated higher total annual precipitation in the UVRB in Central Asia, compared to other reference datasets. We found the largest deviation between the monthly mean actual evapotranspiration CFSR and the observational datasets, compared to other applied datasets in the UVRB. However, from previous studies, we have observed that in some regions, CFSR, when calculating actual evapotranspiration, produced values very similar to conventional weather data. For instance, the CFSR and observational datasets, with the SWAT model, were applied to different watersheds in the Blue Nile Basin in the northwestern Ethiopian Plateau from 1994 to 2008, to estimate actual evapotranspiration [15]. For the Megech sub-basins in Ethiopia, the results showed that the deviation between the monthly mean actual evapotranspiration levels, obtained from CFSR simulations, and observational datasets was less than ± 5 mm in all months besides August and September, when it reached 12 mm and 19 mm, respectively. Dile et al. noted that CFSR weather data can be a valuable option for hydrological prediction where conventional data are not available, such as in remote parts of the Upper Blue Nile Basin [15].

Based on the SWAT model application, our results for the simulations of five different climate data combinations, including APHRODITE_V1101+CFSR, APHRODITE_V1101+CRU TS3.1, CFSR, CRU TS3.1, and observational data showed that approximately 81.06%, 63.12%, 82.79%, 81.66%, and 67.67% of annual runoff was contributed by snowmelt runoff from 1982 to 2006 in the UVRB. The largest contribution of snowmelt runoff to the total runoff appears during the spring and summer periods. The monthly APHRODITE_V1101+CFSR simulations showed that from May to August, the snowmelt runoff contribution to river flow was about 68.10% (March), 80.75% (April), 76.97% (May), 83.41% (June), 84.15% (July), and 82.66% (August); for the APHRODITE_V1101+CRU TS3.1 simulations, it was about 72.49%, 73.63%, 62.24%, 63.36%, 63.86%, and 51.29%; for the CFSR simulations, it was about 77.57%, 86.47%, 82.90%, 84.95%, 86.98%, and 88.72%; for the CRU TS3.1 simulations, it was about 89.05%, 83.17%, 78.33%, 79.93%, 83.41%, and 91.75%; and, for the observational simulations, it was about 83.13%, 85.60%, 67.58%, 64.41%, 69.43%, and 60.49%. Many studies have shown that the application of the SWAT model is quite useful in snowmelt simulations [1,2,24,25]. As far as we are aware, in this study, the use of the SWAT model with various combinations of the respective climate datasets was conducted for the first time as a way to simulate snowmelt runoff contribution to river flow in this mountainous basin.

By using observational climate data in the SWAT model, Duan et al. simulated the snowmelt contribution to total runoff in the Tizinafu River Basin (TRB) in Xinjiang in

Central Asia, from 2013 to 2014 [25]. Duan et al. found that about 44.7% of the total runoff comes from snowmelt runoff in the TRB [25]. Siderius et al. [24] calculated the contribution of snowmelt to river runoff using APHRODITE data with the SWAT model, from 1971 to 2000, in the Ganges River in northern India in the Himalayan arc in South Asia. The SWAT results showed that approximately 1% and 5% can be considered to be indicative of the actual total annual contribution of snowmelt to total runoff. The contribution of seasonal snowmelt to total runoff during the dry season before the summer monsoon (MAM) is estimated to range from 12% to 38% [24]. A similar approach was applied by Chiphang et al. [1], who used the SWAT model in the mountainous Mago River Basin, located in the Eastern Himalayan region of India, from 2006 to 2009 to compute the contribution of snowmelt to streamflow changes in the basin. Their results showed that the contribution of snowmelt runoff to the annual streamflow of the basin was about 8% [1]. Another study was performed by Dhami et al. using the snowmelt runoff model (SRM) and SWAT model with conventional weather data, to compute the water balance components of the Karnali River Basin in Nepal in South Asia and to simulate the contribution of snowmelt to river runoff, from 1993 to 2005 [2]. Dhami et al. reported that, from the comparison of the results obtained from the SWAT model and the SRM model, it is recommended that the results obtained from the SWAT model are used due to its better performance in terms of predicting reality than the SRM model [2]. The results of the SWAT model indicated that in the Karnali River Basin, about 35% of the total runoff is contributed by snowmelt runoff [2]. An accurate representation of the snowmelt process could improve the prediction of streamflow in mountainous catchments [89,94]. The UVRB has good seasonal snow cover at high altitudes. Snowmelt is usually an important source of river flow at high altitudes. In this study, SWAT appropriately demonstrates both the beginning of snowmelt and the peak of spring snowmelt.

The perennial river basin system, when combined with steep slopes, provides enormous hydropower potential, the exploitation of which requires a deep understanding of the hydrological system of the catchment. The capacity of hydropower is determined by the flow rate of the river, so a change in flow will directly lead to changes in hydropower potential [95]. Determining the effect of snowmelt on streamflow in the catchment allows an assessment of the hydrological processes within a river basin in a mountainous region. Snowmelt tends to create regular seasonal patterns of river flow during warmer temperatures, with the melting of snowpacks accumulated over the winter. The impact of melting snow on potential flooding, mainly in the spring, is of concern to many inhabitants across the globe. The performance of the SWAT model presents evidence that it can be applied efficiently in the transboundary Vakhsh River Basin in Central Asia for water resources assessment and management. Semi-distributed hydrological models can be used as an essential feature of the water resources monitoring approach and can play a crucial part in the management of transboundary water resources [96]. Furthermore, well-calibrated models are an important tool for a variety of water management applications, such as for assessing the availability and balance of water resources, modeling water quality and sediment transport, etc.

Uncertainties and Limitations

The SWAT model has been applied in several hydrological modeling studies in various catchments around the world [14,25,97]. There may be a few areas of uncertainty in modeling snow and glacier melt, such as orographic impacts and hydrological model parameterization, as well as heterogeneity in forest cover, slope, and features, which are evident issues in snow and glacier hydrology. All of these aspects are not well represented by a simple temperature index of snowmelt and glacier melt simulation. However, the evaluation of snowmelt based on a temperature index appears to be good enough to compute the physics of snowmelt processes entirely. Snowmelt hydrology is specifically considered as an essential variable in local catchments. Due to the degree-day approach with elevation bands in the SWAT snowmelt module, it is hard to avoid the uncertainties inherent in the model structure, parameters and input data. On the other hand, the absence of an ice-melt module would also cause potential uncertainties. Calibration solely by discharge records might produce good results, as well as potential uncertainties [8]. Parameter uncertainty occurs when certain physical processes in a hydrological or climatic system cannot be explicitly resolved. As a replacement, they must be included via parameterization, which contains some uncertain variables. The UVRB is situated in a mountainous area, within which a wide range of regions subject to permafrost and seasonal cold is spread. The freezing/thawing processes of the soil also affect the accuracy of the simulation [98]. Furthermore, measurement-based flow data that have been used in a comparison with a simulated flow may be exposed to human or tool errors in river-level observations, or inaccuracies in estimated curves, all of which are designed for natural river stretches that are subject to erosion and deposition [99]. The uncertainty of the data source stems from the use of the CRU, which is generated by interpolating data from weather stations in the region. Therefore, the entered climatic data are approximate [100]. The APHRODITE dataset, which includes multiple climate stations, provides high-resolution, gridded, long-term daily precipitation estimates, as commonly used in South Asia. In the UVRB, most of the climate stations are located in valleys and are not in mountains with a large amount of precipitation. The combination of all of these factors can increase the uncertainty of the APHRODITE estimates. Precipitation data from CFSR showed more heterogeneity than the temperature data. The gridded dataset was obtained directly; thus, its applicability in the study region must be assessed prior to its use. The precipitation error characteristics of datasets vary due to climatic regions, elevations, surface conditions, seasons and storm patterns [101]. Similarly, these datasets are inevitably prone to inaccuracies caused by sampling uncertainties, indirect measurements, and exploration algorithms [102]. In future investigations, it might be possible to analyze the effects of bias correction on multiple gridded climate data estimates in SWAT hydrological element simulations.

6. Conclusions

We used the SWAT–CUP Sequential Uncertainty Fitting (SUFI-2) program to assess the water balance components of the basin. APHRODITE_V1101, CFSR, and CRU TS3.1 gridded datasets were tested both independently and in combination with observational datasets for hydrological simulation. The SWAT hydrological model was used to examine the datasets in Central Asia's Vakhsh River Basin. We also tested the SWAT model's applicability to the UVRB hydrology. The authors came to the following conclusions:

(a) The results of the study indicated that the applied gridded datasets, such as CRU TS3.1, AHPRODITE_V1101, and CFSR, can be used as alternative climate data in an assessment of the hydrology in the UVRB.

(b) We observed that the CFSR datasets performed worse than APHRODITE_V1101+CFSR, APHRODITE_V1101+CRU TS3.1, and observational datasets.

(c) Aquifer percolation fraction (RCHRG_DP), baseflow alpha-factor (ALPHA_BF), moist bulk density (SOL_BD), SCS runoff curve number for moisture condition-II (CN2), and effective hydraulic conductivity in main channel alluvium (CH_K2) were determined to be the most sensitive factors for streamflow simulation using the SWAT model in the UVRB.

(d) APHRODITE_V1101+CFSR showed "good" [71] results in simulating the monthly observed streamflow (with NSE = 0.70 and 0.72 for calibration and for overall period, respectively, and PBIAS = 14.8% and 15% for calibration and for overall period, respectively). The APHRODITE_V1101+CRU TS3.1 gave "good" results for calibration and overall periods (NSE = 0.74 and 0.74, respectively), with "satisfactory" [71] (PBIAS = −17.7%) and "good" (PBIAS = −10.80%) underestimation of streamflow. This combination showed "very good" [71] results (NSE = 0.78) for the validation period with a "very good" overestimation of streamflow (PBIAS of 1.30%). The CFSR datasets presented "good" results (NSE = 0.68) with a "very good" (PBIAS of 2%) overestimation of streamflow and, for the validation period, CFSR provided "very good" results (NSE = 0.83) with a "very good" overestimation of streamflow (PBIAS of 9.58%). The CRU TS3.1 datasets gave "very good" results for calibration, validation and overall scales (NSE = 0.91, 0.88 and 0.91, and PBIAS = 0.32%, 6.09%, and 0.29%, respectively). The observational datasets for calibration, validation and overall provided "very good" results (NSE = 0.90, 0.93 and 0.91 and PBIAS = 0.69%, 5.51%, and 5.21%, respectively). The better performance corresponds to observational datasets with higher NSE values, followed by the CRU TS3.1 datasets.

(e) The coefficient of determination (R^2) showed an acceptable (>0.6) [103] correlation between the observed and simulated monthly streamflow during calibration, validation, and overall scales (R^2 ranges from 0.74 to 0.94) for all five datasets.

(f) For APHRODITE_V1101+CFSR, APHRODITE_V1101+CRU TS3.1, CFSR, CRU TS3.1, and observational datasets, the SWAT model simulated mean annual precipitation of the UVRB as 1875.9 mm, 1236.9 mm, 2479 mm, 1215.9 mm, and 1098.5 mm, respectively, with 93.82% (1760.11 mm), 70.41% (870.85 mm), 96% (2379.86 mm), 86.52% (1051.98 mm), and 76.16% (836.65 mm) as snowfall, out of which 29.70% (522.72 mm), 72.19% (628.63 mm), 25.88% (615.80 mm), 68.51% (720.67 mm), and 68.87% (576.19 mm), respectively, melts and facilitates snowmelt runoff in the basin.

(g) The UVR basin evapotranspiration is 9.93% (APHRODITE_V1101+CFSR), 25.52% (APHRODITE_V1101+CRU TS3.1), 2.9% (CFSR), 21.08% (CRU TS3.1), and 27.28% (observational datasets) (186.3 mm, 315.7 mm, 72.1 mm, 256.4 mm, and 299.7 mm out of 1875.9 mm, 1236.9 mm, 2479 mm, 1215.9 mm, and 1098.5 mm, respectively). Overall, less evapotranspiration occurs in December, January, and February.

(h) From 1982 to 2006, the simulation of five different weather products (APHRODITE_V1101 +CFSR, APHRODITE_V1101+CRU TS3.1, CFSR, CRU TS3.1) and observational datasets in the UVR basin showed that snowmelt runoff contributes approximately 81.06%, 63.12%, 82.79%, 81.66%, and 67.67%, respectively, of annual runoff. Snowmelt runoff contributes the most to overall runoff in spring and summer.

(i) The annual flow contribution of rain was estimated at 18.94%, 36.88%, 17.21%, 18.34%, and 32.33% using APHRODITE_V1101+CFSR, APHRODITE_V1101+CRU TS3.1, CFSR, CRU TS3.1, and observational datasets.

The SWAT hydrological simulation of observational datasets outperformed gridded products in the mountainous UVRB. The SWAT model-simulated streamflow variations better than CFSR gridded datasets using CRU TS3.1 and APHRODITE_V1101. Gridded meteorological datasets like CRU TS3.1, APHRODITE_V1101, and CFSR can also be used for hydrologic modeling, especially if observational data is scarce. The SWAT model captured the monthly observed flow patterns and trends well. The model produced reliable monthly streamflow estimates, as evidenced by NSE, R^2, and PBIAS values that were superior for calibration, validation, and overall scales. The high NSE, R^2, and PBIAS values for monthly streamflow during the calibration, validation and overall periods indicate the model's predictive ability. For water management challenges in the basin, the SWAT model offered a baseline understanding of hydrological dynamics. The current study demonstrates that the SWAT model could be a helpful tool for predicting water balance components to assist basin-level policies and decisions.

Author Contributions: All authors were involved in the intellectual elements of this paper. X.C., A.G., and T.L. designed the research. A.G. conducted the research and wrote the manuscript. N.G. and M.G. helped in the data arrangement and data analysis. J.P. and Z.L. helped in the validation. T.L. and Z.K. helped revise the manuscript and also provided many suggestions. All authors have read and agreed to the published version of the manuscript.

Funding: This study was financially supported by the Strategic Priority Research Program of the Chinese Academy of Sciences, the Pan-Third Pole Environment Study for a Green Silk Road (Grant No. XDA20060303), the project of Research Center for Ecology and Environment of Central Asia (Grant No. Y934031),the International Cooperation Project of National Natural Science Foundation of China (Grant No. 41761144079),the CAS PIFI fellowship (Grant No. 2021PC0002), the Xinjiang Tianchi Hundred Talents Program (Grant No. Y848041), the CAS Inter-displinary Innovation Team

(Grant No. JCTD-2019-20), and the Regional Collaborative Innovation Project of Xinjiang Uygur Autonomous Regions (Grant No. 2020E01010).

Institutional Review Board Statement: Not applicable.

Informed Consent Statement: Not applicable.

Data Availability Statement: The data of this research in the Upper Vakhsh River Basin was offered by Ministry of Energy and Water Resources of the Republic of Tajikistan and the Agency of Hydrometeorology of the Committee for Environmental Protection under the Government of the Republic of Tajikistan.

Acknowledgments: The authors are thankful to the Ministry of Energy and Water Resources of the Republic of Tajikistan and the Agency of Hydrometeorology of the Committee for Environmental Protection under the Government of the Republic of Tajikistan for providing the data for this research.

Conflicts of Interest: The authors declare no conflict of interest.

References

1. Chiphang, N.; Bandyopadhyay, A.; Bhadra, A. Assessing the Effects of Snowmelt Dynamics on Streamflow and Water Balance Components in an Eastern Himalayan River Basin Using SWAT Model. *Environ. Model. Assess.* **2020**, *25*, 861–883. [CrossRef]
2. Dhami, B.; Himanshu, S.K.; Pandey, A.; Gautam, A.K. Evaluation of the SWAT model for water balance study of a mountainous snowfed river basin of Nepal. *Environ. Earth Sci.* **2018**, *77*, 21. [CrossRef]
3. Schuol, J.; Abbaspour, K. Using monthly weather statistics to generate daily data in a SWAT model application to West Africa. *Ecol. Model.* **2007**, *201*, 301–311. [CrossRef]
4. Liu, Z.J.; Weller, D.E.; Jordan, T.E.; Correll, D.L.; Boomer, K.B. Integrated Modular Modeling of Water and Nutrients From Point and Nonpoint Sources in the Patuxent River Watershed 1. *JAWRA J. Am. Water Resour. Assoc.* **2008**, *44*, 700–723. [CrossRef]
5. Abbott, M.; Bathurst, J.; Cunge, J.; O'connell, P.; Rasmussen, J. An introduction to the European Hydrological System—Systeme Hydrologique Europeen,"SHE", 2: Structure of a physically-based, distributed modelling system. *J. Hydrol.* **1986**, *87*, 61–77. [CrossRef]
6. Arnold, J.G.; Srinivasan, R.; Muttiah, R.S.; Williams, J.R. Large area hydrologic modeling and assessment part I: Model development 1. *JAWRA J. Am. Water Resour. Assoc.* **1998**, *34*, 73–89. [CrossRef]
7. Holtan, H.N.; Lopez, N.C. *USDAHL-70 Model of Watershed Hydrology*; US Department of Agriculture, U.S. Government Printing Office: Washington, DC, USA, 1971.
8. Fontaine, T.; Cruickshank, T.; Arnold, J.; Hotchkiss, R. Development of a snowfall–snowmelt routine for mountainous terrain for the soil water assessment tool (SWAT). *J. Hydrol.* **2002**, *262*, 209–223. [CrossRef]
9. Abbaspour, K. *User Manual for SWAT-CUP, SWAT Calibration and Uncertainty Analysis Programs*; Swiss Federal Institute of Aquatic Science and Technology: Eawag, Switzerland, 2007.
10. Debele, B.; Srinivasan, R.; Gosain, A. Comparison of process-based and temperature-index snowmelt modeling in SWAT. *Water Resour. Manag.* **2010**, *24*, 1065–1088. [CrossRef]
11. Pradhanang, S.M.; Anandhi, A.; Mukundan, R.; Zion, M.S.; Pierson, D.C.; Schneiderman, E.M.; Matonse, A.; Frei, A. Application of SWAT model to assess snowpack development and streamflow in the Cannonsville watershed, New York, USA. *Hydrol. Process.* **2011**, *25*, 3268–3277. [CrossRef]
12. Troin, M.; Caya, D. Evaluating the SWAT's snow hydrology over a Northern Quebec watershed. *Hydrol. Process.* **2014**, *28*, 1858–1873. [CrossRef]
13. Zhang, L.; Jin, X.; He, C.; Zhang, B.; Zhang, X.; Li, J.; Zhao, C.; Tian, J.; DeMarchi, C. Comparison of SWAT and DLBRM for hydrological modeling of a mountainous watershed in arid northwest China. *J. Hydrol. Eng.* **2016**, *21*, 04016007. [CrossRef]
14. Omani, N.; Srinivasan, R.; Karthikeyan, R.; Reddy, V.; Smith, P.K. Impacts of climate change on the glacier melt runoff from five river basins. *Trans. ASABE* **2016**, *59*, 829–848.
15. Dile, Y.T.; Srinivasan, R. Evaluation of CFSR climate data for hydrologic prediction in data-scarce watersheds: An application in the Blue Nile River Basin. *JAWRA J. Am. Water Resour. Assoc.* **2014**, *50*, 1226–1241. [CrossRef]
16. Tolera, M.B.; Chung, I.-M.; Chang, S.W. Evaluation of the climate forecast system reanalysis weather data for watershed modeling in Upper Awash basin, Ethiopia. *Water* **2018**, *10*, 725. [CrossRef]
17. Adeogun, A.G.; Sule, B.F.; Salami, A.W.; Daramola, M.O. Validation of SWAT model for prediction of water yield and water balance: Case study of upstream catchment of Jebba dam in Nigeria. *Int. J. Math. Comput. Phys. Comput. Eng.* **2014**, *8*, 264–270.
18. Gupta, A.; Himanshu, S.K.; Gupta, S.; Singh, R. Evaluation of the SWAT Model for Analysing the Water Balance Components for the Upper Sabarmati Basin. In *Advances in Water Resources Engineering and Management*; Springer: Singapore, 2020; pp. 141–151.
19. Goswami, S.B.; Kar, S.C. Simulation of water cycle components in the Narmada River basin by forcing SWAT model with CFSR data. *Meteorol. Hydrol. Water Manag. Res. Oper. Appl.* **2018**, *13–25*, 6. [CrossRef]
20. Himanshu, S.K.; Pandey, A.; Shrestha, P. Application of SWAT in an Indian river basin for modeling runoff, sediment and water balance. *Environ. Earth Sci.* **2017**, *76*, 3. [CrossRef]

21. Nasiri, S.; Ansari, H.; Ziaei, A.N. Simulation of water balance equation components using SWAT model in Samalqan Watershed (Iran). *Arab. J. Geosci.* **2020**, *13*, 421. [CrossRef]
22. Yin, Z.; Feng, Q.; Zou, S.; Yang, L. Assessing variation in water balance components in mountainous inland river basin experiencing climate change. *Water* **2016**, *8*, 472. [CrossRef]
23. Pritchard, H.D. Asia's glaciers are a regionally important buffer against drought. *Nature* **2017**, *545*, 169–174. [CrossRef]
24. Siderius, C.; Biemans, H.; Wiltshire, A.; Rao, S.; Franssen, W.; Kumar, P.; Gosain, A.; Van Vliet, M.; Collins, D. Snowmelt contributions to discharge of the Ganges. *Sci. Total Environ.* **2013**, *468*, S93–S101. [CrossRef] [PubMed]
25. Duan, Y.; Liu, T.; Meng, F.; Luo, M.; Frankl, A.; De Maeyer, P.; Bao, A.; Kurban, A.; Feng, X. Inclusion of modified snow melting and flood processes in the swat model. *Water* **2018**, *10*, 1715. [CrossRef]
26. Mohammed, I.N.; Bolten, J.D.; Srinivasan, R.; Lakshmi, V. Satellite observations and modeling to understand the Lower Mekong River Basin streamflow variability. *J. Hydrol.* **2018**, *564*, 559–573. [CrossRef] [PubMed]
27. Tan, M.L.; Samat, N.; Chan, N.W.; Roy, R. Hydro-meteorological assessment of three GPM satellite precipitation products in the Kelantan River Basin, Malaysia. *Remote Sens.* **2018**, *10*, 1011. [CrossRef]
28. Li, D.; Christakos, G.; Ding, X.; Wu, J. Adequacy of TRMM satellite rainfall data in driving the SWAT modeling of Tiaoxi catchment (Taihu lake basin, China). *J. Hydrol.* **2018**, *556*, 1139–1152. [CrossRef]
29. Malsy, M.; aus der Beek, T.; Flörke, M. Evaluation of large-scale precipitation data sets for water resources modelling in Central Asia. *Environ. Earth Sci.* **2015**, *73*, 787–799. [CrossRef]
30. Hajihosseini, H.; Hajihosseini, M.; Morid, S.; Delavar, M.; Booij, M.J. Hydrological assessment of the 1973 treaty on the transboundary Helmand River, using the SWAT model and a global climate database. *Water Resour. Manag.* **2016**, *30*, 4681–4694. [CrossRef]
31. Luo, M.; Meng, F.; Liu, T.; Duan, Y.; Frankl, A.; Kurban, A.; De Maeyer, P. Multi–model ensemble approaches to assessment of effects of local Climate Change on water resources of the Hotan River Basin in Xinjiang, China. *Water* **2017**, *9*, 584. [CrossRef]
32. Liu, J.; Shanguan, D.; Liu, S.; Ding, Y. Evaluation and hydrological simulation of CMADS and CFSR reanalysis datasets in the Qinghai-Tibet Plateau. *Water* **2018**, *10*, 513. [CrossRef]
33. Luo, X.; Fan, X.; Ji, X.; Li, Y. Evaluation of corrected APHRODITE estimates for hydrological simulation in the Yarlung Tsangpo–Brahmaputra River Basin. *Int. J. Climatol.* **2019**, *40*, 4158–4170. [CrossRef]
34. Tan, M.L.; Gassman, P.W.; Cracknell, A.P. Assessment of three long-term gridded climate products for hydro-climatic simulations in tropical river basins. *Water* **2017**, *9*, 229. [CrossRef]
35. Xu, H.; Xu, C.-Y.; Chen, S.; Chen, H. Similarity and difference of global reanalysis datasets (WFD and APHRODITE) in driving lumped and distributed hydrological models in a humid region of China. *J. Hydrol.* **2016**, *542*, 343–356. [CrossRef]
36. Singh, L.; Saravanan, S. Evaluation of various spatial rainfall datasets for streamflow simulation using SWAT model of Wunna basin, India. *Int. J. River Basin Manag.* **2020**, *20*, 1–25. [CrossRef]
37. Shen, Y.J.; Shen, Y.; Fink, M.; Kralisch, S.; Brenning, A. Unraveling the Hydrology of the glacierized Kaidu Basin by integrating multisource data in the Tianshan Mountains, Northwestern China. *Water Resour. Res.* **2018**, *54*, 557–580. [CrossRef]
38. Yatagai, A.; Krishnamurti, T.; Kumar, V.; Mishra, A.; Simon, A. Use of APHRODITE rain gauge–based precipitation and TRMM 3B43 products for improving Asian monsoon seasonal precipitation forecasts by the superensemble method. *J. Clim.* **2014**, *27*, 1062–1069. [CrossRef]
39. Li, L.; Ngongondo, C.S.; Xu, C.-Y.; Gong, L. Comparison of the global TRMM and WFD precipitation datasets in driving a large-scale hydrological model in southern Africa. *Hydrol. Res.* **2013**, *44*, 770–788. [CrossRef]
40. Gulakhmadov, A.; Chen, X.; Gulahmadov, N.; Liu, T.; Davlyatov, R.; Sharofiddinov, S.; Gulakhmadov, M. Long-Term Hydro–Climatic Trends in the Mountainous Kofarnihon River Basin in Central Asia. *Water* **2020**, *12*, 2140. [CrossRef]
41. Vaghefi, S.A.; Abbaspour, N.; Kamali, B.; Abbaspour, K.C. A toolkit for climate change analysis and pattern recognition for extreme weather conditions–Case study: California-Baja California Peninsula. *Environ. Model. Softw.* **2017**, *96*, 181–198. [CrossRef]
42. Touseef, M.; Chen, L.; Yang, K.; Chen, Y. Long-Term Rainfall Trends and Future Projections over Xijiang River Basin, China. *Adv. Meteorol.* **2020**, *2020*, 6852148. [CrossRef]
43. Yatagai, A.; Kamiguchi, K.; Arakawa, O.; Hamada, A.; Yasutomi, N.; Kitoh, A. APHRODITE: Constructing a long-term daily gridded precipitation dataset for Asia based on a dense network of rain gauges. *Bull. Am. Meteorol. Soc.* **2012**, *93*, 1401–1415. [CrossRef]
44. Palazzi, E.; Von Hardenberg, J.; Provenzale, A. Precipitation in the Hindu-Kush Karakoram Himalaya: Observations and future scenarios. *J. Geophys. Res. Atmos.* **2013**, *118*, 85–100. [CrossRef]
45. Fuka, D.R.; Walter, M.T.; MacAlister, C.; Degaetano, A.T.; Steenhuis, T.S.; Easton, Z.M. Using the Climate Forecast System Reanalysis as weather input data for watershed models. *Hydrol. Process.* **2014**, *28*, 5613–5623. [CrossRef]
46. Cuceloglu, G.; Ozturk, I. Assessing the Impact of CFSR and Local Climate Datasets on Hydrological Modeling Performance in the Mountainous Black Sea Catchment. *Water* **2019**, *11*, 2277. [CrossRef]
47. Worqlul, A.W.; Yen, H.; Collick, A.S.; Tilahun, S.A.; Langan, S.; Steenhuis, T.S. Evaluation of CFSR, TMPA 3B42 and ground-based rainfall data as input for hydrological models, in data-scarce regions: The upper Blue Nile Basin, Ethiopia. *Catena* **2017**, *152*, 242–251. [CrossRef]
48. Jarvis, A.; Reuter, H.I.; Nelson, A.; Guevara, E. Hole-Filled SRTM for the Globe Version 4. 2008. Available online: http://srtm.csi.cgiar.org (accessed on 16 July 2020).

49. Food and Agriculture Organization; International Institute for Applied Systems Analysis; International Soil Reference and Information Centre. *JRC: Harmonized World Soil Database (Version 1.2)*; FAO: Rome, Italy; IIASA: Laxenburg, Austria, 2012.
50. Arino, O.; Ramos Perez, J.J.; Kalogirou, V.; Bontemps, S.; Defourny, P.; Van Bogaert, E. *Global Land CoverMap for 2009 (GlobCover 2009)*; European Space Agency: Paris, France; University College London: London, UK, 2012.
51. Yacoub, C.; Foguet, A.P. Slope effects on SWAT modeling in a mountainous basin. *J. Hydrol. Eng.* **2013**, *18*, 1663–1673. [CrossRef]
52. Neitsch, S.L.; Arnold, J.G.; Kiniry, J.R.; Williams, J.R. *Soil and Water Assessment Tool Theoretical Documentation Version 2009*; Texas Water Resources Institute: Temple, TX, USA, 2011.
53. Gassman, P.W.; Reyes, M.R.; Green, C.H.; Arnold, J.G. The soil and water assessment tool: Historical development, applications, and future research directions. *Trans. ASABE* **2007**, *50*, 1211–1250. [CrossRef]
54. Cronshey, R. *Urban Hydrology for Small Watersheds*; US Department of Agriculture, Soil Conservation Service, Engineering Division: Washington, DC, USA, 1986.
55. Hydrology, S. *National Engineering Handbook*; US Soil Conservation Service: Washington, DC, USA, 1972.
56. Hock, R. Temperature index melt modelling in mountain areas. *J. Hydrol.* **2003**, *282*, 104–115. [CrossRef]
57. Rahman, K.; Maringanti, C.; Beniston, M.; Widmer, F.; Abbaspour, K.; Lehmann, A. Streamflow modeling in a highly managed mountainous glacier watershed using SWAT: The Upper Rhone River watershed case in Switzerland. *Water Resour. Manag.* **2013**, *27*, 323–339. [CrossRef]
58. Ruan, H.; Zou, S.; Yang, D.; Wang, Y.; Yin, Z.; Lu, Z.; Li, F.; Xu, B. Runoff simulation by SWAT model using high-resolution gridded precipitation in the upper Heihe River Basin, Northeastern Tibetan Plateau. *Water* **2017**, *9*, 866. [CrossRef]
59. Omani, N.; Srinivasan, R.; Karthikeyan, R.; Smith, P.K. Hydrological modeling of highly glacierized basins (Andes, Alps, and Central Asia). *Water* **2017**, *9*, 111. [CrossRef]
60. Garee, K.; Chen, X.; Bao, A.; Wang, Y.; Meng, F. Hydrological modeling of the upper indus basin: A case study from a high-altitude glacierized catchment Hunza. *Water* **2017**, *9*, 17. [CrossRef]
61. Gulakhmadov, A.; Chen, X.; Gulahmadov, N.; Liu, T.; Anjum, M.N.; Rizwan, M. Simulation of the Potential Impacts of Projected Climate Change on Streamflow in the Vakhsh River Basin in Central Asia under CMIP5 RCP Scenarios. *Water* **2020**, *12*, 1426. [CrossRef]
62. Yang, J.; Reichert, P.; Abbaspour, K.C.; Xia, J.; Yang, H. Comparing uncertainty analysis techniques for a SWAT application to the Chaohe Basin in China. *J. Hydrol.* **2008**, *358*, 1–23. [CrossRef]
63. Xu, C.Y.; Chen, D. Comparison of seven models for estimation of evapotranspiration and groundwater recharge using lysimeter measurement data in Germany. *Hydrol. Process. Int. J.* **2005**, *19*, 3717–3734. [CrossRef]
64. Earls, J.; Dixon, B. A comparison of SWAT model-predicted potential evapotranspiration using real and modeled meteorological data. *Vadose Zone J.* **2008**, *7*, 570–580. [CrossRef]
65. Priestley, C.H.B.; Taylor, R. On the assessment of surface heat flux and evaporation using large-scale parameters. *Mon. Weather Rev.* **1972**, *100*, 81–92. [CrossRef]
66. Hargreaves, G.H.; Samani, Z.A. Reference crop evapotranspiration from temperature. *Appl. Eng. Agric.* **1985**, *1*, 96–99. [CrossRef]
67. Penman, H. Evaporation: An introductory survey. *Neth. J. Agric. Sci.* **1956**, *4*, 9–29. [CrossRef]
68. Monteith, J. The state and movement of water in living organisms. In Proceedings of the 19th Symposia of the Society for Experimental Biology; Cambridge University Press: London, UK, 1965; pp. 205–234.
69. Li, Z.; Yang, Y.; Kan, G.; Hong, Y. Study on the applicability of the Hargreaves potential evapotranspiration estimation method in CREST distributed hydrological model (version 3.0) applications. *Water* **2018**, *10*, 1882. [CrossRef]
70. Gupta, H.V.; Kling, H.; Yilmaz, K.K.; Martinez, G.F. Decomposition of the mean squared error and NSE performance criteria: Implications for improving hydrological modelling. *J. Hydrol.* **2009**, *377*, 80–91. [CrossRef]
71. Moriasi, D.N.; Arnold, J.G.; Van Liew, M.W.; Bingner, R.L.; Harmel, R.D.; Veith, T.L. Model evaluation guidelines for systematic quantification of accuracy in watershed simulations. *Trans. ASABE* **2007**, *50*, 885–900. [CrossRef]
72. KGE Package 'hydroGOF'. 2020. Available online: https://cran.r-project.org/web/packages/hydroGOF/hydroGOF.pdf (accessed on 9 September 2020).
73. Arnold, J.G.; Moriasi, D.N.; Gassman, P.W.; Abbaspour, K.C.; White, M.J.; Srinivasan, R.; Santhi, C.; Harmel, R.; Van Griensven, A.; Van Liew, M.W. SWAT: Model use, calibration, and validation. *Trans. ASABE* **2012**, *55*, 1491–1508. [CrossRef]
74. Abbaspour, K. *SWAT-CUP 2012: SWAT Calibration and Uncertainty Programs-A User Manual, 2012*; Eawag–Swiss Federal Institute of Aquatic Science and Technology: Dübendorf, Switzerland, 2015.
75. Ahmed, E.; Al Janabi, F.; Zhang, J.; Yang, W.; Saddique, N.; Krebs, P. Hydrologic Assessment of TRMM and GPM-Based Precipitation Products in Transboundary River Catchment (Chenab River, Pakistan). *Water* **2020**, *12*, 1902. [CrossRef]
76. Salles, L.; Satgé, F.; Roig, H.; Almeida, T.; Olivetti, D.; Ferreira, W. Seasonal effect on spatial and temporal consistency of the new GPM-based IMERG-v5 and GSMaP-v7 satellite precipitation estimates in Brazil's central plateau region. *Water* **2019**, *11*, 668. [CrossRef]
77. Abbaspour, K.C.; Johnson, C.; Van Genuchten, M.T. Estimating uncertain flow and transport parameters using a sequential uncertainty fitting procedure. *Vadose Zone J.* **2004**, *3*, 1340–1352. [CrossRef]
78. Schuol, J.; Abbaspour, K.C.; Srinivasan, R.; Yang, H. Estimation of freshwater availability in the West African sub-continent using the SWAT hydrologic model. *J. Hydrol.* **2008**, *352*, 30–49. [CrossRef]

79. Anand, J.; Gosain, A.; Khosa, R.; Srinivasan, R. Regional scale hydrologic modeling for prediction of water balance, analysis of trends in streamflow and variations in streamflow: The case study of the Ganga River basin. *J. Hydrol. Reg. Stud.* **2018**, *16*, 32–53. [CrossRef]
80. Fletcher, T.D.; Andrieu, H.; Hamel, P. Understanding, management and modelling of urban hydrology and its consequences for receiving waters: A state of the art. *Adv. Water Resour.* **2013**, *51*, 261–279. [CrossRef]
81. Hu, Z.; Hu, Q.; Zhang, C.; Chen, X.; Li, Q. Evaluation of reanalysis, spatially interpolated and satellite remotely sensed precipitation data sets in central Asia. *J. Geophys. Res. Atmos.* **2016**, *121*, 5648–5663. [CrossRef]
82. Arsenault, R.; Essou, G.R.; Brissette, F.P. Improving hydrological model simulations with combined multi-input and multimodel averaging frameworks. *J. Hydrol. Eng.* **2017**, *22*, 04016066. [CrossRef]
83. Duan, W.; Chen, Y.; Zou, S.; Nover, D. Managing the water-climate-food nexus for sustainable development in Turkmenistan. *J. Clean. Prod.* **2019**, *220*, 212–224. [CrossRef]
84. Duan, W.; Takara, K. *Impacts of Climate and Human Activities on Water Resources and Quality*; Springer Nature: Berlin/Heidelberg, Germany, 2020.
85. Schuol, J.; Abbaspour, K.C.; Yang, H.; Srinivasan, R.; Zehnder, A.J. Modeling blue and green water availability in Africa. *Water Resour. Res.* **2008**, *44*. [CrossRef]
86. Deng, H.; Chen, Y. Influences of recent climate change and human activities on water storage variations in Central Asia. *J. Hydrol.* **2017**, *544*, 46–57. [CrossRef]
87. Li, Z.; Chen, Y.; Fang, G.; Li, Y. Multivariate assessment and attribution of droughts in Central Asia. *Sci. Rep.* **2017**, *7*, 1–12. [CrossRef]
88. Eini, M.R.; Javadi, S.; Delavar, M.; Monteiro, J.A.; Darand, M. High accuracy of precipitation reanalyses resulted in good river discharge simulations in a semi-arid basin. *Ecol. Eng.* **2019**, *131*, 107–119. [CrossRef]
89. Grusson, Y.; Anctil, F.; Sauvage, S.; Sánchez Pérez, J.M. Testing the SWAT model with gridded weather data of different spatial resolutions. *Water* **2017**, *9*, 54. [CrossRef]
90. Pathak, S.; Ojha, C.; Shukla, A.; Garg, R. Assessment of Annual Water-Balance Models for Diverse Indian Watersheds. *J. Sustain. Water Built Environ.* **2019**, *5*, 04019002. [CrossRef]
91. Thapa, B.R.; Ishidaira, H.; Pandey, V.P.; Shakya, N.M. A multi-model approach for analyzing water balance dynamics in Kathmandu Valley, Nepal. *J. Hydrol. Reg. Stud.* **2017**, *9*, 149–162. [CrossRef]
92. Leta, O.T.; El-Kadi, A.I.; Dulai, H.; Ghazal, K.A. Assessment of climate change impacts on water balance components of Heeia watershed in Hawaii. *J. Hydrol. Reg. Stud.* **2016**, *8*, 182–197. [CrossRef]
93. Budhathoki, A.; Babel, M.S.; Shrestha, S.; Meon, G.; Kamalamma, A.G. Climate change impact on water balance and hydrological extremes in different physiographic regions of the West Seti River Basin, Nepal. *Ecohydrol. Hydrobiol.* **2020**, *21*, 79–95. [CrossRef]
94. Ferguson, R. Snowmelt runoff models. *Prog. Phys. Geogr.* **1999**, *23*, 205–227. [CrossRef]
95. Harrison, G.P.; Whittington, H. Vulnerability of hydropower projects to climate change. *IEE Proc. Gener. Transm. Distrib.* **2002**, *149*, 249–255. [CrossRef]
96. Alekseevskii, N.; Zavadskii, A.; Krivushin, M.; Chalov, S. Hydrological monitoring at international rivers and basins. *Water Resour.* **2015**, *42*, 747–757. [CrossRef]
97. Ta, Z.; Yu, Y.; Sun, L.; Chen, X.; Mu, G.; Yu, R. Assessment of Precipitation Simulations in Central Asia by CMIP5 Climate Models. *Water* **2018**, *10*, 1516. [CrossRef]
98. Yang, D.; Gao, B.; Jiao, Y.; Lei, H.; Zhang, Y.; Yang, H.; Cong, Z. A distributed scheme developed for eco-hydrological modeling in the upper Heihe River. *Sci. China Earth Sci.* **2015**, *58*, 36–45. [CrossRef]
99. Shaw, E.M.; Beven, K.J.; Chappell, N.A.; Lamb, R. *Hydrology in Practice*; CRC Press: Boca Raton, FL, USA, 2010.
100. Harris, I.; Jones, P.D.; Osborn, T.J.; Lister, D.H. Updated high-resolution grids of monthly climatic observations–the CRU TS3. 10 Dataset. *Int. J. Climatol.* **2014**, *34*, 623–642. [CrossRef]
101. Sorooshian, S.; AghaKouchak, A.; Arkin, P.; Eylander, J.; Foufoula-Georgiou, E.; Harmon, R.; Hendrickx, J.M.; Imam, B.; Kuligowski, R.; Skahill, B. Advanced concepts on remote sensing of precipitation at multiple scales. *Bull. Am. Meteorol. Soc.* **2011**, *92*, 1353–1357. [CrossRef]
102. Conti, F.L.; Hsu, K.-L.; Noto, L.V.; Sorooshian, S. Evaluation and comparison of satellite precipitation estimates with reference to a local area in the Mediterranean Sea. *Atmos. Res.* **2014**, *138*, 189–201. [CrossRef]
103. Santhi, C., Arnold, J.G.; Williams, J.R.; Dugas, W.A.; Srinivasan, R.; Hauck, L.M. Validation of the swat model on a large rwer basin with point and nonpoint sources 1. *JAWRA J. Am. Water Resour. Assoc.* **2001**, *37*, 1169–1188. [CrossRef]

atmosphere

MDPI

Article

Modifications to Snow-Melting and Flooding Processes in the Hydrological Model—A Case Study in Issyk-Kul, Kyrgyzstan

Solange Uwamahoro [1,2,3], Tie Liu [1,2,3,4,5,*], Vincent Nzabarinda [1,2,3], Jules Maurice Habumugisha [3,6], Theogene Habumugisha [3,7], Barthelemy Harerimana [3,6] and Anming Bao [1,2,3,4,5]

1 State Key Laboratory of Desert and Oasis Ecology, Xinjiang Institute of Ecology and Geography, Chinese Academy of Sciences, Urumqi 830011, China; uwamahoroso@yahoo.com (S.U.); vincentnzabarinda@mails.ucas.ac.cn (V.N.); baoam@ms.xjb.ac.cn (A.B.)
2 Key Laboratory of GIS & RS Application Xinjiang Uygur Autonomous Region, Urumqi 830011, China
3 University of Chinese Academy of Sciences, Beijing 100049, China; hjulesmaurice@yahoo.com (J.M.H.); habumugisha@iue.ac.cn (T.H.); bthelemy@gmail.com (B.H.)
4 China-Pakistan Joint Research Center on Earth Sciences, CAS-HEC, Islamabad 45320, Pakistan
5 Research Center for Ecology and Environment of Central Asia, Chinese Academy of Sciences, Urumqi 830011, China
6 Key Laboratory for Mountain Hazards and Earth Surface Process, Institute of Mountain Hazards and Environment, Chinese Academy of Sciences (CAS), Chengdu 610041, China
7 Institute of Urban Environmental, Chinese Academy of Sciences, 1799 Jimei road, Xiamen 361021, China
* Correspondence: liutie@ms.xjb.ac.cn

Abstract: Streamflow impacts water supply and flood protection. Snowmelt floods occur frequently, especially in mountainous areas, and they pose serious threats to natural and socioeconomic systems. The current forecasting method relies on basic snowmelt accumulation and has geographic limitations that restrict the accuracy and timeliness of flood simulation and prediction. In this study, we clarified the precipitation types in two selected catchments by verifying accumulated and maximum temperatures' influences on snow melting using a separation algorithm of rain and snow that incorporates with the temperatures. The new snow-melting process utilizing the algorithm in the soil and water assessment tool model (SWAT) was also developed by considering the temperatures. The SWAT model was used to simulate flooding and snowmelt in the catchments. We found that the contributions of snowmelt to the river flow were approximately 6% and 7% higher, according to our model compared to the original model, for catchments A and B, respectively. After the model improvement, the flood peaks increased by 49.42% and 43.87% in A and B, respectively. The contributions of snowmelt to stream flow increased by 24.26% and 31% for A and B, respectively. Generally, the modifications improved the model accuracy, the accuracy of snowmelt's contributions to runoff, the accuracy of predicting flood peaks, the time precision, and the flood frequency simulations.

Keywords: Issyk-Kul; SWAT; accumulated temperature; snowmelt

Citation: Uwamahoro, S.; Liu, T.; Nzabarinda, V.; Habumugisha, J.M.; Habumugisha, T.; Harerimana, B.; Bao, A. Modifications to Snow-Melting and Flooding Processes in the Hydrological Model—A Case Study in Issyk-Kul, Kyrgyzstan. *Atmosphere* **2021**, *12*, 1580. https://doi.org/10.3390/atmos12121580

Academic Editors: Ognjen Bonacci and Jimy Dudhia

Received: 13 August 2021
Accepted: 26 November 2021
Published: 27 November 2021

1. Introduction

Water resources are essential for society's long-term development, economic growth, and ecological environment [1–4]. Large amounts of water are stored as snow and glaciers [5,6], and this water can be discharged into catchments [7–9]. Approximately one-sixth of the world's population lives near rivers that originate from snowmelt [10], which can occur in mountainous areas, even in otherwise arid regions. The melt water is used for agricultural, industrial, and municipal purposes [11]; however, its availability may alternate the water levels of lakes and cause floods that can pose serious threats to natural and socioeconomic systems. According to recent statistics, mountain torrents were responsible for approximately 70% of flood-related deaths [12,13], and associated disaster losses accounted for more than 50% of the total deaths [14,15]. Frequent floods may pose a significant threat to certain populations, especially those in high-altitude areas, such as Kyrgyzstan. Indeed, Kyrgyzstan is among the countries affected by such floods.

Kyrgyzstan is a Central Asian country with abundant underground and surface water resources. Changes in the runoff and distribution of its sources, mainly rainfall, snowmelt water, glaciers, and other tributaries, affect the water in Issyk-Kul, Kyrgyzstan's large endorheic lake [16]. The Issyk-Kul basin contains a number of streams. About 123 of them are used for irrigation [17,18]. Water supply and flood protection are both impacted by streamflow [19,20]. Furthermore, Issyk-Kul is fed by rivers, the majority of whose water comes from snow and glaciers, which cover about 509 km^2 of the drainage basin and are located at elevations of 3000 m a.s.l. and higher [21–23]. Precipitation increases from May to August, corresponding with a seasonal increase in agricultural water demand [16]. During the summer (June to September), rivers primarily fed by glacial meltwater experience significant increases in runoff, resulting in floods during the "flood season" [5]. There are four types of rivers. The majority of glacial-snow-type rivers flow into Issyk-Kul [24]. River runoff's shifting properties have long been focuses of hydrological and water resource research [25]. Since Issyk-Kul's rivers affect its water level, it is important to focus on the sources of the rivers, especially the snowmelt.

Snowmelt runoff is a significant source of water and a substantial driving force for catastrophic flooding in inland desert regions during the spring flood season, so it must be carefully analyzed. To mitigate or minimize the tragedies and losses caused by floods, it is vital to analyze the complete snow melting and flood process [26]. Many models have been developed, including empirical models, conceptual models, physical models, and distributed hydrological models with snowmelt modules [27–33]. Water balance is calculated by analyzing evaporation during strong winds, low relative humidity, and low temperatures [34]. More advanced models [35,36] were shown to be flawed when simulating runoff in snowmelt watersheds [37]. The soil water assessment tool (SWAT) [38–40] is a distributed watershed hydrological model developed by the US Department of Agriculture and relies on the Simulator for Water Resources in Rural Basins (SWRRB) model [41–44]. It has good precision when dealing with plains with abundant precipitation and flat terrain, but it has relatively lower accuracy when dealing with mountainous areas with complex terrains [37,45]. The model uses few parameters and little input data. For snow watersheds, the model is used to evaluate dispersed snowmelt and runoff formation [46]. The degree-day factor approach is used in the SWAT model to calculate snowmelt [35,47,48]. In this approach, snowmelt is considered to occur when the average temperature on a given day exceeds the snow-melting temperature threshold. However, the condition for snowmelt is only the daily average temperature, thereby ignoring the influence of cumulative temperature on energy accumulation [49]. Previous researchers, such as Meng et al., Yu et al., and Luo et al., attempted to use the SWAT model with the snowmelt module, but there was confusion about how to distinguish precipitation types. Other studies failed to account for the conditions of mountainous regions, expended immense amounts of effort, or relied on the standard degree-day factor system (which provides limited simulation accuracy) [36,37,50]. Improvements to precipitation type recognition have been ignored in related attempts to enhance snow-related models which omit the standard degree-day factor technique. Therefore, based on recent analysis, the determination of precipitation type was used in this study, and the classic degree-day factor approach was updated. The goal of this research was to differentiate rain from snow in total precipitation by adding accumulated temperature to the temperature condition. Thereby, the accuracy of the original model was improved to raise its ability to determine precipitation type. The temperature condition of the traditional degree-day factor method was improved by adding an accumulated temperature judgment condition and modifying the parameter set. This improved the simulation accuracy of the model [9,37] and improved the calculations of snowmelt capacity and snowmelt's contributions to runoff in the selected catchments around Issyk-Kul.

2. Study Area and Materials

2.1. Study Area

The Issyk-Kul basin is located on the northern slopes of the Tian-Shan Mountains in Kyrgyzstan, which is part of arid Central Asia, between 42°250 N and 77°150 E (Figure 1), at an altitude of 1606 m above sea level (a.s.l.) [51]. Issyk-Kul is the continent's fourth-deepest reservoir. It is an endogenous mountain lake (one of the world's highest saline lakes). It is Central Asia's largest high-altitude lake [52]. It is surrounded by high mountains, such as Teskey Ala-Too, a mountain range to the south with peaks exceeding 4808 m, and Kungey Ala-Too, a mountain range to the north with peaks exceeding 4648 m. Its watershed covers an area of 22,080 km^2, and all major branches of Issyk-Kul originate from 834 glaciers with a volume of 48 km^3 and a total glacial area of 650 km^2 [53,54].

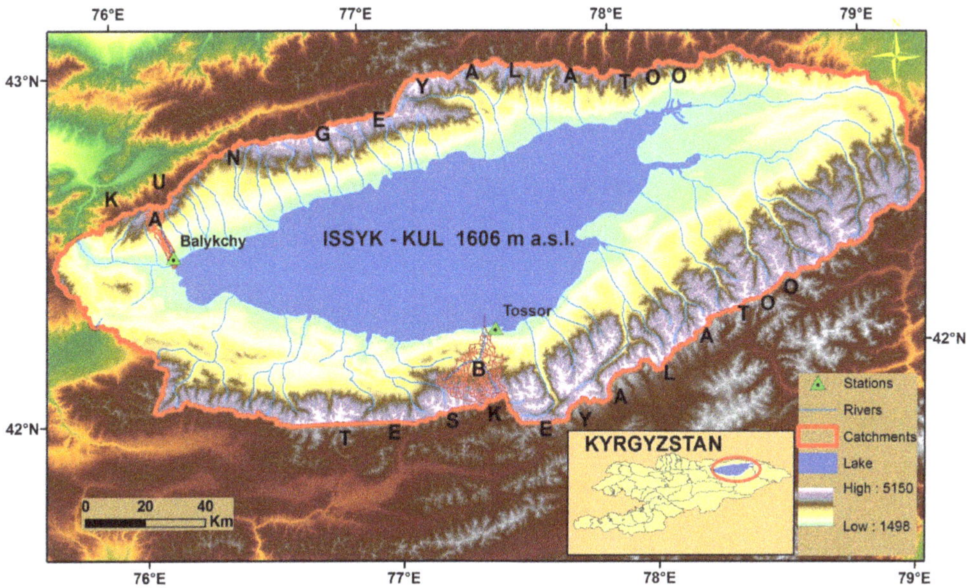

Figure 1. Locations of the selected catchments around the Issyk-Kul basin and digital elevation levels.

One of Kyrgyzstan's most densely populated areas is around Issyk-Kul. With an average annual growth rate of 1.84%, the population of its Oblast increased from 177,300 in 1940 to 448,000 in 2012 [55]. The moderately warm climate of the Issyk-Kul basin is ideal for cereals, crops, and gardening [16,56]. The average temperature in the basin is 19–20 °C in July and 2–3 °C in January, and precipitation ranges from 12.3–35 mm per year [16]. Intensive agriculture has developed throughout the lake basin due to the basin's unusually mild climate [57–59]. Therefore, we selected two catchments in the Issyk-Kul basin, one in the northwest (A) and the other in the south (B), with areas of 2153.46 km^2 and 2254.32 km^2, respectively. The aim of this study was to improve the modelling processes of the selected catchments by incorporating accumulated temperature, allowing for better differentiation of rain and snow in overall precipitation.

2.2. Data and Source

The hydrological model (SWAT) was built with the help of digital elevation models (DEMs), land cover, soil classification, meteorological data, and precipitation type event statistics. A DEM with a resolution of 30 m was downloaded from the Shuttle Radar Topography Mission (SRTM); (http://srtm.csi.cgiar.org/ (accessed on 22 January 2020)). The visual interpretation of the Landsat 8 with 30-meter resolution remote sensing imagery extracted from (https://www.usgs.gov/products/data-and-tools/real-time-data/remote-land-sensing-and-landsat (accessed on 30 December 2019)) provided land use data. The Food and Agriculture Organization (FAO) and the International Institute for Applied Systems Analysis (IIASA) provided the soil data, which included a 1:100,000 soil type map and information on related soil properties. Local Meteorological stations provided precipitation type event statistics. The stations Balykchy and Tossor also provided daily maximum and minimum temperature data (MAX.T and MIN.T, respectively), along with average daily temperatures and precipitation data. The altitude, solar radiation, relative humidity, wind speed, and wind direction data of catchments A and B were collected from the Meteorological Service of Canada (MSC) https://www.canada.ca/en/environment-climate-change/services/climate-change/canadian-centre-climate-services/display-download.html (accessed on 3 February 2020). Model input data (snow cover) were provided by the MODIS snow product MODA10A2.006, with data for 500 m, 8-day, and from 2015 to 2016. We used the data to determine the corresponding temperature thresholds. The Kyrgyzstan hydrological bureau provided daily discharge data from 2015 to 2016, which were used to validate the model.

3. Methods

The daily accumulated temperature was calculated using the temperature integral method. The accumulated temperature inflection points of the precipitation type and snow melting were validated by remote sensing snow data. The traditional degree-day factor, precipitation type determination, and accumulated and maximum temperatures for snow melting were used in this study.

3.1. *SWAT Model Definition*

The SWAT model is a hydrological, physically-based, and distributed model [60,61]. It uses the runoff curve number approach from the soil conservation service (SCS) to quantify surface runoff and the degree-day factor method to measure snowmelt runoff. The model simulates snowmelt runoff, surface runoff [39,62,63]. The average runoff for the whole watershed is calculated using the hydrological model [37,64]. The water balance equation for the SWAT model is the following:

$$\mathrm{sw_t} = \mathrm{sw_0} + \Sigma_{\mathrm{n=1}}^{\mathrm{t}}(Nday - Qsurf - Ea - Rseep - Qgw) \quad (1)$$

The final soil water content is denoted by $\mathrm{sw_t}$ (mm H_2O), $\mathrm{sw_0}$ is the soil water content on day i (mm H_2O), t is the time (days), $Nday$ is the amount of precipitation on day n (mm H_2O), $Qsurf$ is the amount of surface runoff on day n (mm H_2O), the initial soil water content is $\mathrm{sw_0}$, Ea is the quantity of evapotranspiration on day n (mm H_2O), $Rseep$ is the amount of water from the soil profile that enters the vadose region on day n (mm H_2O), and Qgw is the groundwater recharge (water that is not consumed by evapotranspiration) on day n (mm H_2O).

3.2. *Snow Cover in the SWAT Model*

The mean daily air temperature is used in the SWAT model to classify precipitation as rain or freezing rain/snow. The boundary temperature, $k_{s\ r}$, is set by the user and is used to classify precipitation as rain or snow. The mass balance of the snow pack is as follows:

$$SNO_i = SNO_{i-1} + R_{day_i} - E_{sub_i} - SNO_{mlt_i} \quad (2)$$

where SNO_i and SNO_{i-1} are the water content of the snow pack on the current day (i) and previous day ($i-1$), respectively (mm H_2O), R_{day_i} is the amount of precipitation on a given day (added only if $\overline{\mathrm{k}}_{\mathrm{av}} \leq \mathrm{k}$ (mm H_2O)), E_{sub_i} is the amount of sublimation on a given day (mm H_2O), and SNO_{mlt_i} is the amount of snowmelt on a given day (mm H_2O). The equation for the areal depletion curve is

$$SNO_{cov} = \frac{SNO}{SNO_{100}} \cdot \left(\frac{SNO}{SNO_{100}} + \exp\left(cov_1 - cov_2 \cdot \frac{SNO}{SNO_{100}} \right) \right)^{-1} \tag{3}$$

where SNO_{cov} is the fraction of the HRU area covered by snow, SNO is the water content of the snow pack on the current day (mm H_2O), SNO_{100} is the threshold depth of snow at 100% coverage (mm H_2O), and cov_1 and cov_2 are coefficients that define the shape of the curve.

3.3. The Original Degree-Day Factor Algorithm

The snowmelt in the SWAT model is important, as is the source of water during the alpine spring [65,66]. The snowmelt runoff temperature threshold is obtained from the snow cover state and the temperature threshold of the snowmelt runoff [67]. The melting snow equation is given below.

$$SNO_{mlt} = y_{mlt} \cdot Sno_{cov} \cdot \left[\frac{K_{snow} + K_{mx}}{2} - K_{mlt} \right] \tag{4}$$

The total snowmelt on a given day (mm H_2O) is represented by SNO_{mlt}. y_{mlt} stands for the day's melt factor (mm H_2O/°C day^{-1}). The fraction of the HRU region is represented by Sno_{cov}. K_{snow} represents the temperature of the snow pack. K_{mx} is the highest air temperature on a given day (°C), and K_{mlt} is the base temperature threshold (°C) for snow being able to melt. On a daily basis, the classical degree-day model links ice or snowmelt (mm) to air temperature [68]:

$$M = \begin{cases} r \cdot \left(K_{av} - K_{gmlt} \right), & \text{when } K_{av} > K_{gmlt} \\ 0, & \text{otherwise} \end{cases} \tag{5}$$

where M is the melt rate, r (mm day^{-1} °C) is the degree-day factor for snowmelt, K_{av} (°C) is the average air temperature of a given day, and K_{gmlt} is the snowmelt base temperature in °C. The degree-day factor for snowmelt is calculated in the SWAT model with a sinusoidal function to simulate the seasonal shift pattern [69]. The temperature of the snow pack is:

$$K_{snowp} = K_{snow_{p-1}} \cdot (1 - \alpha_{sno}) + K_{av} \cdot \alpha_{sno} \tag{6}$$

K_{snowp} is the temperature (°C) of the snow pack on a given day. α_{sno} is the snow temperature (°C), which takes into account the previous day's snow pack temperature. K_{av} is the average air temperature (°C) on the current day. α_{sno} is used to represent the influence of linkages among factors affecting snow pack temperature. y_{mlt} is the melt factor, which incorporates the seasonal alterations in maximum and minimum temperature values that happen during dry and wet periods, respectively. The snow/ice melt factor is calculated using sinusoidal interpolation in some snow/ice melt–runoff models [70]. The degree-day factor for snowmelt is expressed as follows:

$$y_{mlt} = \frac{y_{mlt6} + y_{mlt12}}{2} + \frac{y_{mlt6} - y_{mlt12}}{2} \cdot \sin\left(\frac{2\pi}{365} \cdot (p - 81) \right) \tag{7}$$

where y_{mlt6} (mm H_2O day^{-1} °C^{-1}) is the melt factor for 21 June, y_{mlt12} (mm H_2O day^{-1} °C^{-1}) is the melt factor for 21 December, and p denotes the number of days in a year. The model divides any sub-basin into several elevation zones [71].

3.4. Model Modifications: Accumulated Temperature and Differentiation of Snowfall and Rainfall

Previous studies, such as [66–69], revealed that the accumulated temperature can only affect the form of precipitation and the determination of snow melting when the minimum air temperature is above 0 °C. The melt factor is calculated using the following equation:

$$T = (T_{max} - T_{min}) \sin t + T_{min} 0 \le t \le \pi \quad (8)$$

T denotes the temperature at any given time of day; T_{max} and T_{min} are the highest and lowest temperatures ever recorded in a single day, respectively; and t is the value of the arc in a single moment. The equation used to calculate the accumulated temperature when the maximum daily air temperature is greater than 0 °C and the minimum daily air temperature is less than 0 °C is as follows:

$$T = \begin{cases} \int_0^{\pi} (T_{max} - T_{min}) \sin t + T_{min} du, u \in (0, \pi) \\ \int_{\sin^{-1}\left(\frac{-T_{min}}{T_{max}-T_{min}}\right)}^{\pi-\sin^{-1}\left(\frac{-T_{min}}{T_{max}-T_{min}}\right)} T_{max} \sin u du, u \in \left(\sin^{-1}\left(\frac{-T_{min}}{T_{max}-T_{min}}\right)\right), \pi - \sin^{-1}\left(\frac{-T_{min}}{T_{max}-T_{min}}\right) \end{cases} \quad (9)$$

The $\sin^{-1}\left(\frac{-T_{min}}{T_{max}-T_{min}}\right)$ and $\pi - \sin^{-1}\left(\frac{-T_{min}}{T_{max}-T_{min}}\right)$ represent Radian values when the temperature is 0 °C. The determination of precipitation type and snowmelt calculations are primarily based on the maximum temperature on a given day, the average temperature on that same day, and the set threshold value, while neglecting the accumulated temperature as an important factor affecting both the precipitation type and the snowmelt process [72,73].

Our modified method of snow melting is mentioned in segments (for the original model, see Figure 2). The modifications were added to distinguish between snow and rain. The modified model requires spatial and weather data, and the types of precipitation are determined by applying accumulated and maximum temperatures to HRUs. If the temperature requirement for rainfall is reached, the precipitation is classed as rainfall, and the contribution of rainfall to river flow is estimated using processes from the original model. Snow can melt when accumulated and maximum temperatures fulfill the prescribed conditions at the same time. The snow pack is formed when there is no release of snowfall and snowfall accumulates. If the type of precipitation is found to be snowfall, snowfall is added to the snow pack. When snow-melting conditions are satisfied, the amount of snow melting and its contribution to river runoff are both measured at the same time; otherwise, there is no snowmelt. In other words, the degree-day factor has been corrected, as snowmelt conditions are calculated by adding accumulated and maximum temperatures, and the model was also improved by making it ignore non-melt periods.

3.5. Comparison of Snow Pack Area and Temperature for the Selected Catchments

Analysis of the temperatures (maximum/accumulated) and remotely sensed snow packing of A and B catchments in 2015 was conducted (Figure 3). The eight-day range was the time taken for the snow pack measurements. The two arrows in Figure 3 indicate how the snow area's turning points change throughout the year. For catchments A and B, the snow pack areas at the turning points from left to right were 1853.68 and 1200 km^2 and 2053.68 and 2164.76 km^2, respectively. For A, the first arrow was between May and June. The snow pack areas of both catchments began to decrease then. The snow-melting process started at this point denoted by the red arrow while the inflection point, which is the second turning point, started in early November. As the time between the first arrow (on the left) and the second arrow (on the right) contained temperatures that fit the conditions for rainfall, the snow pack obviously decreased, and the precipitation took the form of rain. When there was precipitation on a given day, the measured temperature conditions were compared to the values we were given. The daily average values of temperatures (maximum/accumulated) were used to determine the temperature conditions for snowfall during this time. Since the snow pack area increased as time approached the right arrow, for both catchments, snowfall clearly occurred (A and B).

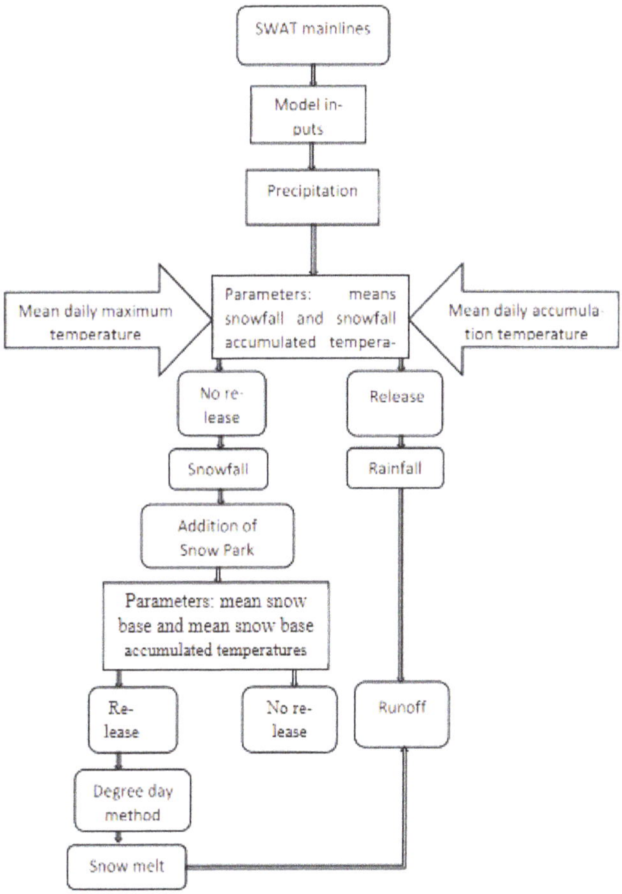

Figure 2. Outline of the degree-day factor method in the SWAT model.

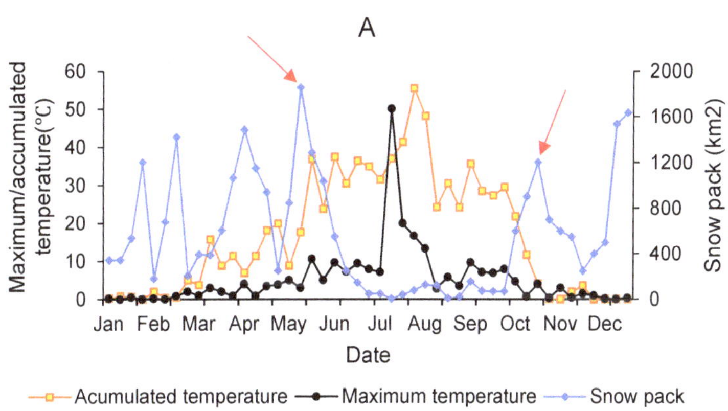

Figure 3. *Cont.*

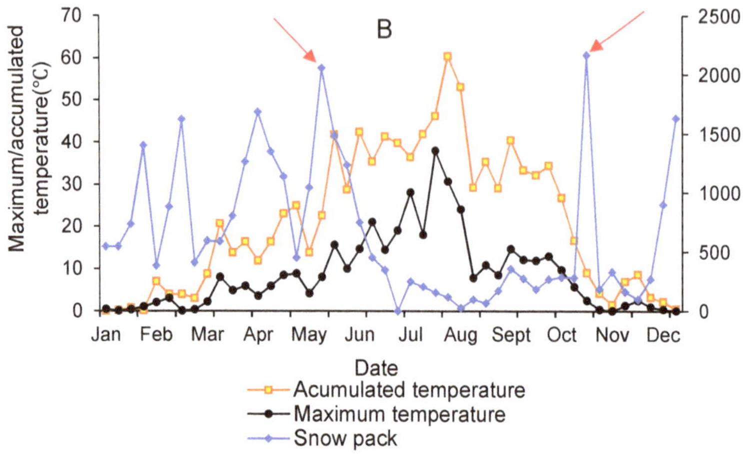

Figure 3. Relationship between snow pack and temperatures (accumulated and maximum) for catchments A (**A**) and B (**B**) in 2015. The red arrows on catchment A and B represent the 2 inflection points of the snow area change during the year.

3.6. Calibration and Validation

To improve the model's simulation accuracy, the optimal parameter set was determined using parameter calibration. The study period was divided into three sections: warm-up period (2013–2014), calibration period (2015), and validation period (2016). The objective function was used to calibrate the simulated daily streamflow based on observations made at the Balykchy and Tossor stations, depending on the catchment's location. The simulated results were evaluated using three statistical coefficients indices: the Nash–Sutcliffe efficiency (NSE) [74], goodness of fit (R^2), and percent bias (PBIAS) [75]. The NSE is a metric that indicates how well the simulated and measured values match. The result is acceptable if the NSE value is between 0 and 1. The model's performance is defined as "very good", "good", "satisfactory", or "unsatisfactory" [75]. The NSE is calculated using the following formula:

$$\mathrm{NSE} = 1 - \frac{\sum_{i=1}^{n}\left(Q_{obs,i} - Q_{sim,i}\right)^2}{\sum_{i=1}^{n}\left(Q_{obs,i} - \overline{Q}_{sim,i}\right)^2}; -\infty \leq \mathrm{NSE} \leq 1 \tag{10}$$

The correlation between the simulated and measured values is expressed as R^2. It is calculated using the following formula:

$$R^2 = \frac{\sum_{i=1}^{n}\left[\left(Q_{sim,i} - \overline{Q}_{sim,i}\right)\left(Q_{obs,i} - \overline{Q}_{0bs,i}\right)\right]^2}{\sqrt{\sum_{i=1}^{n}\left(Q_{sim,i} - \overline{Q}_{sim,i}\right)^2 \sum_{i=1}^{n}\left(Q_{obs,i} - \overline{Q}_{0bs,i}\right)^2}} \tag{11}$$

The PBIAS value measures the average tendency of the simulated results to be higher or lower than the observations. The PBIAS is calculated using the following equation:

$$\mathrm{PBIAS} = \left(\frac{\sum_{i=1}^{n} Q_{sim,i} - \sum_{i=1}^{n} Q_{obs,i}}{\sum_{i=1}^{n} Q_{obs,i}}\right) \tag{12}$$

where $Q_{obs,i}$ is the measured discharge on the ith day ($m^3 \cdot s^{-1}$); $Q_{sim,i}$ is the simulated discharge on the ith day ($m^3 \cdot s^{-1}$); $Q_{sim,i}$ and $Q_{obs,i}$ are the average simulated and measured discharges during the simulation period ($m^3 \cdot s^{-1}$), respectively; and n is the total number of daily flow observations.

4. Results

4.1. Rain and Snow Temperature Differences

Due to the sizes of the watersheds, temperature data from two corresponding stations could represent the temperature situation of the entire catchment (2153.46 km^2 and 2254.32 km^2). The temperature conditions in relation to rainfall and snowfall for the years 2015–2016 have been measured (Table 1). It is clear that rainfall occurs when ACCT reaches 40.59 or 31.97 °C with a MAXT of 16.18 or 13.095 °C for A and B, respectively. Snowfall occurs when ACCT is lower than 31.97 or 26.80 °C with a MAXT of 16.93 or 13.1 °C for A and B, respectively. During snowfall events on catchment A, the ACCT and MAXT cannot exceed 31.9 and 16.93 °C, respectively, and on catchment B, the ACCT and MAXT cannot exceed 26.805 and 13.1 °C, respectively. The two catchments experience differences in temperature conditions. The temperatures were higher in the northern catchment (A) than in the southern catchment (B). Catchments A and B are located in high mountains, Kungey Ala-Too and Teskey Ala-Too, at 3798 and 4762 m, respectively [76]. The temperatures of snowfall in high mountains are lower than those in low mountains [39,77]. From the Balykchy and Tossor stations, the different precipitation events were corrected. Using probability statistics for different precipitation type events at the two stations was considered to indicate the conditions of temperature linked to the precipitation events. Measured temperature conditions, all precipitation types, and simulated temperature conditions were compared. For the correct precipitation type, the temperature conditions needed to be the same. The proportion of the precipitation events that met that condition was used to show the accuracy. The accuracies were 79.45% and 84% for both catchments, given measurements from Balykchy and Tossor stations, respectively.

Table 1. The temperature (°C) conditions of rain and snow in the study area. ACC.T means accumulated temperature and MAX.T means maximum temperature. Balykchy: 42.46 lat, 76.19 long; Tossor: 42.17 lat, 77.44 long.

Location	A				B			
Average	Temperature for Rainfall		Temperature for Snowfall		Temperature for Rainfall		Temperature for Snowfall	
	ACC.T	MAX.T	ACC.T	MAX.T	ACC.T	MAX.T	ACC.T	MAX.T
2015	39.12	16.42	28.51	14.84	27.91	11.94	29.4	12.5
2016	42.06	15.94	35.43	19.02	36.12	14.25	24.21	13.7
Average	40.59	16.18	31.97	16.93	32.01	13.09	26.80	13.1

Sensitivity Analysis

The parameter sensitivity analysis for the original and updated SWAT models in SWAT-CUP, a SWAT model extension, is summarized in Table 2. SNO_SUB represents initial snow water content (mm H_2O), SNOCOVMX stands for minimum snow water content equal to 100% snow cover (mm), SOL_AWC is the soil evaporation compensation factor, SMFMX means melt factor for snow on June 21 (mm H_2O/°C-day), and ESCO stands for the soil evaporation compensation factor. When comparing the performance of the model before and after adjustment, the model was found to be less effective before adjustment. The original model's p-values for catchments A and B were 0.03–0.85 and 0.81–0.08, respectively. The updated model's p-values were 0.57–0.95 and 0.51–0.96, respectively.

Table 2. Parameter sensitivity analysis of the model for the chosen catchments.

	A						B					
	Before Modification			After Modification			Before Modification			After Modification		
No	Parameter	*T*-Test	*p*-Value	Parameter	*T*-Test	*p*-Value	Parameter	*T*-Test	*p*-Value	Parameter	*T*-Test	*p*-Value
1	SMTMP	0.05	0.85	SMTMP	0.04	0.95	SNO_SUB	0.24	0.81	SMTMP_accu	0.04	0.96
2	SNO_SUB	−0.08	0.81	TLAPS	0.04	0.94	SMTMP	−0.68	0.79	TLAPS	0.04	0.94
3	TLAPS	−0.32	0.75	SMTMP_accu	−0.1	0.90	SFTMP	−0.33	0.71	SFTMP_accu	−0.1	0.92
4	SNOCOVMX	−0.34	0.69	SFTMP_accu	0.14	0.89	SMFMX	−0.44	0.69	SNO_SUB	−0.15	0.87
5	SFTMP	−0.37	0.61	SFTMP	0.20	0.85	TLAPS	−0.47	0.61	PLAPS	0.20	0.86
6	SMFMX	0.50	0.59	SNO_SUB	0.24	0.80	SOL_AWC	0.48	0.58	SFTMP	0.24	0.81
7	PLAPS	−0.49	0.57	PLAPS	0.33	0.75	PLAPS	−0.58	0.57	SMTMP	0.33	0.79
8	SOL_AWC	0.83	0.41	SNOCOVMX	0.60	0.63	SNOCOVMX	0.61	0.40	SNOCOVMX	0.	0.71
9	ESCO	2.78	0.39	SMFMX	0.68	0.60	SMFMX	1.58	0.29	SMFMX	0.75	0.69
10	SMFMN	−0.79	0.03	SMFMN	0.77	0.57	SMFMN	−0.80	0.08	SMFMN	0.80	0.51

4.2. Best Parameter Set

The parameters adjusted in the snow-melting route were analyzed (Table 3). The ACCT thresholds for rainfall during model calibration (in 2015) were 39.12 and 27.91 °C for A and B, respectively. The ACCT thresholds for snowmelt were 28.51 and 29.4 °C for A and B, respectively. The parameters were set to different values during the calibration process. Some parameters, such as groundwater and soil, were omitted because they played a minor role in model calibration. The temperature of snow melting was one of the most important parameters.

Table 3. List of important parameters adjusted during calibration.

		CB		CA	
Parameter Description	**Unit**	**CV(A)**	**CV(B)**	**CV(A)**	**CV(B)**
Snowfall temperature (SFTMP)	°C	3.67	4.5	3.89	4.68
Snowfall (SFTMP_accu)	°C	31.97	25.6	31.8	25.5
Snowmelt base temperature (SMTMP)	°C	2.42	3.8	2.69	3.89
Snowmelt base (SMTMP_accu)	°C	28.51	29.4	28.39	28.9
Melt factor for snow on 21 June (SMFMX)	$mmH_2O/°C$-day	6.87	2.93	6.81	2.83
Melt factor for snow on 21 December (SMFMN)	$mmH_2O/°C$-day	9.76	5.62	9.77	5.61
Temperature lapse rate (TPLAS)	$°C.km^{-1}$	−7.46	2.32	−7.44	2.33
Precipitation lapse rate (PLAPS)	$mm.km^{-1}$	24	21	25	22

CB: Calibration before modification, CA: Calibration after modification CV (A): Calibrated values (A), CV (B): Calibrated values (B).

4.3. SWAT Model Performance

The statistical coefficients (NSE, R^2, and PBIAS) were determined and used to indicate the accuracy of the model (during calibration and validation) by comparing the observed and simulated daily streamflow of the catchments before and after modifying the model (Tables 4 and 5). The simulation period was from 2013 to 2016; the warmup period was from 2013 to 2014; and the calibration and validation periods were 2015 and 2016, respectively. The periods were chosen based on observation discharge data.

Table 4. Statistical coefficients calculated to verify the model's accuracy for the catchment A.

	Before Modification			After Modification		
	NSE	**R^2**	**PBIAS (%)**	**NSE**	**R^2**	**PBIAS (%)**
Calibration	0.72	0.73	2.56	0.80	0.84	1.51
Validation	0.67	0.69	1.24	0.79	0.79	−2.2
Overall	0.64	0.72	5.75	0.75	0.87	4.31

After model modification, there was improvement of the statistical values, which were used to verify the accuracy of the model for the catchments (Tables 4 and 5). Overall, differentiation of snowfall and rainfall, along with the modified degree-day factor, may cause PBIAS to increase or decrease [37]. The statistical values improved significantly after the model was updated. As a result, it should be stated that modifications made to the SWAT model improved its performance. In general, the SWAT model was upgraded due to the enhancement of the snowmelt module.

Table 5. Statistical coefficients calculated to verify the model's accuracy for the catchment B.

	Before Modification			After Modification		
	NSE	**R^2**	**PBIAS (%)**	**NSE**	**R^2**	**PBIAS (%)**
Calibration	0.69	0.74	4.71	0.75	0.75	3.79
Validation	0.73	0.73	1.02	0.79	0.81	0.94
Overall	0.61	0.75	6.5	0.69	0.86	4.97

4.3.1. Stream Flow in the Catchments

For both catchments (A and B), a comparison of observed and simulated discharges before and after model adjustment was conducted (Figure 4). During the winter months (October to March), the mean observed daily discharges for catchments A and B were 2.2 and 1.91 $m^3 \cdot s^{-1}$, respectively. The simulated discharges before model modification were 1.89 and 2.4 $m^3 \cdot s^{-1}$ for catchments A and B, respectively; and after model modification, they were 4.3 and 3.9 $m^3 \cdot s^{-1}$. Before, the model underestimated the peaks. After model modifications, some peaks were overestimated, and others remained unchanged, especially during the snowmelt periods (April to September) during the calibration and validation periods. The mean observed daily discharges for catchments A and B were 12.3 and 11.34 $m^3 \cdot s^{-1}$, respectively, and the mean simulated discharges before and after model modification were 10.23 and 14.13 $m^3 \cdot s^{-1}$ for catchment A and 11.67 and 13.43 $m^3 \cdot s^{-1}$ for catchment B during these seasons. In general, the model's simulated discharge was closer to the observed discharge before modification. After modification, the model estimated more flood peaks compared to the peaks estimated before modification (Figure 4), which could indicate the likelihood of flood occurrence due to snow in high mountains causing flooding. The shifting time of peak appearing time (S.T) and error of peak flood (E.F: between observed and simulated discharges during calibration and validation periods) were calculated. The maximum value of the peak flood appeared in July for both catchments. The S.T values were 2.14 and 0.95–1 days for A and B, respectively; E.F accounted for –2.12 and –0.39% for A and B, respectively. This is consistent with the findings of [37,78], who also stated that, occasionally, there is a "one peak, one day" flood phenomenon, which helps peak selection analysis. During calibration and validation, the arrangement of the 16 high peaks for each catchment was analyzed. The average peak flows after model modification in catchments A and B were 12 and 10% higher than the average of peak flows before model modification. It is indicated by the peaks provided in different colors for the simulated, observed, and modified discharge (Figure 4). This made the peaks become closer to their corresponding observation discharges. This suggests that upgrading the snowmelt module can result in more accurate model simulations, which is indicated by statistical values (Tables 4 and 5).

4.3.2. Accumulation Temperature and Snowmelt

The degree-day factor method engaged in SWAT is an important approach for calculating snow and ice melt [79,80]. A comparison of daily snowmelt before and after model modification was conducted (Figure 5). This snowmelt stream flow simulation approach matches the characteristics of the snowmelt floods in the watersheds surrounding the Issyk-Kul Lake. The daily average snowmelt amounts for catchments A and B were 43.01 and 46.14 mm, respectively, and after model modification, they were 37.36 and 45.20 mm reductions of 13.12 and 2.03%. However, prior to model modification, the amount of snowmelt by considering flood peaks were 122.45 and 134.27 mm for A and B, respectively. After model improvement, the snowmelt amounts for A and B were 129.8 and 143.67 mm, respectively. After modifying the model, the increases were about 6 and 7% for A and B, respectively. These results revealed that the simulated snowmelts were significantly higher in the spring and summer when using the modified model (high frequency in spring). Snowmelt is obviously more pronounced in high mountains with snow and ice, causing floods; similar findings have been found by [16] and in neighboring regions [37]. Some exceptions, such as snowmelt obtained using the original model, were neglected. Before the modifications, runoff contributed 45.3 and 40.67% to catchments A and B, respectively, which then became 56.29 and 53.26% for A and B.

The relationship between the amount of eight-day snowmelt and the snow cover during melting periods was investigated. The model's rationality before and after modification and snowmelt calculations were revealed too (Figure 6). For all catchments in August, the volume of snowmelt calculated after modification showed an increasing trend that tended towards strong matching of the peaks, whereas before modification, there was a decreasing trend. In general, the snow cover decreased in summer, owing to the high temperatures. As indicated by the peaks that form a circle for a few days, the snow cover can increase for a short period of time, which corresponds to a rapid decrease and then increase in temperature. According to the findings, the model's performance after modification was better. This is also indicated by how the modified model's trend was an increasing one, whereas the original model's trend decreased in Figure 6, more especially in August. Sudden changes in a short period of time, according to [81], can result in intense runoff, which can cause flooding. The modified model revealed the characteristics of floods in the area.

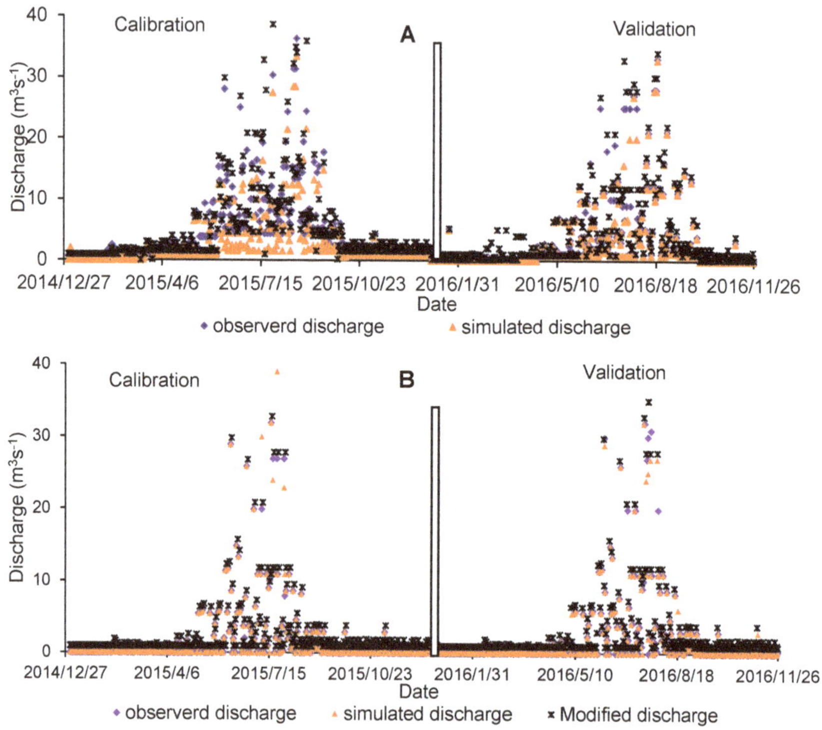

Figure 4. The model calibration (2015) and validation (2016) comparisons for the discharge processes at catchment A (**A**) and catchment B (**B**).

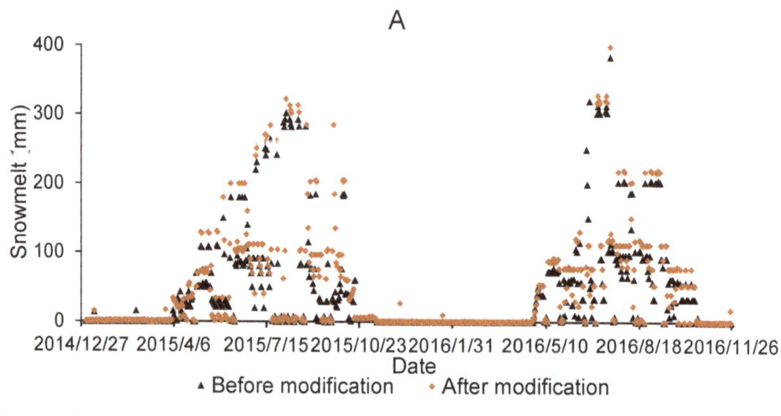

Figure 5. *Cont.*

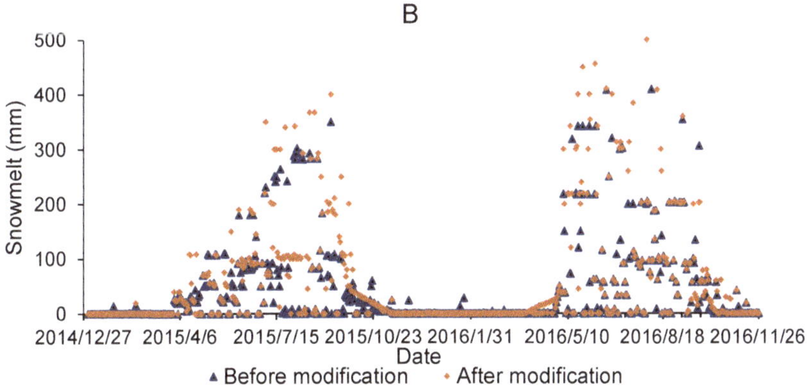

Figure 5. Amounts of snowmelt calculated by the model before and after modification during calibration (2015) and validation (2016). Catchment A (**A**) and catchment B (**B**).

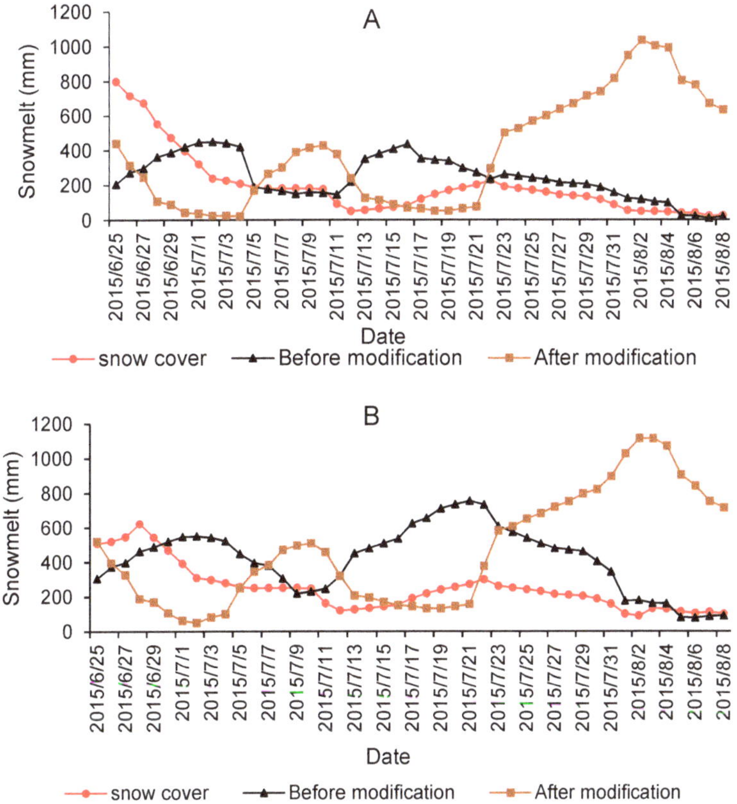

Figure 6. Amount of eight-day snowmelt during the snowmelt season calculated by the modified model, compared with the snow cover area from the original model. Catchment A (**A**) and catchment B (**B**).

The relationships between ACCT and MAXT and monthly snow melting during the calibration (2015) and verification periods (2016) are presented in Figure 7. When the accumulated temperature in the selected catchments reaches 10 °C, the snow melts. The crosses denote non-melting months; normally, there is no snowmelt from November to

February. Prior to the introduction of the degree-day though, the model indicated some amount of snowmelt in the non-melting months; however, after modification, there was no snowmelt found in those months, making snowmelt more accurate and reasonable. The cumulative temperature threshold of snow melting was reached between January and March in 2016 and November and December of the same year.

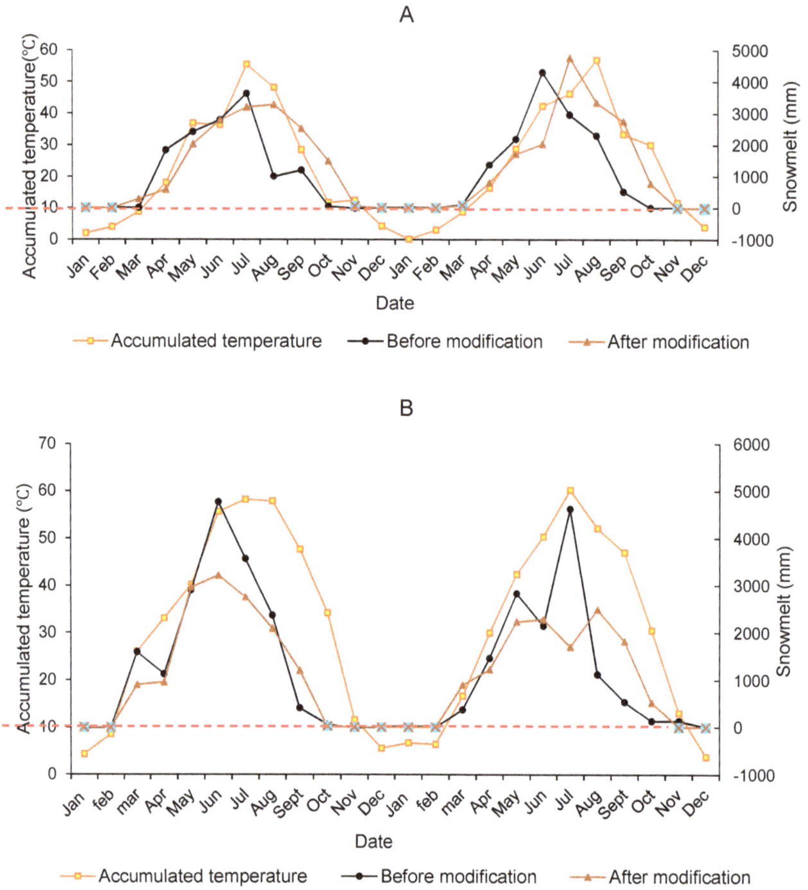

Figure 7. Comparison of accumulated temperature (ACCT), monthly snowmelt before and after modification, and maximum temperature (MAXT) during calibration (2015) and verification (2016) periods. Catchment A (**A**) and catchment B (**B**). The crosses denote non-melting months. The red dummy line represents the amount of accumulated temperature at which snow melt will start.

4.4. Land Surface and Air Temperatures

Simulation of Snow Cover Area

To validate the modeled snow cover, simulated results of snow cover area and data of MODIS (Figure 8) were compared during the calibration and verification periods, 2015 and 2016, respectively. During the calibration period, the snow cover area reached a lower value in summer and the MODIS data overestimated the value in later and earlier spring for catchments A and B, respectively. In general, in spring and summer during calibration and validation, the snow cover area tended to decrease, and in non-melting periods, the area increased.

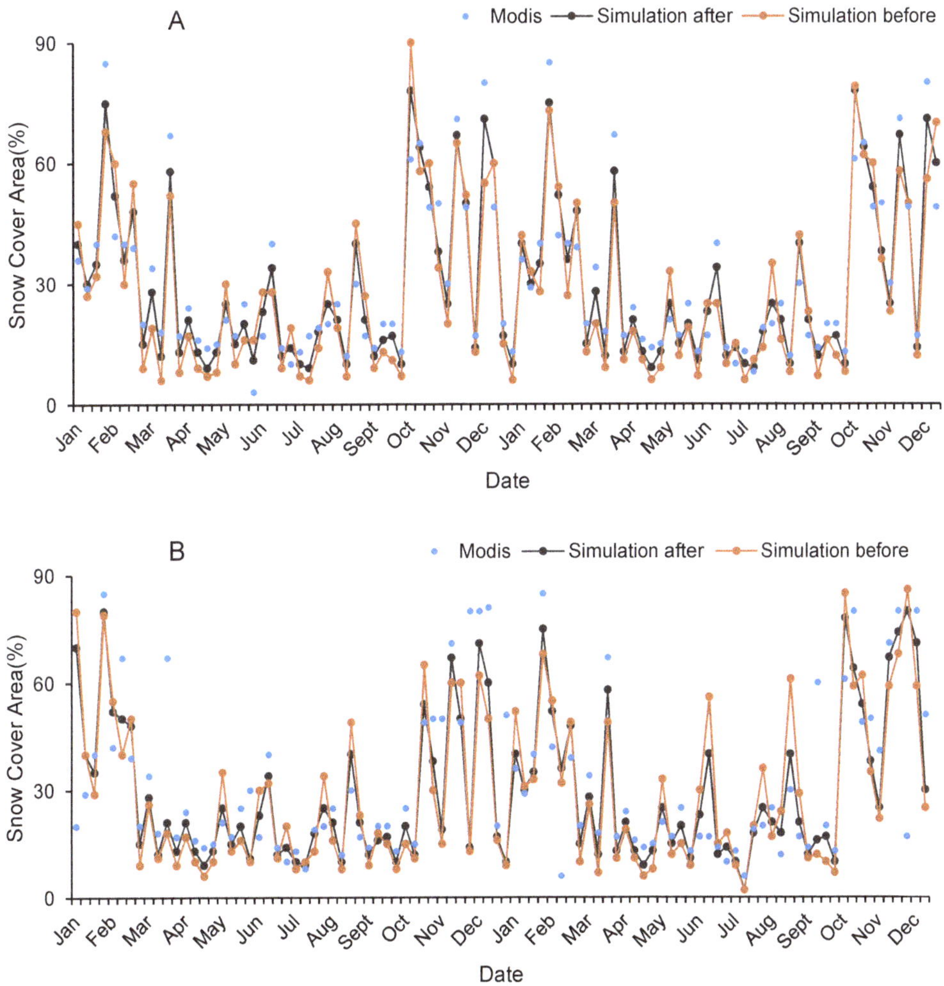

Figure 8. Comparison between (**A**) eight-day simulated snow cover area and MODIS data during model calibration and (**B**) validation periods for catchments.

The results of a linear regression analysis of the relationship between snowmelt and ACCT are presented in Figure 9. Snow is sensitive to temperature, and snow accumulation in the area is concentrated in the high-altitude regions [82]. On the surface of snow cover, the temperature drives the melting process. The modified model performed better with $R^2 = 0.762$ and 0.757 for A and B, respectively. However, after the model was modified, the correction was relatively low, with $R^2 = 0.506$ and 0.515 for A and B, respectively. Based on our results, it can be concluded that modifying the SWAT model, particularly using the degree-day factor method, is highly effective for snowmelt analysis and prediction.

The average monthly land surface temperatures (LST) of MODIS were calculated and compared to air temperatures of the SWAT model from 2015 to 2016. To validate the air temperatures used in the model, the land surface temperatures within elevation bands of the whole catchments were calculated and were compared with the air temperatures. The R^2 values were 0.85 and 0.83 for A and B catchments, respectively, (Figure 10). These statistical values indicate that the used air temperature in the model was sufficient for the simulations. Furthermore, a comparison between discharge and accumulated temperature has been undertaken to better understand the relationship between the two (Figure 11). The findings demonstrated that as the accumulated temperature rises in the spring and summer, so does the overall discharge.

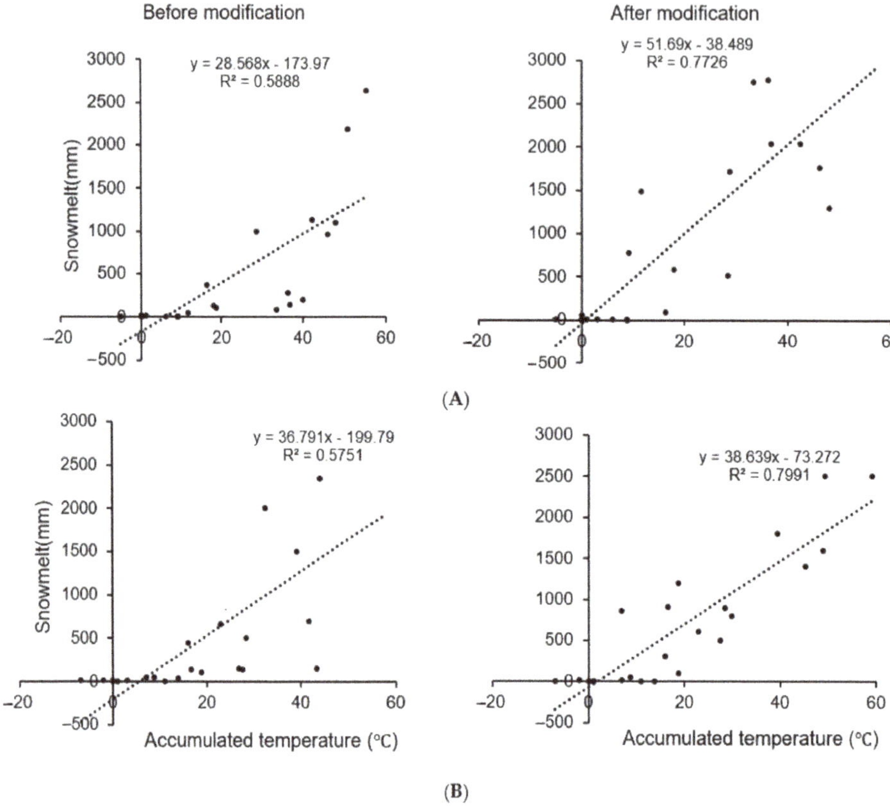

Figure 9. Correlations between monthly snow melting and accumulation temperature before and after model modification for catchments A and B. Catchment A (**A**) and catchment B (**B**).

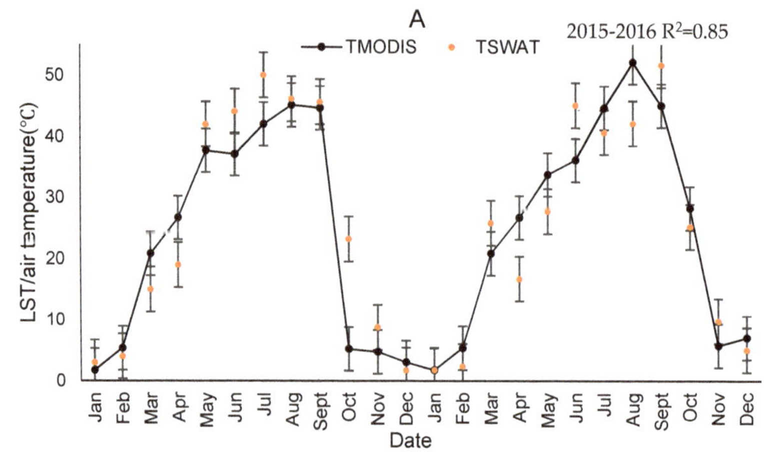

Figure 10. *Cont.*

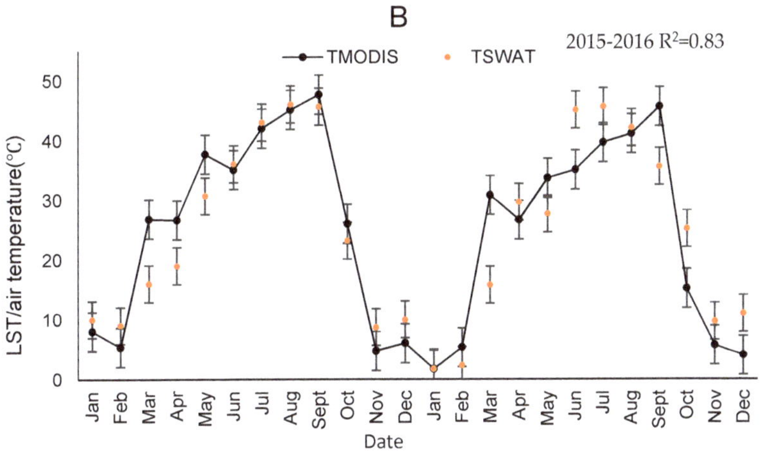

Figure 10. The monthly land surface temperatures of MODIS and air temperature in the SWAT model for the period 2015–2016 in the catchment (**A**,**B**).

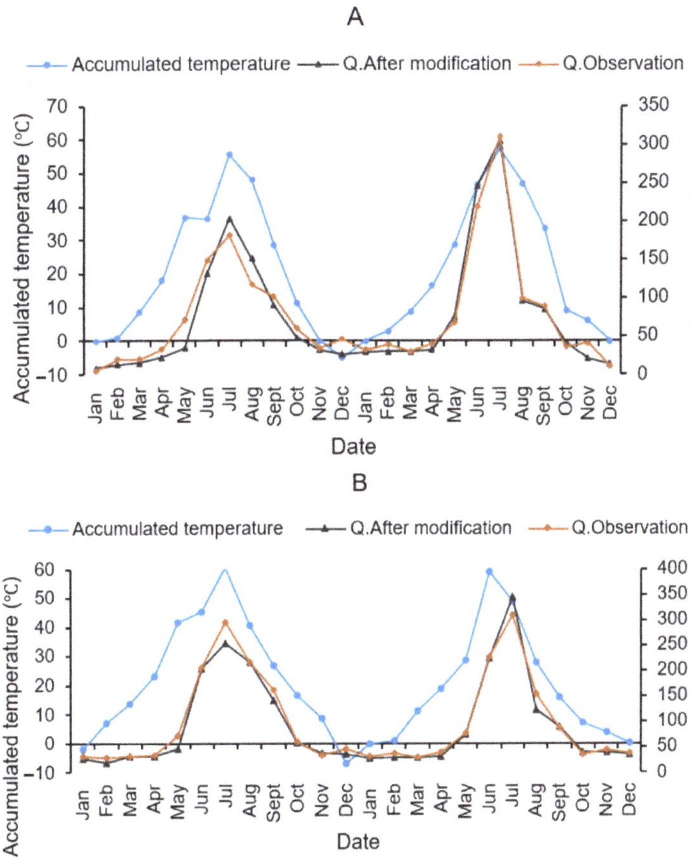

Figure 11. Comparison between accumulated temperature and discharge of catchment (**A**,**B**) for the period 2015–2016.

5. Discussion

This study proposed a method for estimating the snowmelt degree-day factor based on MODIS snow cover data. In the selected catchments, the spatial distribution of the degree-day factor method was estimated and attributed to the interactions of climate conditions, topography, and vegetation. The estimated degree-day factor values were 4.7 and 4.3 $mmH_2/°C$ for A and B, respectively, which are within the ranges of 3.1–5.9 $mmH_2/°C$ [43] and 4.3–5 $mmH_2/°C$ [83]. These values are also close to the range of 5.0–11.6 $mmH_2/°C$ found by studies in the Himalayas [3,4]. The simulations using the modified degree-day factor method on runoff in terms of discharge and volume of snowmelt are more plausible than the original model, pre-modification. They were influenced by a variety of hydrological processes, along with the interactions of the hydrological model parameters [84]. The SWAT model ignores the possibility of snowfall and calculates precipitation as rain [85]. This may reduce snowmelt runoff and increase rainfall. The modeling scale can have a significant influence on the simulations, considering the spatial resolution of MODIS data for precipitation and temperature [86]. The temperature threshold values determine the occurrence of snowmelt in mountain basins. This is in line with the study by [87], who stated that the accumulation temperature is the driving force for triggering the snowmelt. The results revealed that the temperature threshold of the catchment located in the northwest of Issyk-Kul was 4.89% higher than the threshold of the catchment located in the south of the lake, due to elevation differences, topography, and water vapor conditions; this is consistent with [37]. The lower the elevation, the more accurate the rain and snow estimations [88]. Changes in climate conditions in high mountains affect the correctness. In the SWAT model, the daily maximum and minimum temperatures are used, but there is a difference between land surface temperature and air temperature [89]. Temperature variations in a single day can be classified into three main parts: the maximum daily temperature is below 0 °C; the maximum daily temperature is above 0 °C, but the minimum daily temperature is less than 0 °C; and the minimum daily temperature is greater than 0 °C [90]. It is preferable to use land surface temperature to calculate snowmelt [91,92]. In this context, for this study, the differences in air and ground temperatures were taken into account [37]. When observations of precipitation are corrected from meteorological stations, the precipitation types are always the same [93], except that the temperature threshold is set to determine the precipitation type when passing observation values into the SWAT model [94,95]. The catchment's area is enough for their components to be verified. The accumulated and maximum temperatures were improved, and then the snow and rain accuracies were verified, as part of a model modification process, to improve its performance. Furthermore, the optimal accumulated and maximum temperatures were achieved, and the accuracies of precipitation type and snowmelt measurements were modified. The use of average temperature as a criterion for snowmelt has an impact on snowmelt volume computation [37,96]. The process of surface temperature accumulation can then be well represented, improving calculation accuracy, by using both the accumulated and maximum temperatures in the snow-melting calculation.

5.1. Model Performance

The statistical coefficients before and after model modification were calculated (Tables 4 and 5). For all catchments, the simulation results obtained after modification were better than before. The PBIAS values from verification using the modified model were better than that of the model before (Tables 4 and 5). The temperatures in Issyk-Kul's northern and southern catchments begin to rise in early April, snow begins to melt into stream flow, and the water levels in the catchments begin to rise rapidly [25], eventually leading to floods [97–99]. Our results showed that the participation of snowmelt from October to March was limited to the modified degree-day method, which increased from April to September. In fact, the method brought the observed discharge closer to the simulated values (Figure 2). During simulations, the differentiation between rain and snow in terms of precipitation for all catchments increased snowmelt and decreased rainfall in proportion to runoff, which matches our findings from a simulation conducted in the Xinjiang region's Kunlun Mountains [40]. Rainfall in the catchments is scarce, and the increase in the runoff from April to September is due to snowmelt, as described by [16,48]. Flooding was predicted by the modified model to occur during the spring. The temperature, which causes snow to melt, increases in summer [100]. To evaluate the discharge, the shifting time of peak appearing time and the error of peak flood were considered. Most of the high peaks occurred during spring and winter.

The modified model performed better than the original model; the peaks of flood flow increased, bringing the discharge closer to the measured values. The northwestern catchment of Issyk-Kul indicated lower discharge compared to the southern catchment for the study period, probably due to differences in location, elevation, and catchment area. For catchment A, the average daily flow in 2015 was 25.72 $m^3{\cdot}s^{-1}$, which was lower than 35.12 $m^3{\cdot}s^{-1}$ in 2016. In 2015, the average daily flow into catchment B was 39.04 $m^3{\cdot}s^{-1}$, which was lower than 41.83 $m^3{\cdot}s^{-1}$ in 2016. These differences could be due to both climate change and human activities [16]. Furthermore, variations in runoff in catchments are primarily caused by changes in climate [76]. The contributions of snowmelt to the water flow were approximately 6 and 7% greater for catchments A and B, respectively, according to the modified model compared to the original model. The Sequential Uncertainty Fitting (SUFI-2) algorithm [101] considers uncertainties in observation data, parameters, model structure, and input data. After consideration, the algorithm determines the impacts of the parameters and puts them in ranges. The sensitivity parameters were ranked before and after modification for both models. To improve the model's performance, the parameters related to snowmelt were modified. In the traditional and modified models, the parameters of snowmelt base temperature SMTMP for catchment A, initial snow water content (mm H_2O) (SNO_SUB) for catchment B, (SMTMP_accu) for catchment A, and snowmelt base temperature (SMTMP_accu) for catchment B showed strong sensitivity. Before and after model modification, the parameter of melt factor for snow on December 21 (SMFMN) showed a low sensitivity for both catchments. Because of differences in input data, the snowmelt module in the modified model of catchment A performed better than the modified model for catchment B. Generally, the entire snowmelt module in both selected catchments' original models indicated a low response during the calibration process. Thus, model modification for the snowmelt module improved the model's results by increasing accuracy and reliability.

5.2. Accumulation Temperature and Snowmelt

The precipitation within the HRU is labeled as snow if the mean daily air temperature falls below the boundary temperature and the snow equivalent is applied to the snow pack. When snowfall accumulates on the ground surface, it forms a snow pack. The amount of water contained in the snow pack is measured in snow water equivalents. Additionally, snowfall will increase the snow pack, whereas snowmelt or sublimation will reduce it. During simulation, the traditional degree-day factor method only uses average temperature in calculations [99], but after adding accumulation temperature, the model performed better in melt seasons simulations. The temperatures were higher in the northern catchment (A) than in the southern catchment (B). This is due to altitude differences and the wet air from Issyk-Kul. The increase in elevation causes air pressure to decrease; hence, air molecules are further apart, causing the temperature decreases [1–3]. The snow melts when certain accumulated and maximum temperatures are reached, which can happen in a single day. The modified model simulated more flood peaks in the simultaneous snowmelt floods and rainfall floods that occur during melting seasons. The accumulated threshold for all catchments was 10 °C (Figure 7), which improved the snowmelt calculations. There was some inconsistency between the simulated discharge before the model modification during melt seasons and non-melt seasons for all catchments, but the model modification eliminated the errors. After model modification, the results showed higher snowmelt peaks during snowmelt seasons (summer), and the amount of snowmelt was increased for all catchments (Figure 5). This means that upgrading the snow melt module led to an increase in the amount of snowmelt, which shows that the model made a significant improvement after the modifications. Our results look quite similar to the results obtained by [37].

6. Conclusions

In this study, a new algorithm was used to calculate snowmelt in the catchments around Issyk-Kul in SWAT. The model was able to simulate the discharge and snowmelt. It needs to be calibrated and validated. The existence of the degree-day factor method and the precipitation type differentiation in the model were highlighted. The modifications seemed to improve the snowmelt module of the model. Through the improvement of the module, the sensitivity of the snowmelt-related parameters was been greatly improved. To improve the accuracy of the original model, the accumulation of temperature conditions was determined. During the calibration process, the modified model was used to calculate the discharges in the catchments. Through model modification, the degree day factor method calculated snow melt, and the snow-related parameters were calibrated manually; the statistical values such as NSE and R^2 were improved. The NSE increased from 0.72 to 0.80 and from 0.69 to 0.75 for catchments A and B, respectively. R^2 increased from 0.73 to 0.84 and from 0.74 to 0.75 for A and B, respectively (Tables 4 and 5). The shifting times were 2.14 and 0.95–1 days for A and B, respectively. Errors of peaks accounted for −2.12 and −0.39% for A and B, respectively. These results indicated that the model was good enough to simulate the discharge. After modifying the model, the flood peaks increased for A and B by 49.42 and 43.87%, respectively, bringing them closer to the reality of the observation discharge. This was more consistent with the characteristics of floods caused by snowmelt in the catchments around Issyk-Kul. The contributions of snowmelt to stream flow increased by 24.26 and 31% for A and B, respectively. Therefore, it is possible to limit snowmelt to reasonable times of the year by improving the traditional degree-day factor method with the addition of accumulated temperature conditions, while avoiding non-melting seasons. Finally, when using the model for mountainous regions, it is best to improve the degree-day factor method to improve accuracy. Differentiating precipitation and flood types may provide reliable information for flood forecasting, allowing for the identification of hazards and appropriate adaptive measures.

Author Contributions: Conceptualization, S.U., T.L., V.N., J.M.H., T.H. and B.H.; methodology, S.U., T.L. and V.N.; software, S.U. and V.N.; validation, S.U.; formal analysis, S.U., T.L. and V.N.; investigation and resources, S.U. and V.N.; data curation, T.L; writing—original draft preparation, S.U. and V.N.; writing—review and editing, S.U.; visualization, V.N.; supervision, T.L.; project administration, T.L., A.B.; funding acquisition, T.L. All authors have read and agreed to the published version of the manuscript.

Funding: This study was supported by the Strategic Priority Research Program of the Chinese Academy of Sciences, Pan-Third Pole Environment Study for a Green Silk Road (grant number XDA20060303), the National Natural Science Foundation of China (grant number 41761144079), the International Partnership Program of the Chinese Academy of Sciences (grant number 131551KYS B20160002), CAS Interdisciplinary Innovation Team (grant number JCTD-2019-20), CAS Research Center for Ecology and Environment of Central Asia (grant number Y934031), K.C. Wong Education Foundation (GJTD-2020-14), and Regional Collaborative Innovation Project of Xinjiang Uygur Autonomous Regions (grant number 2020E01010).

Institutional Review Board Statement: Not applicable.

Informed Consent Statement: Not applicable.

Data Availability Statement: The source of all the data used in this study is provided in the manuscript.

Acknowledgments: We acknowledge the United States Geological Survey and the NASA team, the ESA CCI Land Cover project, the Food and Agriculture Organization (FAO), and the Meteorological Service of Canada (MSC) for providing the necessary data. We also thank the Hydrological Bureau of Kyrgyzstan for the production and provision of their hydrological data. In addition, the authors would like to thank the Chinese government, the Xinjiang Institute of Ecology and Geography, and the University of the Chinese Academy of Sciences (UCAS) for their financial assistance and laboratory facilities. Special acknowledgement goes to the China–Pakistan Joint Research Center on Earth Sciences, which supported the implementation of this study.

Conflicts of Interest: The authors declare that there is no conflict of interests regarding the publication of this paper.

References

1. Rasool, S. Climate change, Global change: What is the difference? *Eos Trans. Am. Geophys. Union* **1988**, *69*, 668. [CrossRef]
2. Lin, N.-F.; Tang, J.; Han, F.-X. Eco-environmental problems and effective utilization of water resources in the Kashi Plain, western Terim Basin, China. *Hydrogeol. J.* **2001**, *9*, 202–207. [CrossRef]
3. Ji, X.; Kang, E.; Chen, R.; Zhao, W.; Zhang, Z.; Jin, B. The impact of the development of water resources on environment in arid inland river basins of Hexi region, Northwestern China. *Environ. Geol.* **2006**, *50*, 793–801. [CrossRef]
4. Michel-Guillou, E. Water resources and climate change: Water managers' perceptions of these related environmental issues. *J. Water Clim. Chang.* **2015**, *6*, 111–123. [CrossRef]
5. Chen, Y.; Li, Z.; Fang, G.; Deng, H. Impact of climate change on water resources in the Tianshan Mountians. *Cent. Asia. Acta Geogr. Sin.* **2017**, *72*, 18–26.
6. Fischer, A. Glaciers and climate change: Interpretation of 50 years of direct mass balance of Hintereisferner. *Glob. Planet. Chang.* **2010**, *71*, 13–26. [CrossRef]
7. Gascoin, S.; Kinnard, C.; Ponce, R.; Lhermitte, S.; Macdonell, S.; Rabatel, A. Glacier contribution to streamflow in two headwaters of the Huasco River, Dry Andes of Chile. *Cryosphere* **2011**, *5*, 1099–1113. [CrossRef]
8. Pelto, M.S. Quantifying Glacier Runoff Contribution to Nooksack River, WA in 2013-15. In Proceedings of the 2015 AGU Fall Meeting Abstracts, San Francisco, CA, USA, 14–18 December 2015.
9. Swick, M.; Kaspari, S. Partitioning the Contribution of Light Absorbing Aerosols to Snow and Glacier Melt Using a Novel Hyperspectral Microscopy Method. In Proceedings of the 2017 AGU Fall Meeting Abstracts, New Orleans, LA, USA, 11–15 December 2017.
10. Hock, R.; Rees, G.; Williams, M.W.; Ramirez, E. Contribution from glaciers and snow cover to runoff from mountains in different climates. *Hydrol. Process.* **2006**, *20*, 2089–2090. [CrossRef]
11. Luo, Y.; Arnold, J.; Liu, S.; Wang, X.; Chen, X. Inclusion of glacier processes for distributed hydrological modeling at basin scale with application to a watershed in Tianshan Mountains, northwest China. *J. Hydrol.* **2013**, *477*, 72–85. [CrossRef]
12. Li, X.; Ma, Y.; Sun, Y.; Gong, H.; Li, X. Flood hazard assessment in Pakistan at grid scale. *J. Geo-Inf. Sci.* **2013**, *15*, 314–320. [CrossRef]
13. Lian, J.; Gong, H.; Li, X.; Zhao, W.; Hu, Z. Design and development of flood/waterlogging disaster risk model based on Arcobjects. *J. Geo-Inf. Sci.* **2009**, *11*, 376–381. [CrossRef]
14. Long, H.; Liu, Y.; Li, X.; Chen, Y. Building new countryside in China: A geographical perspective. *Land Use Policy* **2010**, *27*, 457–470. [CrossRef]
15. Zhao, G.; Pang, B.; Xu, Z.; Wang, Z.; Shi, R. Assessment on the hazard of flash flood disasters in China. *J. Hydraul. Eng.* **2016**, *47*, 1133–1142.
16. Alymkulova, B.; Abuduwaili, J.; Issanova, G.; Nahayo, L. Consideration of water uses for its sustainable management, the case of Issyk-Kul Lake, Kyrgyzstan. *Water* **2016**, *8*, 298. [CrossRef]
17. Giralt, S.; Klerkx, J.; Riera, S.; Julia, R.; Lignier, V.; Beck, C.; De Batist, M.; Kalugin, I. Recent paleoenvironmental evolution of Lake Issyk-Kul. In *Lake Issyk-Kul: Its Natural Environment*; Springer: Berlin/Heidelberg, Germany, 2002; pp. 125–145.
18. Vollmer, M.K.; Weiss, R.F.; Schlosser, P.; Williams, R.T. Deep-water renewal in Lake Issyk-Kul. *Geophys. Res. Lett.* **2002**, *29*, 124-1–124-4. [CrossRef]
19. Dawadi, S.; Ahmad, S. Changing climatic conditions in the Colorado River Basin: Implications for water resources management. *J. Hydrol.* **2012**, *430*, 127–141. [CrossRef]
20. Abadi, L.S.K.; Shamsai, A.; Goharnejad, H. An analysis of the sustainability of basin water resources using Vensim model. *KSCE J. Civ. Eng.* **2015**, *19*, 1941–1949. [CrossRef]
21. De Batist, M.; Imbo, Y.; Vermeesch, P.; Klerkx, J.; Giralt, S.; Delvaux, D.; Lignier, V.; Beck, C.; Kalugin, I.; Abdrakhmatov, K. Bathymetry and sedimentary environments of Lake Issyk-Kul, Kyrgyz Republic (Central Asia): A large, high-altitude, tectonic lake. In *Lake Issyk-Kul: Its Natural Environment*; Springer: Berlin/Heidelberg, Germany, 2002; pp. 101–123.
22. Jailoobayev, A.; Neronova, T.; Nikolayenko, A.; Mirkhashimov, I. *Water Quality Standards and Norms in Kyrgyz Republic*; Regional Environmental Centre for Central Asia (CAREC): Almaty, Kazakhstan, 2009.

23. Wang, G.; Shen, Y.; Wang, N.; Wu, Q. The effects of climate change and human activities on the lake level of the Issyk-Kul during the past 100 years. *J. Glaciol. Geocryol.* **2010**, *32*, 1097–1105.
24. Narama, C.; Shimamura, Y.; Nakayama, D.; Abdrakhmatov, K. Recent changes of glacier coverage in the western Terskey-Alatoo range, Kyrgyz Republic, using Corona and Landsat. *Ann. Glaciol.* **2006**, *43*, 223–229. [CrossRef]
25. Alifujiang, Y.; Abuduwaili, J.; Ge, Y. Trend Analysis of Annual and Seasonal River Runoff by Using Innovative Trend Analysis with Significant Test. *Water* **2021**, *13*, 95. [CrossRef]
26. Jost, G.; Moore, R.D.; Weiler, M.; Gluns, D.R.; Alila, Y. Use of distributed snow measurements to test and improve a snowmelt model for predicting the effect of forest clear-cutting. *J. Hydrol.* **2009**, *376*, 94–106. [CrossRef]
27. Martinec, J.; Rango, A. Parameter values for snowmelt runoff modelling. *J. Hydrol.* **1986**, *84*, 197–219. [CrossRef]
28. Anderson, E.A. *A Point Energy and Mass Balance Model of a Snow Cover*; US Department of Commerce, National Oceanic and Atmospheric Administration: Silver Spring, MD, USA, 1976; Volume 19.
29. He, Z.; Parajka, J.; Tian, F.; Blöschl, G. Estimating degree-day factors from MODIS for snowmelt runoff modeling. *Hydrol. Earth Syst. Sci.* **2014**, *18*, 4773–4789. [CrossRef]
30. Jones, H.; Sochanska, W.; Stein, J.; Roberge, J.; Plamondon, A.; Charette, J. Snowmelt in a boreal forest site: An integrated model of meltwater quality (SNOQUAL1). In *Acidic Precipitation*; Springer: Berlin/Heidelberg, Germany, 1986; pp. 1485–1493.
31. Smith, M.B.; Koren′, V.; Zhang, Z.; Reed, S.M.; Seo, D.; Moreda, F.; Kuzmin, V.A. *NOAA NWS Distributed Hydrologic Modeling Research and Development*; NOAA Institutional Repository: Silver Spring, MD, USA, 2004.
32. Shimamura, Y.; Izumi, T.; Matsuyama, H. Remote sensing of areal distribution of snow cover and snow water resources in mountains based on synchronous observations of Landsat-7 satellite-A case study around the Joetsu border of Niigata prefecture in Japan. In *Proceedings of the General Meeting of the Association of Japanese Geographers Annual Meeting of the Association of Japanese Geographers, Spring 2004, 29 July 2004*; The Association of Japanese Geographers: Tokyo, Japan, 2004.
33. Herrero, J.; Polo, M.; Moñino, A.; Losada, M. An energy balance snowmelt model in a Mediterranean site. *J. Hydrol.* **2009**, *371*, 98–107. [CrossRef]
34. Feng, T.; Feng, S. An Energy Balance Snowmelt Model for Application at a Continental Alpine Site. *Procedia Eng.* **2012**, *37*, 208–213. [CrossRef]
35. Jost, G.; Moore, R.D.; Smith, R.; Gluns, D.R. Distributed temperature-index snowmelt modelling for forested catchments. *J. Hydrol.* **2012**, *420*, 87–101. [CrossRef]
36. Yu, W.; Zhao, Y.; Nan, Z.; Li, S. Improvement of snowmelt implementation in the SWAT hydrologic model. *Acta Ecol. Sin.* **2013**, *33*, 6992–7001.
37. Duan, Y.; Liu, T.; Meng, F.; Luo, M.; Frankl, A.; De Maeyer, P.; Bao, A.; Kurban, A.; Feng, X. Inclusion of modified snow melting and flood processes in the swat model. *Water* **2018**, *10*, 1715. [CrossRef]
38. Arnold, J.G.; Fohrer, N. SWAT2000: Current capabilities and research opportunities in applied watershed modelling. *Hydrol. Process. Int. J.* **2005**, *19*, 563–572. [CrossRef]
39. Fontaine, T.; Cruickshank, T.; Arnold, J.; Hotchkiss, R. Development of a snowfall–snowmelt routine for mountainous terrain for the soil water assessment tool (SWAT). *J. Hydrol.* **2002**, *262*, 209–223. [CrossRef]
40. Xu, C.; Chen, Y.; Hamid, Y.; Tashpolat, T.; Chen, Y.; Ge, H.; Li, W. Long-term change of seasonal snow cover and its effects on river runoff in the Tarim River basin, northwestern China. *Hydrol. Process. Int. J.* **2009**, *23*, 2045–2055. [CrossRef]
41. Zhang, X.-Y; Li, J.; Yang, Y.-Z.; You, Z. Runoff Simulation of the Catchment of the Headwaters of the Yangtze River Based on SWAT Model. *J. Northwest For. Univ.* **2012**, 5.
42. Arnold, J.G.; Muttiah, R.S.; Srinivasan, R.; Allen, P.M. Regional estimation of base flow and groundwater recharge in the Upper Mississippi river basin. *J. Hydrol.* **2000**, *227*, 21–40. [CrossRef]
43. Zhang, Y.; Liu, S.; Ding, Y. Spatial variation of degree-day factors on the observed glaciers in western China. *Acta Geogr. Sin.* **2006**, *61*, 89. [CrossRef]
44. Zhang, Y.; Suzuki, K.; Kadota, T.; Ohata, T. Sublimation from snow surface in southern mountain taiga of eastern Siberia. *J. Geophys. Res. Atmos.* **2004**, *109*, D21. [CrossRef]
45. Arnold, J.; Allen, P.; Volk, M.; Williams, J.; Bosch, D. Assessment of different representations of spatial variability on SWAT model performance. *Trans. ASABE* **2010**, *53*, 1433–1443. [CrossRef]
46. Wang, X.; Melesse, A. Evaluation of the SWAT model's snowmelt hydrology in a northwestern Minnesota watershed. *Trans. ASAE* **2005**, *48*, 1359–1376. [CrossRef]
47. Kumar, M.; Marks, D.; Dozier, J.; Reba, M.; Winstral, A. Evaluation of distributed hydrologic impacts of temperature-index and energy-based snow models. *Adv. Water Resour.* **2013**, *56*, 77–89. [CrossRef]
48. Hock, R. A distributed temperature-index ice-and snowmelt model including potential direct solar radiation. *J. Glaciol.* **1999**, *45*, 101–111. [CrossRef]
49. Cao, K.; Long, A.; Wang, J.; Liu, Y.; Cai, S.; Li, Y. Research and Application on Basin Accumulated Temperature Distribution (Atd) Model at the Snowmelt Flood Magnitude. *J. North China Univ. Water Resour. Electr. Power* **2017**, *38*, 10–18.
50. Meng, X.; Ji, X.; Liu, Z.; Xiao, J.; Chen, X.; Wang, F. Research on improvement and application of snowmelt module in SWAT. *J. Nat. Resour.* **2014**, *29*, 528–539.

51. Ferronskii, V.; Polyakov, V.; Brezgunov, V.; Vlasova, L.; Karpychev, Y.A.; Bobkov, A.; Romaniovskii, V.; Johnson, T.; Ricketts, D.; Rasmussen, K. Variations in the hydrological regime of Kara-Bogaz-Gol Gulf, Lake Issyk-Kul, and the Aral Sea assessed based on data of bottom sediment studies. *Water Resour.* **2003**, *30*, 252–259. [CrossRef]
52. Romanovsky, V. Water level variations and water balance of Lake Issyk-Kul. In *Lake Issyk-Kul: Its Natural Environment*; Springer: Berlin/Heidelberg, Germany, 2002; pp. 45–57.
53. uulu Salamat, A.; Abuduwaili, J.; Shaidyldaeva, N. Impact of climate change on water level fluctuation of Issyk-Kul Lake. *Arab. J. Geosci.* **2015**, *8*, 5361–5371. [CrossRef]
54. Shabunin, G.; Shabunin, A. Climate and physical properties of water in Lake Issyk-Kul. In *Lake Issyk-Kul: Its Natural Environment*; Springer: Berlin/Heidelberg, Germany, 2002; pp. 3–11.
55. Alifujiang, Y.; Abuduwaili, J.; Ma, L.; Samat, A.; Groll, M. System Dynamics Modeling of Water Level Variations of Lake Issyk-Kul, Kyrgyzstan. *Water* **2017**, *9*, 989. [CrossRef]
56. Propastin, P. Assessment of climate and human induced disaster risk over shared water resources in the Balkhash Lake drainage basin. In *Climate Change and Disaster Risk Management*; Springer: Berlin/Heidelberg, Germany, 2013; pp. 41–54.
57. Romanovsky, V.V.; Tashbaeva, S.; Crétaux, J.-F.; Calmant, S.; Drolon, V. The closed Lake Issyk-Kul as an indicator of global warming in Tien-Shan. *Nat. Sci.* **2013**, *5*, 32106. [CrossRef]
58. Dong, X.; Wang, Y.; Ding, Y.; Wang, C.; Sun, H.; Qin, X.; Jiang, C.; He, F. Assessment of impact of unbalancing power allocation on calculating maximum loading point. In Proceedings of the 2016 IEEE PES Asia-Pacific Power and Energy Engineering Conference (APPEEC), Xi'an, China, 25–28 October 2016; IEEE: Piscataway, NJ, USA, 2016.
59. Wu, L.; Wang, S.; Bai, X.; Luo, W.; Tian, Y.; Zeng, C.; Luo, G.; He, S. Quantitative assessment of the impacts of climate change and human activities on runoff change in a typical karst watershed, SW China. *Sci. Total Environ.* **2017**, *601*, 1449–1465. [CrossRef] [PubMed]
60. Braud, I.; Roux, H.; Anquetin, S.; Maubourguet, M.-M.; Manus, C.; Viallet, P.; Dartus, D. The use of distributed hydrological models for the Gard 2002 flash flood event: Analysis of associated hydrological processes. *J. Hydrol.* **2010**, *394*, 162–181. [CrossRef]
61. Vincendon, B.; Ducrocq, V.; Saulnier, G.-M.; Bouilloud, L.; Chancibault, K.; Habets, F.; Noilhan, J. Benefit of coupling the ISBA land surface model with a TOPMODEL hydrological model version dedicated to Mediterranean flash-floods. *J. Hydrol.* **2010**, *394*, 256–266. [CrossRef]
62. Fuka, D.R.; Easton, Z.M.; Brooks, E.S.; Boll, J.; Steenhuis, T.S.; Walter, M.T. A Simple Process-Based Snowmelt Routine to Model Spatially Distributed Snow Depth and Snowmelt in the SWAT Model 1. *JAWRA J. Am. Water Resour. Assoc.* **2012**, *48*, 1151–1161. [CrossRef]
63. Green, C.; Van Griensven, A. Autocalibration in hydrologic modeling: Using SWAT2005 in small-scale watersheds. *Environ. Model. Softw.* **2008**, *23*, 422–434. [CrossRef]
64. Ahl, R.S.; Woods, S.W.; Zuuring, H.R. Hydrologic calibration and validation of swat in a snow-dominated rocky mountain watershed, montana, USA 1. *JAWRA J. Am. Water Resour. Assoc.* **2008**, *44*, 1411–1430. [CrossRef]
65. Haq, M. *Snowmelt Runoff Investigation in River Swat Upper Basin Using Snowmelt Runoff Model, Remote Sensing and GIS Techniques*; ITC: Hudsonville, MI, USA, 2008.
66. Dudley, R.W.; Hodgkins, G.A.; Mchale, M.; Kolian, M.J.; Renard, B. Trends in snowmelt-related streamflow timing in the conterminous United States. *J. Hydrol.* **2017**, *547*, 208–221. [CrossRef]
67. Stigter, E.E.; Wanders, N.; Saloranta, T.M.; Shea, J.M.; Bierkens, M.F.; Immerzeel, W.W. Assimilation of snow cover and snow depth into a snow model to estimate snow water equivalent and snowmelt runoff in a Himalayan catchment. *Cryosphere* **2017**, *11*, 1647–1664. [CrossRef]
68. Hock, R.; Jansson, P.; Braun, L.N. Modelling the response of mountain glacier discharge to climate warming. In *Global Change and Mountain Regions*; Springer: Berlin/Heidelberg, Germany, 2005; pp. 243–252.
69. Luo, Y.; He, C.; Sophocleous, M.; Yin, Z.; Hongrui, R.; Ouyang, Z. Assessment of crop growth and soil water modules in SWAT2000 using extensive field experiment data in an irrigation district of the Yellow River Basin. *J. Hydrol.* **2008**, *352*, 139–156. [CrossRef]
70. Braun, L.; Grabs, W.; Rana, B. Application of a conceptual precipitation-runoff model in the Langtang Khola basin, Nepal Himalaya. *IAHS Publ.-Publ. Int. Assoc. Hydrol. Sci.* **1993**, *218*, 221–238.
71. Pomeroy, J.W.; Marks, D.; Link, T.; Ellis, C.; Hardy, J.; Rowlands, A.; Granger, R. The impact of coniferous forest temperature on incoming longwave radiation to melting snow. *Hydrol. Process. Int. J.* **2009**, *23*, 2513–2525. [CrossRef]
72. Wang, X.; Luo, Y.; Sun, L.; Zhang, Y. Assessing the effects of precipitation and temperature changes on hydrological processes in a glacier-dominated catchment. *Hydrol. Process.* **2015**, *29*, 4830–4845. [CrossRef]
73. Xiao, Y. A Method of Calculating Effective Accumulated Temperature Is Introduced Based on Daily Maximum and Minimum Temperature. *Plant Prot.* **1983**, *9*, 43–45.
74. Nash, J.E.; Sutcliffe, J.V. River flow forecasting through conceptual models part I—A discussion of principles. *J. Hydrol.* **1970**, *10*, 282–290. [CrossRef]
75. Moriasi, D.N.; Arnold, J.G.; Van Liew, M.W.; Bingner, R.L.; Harmel, R.D.; Veith, T.L. Model evaluation guidelines for systematic quantification of accuracy in watershed simulations. *Trans. ASABE* **2007**, *50*, 885–900. [CrossRef]
76. Alifujiang, Y.; Abuduwaili, J.; Groll, M.; Issanova, G.; Maihemuti, B. Changes in intra-annual runoff and its response to climate variability and anthropogenic activity in the Lake Issyk-Kul Basin, Kyrgyzstan. *Catena* **2021**, *198*, 104974. [CrossRef]

77. Deng, H.; Pepin, N.; Chen, Y. Changes of snowfall under warming in the Tibetan Plateau. *J. Geophys. Res. Atmos.* **2017**, *122*, 7323–7341. [CrossRef]
78. Ficklin, D.L.; Barnhart, B.L. SWAT hydrologic model parameter uncertainty and its implications for hydroclimatic projections in snowmelt-dependent watersheds. *J. Hydrol.* **2014**, *519*, 2081–2090. [CrossRef]
79. Dahri, Z.H.; Ahmad, B.; Leach, J.H.; Ahmad, S. Satellite-based snowcover distribution and associated snowmelt runoff modeling in Swat River Basin of Pakistan. *Proc. Pak. Acad. Sci.* **2011**, *48*, 19–32.
80. Engelhardt, M.; Schuler, T.; Andreassen, L. Contribution of snow and glacier melt to discharge for highly glacierised catchments in Norway. *Hydrol. Earth Syst. Sci.* **2014**, *18*, 511–523. [CrossRef]
81. Rosenwinkel, S.; Landgraf, A.; Schwanghart, W.; Volkmer, F.; Dzhumabaeva, A.; Merchel, S.; Rugel, G.; Preusser, F.; Korup, O. Late Pleistocene Outburst Floods from Issyk Kul, Kyrgyzstan? *Earth Surf. Process. Landf.* **2017**, *42*, 1535–1548. [CrossRef]
82. Deng, H.; Chen, Y.; Wang, H.; Zhang, S. Climate change with elevation and its potential impact on water resources in the Tianshan Mountains, Central Asia. *Glob. Planet. Chang.* **2015**, *135*, 28–37. [CrossRef]
83. Zhang, F.; Zhang, H.; Hagen, S.C.; Ye, M.; Wang, D.; Gui, D.; Zeng, C.; Tian, L.; Liu, J.S. Snow cover and runoff modelling in a high mountain catchment with scarce data: Effects of temperature and precipitation parameters. *Hydrol. Process.* **2015**, *29*, 52–65. [CrossRef]
84. Merz, R.; Parajka, J.; Blöschl, G. Time stability of catchment model parameters: Implications for climate impact analyses. *Water Resour. Res.* **2011**, *47*. [CrossRef]
85. Sexton, A.; Sadeghi, A.; Zhang, X.; Srinivasan, R.; Shirmohammadi, A. Using NEXRAD and rain gauge precipitation data for hydrologic calibration of SWAT in a northeastern watershed. *Trans. ASABE* **2010**, *53*, 1501–1510. [CrossRef]
86. Zhou, Z.; Bi, Y. Improvement of Swat Model and Its Application in Simulation of Snowmelt Runoff. In Proceedings of the National Symposium on Ice Engineering, Hohhot, China, 1 July 2011.
87. Yanmei, T.; Weize, M.; Xinjian, L.; Hui, W. Study on the models of predicting the annual accumulated temperature in the main cotton-production regions in Xinjiang. *Arid. Zone Res.* **2005**, *22*, 259–263.
88. Gafurov, A.; Kriegel, D.; Vorogushyn, S.; Merz, B. Evaluation of remotely sensed snow cover product in Central Asia. *Hydrol. Res.* **2013**, *44*, 506–522. [CrossRef]
89. Richard, C.; Gratton, D. The importance of the air temperature variable for the snowmelt runoff modelling using the SRM. *Hydrol. Process.* **2001**, *15*, 3357–3370. [CrossRef]
90. Duan, Y.; Liu, T.; Meng, F.; Yuan, Y.; Luo, M.; Huang, Y.; Xing, W.; Nzabarinda, V.; De Maeyer, P. Accurate simulation of ice and snow runoff for the mountainous terrain of the kunlun mountains, China. *Remote Sens.* **2020**, *12*, 179. [CrossRef]
91. Mernild, S.H.; Liston, G.E. The influence of air temperature inversions on snowmelt and glacier mass balance simulations, Ammassalik Island, Southeast Greenland. *J. Appl. Meteorol. Climatol.* **2010**, *49*, 47–67. [CrossRef]
92. Cazorzi, F.; Dalla Fontana, G. Snowmelt modelling by combining air temperature and a distributed radiation index. *J. Hydrol.* **1996**, *181*, 169–187. [CrossRef]
93. Lu, X.; Xie, G.; Li, Y.; Chen, S. Variation characteristics of snow cover and the relation to air temperature and precipitation in Manasi River Basin. *Desert Oasis Meteorol.* **2010**, *4*, 35–39.
94. Zhao, C.; Yan, X.; Li, D.; Wang, Y.; Luo, Y. The variation of snow cover and its relationship to air temperature and precipitation in Liaoning Province during 1961–2007. *J. Glaciol. Geocryol.* **2010**, *32*, 461–468.
95. Xu, L.; Wu, B. Relationship between Eurasian snow cover and late-spring and early-summer rainfall in China in 2010. *Plateau Meteorol.* **2012**, *31*, 706–714.
96. Wang, P.; Mu, Z. Study on Relationship of Snowmelt Runoff with Snow Area and Temperature in Km River Basin. *J. Water Resour. Water Eng.* **2013**, *24*, 28–31.
97. Aizen, V.; Aizen, E.; Nesterov, V.; Sexton, D. A study of glacial runoff regime in Central Tien Shan during 1989–1990. *J. Glaciol. Geocryol.* **1993**, *3*, 442–459.
98. Zhang, Y.; Luo, Y.; Sun, L. Quantifying future changes in glacier melt and river runoff in the headwaters of the Urumqi River, China. *Environ. Earth Sci.* **2016**, *75*, 770. [CrossRef]
99. Rulin, O.; Liliang, R.; Weiming, C.; Zhongbo, Y. Application of hydrological models in a snowmelt region of Aksu River Basin. *Water Sci. Eng.* **2008**, *1*, 1–13.
100. Hock, R. Temperature index melt modelling in mountain areas. *J. Hydrol.* **2003**, *282*, 104–115. [CrossRef]
101. Iwata, Y.; Nemoto, M.; Hasegawa, S.; Yanai, Y.; Kuwao, K.; Hirota, T. Influence of rain, air temperature, and snow cover on subsequent spring-snowmelt infiltration into thin frozen soil layer in northern Japan. *J. Hydrol.* **2011**, *401*, 165–176. [CrossRef]

atmosphere MDPI

Article

Geothermal Energy Potential for Cooling/Heating Greenhouses in Hot Arid Regions

Ibrahim Al-Helal [1], Abdullah Alsadon [2], Samy Marey [3], Abdullah Ibrahim [2], Mohamed Shady [1] and Ahmed Abdel-Ghany [1,4,*]

1 Department of Agricultural Engineering, College of Food and Agriculture Sciences, King Saud University, P.O. Box 2460, Riyadh 11451, Saudi Arabia; imhelal@ksu.edu.sa (I.A.-H.); mshady@ksu.edu.sa (M.S.)
2 Department of Plant Production, College of Food and Agriculture Sciences, King Saud University, P.O. Box 2460, Riyadh 11451, Saudi Arabia; alsadon@ksu.edu.sa (A.A.); adrahim@ksu.edu.sa (A.I.)
3 Sciences, Technology and Innovation Unit, King Saud University, P.O. Box 2454, Riyadh 11451, Saudi Arabia; samarey@ksu.edu.sa
4 Mechanical Power Engineering Department, Faculty of Energy Engineering, Aswan University, Aswan 8510, Egypt
* Correspondence: aghany@ksu.edu.sa or aghany@energy.aswu.edu.eg

Abstract: In arid regions, drastic seasonal variations in the climatic parameters are common; thus, a high potential of geothermal effects for heating/cooling applications is expected. However, such applications are very limited in these regions due to the lack of information about underground temperature profiles of the surface and shallow zones. Therefore, this study aims to (i) measure the underground temperature profile for one year to determine the optimum depth for burying EAHE pipes; (ii) examine the possibility of water vapour condensation occurring in the buried EAHE pipes, if the air let into the pipes was humid; and (iii) quantify the maximum cooling/heating capacity, if an EAHE was implemented. The results show that a 3-m depth is optimal to bury EAHE pipes, where the ground temperature is 32 °C in the summer and 29 °C in the winter. These temperatures would provide a maximum cooling/heating capacity of 1000/890 MJ day^{-1} for each 1 m^3 of humid air exhausted from a greenhouse. If the EAHE were to operate in a closed loop with a greenhouse, the condensation of water vapour in the EAHE pipes would be impossible during the cooling process. The results of this study are useful for designers using geothermal effects for indoor space cooling and heating in arid regions.

Keywords: arid climate; geothermal energy; underground temperature; greenhouse; heat exchanger

Citation: Al-Helal, I.; Alsadon, A.; Marey, S.; Ibrahim, A.; Shady, M.; Abdel-Ghany, A. Geothermal Energy Potential for Cooling/Heating Greenhouses in Hot Arid Regions. *Atmosphere* **2022**, *13*, 105. https://doi.org/10.3390/atmos13010105

Academic Editors: Zengyun Hu, Xuguang Tang and Qinchuan Xin

Received: 10 December 2021
Accepted: 5 January 2022
Published: 10 January 2022

1. Introduction

The use of greenhouses in arid areas, such as in the Arabian Peninsula, faces serious obstacles due to the extremely harsh environment; daytime overheating caused by intensive solar irradiance and night-time unfavourable undercooling are common weather conditions in desert areas [1]. By solving greenhouse operational problems (cooling in summer days and heating in winter nights), farmers can grow protected crops throughout the year.

Evaporative cooling (wet pad–fan systems) and fuel burners are commonly used for cooling and heating greenhouse air, whereby cold humid air (in summer) and warm air (in winter) are exhausted from the greenhouse and lost to the surrounding (useless air) during ventilation processes. Moreover, heating and evaporative cooling systems are an additional burden of energy and water consumption in greenhouses [2]. To overcome these difficulties, a semi-closed greenhouse has been suggested, where the inside greenhouse air is circulated in a closed loop and its cooling or heating are obtained by using sustainable energy resources. Looking for low-cost, energy- and water-saving techniques for heating/cooling inside greenhouse air is an essential priority for sustainable agriculture and the development of the rural areas in arid regions [3]. Therefore, a semi-closed greenhouse coupled with an earth–air heat exchanger (EAHE) was suggested in the Kingdom of Saudi Arabia (KSA). The EAHE is expected to be an effective way for heating/cooling greenhouses in the deserts characterised by hot arid climates. This is because the extracted low-grade geothermal energy (fluid temperature < 100 °C) as a sustainable energy resource is applicable for different purposes, such as desalination; drying technology; heating and cooling load production, using single-effect lithium bromide absorption chillers; agriculture greenhouses; and domestic use [4]. On the other hand, high-grade geothermal energy is applicable for power generation based on the Rankin cycle concept [5,6].

Even though the KSA is considered one of the most geothermally active countries in the Middle East [7], there is a lack of studies that address the issues of geothermal energy potential and its applications in the KSA [8]. Very few studies can be found that discuss the geothermal activity in the KSA from a geological science point of view [7,8]. Moreover, the geothermal energy potential, as well as the underground temperature profile in the shallow zone (0~8 m), has never been measured or evaluated in the KSA desert. Specifically, accurate information about the vertical distribution of soil temperature and geothermal energy potential in the shallow zone are still missing relatively to the KSA desert.

It is well known that the earth strongly absorbs solar energy during the daytime and stores a considerable amount of thermal energy at a particular depth. This is mainly attributable to the high solar irradiance intensity and large heat capacity of the soil. According to the diurnal variation in solar irradiance and ambient air temperature, the maximum temperature oscillation occurs at the Earth's surface, and it varies with the depth inside the Earth. Many researchers have measured the temperature distribution with depth and found that the oscillation of temperature below the ground attenuated with depth and became constant at a particular depth. For example, at a 0.18-m depth, the daily average soil temperature was 17.84 °C in the winter and 32.87 °C in the summer in New Delhi, India [9]. Moreover, the annual average temperatures of soil were measured at a 4-m depth and under different conditions of the ground surface in New Delhi, India. It was 29 °C when the ground surface was exposed to solar radiation, 19 °C when the ground surface was wetted, and 17 °C when the ground surface was wetted and shaded [9]. Different values of the annual average temperature of different soils in different places worldwide have been measured at different depths and reported in the literature, e.g., 23.45 °C at a 4-m depth in Las Vegas, USA [10]; 27–28 °C at a 2.5-m depth under warm humid weather conditions in Mexico City [11]; 18.7 °C at a 2-m depth and 20 °C at a 4-m depth in Brazil [12]; 25 °C at a 2-m depth in Bhopal, India [13]; 11.5 °C in winter and 17.5 °C in summer at a 2-m depth in Tianjin, China [14]; and 17.6 °C at a 3.6-m depth in Shouguang City, China [15]. In a typical arid climate (the desert in the south part of Algeria), the underground temperature was measured, in the hottest month in summer, to be 30 °C at a 2-m depth and 27 °C at a 5-m depth [16]. These temperatures are usually defined as ground undisturbed temperatures (GUTs). The GUT at a certain depth in the ground is nearly constant during the day and night and throughout the year. This depends on the groundwater level, the physical/chemical properties of soil and the ground surface conditions (mainly solar radiation flux and ambient air temperature). For greenhouse applications, the GUTs reported in the literature are reasonable for heating and cooling greenhouses; further, they can be used for indoor space heating and cooling (e.g., residential buildings, poultry houses, livestock houses, etc.) [17]. However, the cost of digging to the optimum depth for availing free geothermal energy should be considered; this depends on the type and nature of the soil. Besides the sustainability and low-cost and/or free geothermal energy for heating/cooling greenhouses, it can also be recognized that the thermal load levelling of the underground temperature at a depth from 2.5 m to 4 m is very low and is required for plants' healthy growth, irrespective of any climatic condition [17,18]. To use the geothermal energy potential for cooling/heating greenhouses in a closed loop, Polyvinyl Chloride (PVC) or High-density Polyethylene (HDPE) pipes are an optimum choice due to their low cost, low heat capacity and physical/chemical properties when they are buried at an optimum depth in the ground. In greenhouse applications, buried pipes are to carry the exhausted humid air from the greenhouse; then, the air flowing through the pipes is either heated or cooled according to the requirement. This depends on the temperature difference between the inside greenhouse air and the inside surface temperature of the buried PVC or HDPE pipes. The mass flow rate, which depends on the cross-sectional area of the pipe and air velocity, can be optimized for a given greenhouse volume. This arrangement is generally referred to as the closed-loop earth–air heat exchanger (CL-EAHE). Such an arrangement has been implemented and evaluated for heating/cooling greenhouses by many researchers [17–23]; however, a survey of the literature revealed that there is a lack of and unclear information about CL-EAHEs operated in gravel–sand soil and arid climates (such as in the KSA). Moreover, because the buried pipes of CL-EAHEs carry humid greenhouse air, water vapour condensation inside the pipes is possible during the cooling process. This possibility needs to be examined for greenhouse applications. Another arrangement, called open-loop earth–air heat exchanger (OL-EAHE), is out of the scope of the present study. This is because, in the OL-EAHE, the preconditioned greenhouse air is usually discharged to the surrounding area outside the greenhouse (i.e., energy and water vapor losses) during the heating and cooling processes of inside greenhouse air. Accordingly, the main objectives of this study are (i) to experimentally measure the temperature profile at different depths inside the ground at the King Saud University campus (as a desert in an arid climate) to determine the optimum depth for burying EAHE pipes; (ii) to examine the possibility of condensation inside the buried pipes; and (iii) to evaluate the maximum heating and cooling potential of the geothermal energy in the KSA for possible use in greenhouses and other domestic and residential building applications.

2. Methodology

2.1. Study Area

The experiment was conducted at the Agricultural Research and Experiment Station, Agriculture Engineering Department, King Saud University (Riyadh, Saudi Arabia; 46″43′ E longitude and 24″38′ N latitude). The soil at the site of the experiment, as well as that of most of arid regions such as the Arabian Peninsula, is composed of gravelly sand and dries up to a 5-m depth or more. Therefore, mechanical digging was used to prepare a hole with a surface area of about 1 m^2 and a depth of 3.5 m. In order to protect the cables of temperature sensors in the soil, these cables were collected to pass through a 5-cm diameter PVC pipe, installed vertically and fixed at the bottom of the hole with a concert block (Figure 1).

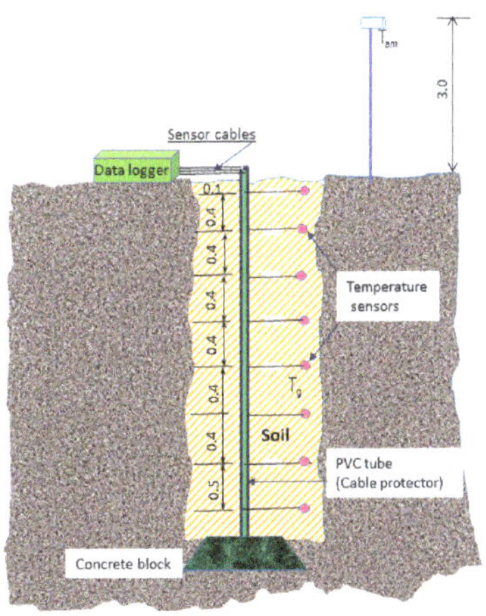

Figure 1. Schematic diagram showing the hole dug into the ground and the locations of thermocouples used to measure underground temperatures (T_g) and ambient temperature (T_{am}).

2.2. Measuring Underground Temperature Profiles

To determine the optimum depth at which the CL-EAHE pipes should be buried, the temperature profile of underground soil was measured at different depths. The optimum location for burying pipes is where the soil temperature became constant, unaffected by the diurnal and seasonal variations in the climatic parameters (GUT). The experiment was conducted during the period from 1 November 2020 to 31 October 2021, to measure the temperatures at the depths of 0.1, 0.5, 0.9, 1.3, 1.7, 2.1, 2.5 and 3 m, in addition to ambient air temperature. Digging up to only a 3.5-m depth was considered because more digging is quite hard and expensive due to the hard nature of soil in the desert of the KSA (i.e., mixture of gravelly sand and rocks). Moreover, the annual variation in soil temperature, T_g ($T_{g,max}$−$T_{g,min}$) at the 3-m depth was expected to be low. Eight small holes were made vertically in the pipe wall (Figure 1) at the levels of measurement (3, 2.5, 2.1, 1.7, 1.3, 0.9, 0.5 and 0.1 m below the ground surface). These holes were to create a path for the allocation of the cables of the temperature sensors and to protect the cables to reach the data logger above the ground via the pipe in a safe (Figure 1). The sensors were inserted into the soil 0.8 m apart from the vertical pipe. The backfilling of the ground hole was performed in steps. The first filling was up to a 3-m depth; then, the temperature sensor was fixed carefully on the soil surface; after that, the second backfilling was made at the level of a 2.5-m depth; then, the temperature sensor was inserted, etc. Thermocouple sensors (wzp-035 Pt100/k; Shenzhen More-Suns Electronics Co., Ltd., Shenzhen, China) were used to measure the underground temperature at the specified locations. The thermocouple had a precision of ±0.1 °C and a temperature range of 0~85 °C. Measurements were taken every 5 min, averaged at every 15 min and recorded in a COMBILOG-1022 data logger (32 channels; Theoder Friedrichs & Co., Schenefeld, Germany).

Due to difficulties and to the high cost of digging, the vertical variation in T_g at different depths and at any time was calculated using the Kasuda formula [24]. Assuming homogeneous soil with a constant thermal diffusivity (α_s), the monthly averaged soil temperature (T_g) at any depth (z) and month number (t) can be estimated by using the following formula:

$$T_g(z,t) = T_m - T_{amp} \times Exp\left[-z\sqrt{\left(\frac{\pi}{12\alpha_s}\right)}\right] \times cos\left\{\frac{2\pi}{12}\left[t - t_o - \frac{z}{2}\sqrt{\frac{12}{\pi\alpha_s}}\right]\right\} \quad (1)$$

where T_g (z,t) is the soil temperature at depth z and month number t; T_m is the annual mean soil surface temperature (°C); T_{amp} is the amplitude of the soil surface temperature [(max − min)/2] in (°C); z is the depth below the ground surface (m); α_s is the ground soil thermal diffusivity (m^2/month); t is time (the month number that the ground temperature is calculated for); t_o is the time shift (month of the year of the lowest ground surface temperature). For the desert of the KSA, the value of α_s was estimated as 2.736 (m^2/month) and thermal conductivity of soil as 2.2–2.8 (W m^{-1} °C^{-1}) [7,8]. From 1 November 2020 to 31 October 2021, the values of T_m and T_{amp} were estimated (from measurements) to be 30 and 12 °C, respectively. Moreover, in January, the lowest ground surface temperatures were recorded; then, t_o, in Equation (1), is equal to one.

2.3. Possibility of Condensation in the Buried CL-EAHE Pipes

In the summer months, the inner surface temperature of the buried EAHE pipes is expected to be much lower than the temperature of the humid air exhausted from the greenhouse. This makes the geothermal effect for cooling inside greenhouse air possible. In summer, when humid air flowing through the buried pipes may cool to its dew-point temperature (T_{dp}) through contact with the inner surface of the pipe, which is cooler than the air, water vapour condenses on the pipe surface. In this case, design considerations should be taken into account to collect the condensed water from the buried EAHE pipes. This may increase the EAHE's cost; therefore, an accurate calculation for T_{dp} is important for the appropriate design of a low-cost CL-EAHE. To examine the possibility of condensation, an experiment was conducted to measure the dry-bulb temperature (T_d) and relative humidity (*RH*) of the ventilated humid air at the greenhouse outlet (exactly before the exhaust fans). A crop-free greenhouse with a floor area of 165 m^2 was used; it was covered with a double-layered polycarbonate sheet of an 8.15-mm thickness. The greenhouse was mechanically ventilated using two exhaust fans, each with an airflow rate of 350 m^3 min^{-1}, operated in summer (July 2021) and in winter (December 2020) to obtain data for two extreme weather conditions, in Riyadh, KSA. T_d and *RH* were measured using a DMA033 Thermo-hygrometer (LSI-Lastem, Milano, Italy). The parameters were measured every 1 min, averaged at every 15 min and recorded in a data logger (CR23X Micro logger, Campbell Scientific, Inc., Oldenburg, Germany). The well-known approximate formulation used to calculate the dew-point temperature (T_{dp}) is based on Magnus' formula [25], for which the measured T_d (in °C) and *RH* (%) are required; then, the saturated water vapour pressure (P_s in Pascal) corresponding to T_d is given, by [25], as follows:

$$P_s = 610.78 \times Exp\left(\frac{17.2694 \times T_d}{(T_d + 238.3)}\right) \tag{2}$$

The actual water vapour pressure (P_a, in Pascal) is given by the following:

$$P_a = P_s \times RH(\%)/100 \tag{3}$$

$$S = ln(P_a/610.78) \tag{4}$$

$$T_{dp} = S \times 238.3/(17.294 - S) \tag{5a}$$

Another simple approximation is used to calculate T_{dp} (°C) for $RH > 50\%$ and ±1 °C error [26], in the following form:

$$T_{dp} = T_d - \left(\frac{100 - RH}{5}\right) \tag{5b}$$

These approximations are commonly used to calculate T_{dp}; however, in arid climates (where T_d is very high and *RH* is very low), the validity of these approximations ((Equation (5a,b)) needs to be examined specifically for the arid climate.

For greater accuracy, the saturation water vapour pressure (P_s) has been modified and Equation (5a) enhanced, becoming known as an Arden Buck equation [27], by which the modified saturation water vapour pressure (P_{sm}, in Pascal) corresponding to a dry bulb temperature of air (T_d, in °C) is given by the following:

$$P_{sm} = a \times Exp\left\{\left(b - \frac{T_d}{D}\right)\left(\frac{T_d}{c + D}\right)\right\} \tag{6}$$

$$T_{dp} = \frac{c \times ln\left(\frac{RH}{100} \times \frac{P_{sm}}{a}\right)}{b - ln\left(\frac{RH}{100} \times \frac{P_{sm}}{a}\right)} \tag{7}$$

The constants in Equations (6) and (7) were provided with a maximum error of ≤0.05% and T_d from 0 up to 50 °C as follows: a = 611.21 Pa, b = 17.368, c = 238.88 °C and D = 234.5 °C [27].

2.4. Geothermal Cooling/Heating Capacity

In an EAHE, the PVC or HDPE pipes are buried permanently under the ground at a specified location, where the ground temperature (T_g) is annually stable and nearly constant; therefore, the thermal conditions can be characterized as steady state and equilibrium. Moreover, the thickness of an EAHE pipe is very modest (e.g., 4.25 mm) compared to the pipe surface area or the pipe diameter (203.2 mm). Hence, for simplicity, we assumed that the thermal resistance of the pipe material was negligible, and the inner surface temperature of the pipe (T_s) was almost uniform and constant along the axial direction of pipe and equal, in most cases, to the ground temperature (T_g) at the pipe location. The cooling and heating capacity of an EAHE ($\dot{Q}_{c/h}$) is defined as follows:

$$\dot{Q}_{c/h} = \dot{m}_a C_p \left(T_{out} - T_{in}\right) \tag{8}$$

where $\dot{m}_a$ is the mass flow rate of air exhausted from the greenhouse (ventilation rate at T_{ex}) and flowing through the EAHE pipes (kg s^{-1}); C_p is the specific heat capacity of air (J kg^{-1} °C^{-1}); and T_{in} and T_{out} are the temperatures of the inlet and outlet air (°C) for the EAHE pipes. T_{in} is assumed to be the temperature of air exhausted from the greenhouse (T_{in} = T_{ex} in °C) and it can be obtained directly from measurements. For a very long pipe, T_{out} is assumed to be equal to the inner surface temperature of the EAHE pipes (i.e., T_{out} = T_s = T_g). In this case, Equation (8) represents the maximum cooling/heating capacity that can be provided by an EAHE ($Q_{c/h,max}$). In a similar manner, the cooling/heating potential ($Q_{c/h}$), in Joules, over a specified period is defined, by [17,18], as follows:

$$Q_{c/h} = \sum_{time} \dot{m}_a C_p (T_{out} - T_{in}) \Delta t \tag{9}$$

where Δt is the time interval (s). In fact, the outlet air temperature of EAHE pipes (T_{out}) is lower than the ground temperature (T_g) or the inner surface temperature of the pipes (T_s); this depends on the effectiveness of the EAHE. Therefore, an expression for T_{out} in terms of underground temperature (T_g) and inlet temperature of air (T_{in}) is given, by [13], as follows:

$$T_{out} = T_g \left\{ 1 - exp\left(-\frac{A_s h_f}{\dot{m}_a C_p} \right) \right\} + T_{in} exp\left(-\frac{A_s h_f}{\dot{m}_a C_p} \right) \tag{10}$$

where A_s is the inner surface area of a single pipe in the EAHE; $\dot{m}_a$ and C_p are the mass flow rate and specific heat of air flowing in a single pipe (kg s^{-1} and J kg^{-1} °C^{-1}); and h_f is the convective heat transfer coefficient between the inner surface of the pipe and the flowing air (W m^{-2} °C^{-1}). h_f is a critical parameter in the EAHE design; in such a case, it is reasonable to assume that the air flow in the EAHE pipes is fully developed. To adapt an appropriate correlation to calculate h_f, Bisoniya [13] examined eight Nusselt number (Nu) correlations and recommended the following correlation for turbulent flows in a pipe with a smooth internal surface [13]:

$$N_u = \frac{f/8(R_e - 1000)P_r}{1 + 12.7\sqrt{(f/8)}\left(P_r^{2/3} - 1\right)} \tag{11}$$

The friction coefficient f for turbulent flow is given by the following:

$$f = (1.82\, logR_e - 1.64)^{-2}, \left(2300 \le R_e < 5 \times 10^6\right) \text{ and } \left(0.5 < P_r < 10^6\right) \tag{12}$$

where the Reynolds number ($R_e = dv\rho/\mu$) and Prandtl number ($P_r = \mu C_p/k$) are calculated for the air flowing through a pipe; d is the inner diameter of the pipe; v is the air velocity in the pipe; ρ is the air density; μ is the air viscosity; and k is the thermal conductivity of air. The unit of each parameter was adjusted to give dimensionless numbers (Nu, Re and Pr).

By determining the value of h_f, Equation (10) can be used as a design tool to optimize the number of pipes and the length and diameter of each pipe for the EAHE design.

3. Results and Discussion

3.1. Underground Soil Temperature

It is well known that the ground temperature is affected by the cyclic variation in climatic parameters such as solar radiation, ambient air temperature, wind speed, humidity, rainfall and snow, if any, etc. The interaction between the climatic parameters and the ground is mainly at the ground surface; then, the cyclic variation in the ground temperature (T_g) decreases as the ground depth increases. At a certain depth, according to the chemical/physical properties of soil, T_g becomes nearly constant, unaffected by the diurnal and seasonal variations in the climatic parameters, and it is defined as the ground undisturbed temperature (GUT). The GUT is an essential parameter in designing an EAHE system. It is quite difficult to theoretically calculate the GUT correctly, because it depends on the soil parameters and the climatic conditions affecting the ground surface. Therefore, an experiment was conducted (Section 2) for measuring T_g at different ground depths during different seasons to determine the depth at which T_g became nearly constant throughout the year (GUT). To illustrate the daily cyclic variation in ambient air and soil temperatures during the period of the experiment (from 1 November 2020 to 31 October 2021), four days were selected to represent the extreme weather conditions, two days in cold winter (29–30 December) and two days in hot summer (30–31 July). The measured parameters (i.e., T_{am} and T_g at different depths) are depicted in Figure 2 (for winter days) and in Figure 3 (for summer days). In winter, the ground temperature (T_g) increased as the ground depth increased (Figure 2) and the opposite observation was noted in summer (Figure 3). In Figures 2 and 3, the cyclic variation in T_g is significant at a 0.1-m depth as affected by the diurnal variation in the climatic parameters (mainly the solar irradiance and T_{am}) in summer and winter. At a depth from $\geq$0.5 m up to 3 m, T_g became nearly constant in summer and in winter and was not affected by the diurnal variation in the climatic parameters. At a 3-m depth, T_g was about 29 °C in winter (Figure 2) and about 32 °C in summer (Figure 3) and a difference of 3 °C in the T_g value is acceptable between the hot summer and cold winter seasons. For more clarification and to find the optimum depth for burying the EAHE pipes, the vertical variation in the monthly average soil temperature (T_g) at different depths for the experimental site (Riyadh, KSA) is plotted in Figure 4 and the annual variation in temperatures of ambient air (T_{am}) and soil (T_g) at the 0.5-m and 3-m depths is plotted in Figure 5. As illustrated in Figures 4 and 5, at the 3-m depth, the maximum annual variation (max – min) in T_g was about 5 °C. The lowest value of T_g was around 27 °C in February and March and the highest value of T_g was around 32 °C in September and October. This annual variation in T_g (29–32 °C) is adequate for heating and cooling purposes and can be considered as the GUT and the depth of 3 m is adequate for burying the EAHE pipes. The annual variation in T_g is considerable at the 0.5-m depth; more stability of T_g was observed at the 3-m ground depth (Figure 5). According to Figure 5, the heating effect of a supposed EAHE in winter is expected to be relatively higher than its cooling effect in summer; this depends on the temperature difference between T_{am} and GUT (T_g at the 3-m depth).

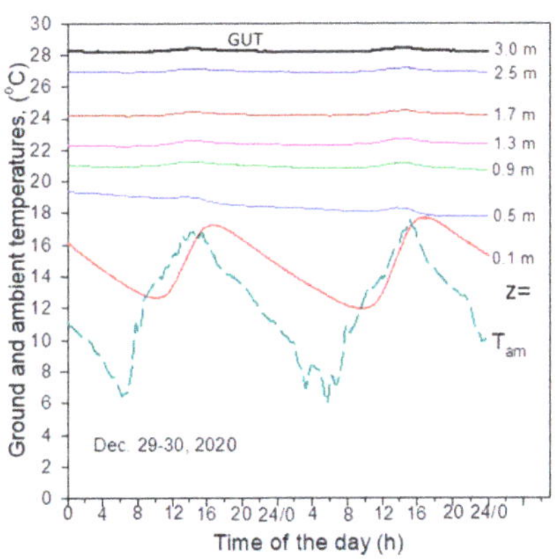

Figure 2. Diurnal variation in ambient temperature (T_{am}) and underground soil temperature (T_g) at different depths (z) in winter season (29–30 December 2020).

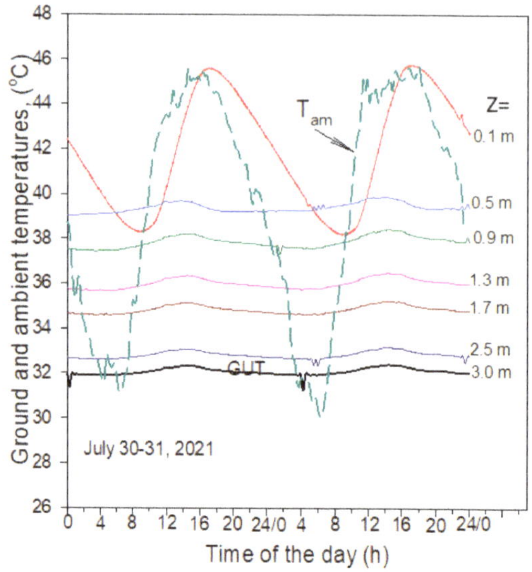

Figure 3. Diurnal variation in ambient temperature (T_{am}) and underground soil temperature (T_g) at different depths (z) in summer season (30–31 July 2021).

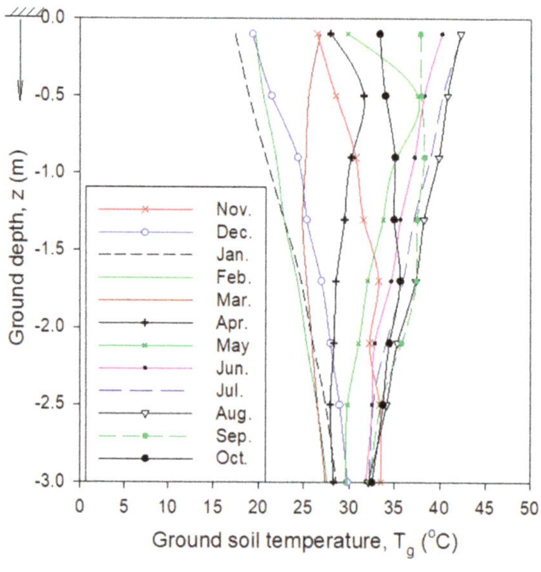

Figure 4. Vertical variation in the monthly average ground temperature (T_g) at different depths (z) for Riyadh region, KSA, in the years 2020–2021.

As farther digging in the desert of arid regions is quite difficult and expensive, we used Equation (1) to predict the monthly average values of T_g at different depths. To calculate the values of T_{m} and T_{amp} in Equation (1), the measured T_g at the 0.1-m depth was considered as the soil surface temperature. The predicted results for 12 months and depths of >3 m are plotted in Figure 6. As illustrated in Figure 6, at a depth (z) greater than 13 m, T_g becomes constant at 30 °C throughout the year. As the arid climatic condition is almost similar every year in the deserts of KSA, the GUT can be considered constant at 30 °C throughout the year at a depth greater than 13 m.

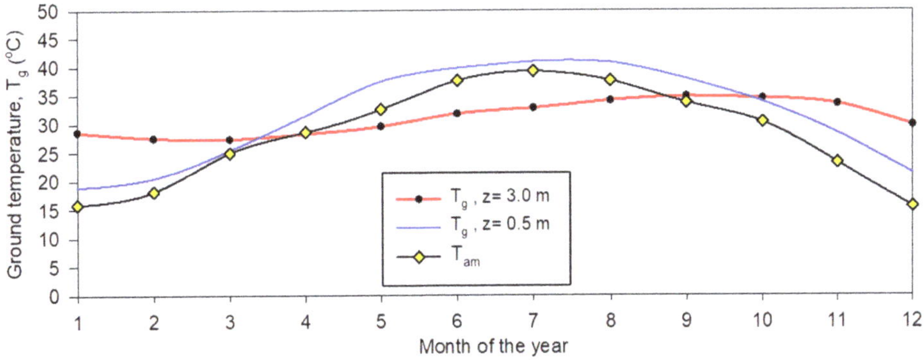

Figure 5. Annual variation in temperatures of ambient air (T_{am}) and ground soil (T_g) at 0.5- and 3.0-m depths for Riyadh, Saudi Arabia.

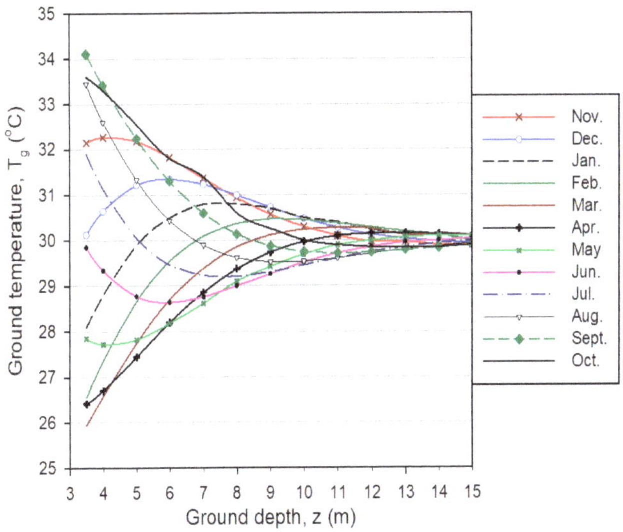

Figure 6. Ground temperature variation (monthly average) with depth, predicted using Equation (1) for the KSA desert.

3.2. Possibility of Condensation in the EAHE Pipes

The proposed CL-EAHE was applied in a greenhouse for cooling/heating purposes in the summer and winter seasons. In the cooling/heating processes, humid air is exhausted from the greenhouse (at T_{ex} and RH_{ex}); it flows in the buried pipes, then the condensation of water vapour in the pipes is possible during cooling. The occurrence of condensation requires extra design consideration to collect the condensed water from the pipes. The most accurate correlation (Equation (7)) was used to calculate the dew-point temperature (T_{dp}) of the exhausted air (having T_{ex} and RH_{ex}) for the four previously selected days (hot and dry in summer and humid and cold in winter). No heating or cooling was applied to the greenhouse to represent the actual practical situations. Figures 7 and 8 illustrate the diurnal variation in the measured T_{ex} and RH_{ex} for the exhausted air from the greenhouse, the measured GUT (T_g at the 3-m depth) and the calculated T_{dp} (using Equation (7)). A considerable difference could be observed between T_{ex} and T_{dp} in summer (Figure 8) and in winter (Figure 7). For winter heating (Figure 7), T_{ex} increased, as affected by the warm inner surface of the EAHE pipes; if the upper limit of T_{ex} was to reach the GUT, then the gap between the T_{ex} and T_{dp} of the air would increase and the condensation of water vapour would be impossible. For summer cooling (Figure 8), if T_{ex} was to decrease to its lower limit GUT (i.e., T_g at the 3-m depth), there would still be a large difference between T_{ex} and T_{dp} (Figure 8); therefore, condensation would never take place or would be impossible. Moreover, a large number of combinations for T_{ex} and RH_{ex} was used to calculate T_{dp} using Equation (7) and the resulting values of T_{dp} do not exceed 20 °C; in each combination, a large difference between T_{ex} and T_{dp} remained. Accordingly, condensation in CL-EAHE pipes would be impossible throughout the year for the pipes buried at a 3-m depth in the Riyadh area, KSA.

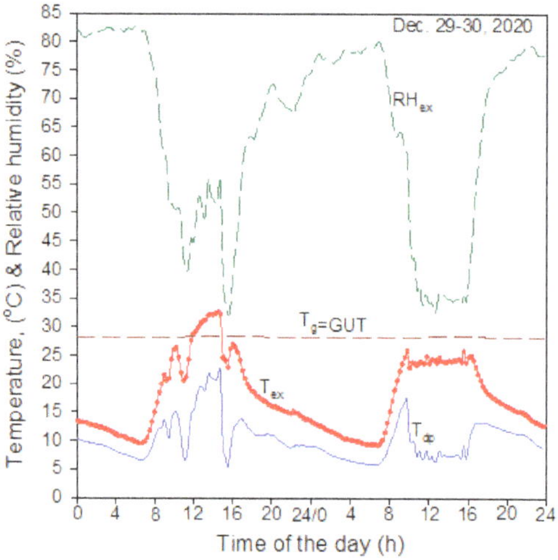

Figure 7. Diurnal variation in estimated dew-point temperature (T_{dp}) of air ventilated from a greenhouse at T_{ex} and RH_{ex} and flowed through EAHE pipes buried at the GUT in winter season (29–30 December 2020).

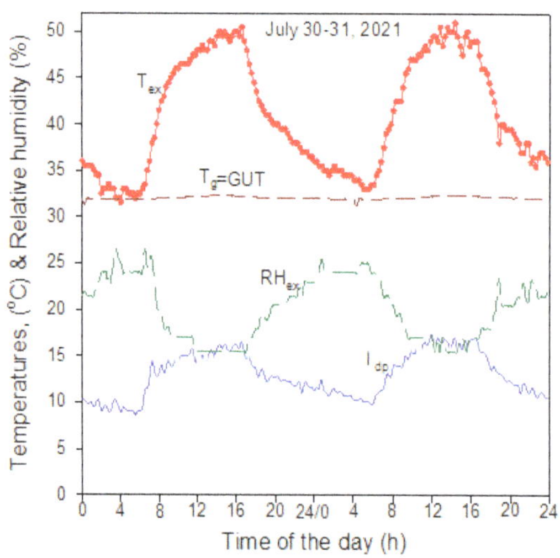

Figure 8. Diurnal variation in estimated dew-point temperature (T_{dp}) of air ventilated from a greenhouse at T_{ex} and RH_{ex} and flowed through EAHE pipes buried at the GUT in summer season (30–31 July 2021).

Similarly, T_{dp} was calculated using Equation (5a,b) to check the validity of these approximations to be used to calculate T_{dp} for an arid climate. The percentage error, E_R (%), was calculated in each case as E_R (%) = abs{($T_{accurate}$-$T_{approximate}$)/$T_{accurate}$} × 100. The values of E_R (%) were calculated for a wide range of T_d and RH and are plotted in Figure 9a,b for Equation (5a) and in Figure 10 for Equation (5b). Based on Figure 9a,b, Equation (5a) can be used to calculate T_{dp} with an error less than 1% for $RH > 40\%$ and $T_d < 30$ °C. However, Equation (5b) can be used only for $RH > 60\%$ with an error of about 3–4%; for low RH values (as in the arid climate), such an approximation (Equation (5b)) cannot be used. Accordingly, for arid climatic conditions (as in the Arabian Peninsula region), such approximations (Equation (5a,b)) are not recommended and Equation (7) is the appropriate correlation to determine T_{dp} correctly.

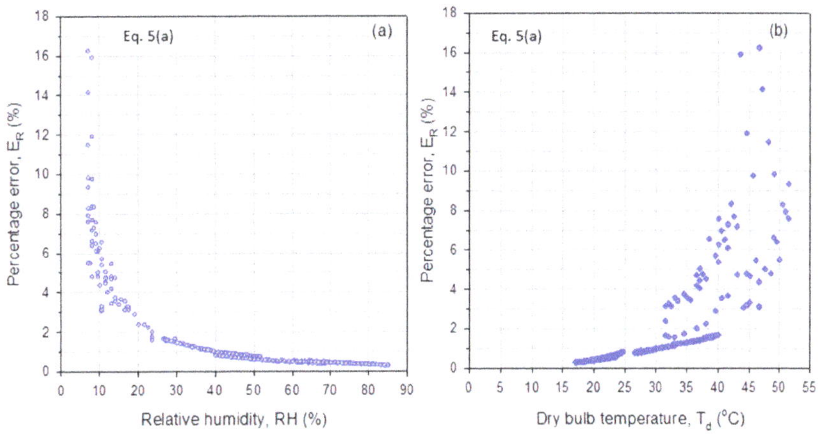

Figure 9. Percentage error, E_R (%), in the approximated value of T_{dp} (estimated by using Equation (5a)) as affected by (**a**) relative humidity (RH) and (**b**) dry bulb temperature (T_d).

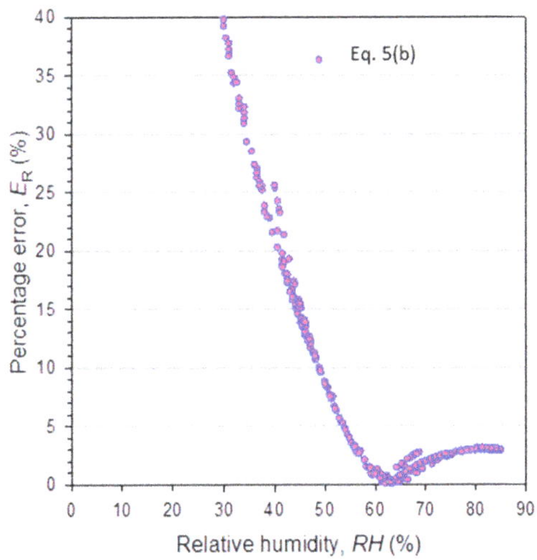

Figure 10. Percentage error, E_R (%), in the approximated value of T_{dp} (estimated by using Equation (5b)) as affected by relative humidity (RH).

3.3. Maximum Cooling/Heating Capacity

An EAHE's geothermal energy potential, as well as its cooling/heating capacity, depends mainly on the climatic parameters and the underground soil conditions. For EAHE pipes buried at a specified optimum depth, T_g equals the GUT. Under the steady-state thermal condition, the inner surface temperature of the pipe (T_s) is assumed to be uniform in the axial flow direction and by assuming the thermal resistance of the pipe material is negligible, the equality $T_s = T_g =$ GUT can be assumed with insignificant error. If the pipe is long enough, the outlet air temperature (T_{out}) from the pipe can be assumed to be equal to T_s ($T_{out} = T_s = T_g =$ GUT). In this case, the EAHE provides its maximum possible cooling/heating capacity. Therefore, in Equation (8), T_{in} and T_{out} were taken as T_{ex} (exit air from the greenhouse) and GUT, respectively; $\dot{Q}_{c/h,max}$ was estimated per cubic meter of airflow. For winter cooling, a considerable amount of heat is expected to be added to the flowing air at around midnight (Figure 11). In addition, at around noon, in winter, operating the EAHE is not necessary and the transmitted solar radiation into the greenhouse is enough for warming up the inside air. In summer, a considerable amount of heat is expected to be removed from the flowing air at around noon; however, at around midnight, it is not necessary to operate the EAHE (Figure 11). Under the presumed ideal conditions and to estimate the maximum possible cooling/heating potential that the ground (at a 3-m depth) can provide in the cold winter and hot summer in the Riyadh region, the results are integrated in Figure 11. The expected maximum possible cooling and heating potential are 890 and 1000 MJ m^{-3} day^{-1}, respectively. These values are promising to use EAHEs for different applications in the KSA and the Arabian Peninsula regions.

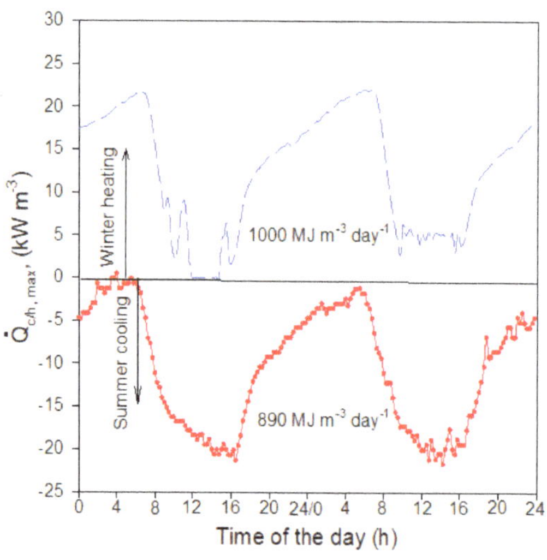

Figure 11. Diurnal variation in maximum cooling/heating capacity ($\dot{Q}_{c/h,max}$) that an EAHE can provide and the maximum cooling/heating potential per day.

4. Conclusions

This study is an attempt to provide critical information for EAHE designers and to evaluate the geothermal energy potential for possible cooling/heating greenhouses and indoor spaces in hot arid regions, for which two experiments were conducted. The main conclusion is summarized below.

The daily cyclic variation in the climatic parameters affected the ground temperature (T_g) up to a 0.5-m depth, after which T_g was constant daily; it increased with depth in winter and decreased with depth in summer.

At a 3-m depth, the annual variation in T_g was minor and it can be considered as the ground undisturbed temperature, GUT. The GUT value was 29 °C in winter and 32 °C in summer; these values are adequate for cooling/heating purposes in a climate in which the ambient air temperature drops below 10 °C on winter nights and exceeds 47 °C on summer days. For an EAHE operating under ideal conditions, the geothermal energy level can provide a maximum cooling/heating potential of 890/1000 MJ per m^3 of flowing air per day.

In summer seasons, during the cooling process of greenhouse air in EAHE pipes distributed at a 3-m depth, the condensation of water vapor in the pipes could never take place.

In arid climates, approximate correlations (Equation (5a,b)) are not recommended to calculate the dew-point temperature of air (T_{dp}). However, Equation (7) is the appropriate correlation to determine T_{dp} correctly.

The geothermal energy potential is promising for cooling/heating applications in arid regions such as the Arabian Peninsula for sustainable development and environmental protection. Further research should be conducted to design, construct and operate an EAHE connected to a greenhouse and evaluate the system performance under different operating and climate conditions.

Author Contributions: Conceptualization and methodology, A.A.-G. and I.A.-H.; validation, A.A., S.M. and A.I.; formal analysis, A.A.-G., S.M. and I.A.-H.; investigation, A.A.-G.; resources, M.S.; data creation, M.S., S.M. and A.I.; writing—original draft preparation, I.A.-H., S.M. and A.I.; writing—review and editing, A.A.-G.; visualization, M.S., S.M. and A.I.; supervision, A.A.-G.; project administration, I.A.-H. All authors have read and agreed to the published version of the manuscript.

Funding: This research study received no external funding.

Acknowledgments: The authors extend their appreciation to the Deputyship for Research and Innovation, Ministry of Education in Saudi Arabia, for funding this research work through the project number DRI-KSU-542.

Conflicts of Interest: The authors declare no conflict of interest.

Nomenclature

Symbol	*Description*	*(unit)*
A_s	Inner surface area of EAHE pipe	(m^2)
C_p	Specific heat of flowing air through EAHE pipes	(J kg^{-1} °C^{-1})
d	Inner diameter of the EAHE pipes	(m)
E_R	Percentage error	(%)
h_f	Convective heat transfer coefficient between inner surface of EAHE pipe and flowing air	(W m^{-2} °C^{-1})
k	Thermal conductivity of flowing air	(W m^{-1} °C^{-1})
L	Length of one EAHE pipe	(m)
$\dot{m}_a$	Mass flow rate of flowing air through EAHE pipes	(kg s^{-1})
N_u	Nusselt number	($N_u = dh/k$)
P_a	Actual water vapor pressure	(Pa)
P_r	Prandtl number	($P_r = \mu C_{pa}/k$)
P_s	Saturation water vapor pressure	(Pa)
$\dot{Q}_{c/h}$	Cooling/heating capacity	(W)
$Q_{c/h}$	Cooling/heating potential	(J)
R_e	Reynolds number	($R_e = dv\rho/\mu$)
RH	Relative humidity of air	(%)
RH_{ex}	Relative humidity of air exhausted from the greenhouse	(%)
t	Time	(s; day; month)
T_{amp}	Amplitude of the annual ground surface temperature	(°C)
T_d	Dry bulb temperature of air or ambient temperature ($T_{am} = T_d$)	(°C)
T_{dp}	dew point temperature of flowing air in the EAHE pipe	(°C)
T_{ex}	Temperature of air exhausted from the greenhouse	(°C)

T_g	Underground soil temperature	(°C)
T_{in}	Inlet hot/cold air temperature to the EAHE pipes	(°C)
T_m	Mean ground surface temperature, annual average	(°C)
T_{out}	Outlet hot/cooled air temperature from the EAHE pipes	(°C)
T_s	Outlet hot/cooled air temperature from the EAHE pipes	(°C)
v	Velocity of flowing air through EAHE Pipe	($m\ s^{-1}$)
z	Depth below the soil surface	(m)
Greek letter		
α_s	Thermal diffusivity of soil	(m^2/day)
ρ	Density of flowing air through EAHE pipe	($kg\ m^{-3}$)
μ	Dynamic viscosity of flowing air through EAHE pipe	($kg\ m^{-1}\ s^{-1}$)
Δt	Interval of time	(s, h, day, etc.)
Abbreviations		
CL-EAHE	Closed-loop earth to air heat exchanger	
EAHE	Earth to air heat exchanger	
KSA	the Kingdom of Saudi Arabia	
OL-EAHE	Open loop earth to air heat exchanger	
GUT	Ground undisturbed temperature	(°C)

References

1. Abdel-Ghany, A.M.; Picuno, P.; Al-Helal, I.M.; Alsadon, A.; Ibrahim, A.; Shady, M. Radiometric characterization, solar and thermal radiation in a greenhouse as affected by shading configuration in an arid climate. *Energies* **2015**, *8*, 13928–13937. [CrossRef]
2. Al-Helal, I.M.; Al-Hamdan, A.M. Effect of arid environment on radiative properties of greenhouse polyethylene cover. *Sol. Energy* **2009**, *83*, 790–798. [CrossRef]
3. Alkoaik, F.N.; Al-Faraj, A.A.; Al-Helal, I.M.; Ronnel, B.F.; Mansour, N.I.; Abdel-Ghany, A.M. Toward sustainability in rural areas: Composting palm tree residues in rotation bioreactors. *Sustainability* **2019**, *12*, 201. [CrossRef]
4. El Haj, A.M.; Sadeghzadeh, M.; Ahmadi, M.H.; Al-Shabi, M.; Albawab, M.; Anvari-Moghaddam, A.; Bani-Hani, E. Space cooling using geothermal single-effect water/lithium bromide absorption chiller. *Energy Sci. Eng.* **2021**, *9*, 1747–1760. [CrossRef]
5. El Haj, A.M.; Aryanfar, Y.; Radman, S.; Yousef, B.; Pakatchian, M. Energy and exergy analyses of single flash geothermal power plant at optimum separator temperature. *Int. J. Low-Carbon Technol.* **2021**, *16*, 873–881.
6. El Haj, A.M.; Aryanfar, Y.; Javaherian, A.; Khosravi, A.; Aghaei, K.; Hosseinzadeh, S.; Pabon, J.; Mahmoudi, S.M.S. Energy, exergy, economic and exergoenvironmental analyses of transcritical CO_2 cycle powered by single flash geothermal power plant. *Int. J. Low-Carbon Technol.* **2021**, *16*, 1504–1518. [CrossRef]
7. Demirbas, A.; Alidrisi, H.; Ahmed, W.; Sheikh, M.H. Potential of geothermal energy in the Kingdom of Saudi Arabia. *Energy Sources Part A Recovery Util. Environ. Eff.* **2016**, *38*, 2238–2243. [CrossRef]
8. Lashin, A.; Al Arifi, N. Geothermal energy potential of southwestern of Saudi Arabia "exploitation and possible power generation": A case study at AlKhouba area-Jazan. *Renew. Sustain. Energy Rev.* **2014**, *30*, 771–789. [CrossRef]
9. Bharadwaj, S.S.; Bansal, N.K. Temperature distribution inside ground for various surface conditions. *Build. Environ.* **1981**, *16*, 183–192. [CrossRef]
10. Vidhi, R. A review of underground soil and night sky as passive heat sink: Design configurations and models. *Energies* **2018**, *11*, 2941. [CrossRef]
11. Díaz-Hernández, H.P.; Macias-Melo, E.V.; Aguilar-Castro, K.M.; Hernández-Pérez, I.; Xamán, J.; Serrano-Arellano, J.; López-Manrique, L.M. Experimental study of an earth to air heat exchanger (EAHE) for warm humid climatic conditions. *Geothermics* **2020**, *84*, 101741. [CrossRef]
12. Hermes, V.F.; Ramalho, J.V.A.; Rocha, L.A.O.; Santos, E.D.; Marques, W.C.; Costi, J.; Rodrigues, M.K.; Isoldi, L.A. Further realistic annual simulations of earth-air heat exchangers installations in a coastal city. *Sustain. Energy Technol. Assess.* **2020**, *37*, 100603. [CrossRef]
13. Bisoniya, T.S. Design of earth–air heat exchanger system. *Geotherm Energy* **2015**, *3*, 18. [CrossRef]
14. Cao, S.; Li, F.; Li, X.; Yang, B. Feasibility analysys of earth-air heat exchanger (EAHE) in a sports and culture centre in Tianjin, China. *Case Stud. Therm. Eng.* **2021**, *26*, 101654. [CrossRef]
15. Le, A.T.; Wang, L.; Wang, Y.; Li, D. Measurement investigation on the feasibility of shallow geothermal energy for heating and cooling applied in agricultural greenhouses of Shouguang City: Ground temperature profiles and geothermal potential. *Inf. Process. Agric.* **2021**, *8*, 251–269. [CrossRef]
16. Hebbal, B.; Marif, Y.; Hamdani, M.; Belhadj, M.M.; Bouguettaia, H.; Bechki, D. The geothermal potential of underground buildings in hot climates: Case of southern Algeria. *Case Stud. Therm. Eng.* **2021**, *28*, 101422. [CrossRef]
17. Tiwari, G.N.; Akhtar, M.A.; Shukla, A.; Emran Khan, M. Annual thermal performance of greenhouse with an earth-air heat exchanger: An experimental validation. *Rene Energy* **2006**, *31*, 2432–2446. [CrossRef]

18. Ghosal, M.K.; Tiwari, G.N.; Das, D.K.; Pandey, K.P. Modeling and comparative thermal performance of ground air collector and earth air heat exchanger for heating of greenhouse. *Energy Build.* **2005**, *37*, 613–621. [CrossRef]
19. Stanciu, D.; Stanciu, C.; Paraschiv, I. Mathematical links between optimum solar collector tilts in isotropic sky for intercepting maximum solar irradiance. *J. Atmos. Sol.-Terr. Phys.* **2016**, *137*, 58–65. [CrossRef]
20. Ozgener, O.; Ozgener, L.; Goswami, D.Y. Experimental prediction of total thermal resistance of a closed loop EAHE for greenhouse cooling system. *Int. Commun. Heat Mass Transf.* **2011**, *38*, 711–717. [CrossRef]
21. Ozgener, O.; Ozgener, L.; Goswami, D.Y. Seven years energetic and exergetic monitoring for vertical and horizontal EAHE assisted agricultural building heating. *Renew. Sustain. Energy Rev.* **2017**, *80*, 175–184. [CrossRef]
22. Hepbasli, A. Low exergy modelling and performance analysis of greenhouses coupled to closed earth-to-air heat exchangers (EAHEs). *Energy Build.* **2013**, *64*, 224–230. [CrossRef]
23. Bisoniya, T.S.; Kumar, A.; Baredar, P. Experimental and analytical studies of earth-air heat exchanger (EAHE) systems in India: A review. *Renew. Sustain. Energy Rev.* **2013**, *19*, 238–246. [CrossRef]
24. Kasuda, T.; Archenbach, P.R. Earth temperature and thermal diffusivity at selected stations in the United States. *ASHRAE Trans.* **1965**, *71*, 61–74.
25. NASA Technical Note, NASA TN D-8401, Equations for the Determination of Humidity from Dew Point Ad Psychometric Data. Available online: https://www.nasa.gov/centers/dryden/pdf/87878main_H937.pdf (accessed on 14 August 2021).
26. Dew Point Temperature—An Overview, Science Direct Topics. Available online: https://www.sciencedirect.com/topics/engineering/dewpoint-temperature (accessed on 17 July 2021).
27. Buck, A.L. New equations for computing vapour pressure and enhancement factor. *J. Appl. Meteorol.* **1981**, *20*, 1527–1532. [CrossRef]

atmosphere

Article

Decomposition and Decoupling Analysis between HDI and Carbon Emissions

Dongju Li [1], Tongtong Shen [2,*], Xi Wei [1] and Jie Li [1]

[1] School of Statistics and Big Data, Henan University of Economics and Law, Zhengzhou 450046, China; 20100452@huel.edu.cn (D.L.); 20202300058@stu.huel.edu.cn (X.W.); 20212000653@stu.huel.edu.cn (J.L.)
[2] School of Statistics, Dongbei University of Finance and Economics, Dalian 116021, China
* Correspondence: shenttdufe@163.com

Abstract: The concept of low carbon is extended to the welfare dimension by considering the relationship between carbon emissions and the Human Development Index (HDI). This paper examines the decoupling between carbon emissions per capita and HDI and the welfare output of carbon emissions by using the data from 189 countries, from 1990 to 2019, as well as decomposes the drivers of the decoupling index and carbon emissions performance (CEP) in the example countries. The results show that most countries that achieve strong decoupling have very high human development, while the worst case is that a few countries with an extremely low human development achieved strong decoupling. Moreover, the status of strong decoupling in most countries is not stable, and there is a risk of transformation to another decoupling status. Although the CEP of most countries has gradually improved, very few countries have high CEP. Economic development effects are the primary inhibitor to achieving and maintaining strong decoupling in example countries. The main drivers of CEP improvement are the carbon productivity effects in the Czech Republic, Germany, and the United Kingdom, and the economic development effects in South Korea and Turkey. The main factors inhibiting the increase of CEP are the energy intensity effect in the Czech Republic, Germany, and the UK, and the welfare effect in South Korea and Turkey. These effects are all related to GDP. Economic activity broadly affects the decoupling index and CEP. Recommendations for maintaining HDI growth and reducing carbon emissions are made for countries with different human development.

Keywords: HDI; decoupling index; carbon emission performance; LMDI

Citation: Li, D.; Shen, T.; Wei, X.; Li, J. Decomposition and Decoupling Analysis between HDI and Carbon Emissions. *Atmosphere* **2022**, *13*, 584. https://doi.org/10.3390/atmos13040584

Academic Editors: Zengyun Hu, Xuguang Tang and Qinchuan Xin

Received: 25 February 2022
Accepted: 2 April 2022
Published: 5 April 2022

1. Introduction

In the context of global warming, the low-carbon economy with low energy consumption, low pollution, and low emissions have become the focus of global attention. In compliance with the *Kyoto Protocol* and the *Paris Agreement*, governments have taken actions to reduce carbon emissions, improve energy efficiency, and enhance their ability to cope with climate change. Despite this, global energy demand and CO_2 emissions continue to increase, except in 2020 [1]. Ideally, we would like to see economic growth that is not dependent on the increase in carbon emissions, with decoupling between economic growth and carbon emissions. Therefore, revealing the decoupling process is conducive to finding critical breakthroughs in the path of global low-carbon economic development [2].

Low-carbon development is being achieved at the economic level, but it should also be extended to the welfare dimension. Based on Sen's capability approach, the Human Development Index (HDI) is constructed to measure the combined performance of a country in health, education, and income [3]. The increase in HDI is taken as a starting point for human welfare. To achieve higher human development is the pursuit of each country. However, realizing the right to human development requires the guarantee of energy consumption and the demand for carbon emissions [4]. The 30th anniversary

Human Development Report introduced a new index—Planetary pressures-adjusted HDI (PHDI), which emphasizes how human activities significantly influence the environment and ecology and place great stress on the planet. These pressures will react to human development in the form of extreme climate or species extinction, resulting in the decline or loss of HDI [5]. Therefore, studying the relationship between CO_2 emissions and human development will provide policymakers with more evidence for achieving sustainable development [6].

This paper aims to analyze the decoupling between carbon emissions and HDI and the carbon welfare output of 189 countries from 1990 to 2019 and understand the drivers. First, the overall decoupling performance of 189 countries from 1990 to 2019 is analyzed. Furthermore, the evolution of decoupling trends is shown over six periods. However, the decoupling index and decoupling status are not comprehensive enough to explain the coupling relationship between carbon emissions and HDI. Therefore, the total carbon emissions and carbon emissions per capita were combined to analyze the welfare output of carbon emissions in 189 countries, namely carbon emission performance (CEP). Finally, several countries that maintained continuous decoupling and expansive negative decoupling over six periods are identified in the decoupling analysis. And LMDI analyzed the driving factors of the decoupling index and CEP.

This paper consists of five sections and is arranged as follows. The second section reviews and summarizes the relevant literature. The appropriate methods and data are described in Section 3. Section 4 shows the results of a series of methods. Section 5 summarizes the main conclusions and makes recommendations on maintaining a stable decoupling status.

2. Literature Review

2.1. Review of Decoupling Model

The decoupling model is widely used to analyze the coupling relationship between economy and resource consumption or environmental pollution. The two main decoupling models are the OECD decoupling model and the Tapio decoupling model. OECD analyzed the decoupling status of 30 member countries and confirmed the widespread decoupling between environment and economy in OECD countries [7]. Tapio adopted the concept of elasticity to dynamically reflect the decoupling relationship between variables and determined eight reasonable decoupling statuses, overcoming the difficulties in the base period selection of the OECD decoupling model [8]. Lin et al. used the OECD and Tapio decoupling model to evaluate the decoupling status between CO_2 emissions and GDP in South Africa during 1990–2012 and found that the decoupling effect was gradually increasing [9]. The results of the two decoupling models have a good correlation, but the OECD decoupling model has limitations for analyzing the decoupling of sub-periods. To compare the decoupling trend between economic growth and carbon dioxide emissions in typical developed and developing countries from 1965 to 2015, Wu et al. used three decoupling models—the OECD decoupling model, the Tapio decoupling model, and the IGTX decoupling model [10]. The research has shown that the decoupling effect is stronger in developed countries than in developing countries, and the accuracy of the Tapio decoupling model is not easily limited by the length of the research period among the three decoupling models.

In the decoupling analysis between carbon dioxide and economy, the Tapio model and LMDI decomposition method are generally combined to discuss the driving factors affecting the decoupling status. Chen et al. embedded the LMDI decomposition formula in the Tapio decoupling model to study the driving factors affecting CO_2 emissions in OECD countries from 2001 to 2015 [11]. The results showed that energy intensity and GDP per capita are the main factors affecting CO_2 emissions, and the impact of technological factors is greater than that of non-technological factors. Yang et al. combined the LMDI decomposition model and Tapio decoupling model to analyze the determinants of decoupling between global carbon emissions and global economic growth during 2000–2017 [2]. It is found that energy conservation, production efficiency, and energy structure optimization played a positive role in the global decoupling process, while GDP per capita and population expansion played a restraining role. In general, the application of the Tapio model and LMDI decomposition model is wide. Therefore, the Tapio model and LMDI decomposition model are used in this paper to identify each country's decoupling status and driving factors.

2.2. Review of Carbon Emissions and HDI

Most literature focuses on the relationship between energy consumption/environmental pollution and economic growth but ignores the relationship between energy/environment and human welfare. Economic growth is only one of the goals of human development. In essence, promoting human development is the ultimate goal. In the context of limited global energy resources, ensuring the effective use of energy and promoting human development is particularly important. Martínez and Ebenhack found a significant correlation between HDI and energy consumption per capita in 120 countries [12]. For some energy-poor countries with a low HDI, a small amount of energy assistance can make great strides in human development. Akizu-Gardoki et al. conducted a decoupling analysis of 126 countries from 2000 to 2014, based on Total Primary Energy Footprint and HDI, and the results showed that human welfare could be increased by reducing energy consumption [13]. According to the analytical framework of HDI, there is also research on how to formulate and implement global emission reduction policies. Pan believes that countries with a low HDI should be given priority in international negotiations on global climate change mitigation to guarantee their rights in achieving human development [4]. Hu proposed to implement global emission reduction policies according to the HDI of countries/regions and the principle that major carbon emitters take the lead in emission reduction [14]. Costa et al. used the positive correlation between HDI and CO_2 emissions per capita to evaluate the cumulative CO_2 emissions required to achieve high human development in developing countries and determine their respective emission reduction according to the HDI [15].

The decoupling relationship between carbon emissions and HDI is mainly studied at the levels of nation and district. Based on the Tapio decoupling model and LMDI decomposition, Chen et al. discussed the decoupling status between carbon emissions per capita and the HDI of four provinces in southwest China from 2000 to 2015 [16]. The results showed that the overall decoupling effect of southwest China was gradually increasing. Hossain and Chen demonstrated weak and strong decoupling between HDI and CO_2 emissions in Bangladesh from 1990 to 2018, with economic activity being the primary driver of CO_2 emissions, compared to other influencing factors [6]. Therefore, this paper expands the research perspective to the global dimension instead of studying a specific country or region. The decoupling relationship between carbon emissions and HDI can be studied from the perspective of multiple countries, and comparative analysis can be conducted among countries with the same data source.

2.3. Review of Carbon Emission Performance

Carbon productivity originated from the resource productivity in ecological economics and was later measured by consultants McKinsey & Company, in terms of GDP output per unit of CO_2 emissions [17]. According to the definition of carbon productivity, carbon emission performance (CEP) measures the economic and social welfare output per unit of carbon emission [18,19]. Therefore, the welfare output of carbon emissions should be considered from the perspective of human well-being to broaden the research horizon of low-carbon development. Zhu and Liu measured the CEP of G20 countries, expressed by the ratio of HDI to carbon emission index—the average of the proportion of total carbon emission and carbon emission per capita after standardization [18]. Liu and Zhu and Hu et al., respectively, measured the carbon emission performance of G20 countries and 30 Provinces in China, but the carbon emission index only included carbon emissions per capita, without the important information of total carbon emissions [19,20]. The PHDI in the Human Development Report 2020 adjusts HDI through carbon emissions per capita and material footprint per capita to explain the pressure on the earth caused by human activities [5]. In this paper, the calculation method of Zhu and Liu and PHDI was used to construct the carbon emission performance index.

3. Methods and Data

3.1. Analysis of Decoupling

In economics, the decoupling model is usually used to study the coupling relationship between economic growth and resource consumption or environmental pollution, reflecting the sensitivity of resources and environmental pressure changes to economic changes [21]. This paper uses the Tapio decoupling model to analyze the decoupling relationship between carbon emissions per capita and HDI. Since HDI is a per capita indicator, carbon emissions per capita are selected instead of total carbon emissions. The equation is expressed as follows.

$$D(PC, HDI) = \frac{\%\Delta PC}{\%\Delta HDI} = \frac{\Delta PC / PC_0}{\Delta HDI / HDI_0} \tag{1}$$

In Equation (1), $D(PC, HDI)$ represents the decoupling index between carbon emissions per capita (PC) and the Human Development Index (HDI), and $\%\Delta PC$ and $\%\Delta HDI$ represent the growth rate of carbon emissions per capita and HDI, respectively. Both ΔPC and ΔHDI represent the change of carbon emissions per capita and HDI from the base period to the end period respectively. Lastly, PC_0 and HDI_0 represent carbon emissions per capita and HDI at the base period, respectively.

According to the decoupling index, the decoupling status is divided into eight statuses: strong decoupling, weak decoupling, expansive coupling, expansive negative decoupling, weak negative decoupling, strong negative decoupling, recessive coupling, and recessive decoupling. The meanings of the eight decoupling statuses are detailed in the other literature [22].

3.2. Construction of Carbon Emission Performance (CEP)

According to the definition of CEP, HDI is used to represent welfare. The HDI data provided by the Human Development Data Center is non-continuous, and direct use may cause errors. According to the calculation method, the HDI of 189 countries from 1990 to 2019 was recalculated, but the calculation steps were omitted. Carbon emissions are expressed by Carbon Emission Index (CEI). CEI needs to consider two factors: total carbon emissions and carbon emissions per capita. Zhu and Liu standardized the ratio of total carbon emissions to the total global emissions as the total Carbon Emission Index [18]. However, the total global emissions are determined by the total carbon emissions of the sample countries, and the proportion of total carbon emissions is not stable. In this paper, the total carbon emissions and carbon emissions per capita are directly standardized. Construction of carbon emission performance is shown in Equation (2):

$$CEP = \frac{HDI}{CEI} \tag{2}$$

$$CEI = \frac{1}{2}\left[\frac{Ln(C) - Ln(0)}{Ln(10{,}490{,}000) - Ln(0)} + \frac{Ln(PC) - Ln(0)}{Ln(68{,}720) - Ln(0)}\right] \tag{3}$$

In Equation (3), C is carbon emission and PC is carbon emission per capita. The standardized method for calculating PHDI is used [5]. Zero is set as the minimum value. The maximum value is the highest value in all countries from 1990 to 2019. Therefore, the minimum value of total carbon emission and carbon emission per capita is 0. The maximum value of total carbon emission is 10.49 billion tons, observed in China in 2019, and the maximum value of carbon emission per capita is 68.72 tons, observed in Qatar in 1997. It should be noted that logarithmic standardization is meaningful only if the value of total carbon emission and carbon emission per capita is greater than 1. Therefore, the total carbon emission is expressed in 10,000 tons, and the carbon emission per capita is expressed in kilograms.

3.3. Analysis of Decomposition

Logarithmic Mean Divisia Index (LMDI) decomposes the variables into several effects and analyzes the contribution of each effect to the change in the variable [23]. This method can show the decomposition results as additivity, complete decomposition, and no residual error. The LMDI method has been widely used to decompose drivers of carbon emission due to its variety of forms and simplicity of calculation [24,25]. This paper decomposes the decoupling index and CEP by combining the Kaya Identity and the LMDI method.

3.3.1. Decomposition of Decoupling Index

Based on the idea of existing studies, the decomposition process of drivers affecting CO_2 emission is determined as follows [6,16]:

$$PC = \frac{C}{F} \times \frac{F}{E} \times \frac{E}{GDP} \times \frac{GDP}{P} \tag{4}$$

where, $CF = \frac{C}{F}$, $FE = \frac{F}{E}$, $EG = \frac{E}{GDP}$, $GP = \frac{GDP}{P}$

In Equation (4), C represents total carbon emissions, F represents fossil fuel consumption (oil, natural gas, and coal), E represents primary energy consumption, GDP represents total economic output, and P represents the size of the population. CF represents the carbon emission intensity of fossil fuel. FE represents the structure of energy consumption. EG represents energy intensity. GP represents GDP per capita. Equation (4) can be transformed into:

$$PC = CF \times FE \times EG \times GP \tag{5}$$

According to the LMDI model, the change of carbon emissions per capita from the base period (0 period) to the reporting period (T period) is shown in Equation (6).

$$\begin{aligned} \Delta PC = PC_t - PC_0 &= CF_t \times FE_t \times EG_t \times GP_t - CF_0 \times FE_0 \times EG_0 \times GP_0 \\ &= \Delta CF + \Delta FE + \Delta EG + \Delta GP \end{aligned} \tag{6}$$

$$\Delta CF = Ln\left(\frac{CF_t}{CF_o}\right) \times \frac{PC_t - PC_0}{Ln(PC_t) - In(PC_0)} \tag{7}$$

$$\Delta FE = Ln\left(\frac{FE_t}{FE_0}\right) \times \frac{PC_t - PC_0}{Ln(PC_t) - In(PC_0)} \tag{8}$$

$$\Delta EG = Ln\left(\frac{EG_t}{EG_0}\right) \times \frac{PC_t - PC_0}{Ln(PC_t) - In(PC_0)} \tag{9}$$

$$\Delta GP = Ln\left(\frac{GP_t}{GP_0}\right) \times \frac{PC_t - PC_0}{Ln(PC_t) - In(PC_0)} \tag{10}$$

In Equation (6), ΔCF represents the carbon intensity of fossil fuel effect, ΔFE represents the energy structure effect, ΔEG represents energy intensity of fossil fuel effect, and ΔGP represents economic development effect. The four effects from the base period to T period are calculated based in Equations (7)–(10).

By combining the Tapio decoupling model with the LMDI method, the decoupling index from the base period (0 period) to the reporting period (T period) can be expressed by Equation (11).

$$\begin{aligned} D(PC, HDI) &= \frac{\%\Delta PC}{\%\Delta HDI} = \frac{\Delta PC/PC_0}{\Delta HDI/HDI_0} = \Delta PC \times \frac{HDI_0}{\Delta HDI \times PC_0} \\ &= (\Delta CF + \Delta FE + \Delta EG + \Delta GP) \times \frac{HDI_0}{\Delta HDI \times PC_0} \end{aligned}$$

$$= \Delta CF \times \frac{HDI_0}{\Delta HDI \times PC_0} + \Delta FE \times \frac{HDI_0}{\Delta HDI \times PC_0} + \Delta EG \times \frac{HDI_0}{\Delta HDI \times PC_0} + \Delta GP \times \frac{HDI_0}{\Delta HDI \times PC_0}$$
$$= D_{CF} + D_{FE} + D_{EG} + D_{GP} \quad (11)$$

In Equation (11), the decoupling index of carbon emissions per capita and HDI is divided into four components. D_{CF}, D_{FE}, D_{EG}, and D_{GP} represent the decoupling index of ΔCF, ΔFE, ΔEG, and ΔGP, respectively.

3.3.2. Decomposition of CEP

Similarly, the Kaya Identity and the LMDI method are used to decompose the driving factors of CEP. The decomposition form is shown in Equation (12).

$$CEP = \frac{CEP}{GDP} \times \frac{GDP}{C} \times \frac{C}{F} \times \frac{F}{E} \times \frac{E}{GDP} \times \frac{GDP}{P} \times P \quad (12)$$

where, $CG = \frac{CEP}{GDP}$, $GC = \frac{GDP}{C}$, $CF = \frac{C}{F}$, $FE = \frac{F}{E}$, $EG = \frac{E}{GDP}$, $GP = \frac{GDP}{P}$, $P = POP$.

In Equation (12), CG represents the contribution to CEP per unit of economic growth. GC represents the economic output per unit of carbon emissions, i.e., carbon productivity. CF, FE, EG, and GP are the same as in Equation (4), respectively. P denotes the size of the population. Equation (12) is simplified into the following form:

$$CEP = CG \times GC \times CF \times FE \times EG \times GP \times P \quad (13)$$

The change of CEP from 0 period to T period is expressed as Equation (14).

$$\Delta CEP = CEP_t - CEP_0 = \Delta CG + \Delta GC + \Delta CF + \Delta FE + \Delta EG + \Delta GP + \Delta P \quad (14)$$

$$\Delta CG = Ln\left(\frac{CG_t}{CG_o}\right) \times \frac{CEP_t - CEP_0}{Ln(CEP_t) - ln(CEP_0)} \quad (15)$$

$$\Delta GC = Ln\left(\frac{GC_t}{GC_0}\right) \times \frac{CEP_t - CEP_0}{Ln(CEP_t) - ln(CEP_0)} \quad (16)$$

$$\Delta CF = Ln\left(\frac{CF_t}{CF_0}\right) \times \frac{CEP_t - CEP_0}{Ln(CEP_t) - ln(CEP_0)} \quad (17)$$

$$\Delta FE = Ln\left(\frac{FE_t}{FE_o}\right) \times \frac{CEP_t - CEP_0}{Ln(CEP_t) - ln(CEP_0)} \quad (18)$$

$$\Delta EG = Ln\left(\frac{EG_t}{EG_0}\right) \times \frac{CEP_t - CEP_0}{Ln(CEP_t) - ln(CEP_0)} \quad (19)$$

$$\Delta GP = Ln\left(\frac{GP_t}{GP_0}\right) \times \frac{CEP_t - CEP_0}{Ln(CEP_t) - ln(CEP_0)} \quad (20)$$

$$\Delta P = Ln\left(\frac{P_t}{P_0}\right) \times \frac{CEP_t - CEP_0}{Ln(CEP_t) - ln(CEP_0)} \quad (21)$$

In Equation (14), ΔCG represents the welfare effect, ΔGC represents the carbon productivity effect, and ΔP represents the population effect. The terms ΔCF, ΔFE, ΔEG, and ΔGP are the same as in Equation (6), respectively. The seven effects from the base period to T period are calculated based on Equations (15)–(21).

3.4. Data

Frequently used data on carbon emissions are taken from BP statistics. But data for just over 70 countries is not enough to give a complete picture of global carbon emissions. The carbon emissions data used in this paper are from the Global Carbon Project, which mainly refers to the carbon emissions generated by burning fossil fuels within the national territory [26]. UNDP also uses carbon emissions data from the Global Carbon Project to calculate PHDI [5]. *BP Statistical Review of World Energy 2021* provides a long period of energy consumption data for the decomposition analysis of countries with certain characteristics [1]. The GDP data derived from World Development Indicators is updated in March 2021 [27]. Population data were derived from the World Population Prospects: The 2019 Revision [28]. Raw data on health, education, and income were obtained from the Human Development Data Centre to recalculate the HDI [29]. The variable definitions and sources are shown in Table 1.

Table 1. Variable definition and source.

Variable	Definition	Source
C	Total carbon emissions (million tons)	Global Carbon Project
F	Fossil fuel consumption (Twh)	BP Statistical Review of World Energy
E	primary energy consumption (Twh)	BP Statistical Review of World Energy
GDP	Total GDP (constant 2017 PPP $)	World Development Indicators
P	Total population (thousand)	World Population Prospects
HDI	The geometric average of health, education, and income index	Human Development Data Center

4. Result

4.1. Relationship between Carbon Emissions Per Capita and HDI

Figure 1 depicts the corresponding relationship between CO_2 emissions per capita and the HDI for four years. There is a strong correlation between CO_2 emissions per capita and the HDI. The figures show a power–exponential relationship, that is, with the increase of HDI, the increase of carbon emissions per capita is gradually accelerating. Carbon emissions per capita of countries with extremely low HDI are minor, and the growth is flat. When countries achieve medium, high, and very high human development, the growth rate of carbon emissions per capita begins to accelerate. This indicates that the demand for carbon emissions and the carbon emissions per capita will gradually increase for the country to achieve a higher level of human development. The non-linear relationship between CO_2 emissions per capita and HDI is consistent with Costa et al., who found an exponential relationship between CO_2 emissions from fossil fuels and HDI [15].

It can also be seen from Figure 1 that the CO_2 emissions per capita of countries with extremely low HDI are relatively close, with little difference between countries. As the HDI of countries gradually increases, differences in carbon emissions per capita become apparent. The differences in emissions per capita among countries are most remarkable at very high human development. Since 1990, emissions per capita by the poorest half of the world's population have only slightly increased, from 1.2 tons to 1.6 tons [30]. For countries with high levels of human development, their carbon emissions per capita are different, and they follow different development paths. Some countries will achieve high human development at the cost of large carbon emissions. Some countries will aim at low-carbon development, improving emission-reduction technologies, and developing new energy sources to achieve high human development.

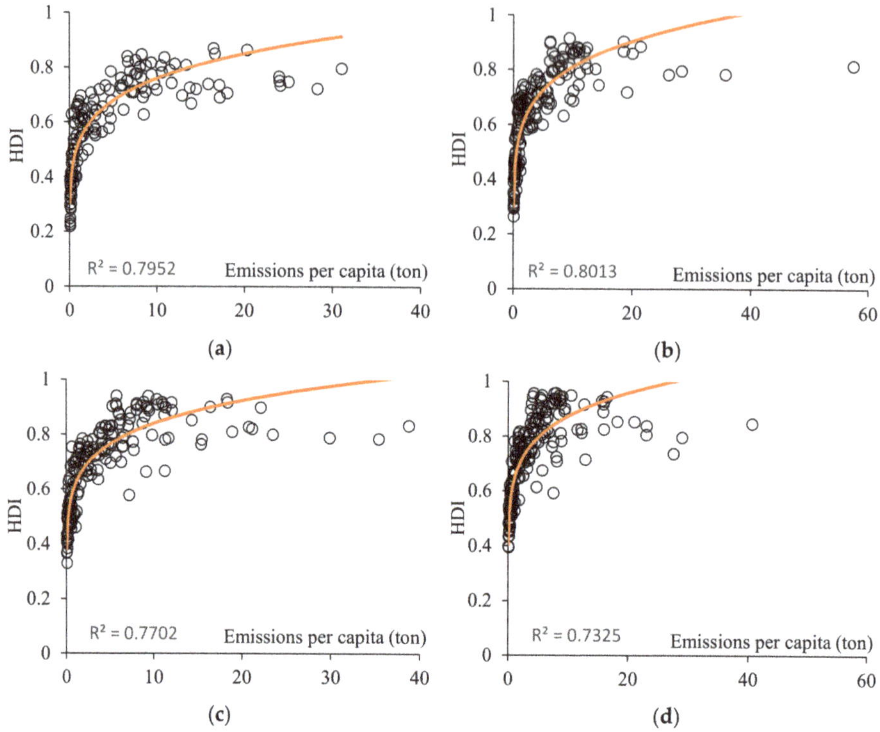

Figure 1. Relationship between HDI and CO_2 per capita emissions. Source: Authors' own construction. Note: The *p*-values for the four exponential functions are all less than 0.01 and pass the significance test. (**a**) 1990; (**b**) 2000; (**c**) 2010; (**d**) 2019.

4.2. Analysis of Decoupling between Carbon Emissions Per Capita and HDI

4.2.1. Decoupling between Carbon Emissions Per Capita and HDI, 1990–2019

It is the right of every country to achieve high human development. Carbon emissions should be targeted to increase the HDI. If carbon emissions per capita increase without increasing the HDI, this may indicate that no effort has been made to reduce carbon emissions. Countries with strong decoupling and very high human development are considered to have achieved positive decoupling. However, countries that do not achieve very high HDI and strong decoupling are referred to as negative decoupling.

As shown in Figure 2, no country experienced negative growth HDI from 1990 to 2019. HDI data for 1990 are missing for 46 countries, and 59 countries out of the 143 are strongly decoupling. In addition, 40 countries are positively decoupling whereas 19 countries are negative decoupling. Negative decoupling is not a good thing for countries with extremely low HDI. These countries are the Central African Republic, the Democratic Republic of Congo, and Yemen. Their HDI is still at low human development after 29 years, due to constant war and political unrest. Because of their weak industrial base and poor infrastructure, they are strongly decoupling with low carbon emissions and low energy consumption. Much investment is needed to develop in the future due to the low starting point.

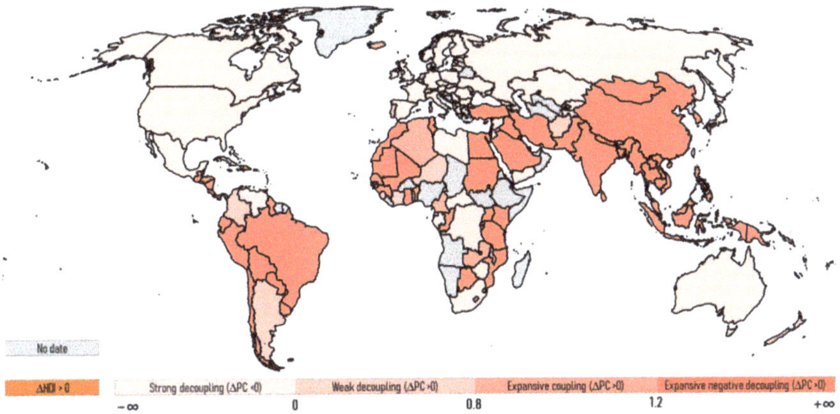

Figure 2. Decoupling between carbon emissions per capita and HDI, 1990–2019. Source: Authors' own construction using Excel.

In total, 63 countries are expansive negative decoupling, meaning carbon emissions per capita are growing much faster than HDI. Most of the 63 countries are at high or medium human development, and 11 countries with very high HDI are expansive negative decoupling, such as South Korea, Kuwait, and Qatar. Achieving high HDI growth comes at the cost of high carbon emissions per capita.

4.2.2. Evolution of the Decoupling between Carbon Emissions Per Capita and HDI

The whole observation period from 1990 to 2019 is divided into six sub-periods. The decoupling index between carbon emissions per capita and HDI is calculated to analyze the change in the decoupling status in each period. As shown in Figure 3, the decoupling status of countries changed significantly over the six periods. Most countries have maintained positive HDI growth. In general, the main decoupling statuses are strong decoupling, weak decoupling, expansive coupling, and expansive negative decoupling. The proportion of countries with these four decoupling statuses is between 80% and 97%.

Four undesirable decoupling status will emerge if some countries experience negative HDI growth. The worst status is recession decoupling. A significant decline in HDI has accompanied the reduction in carbon emissions per capita. There were more countries with recessive decoupling in the first period. The main countries are the Russian Federation, Cuba, and Mongolia, which were more affected by the collapse of the Soviet Union. Other countries are in Sub-Saharan Africa. In the later periods, the countries with recessive decoupling were mainly in Sub-Saharan Africa, the Middle East, and Latin America. Based on the experience of most countries, the decoupling status between carbon emissions per capita and HDI experienced an evolutionary trend: expansive negative decoupling, expansive coupling, weak decoupling, and strong decoupling. Strong decoupling in the early period is unsuitable for countries with extremely low HDI, which requires vast investments later. However, there has been a shift from recessive decoupling to expansive coupling and expansive negative decoupling for some countries with extremely low HDI (Rwanda, Tanzania, and Zambia).

Overall, the proportion of countries with strong decoupling and weak decoupling continues to rise, while the proportion of expansive coupling and expansive negative decoupling is declining. Although the relationship between HDI and energy consumption has entered a new phase in most countries, the status of strong and weak decoupling is unstable. Strong decoupling and weak decoupling carry the risk of shifting to other status of decoupling. In addition, economic, social, political, and resource factors vary among countries, and the unbalanced trend of economic growth and carbon emissions will continue in the future. Solving these problems requires the concerted efforts of the whole world.

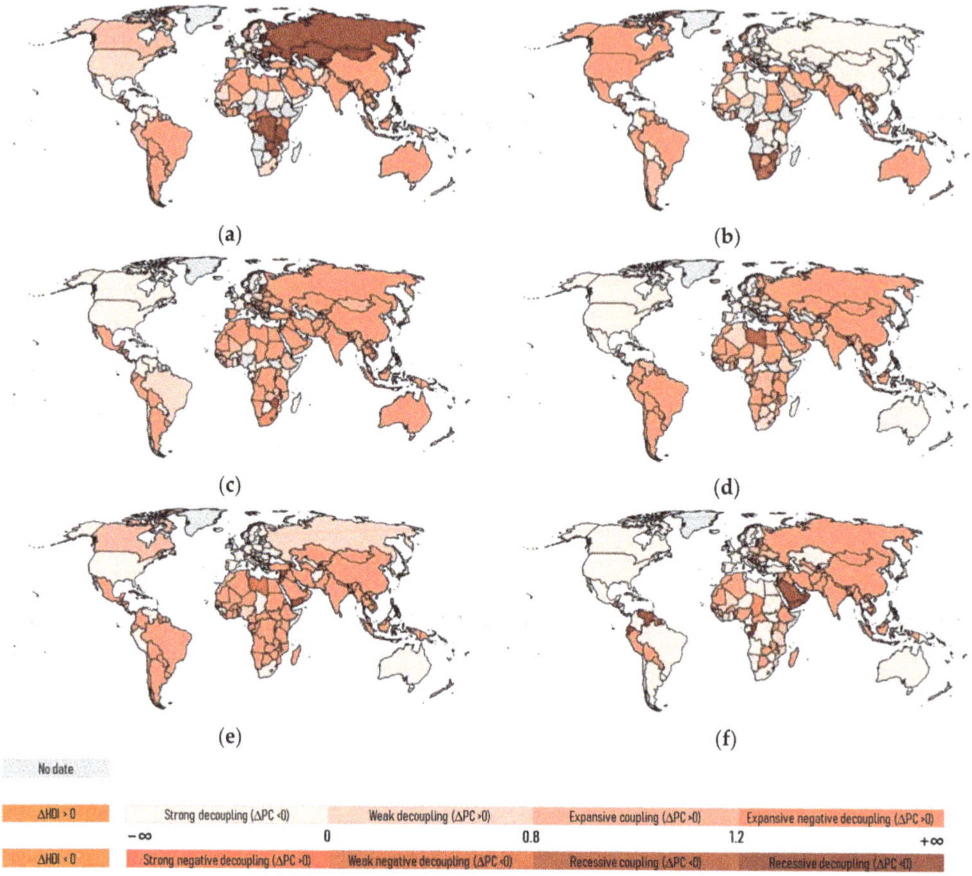

Figure 3. Spatial distribution of the decoupling index over six periods. Source: Authors' own construction using Excel. (**a**) 1990–1995; (**b**) 1995–2000; (**c**) 2000–2005; (**d**) 2005–2010; (**e**) 2010–2015; (**f**) 2015–2019.

Only three countries (the Czech Republic, Germany, and the United Kingdom) have sustained positive decoupling over the six periods. They reduce carbon emissions while maintaining relatively high HDI. Five countries (Mauritius, Panama, South Korea, Turkey, and Uruguay) with high HDI are expansive negative decoupling over at least five periods. Their growth rate of carbon emissions per capita is larger than their HDI. These countries emit more carbon dioxide without a corresponding increase in the HDI.

4.3. Evolution and Two-Dimensional Analysis of Carbon Emission Performance

4.3.1. Results of Carbon Emission Performance Measurement from 1990–2019

Figure 4 shows the evolution of countries' carbon emission performance in 1990, 1996, 2002, 2008, 2014, and 2019. As shown in Figure 4, there are more and more countries with dark colors. The carbon emission performance of countries is gradually increasing.

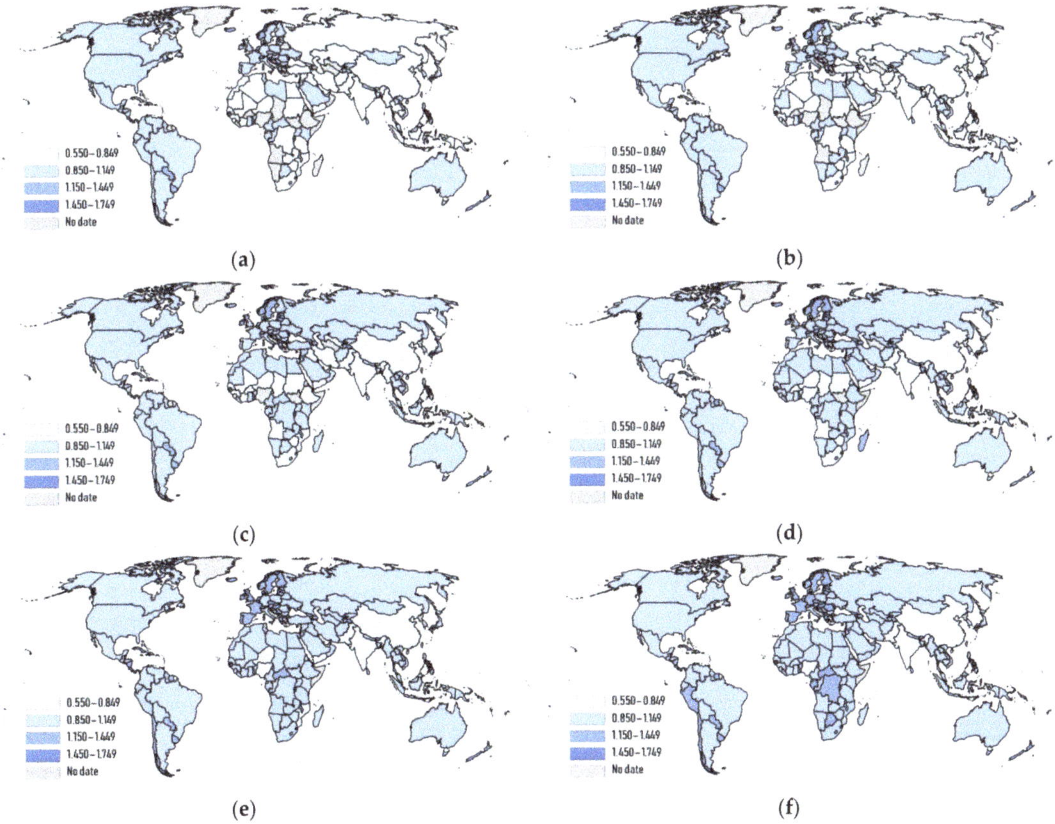

Figure 4. Evolution of carbon emission performance, 1990–2019. Source: Authors' own construction using Excel. (**a**) 1990; (**b**) 1996; (**c**) 2002; (**d**) 2008; (**e**) 2014; (**f**) 2019.

Countries showing significant increases in carbon emissions performance (50% above from 1990 to 2019) include the Democratic Republic of Congo, Mozambique, Niger, Rwanda, and Sierra Leone, all in sub-Saharan Africa. The most considerable increase was in Rwanda, where CEP increased by 99.89%. In 1990, the total and carbon emissions per capita in Rwanda were 0.52 million tons and 0.0719 tons, respectively. The total and per capita emissions in 2019 were 1.19 million tons and 0.094 tons in Rwanda, respectively. The carbon emission index increased by only 9.69%, whereas the HDI doubled by 119.26%. This result is consistent with the transition of Rwanda's decoupling status. Rwanda started with a deficient level of economic and human development. In the long term, economic growth has accelerated, life expectancy has improved significantly, and primary education has developed significantly in Rwanda. Thus, the increase in HDI results from the overall improvement in all aspects.

Carbon emissions performance declined in 14 countries. The most significant decrease in CEP was recorded in Laos. Total carbon emissions were 0.51 million tons, and carbon emissions per capita were 0.1205 tons in 1990. Total and carbon emissions per capita were 33.93 million tons and 4.7328 tons in 2019, respectively. The carbon emission index of Laos increased by 72.15%, and the human development index increased by 51.35%. The growth of HDI is less than that of carbon emissions. Laos achieved an increase in HDI at the cost of environmental pollution.

The carbon emission performance for the six years was a normal distribution. The majority of countries (73–96%) are at the middle intervals (0.85–1.449). Only a tiny proportion of countries (4–27%) are at the two ends of the interval. The number of countries with CEP at 0.85–1.149 is the greatest, but the proportion has gradually decreased, from 63.64% in 1990 to 53.44% in 2019. The proportion of countries with CEP at 1.15–1.449 gradually increased, rising from 9.79% in 1990 to 42.33% in 2019. The proportion of countries with CEP at 0.55–0.849, mainly in South Asia and Sub-Saharan Africa, fell from 25.17% in 1990 to 2.12% in 2019. Carbon emission performance has been significantly improved, and there are fewer and fewer countries with low carbon emission performance. Although the proportion of countries with CEP at 1.45–1.749 increased to 2.12% in 2019, the number of countries with high CEP is still minimal. There are only four countries (Kiribati, Liechtenstein, Micronesia, and Tonga) with CEP at 1.45–1.79, and their CEP is relatively stable.

4.3.2. Two-Dimensional Analysis Based on HDI and Carbon Emission Performance

The HDI ranges from 0 to 1. The UNDP considers countries with HDI above 0.8 to be very high in human development. The country with the highest CEP is Liechtenstein, with CEP at 1.7471 in 2019. Referring to the analysis method of Zhu and Liu, the upper limit of CEP is set as 2 [18]. According to the classification of HDI, countries with CEP at the top 20% (above 1.6) are regarded as countries with high CEP. The two-dimensional analysis of HDI and CEP in 2019 was performed. Countries can be divided into four types based on the set thresholds: (a) high HDI and high CEP; (b) high HDI and low CEP; (c) low HDI and low CEP; (d) low HDI and high CEP.

As shown in Figure 5, only Liechtenstein has a high HDI and high CEP, which is an ideal state. There are 65 countries with high HDI and low CEP, such as the United States, Norway, Canada, and other developed countries, mainly centered in Europe, Central Asia, and North America. These countries with a very high HDI enjoy better economic and social development. Development should focus on maintaining economic growth while reducing environmental pollution. These countries need to improve emission reduction technologies, develop new renewable energy sources, and optimize industrial structures to achieve energy conservation and emission reduction. The number of countries with low HDI and low CEP is the largest, over 65%, and includes developing countries such as China and India, and least-developed countries, such as Lesotho and Yemen. These countries are in the stage of comprehensive development. They not only need to meet the energy needs to promote the economy and society, but also need to control carbon emissions, and face the dual challenges of economic and social development and environmental sustainability. There are currently no countries with low HDI and high CEP. The fourth type will not appear based on the exponential relationship between carbon emissions per capita and HDI.

4.4. *Decomposition of Drivers*

As known from Section 4.2, only the Czech Republic, Germany, and the United Kingdom have maintained continuous strong decoupling over the six periods. The status of decoupling is unstable for most countries, and it is a challenge to sustain strong decoupling for a long time. Therefore, it is necessary to understand the drivers of the decoupling index and CEP of exemplary countries. In addition to the three countries with continuous positive decoupling, the carbon emission paths of five countries with expansive negative decoupling were studied. However, the data for Mauritius, Panama, and Uruguay were not available, and so the decomposition analysis is only for South Korea and Turkey, to explore why they have not yet achieved strong decoupling.

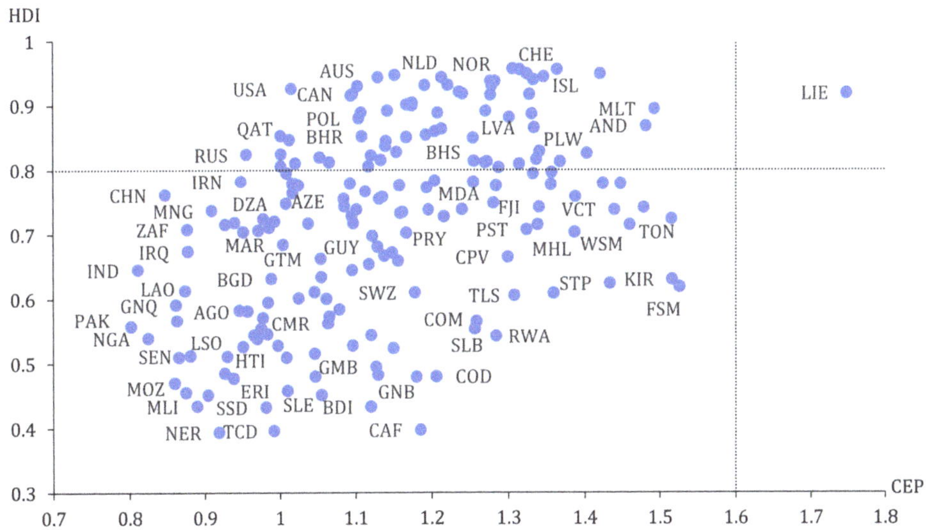

Figure 5. Two-dimensional analysis based on HDI and CEP in 2019. Source: Authors' own construction using Excel.

4.4.1. Decomposition of the Decoupling Index

Combining the Tapio decoupling model with the LMDI decomposition model, the decoupling index between carbon emissions per capita and HDI can be understood from the carbon intensity of fossil energy effect, energy structure effect, energy intensity effect, and economic development effect. The smaller decoupling index indicates the lower dependence of HDI growth on energy consumption for strong decoupling status. The strong decoupling index gets smaller and smaller only when the four effects are negative. That is, the contribution rate of the four effects is positive. As shown in Table 2, the decoupling index is becoming smaller, and the strong decoupling effect is increasing in the Czech Republic, Germany, and the UK. The main factors influencing the decoupling index are the same for all three countries. The main contributing factor is the energy intensity effect, indicating high energy use efficiency. On the other hand, the economic development effect is the main inhibiting factor, indicating an increase in carbon emissions in daily life with the popularization of energy-consuming consumer goods. But the contribution of the economic development effect tends to decrease.

In order to transform from expansive negative decoupling to strong decoupling, South Korea and Turkey need to make the decoupling index as small as possible under the premise of positive HDI growth. In other words, all four driving factors are guaranteed to be negative, i.e., the contribution rate is positive. As seen in Table 1, the decoupling index for Korea has decreased in a fluctuating way but has not yet reached weak and strong decoupling. For Korea, the carbon intensity of fossil energy was the main factor contributing to the decline in the decoupling index during the 1990–1995 and 1995–2000 periods. The main factors in the last four periods are transformed into energy intensity. Economic development effects have been the primary inhibitor to transition to weak and strong decoupling for South Korea. Similarly, Turkey's decoupling index has declined, even to the threshold of strong decoupling in the 2015–2019 period. However, the main factors affecting the decline in the decoupling index are different in each period, indicating that some efforts have been made to optimize energy consumption structure and improve energy efficiency. The factor inhibiting the decrease of the decoupling index is always the effect of economic development.

Table 2. Decomposition of the decoupling index for the example countries.

Country	Phase	$D(PC,HDI)$	D_{CF}/%	D_{FE}/%	D_{EG}/%	D_{GP}/%
CZE	1990–1995	−6.49	39.34	6.09	37.56	17.01
	1995–2000	−0.5	−5.4	35.33	401.95	−331.88
	2000–2005	−0.17	519.75	687.84	1204.19	−2311.78
	2005–2010	−2.78	13.1	32.73	156.96	−102.79
	2010–2015	−4.61	−7.96	17.4	154.99	−64.43
	2015–2019	−4.7	71.07	38.41	245.67	−255.15
DEU	1990–1995	−3.22	33.33	9.43	108.6	−51.36
	1995–2000	−1.11	47.91	49.97	198.52	−196.4
	2000–2005	−1.09	31.5	27.25	101.63	−60.38
	2005–2010	−1.33	−47.16	41.75	331.54	−226.13
	2010–2015	−4.94	−7.48	36.5	195.64	−124.66
	2015–2019	−12.67	38.59	26.05	69.22	−33.86
GBR	1990–1995	−0.83	65.59	37.45	89.43	−92.47
	1995–2000	−0.45	313.71	−47.44	814.77	−981.04
	2000–2005	−0.7	80.7	−22.53	664.23	−622.4
	2005–2010	−8.96	11.57	−1.33	73.01	16.75
	2010–2015	−17.93	0.48	35.1	91.95	−27.53
	2015–2019	−15.34	29.39	23.28	72.66	−25.33
KOR	1990–1995	6.55	−32.13	12.28	24.43	95.42
	1995–2000	1.92	−60.44	−38.77	−51.55	250.76
	2000–2005	2.18	−16.17	−17.38	−88.48	222.03
	2005–2010	5.56	1.85	15.33	−39.42	122.24
	2010–2015	1.92	−26.98	−10.22	−183.35	320.55
	2015–2019	1.15	−288.26	117.59	−566.1	836.77
TUR	1990–1995	2.02	−27.03	−30.24	77.32	79.95
	1995–2000	2.21	−3.73	28.58	−1.66	76.81
	2000–2005	1.28	8.71	−12.29	−141.63	245.21
	2005–2010	1.95	−45.58	−7.35	69.11	83.82
	2010–2015	1.39	−24.29	−14.67	−95.17	234.13
	2015–2019	−0.58	173.68	357.1	145.12	−575.9

Source: Authors' own calculation.

4.4.2. Decomposition of the CEP

As shown in Table 3, the carbon productivity effect, economic development effect, and population effect are positive in most periods for these five countries, indicating that the three effects positively impact the improvement of CEP. But the role of the population effect is relatively small. In most periods, the welfare effect, the carbon intensity of fossil fuels effect, the energy structure effect, and the energy intensity effect are negative. The welfare effect is negative, which means the low CEP is generated by unit economic output growth. The latter three effects are related to the energy utilization, energy structure, and efficiency of fossil fuels, and the negative effects indicate the need to adjust the energy structure further and reduce energy consumption. However, the main influencing factors are different for different countries.

The CEP of the Czech Republic, Germany, and the United Kingdom are primarily similar and gradually increased from around 0.95 in 1990 to 1.2 in 2019, with a growth rate of around 25%. The main influencing factors for the three countries are also the same. The carbon productivity effect is the primary driver of the improvement in CEP. The higher the carbon productivity, the greater the economic output per unit of carbon emissions, representing the progress of emission-reduction technologies in economic development. The main driving factor that inhibits CEP is the energy intensity effect, which indicates that economic growth is highly dependent on primary energy, and new energy needs to be developed to improve energy efficiency.

Table 3. Decomposition of CEP for the example countries.

Country	Phase	Δ*CG*/%	Δ*GC*/%	Δ*CF*/%	Δ*FE*/%	Δ*EG*/%	Δ*GP*/%	Δ*P*/%
CZE	1990–1995	170.42	359.34	−170.32	−26.37	−162.64	−73.68	3.25
	1995–2000	−51.04	211.79	2.65	−17.33	−197.11	162.75	−11.71
	2000–2005	−292.69	416.17	−89.69	−118.69	−207.79	398.91	−6.22
	2005–2010	−210.7	479.05	−30.96	−77.31	−370.79	242.83	67.88
	2010–2015	−133.4	552.1	26.72	−58.41	−520.4	216.33	17.06
	2015–2019	−813.48	1190.71	−238.29	−128.76	−823.66	855.44	58.04
DEU	1990–1995	−90.67	412.67	−90.88	−25.71	−296.09	140.04	50.64
	1995–2000	−113.46	310.98	−50.26	−52.43	−208.28	206.06	7.39
	2000–2005	31.65	164.67	−32.35	−27.98	−104.35	62	6.36
	2005–2010	−132.59	390.05	56.4	−49.93	−396.52	270.46	−37.87
	2010–2015	−420.54	805.75	26.81	−130.88	−701.68	447.09	73.45
	2015–2019	−211.56	840.44	−242.27	−163.58	−434.59	212.59	98.97
GBR	1990–1995	6.57	160.86	−54.82	−31.29	−74.75	77.28	16.15
	1995–2000	−389.36	484.07	−140.47	21.24	−364.84	439.29	50.07
	2000–2005	−414.51	499.08	−55.75	15.56	−458.89	430	84.51
	2005–2010	15.64	454.58	−63.15	7.25	−398.68	−91.46	175.82
	2010–2015	−217.43	925.28	−3.48	−254.7	−667.1	199.76	117.67
	2015–2019	−173.28	835.77	−195.96	−155.27	−484.55	168.94	104.35
KOR	1990–1995	−1323.83	−59.34	−416.45	159.15	316.63	1236.61	187.23
	1995–2000	−612.13	358.88	−143.88	−92.29	−122.71	596.94	115.19
	2000–2005	−580.58	332.14	−44	−47.31	−240.83	604.34	76.24
	2005–2010	−1281.23	230.83	19.24	159.04	−409.11	1268.73	112.5
	2010–2015	−838.08	536.13	−65.57	−24.84	−445.71	779.21	158.86
	2015–2019	−1096.92	977.66	−382.51	156.03	−751.19	1110.36	86.57
TUR	1990–1995	−347.94	−54.58	−73.58	−82.32	210.48	217.65	230.29
	1995–2000	−258.08	−66.25	−10.67	81.65	−4.73	219.46	138.62
	2000–2005	−441.04	224.96	13.49	−19.04	−219.41	379.88	161.16
	2005–2010	−248.82	−40.09	−112.93	−18.2	171.21	207.68	141.15
	2010–2015	−412.24	222.78	−40.34	−24.36	−158.08	388.89	123.35
	2015–2019	−513.97	416.24	−106.95	−219.91	−89.37	354.65	259.31

Source: Authors' own calculation.

The CEP of South Korea was 0.95 in 1990. It was 1.09 in 2019, an increase of only 14.74%. The CEP of Turkey was 0.8 in 1990, a low starting point, and then increased significantly to 1.05, at an increase of 31.25%. The main drivers in the two countries are slightly different. In South Korea, it is clear that the most significant positive effect of increasing CEP is the economic development effect, and the most significant negative effect is the welfare effect. Other effects are not significant. The economic development and the living standard in South Korea have greatly improved, which has played a leading role in the improvement of CEP. However, South Korea's economic growth has not promoted the overall welfare improvement very well, resulting in a potent inhibitory effect on the welfare effect. Although the main positive and negative effects in Turkey are also economic development effects and welfare effects, respectively, the contribution rate is not significantly prominent. The facilitation of carbon productivity effects and the inhibition of energy intensity effects also play a role. As a result, CEP in Turkey has been greatly improved. Overall, South Korea and Turkey should focus on energy efficiency and improve the development of emission-reduction technologies.

5. Conclusions and Recommendations

5.1. Conclusions

This paper examines the decoupling between carbon emissions and HDI and the CEP. Combining the Tapio decoupling model and the LMDI decomposition model, it is found that the decoupling status and CEP of various human development countries are different.

Due to data availability, only the drivers of the decoupling index and CEP for the example countries are decomposed. The main findings of the study are as follows:

1. There are noticeable differences in the decoupling status of countries with different human development. The countries that achieve strong decoupling mostly have very high human development. A few countries with extremely low human development have achieved strong decoupling, which is not an ideal decoupling status. Only three countries are sustaining strong decoupling. The strong decoupling status in most countries is unstable, and there is a risk of transition to another decoupling status.
2. Overall, the CEP of most countries shows a gradual upward trend. Countries with high human development and low CEP are mainly in Europe, Central Asia, and North America. Most countries with low human development and low CEP face the dual challenges of welfare growth and environmental sustainability.
3. The main contributing factor of strong decoupling in the Czech Republic, Germany, and the United Kingdom is the energy intensity effect, while the main inhibitory factor is the economic development effect. The economic development effect is the main inhibiting factor for South Korea and Turkey, which causes South Korea and Turkey to be unable to shift from expansive negative decoupling to strong decoupling. For the Czech Republic, Germany, and the United Kingdom, the main driving force of improvement in CEP is the carbon productivity effect, and the main inhibitory effect is the energy intensity effect. The main positive effect of promotion in South Korea and Turkey is the economic development effect, and the main inhibitory factor is the welfare effect.

5.2. Recommendations

Some targeted recommendations need to be put forward to achieve decoupling between carbon emissions and HDI and improving CEP.

1. For countries with a very high HDI, reduce carbon emissions while maintaining the growth of their HDI. Following the commitments of the *Paris Agreement*, developed countries continue to take the lead in emission-reduction actions, improve the emission-reduction technologies, and provide developing countries with technical and financial support for emission reduction. In daily life, developed countries continue to implement the concept of environmental protection and achieve low-carbon life.
2. Most countries with high and medium human development are the major carbon emitters. A synchronized increase in carbon emissions has accompanied their HDI growth. The first thing to do is increase their HDI to very high human development. Economic growth has made remarkable achievements, and more attention needs to be paid to developing health and education, especially the improvement of quality healthcare and higher education (UNDP, 2019). The improvement of population quality will be conducive to the transmission of low-carbon concepts, as carbon emissions from living are gradually increasing. It is also necessary to reduce carbon emissions, learn advanced emission-reduction technologies, improve energy efficiency, and adjust the energy structure. If new energy sources are developed and new energy industries are encouraged, it is possible to surpass developed countries, such as China's electric vehicle industry.
3. Low human development countries have the worst decoupling status. The most urgent thing for these countries is to maintain a stable political environment. Then there is the construction of infrastructure, including medical, educational, and industrial, to improve the HDI.

Author Contributions: D.L. conceptualized and designed the research. T.S. drafted the paper and conducted the data interpretation. X.W. and J.L. collected data and obtained preliminary data results. All authors have read and agreed to the published version of the manuscript.

Funding: This paper is supported by the Key Program of the National Philosophy and Social Science Foundation of China (Grant No. 21ATJ003), Innovation Team of Philosophy and Social Sciences in Henan Colleges and Universities (2017–CXTD-07), Major Projects in Basic Research of Philosophy and Social Sciences in Henan Colleges and Universities (2019-JCZD-002), and Humanities and Social Sciences Project of Henan Educational Administration (2022-ZZJH-416).

Institutional Review Board Statement: Not applicable.

Informed Consent Statement: Not applicable.

Data Availability Statement: The raw data presented in this study are openly available on some platforms of international organizations, reference numbers 1, 26, 27, 28 and 29.

Conflicts of Interest: The authors declare no conflict of interest.

References

1. BP. BP Statistical Review of World Energy 2021. 2021. Available online: http://www.bp.com/Statisticalreview (accessed on 9 November 2021).
2. Yang, J.; Hao, Y.; Feng, C. A race between economic growth and carbon emissions: What play important roles towards global low-carbon development? *Energy Econ.* **2021**, *100*, 105327. [CrossRef]
3. United Nation Development Program (UNDP). *Human Development Report 1990: Concept and Measurement of Human Development*; UNDP: New York, NY, USA, 1990.
4. Pan, J. A Conceptual Framework for Understanding Human Development Potential—With Empirical Analysis of Global Demand for Carbon Emissions. *Soc. Sci. China* **2002**, *6*, 15–25.
5. United Nation Development Program (UNDP). *Human Development Report 2020 the Next Frontier Human Development and the Anthropocene*; UNDP: New York, NY, USA, 2020.
6. Hossain, M.A.; Chen, S. Nexus between Human Development Index (HDI) and CO_2 emissions in a developing country: Decoupling study evidence from Bangladesh. *Environ. Sci. Pollut. Res.* **2021**, *28*, 58742–58754. [CrossRef] [PubMed]
7. OECD. *Indicators to Measure Decoupling of Environmental Pressure from Economic Growth*; OECD: Paris, France, 2002.
8. Tapio, P. Towards a theory of decoupling: Degrees of decoupling in the EU and the case of road traffic in Finland between 1970 and 2001. *Transp. Policy* **2005**, *12*, 137–151. [CrossRef]
9. Lin, S.J.; Beidari, M.; Lewis, C. Energy consumption trends and decoupling effects between carbon dioxide and gross domestic product in South Africa. *Aerosol Air Qual. Res.* **2015**, *15*, 2676–2687. [CrossRef]
10. Wu, Y.; Zhu, Q.; Zhu, B. Decoupling analysis of world economic growth and CO_2 emissions: A study comparing developed and developing countries. *J. Clean. Prod.* **2018**, *190*, 94–103. [CrossRef]
11. Chen, J.; Wang, P.; Cui, L.; Huang, S.; Song, M. Decomposition and decoupling analysis of CO_2 emissions in OECD. *Appl. Energy* **2018**, *231*, 937–950. [CrossRef]
12. Martínez, D.M.; Ebenhack, B.W. Understanding the role of energy consumption in human development through the use of saturation phenomena. *Energy Policy* **2008**, *36*, 1430–1435. [CrossRef]
13. Akizu-Gardoki, O.; Bueno, G.; Wiedmann, T.; Lopez-Guede, J.M.; Arto, I.; Hernandez, P.; Moran, D. Decoupling between human development and energy consumption within footprint accounts. *J. Clean. Prod.* **2018**, *202*, 1145–1157. [CrossRef]
14. Hu, A. A Global Roadmap for Emission Reductions on the Road to Copenhagen. *J. Contemp. Asia-Pac. Stud.* **2008**, *162*, 22–38.
15. Costa, L.; Rybski, D.; Kropp, J.P. A Human Development Framework for CO_2 Reductions. *PLoS ONE* **2011**, *6*, e29262. [CrossRef]
16. Chen, L.; Cai, W.; Ma, M. Decoupling or delusion? Mapping carbon emission per capita based on the human development index in Southwest China. *Sci. Total Environ.* **2020**, *741*, 138722. [CrossRef]
17. Beinhocker, E.; Oppenheim, J.; Irons, B.; Lahti, M. *The Carbon Productivity Challenge: Curbing Climate Change and Sustaining Economic Growth*; McKinsey & Company: New York, NY, USA, 2008.
18. Zhu, D.; Liu, G. Human Development Performance Index and Empirical Analysis of Carbon Emission. *China Popul. Resour. Environ.* **2010**, *21*, 73–79.
19. Liu, G.; Zhu, Y. Research on well-being performance of carbon emissions: Based on the data of G20. *Stud. Sci. Sci.* **2011**, *29*, 1504–1510.
20. Hu, J.; Gao, P.; Zuo, P. Study on the Regional Differences and Contribution Rate of China's Carbon Emissions Welfare Performance. *Econ. Probl.* **2018**, *11*, 54–60.
21. Pan, J.; Zhang, Y. Spatiotemporal patterns of energy carbon footprint and decoupling effect in China. *Acta Geogr. Sin.* **2021**, *76*, 206–222.
22. Wang, H.; Hu, Z. The measurement research on the decoupling relationship between manufacturing sector economic growth and carbon dioxide in China. *Stud. Sci. Sci.* **2012**, *30*, 1671–1675.
23. Ang, B.W. Decomposition analysis for policymaking in energy: Which is the preferred method? *Energy Policy* **2004**, *32*, 1131–1139. [CrossRef]
24. Wang, C.; Chen, J.; Zou, J. Decomposition of energy-related CO2 emission in China: 1957–2000. *Energy* **2005**, *30*, 73–83. [CrossRef]
25. Liu, Y.; Jin, S. Temporal and Spatial Evolution Characteristics and Influencing Factors of Energy Consumption Carbon Emissions in Six Provinces of Central China. *Econ. Geogr.* **2019**, *39*, 182–191.

26. Global Carbon Project. Supplemental Data of Global Carbon Budget 2021 (Version 1.0) [Data Set]. 2021. Available online: https://www.icos-cp.eu/science-and-impact/global-carbon-budget/2021 (accessed on 9 November 2021).
27. World Bank. The World Development Indicators [Data Set]. 2021. Available online: https://datacatalog.worldbank.org/home (accessed on 15 April 2021).
28. United Nations, Department of economic and social affairs (UNDESA). World Population Prospects: The 2019 Revision. Rev 1. 2020. Available online: https://population.un.org/wpp/ (accessed on 10 November 2021).
29. United Nation Development Program (UNDP). Human Development Data Center [Data Set]. 2021. Available online: https://hdr.undp.org/ (accessed on 10 November 2021).
30. Lucas Chancel. *Climate Change & the Global Inequality of Carbon Emissions, 1990–2020*; World Inequality Lab: Paris, France, 2021.

atmosphere

MDPI

Article

Study on the Associations between Meteorological Factors and the Incidence of Pulmonary Tuberculosis in Xinjiang, China

Chunjie Gao [1,†], Yingdan Wang [1,†], Zengyun Hu [2,3], Haiyan Jiao [1] and Lei Wang [4,*]

1 College of Public Health, Xinjiang Medical University, Urumqi 830017, China; gaochunjie98@163.com (C.G.); wyd2248@163.com (Y.W.); 18893718038@163.com (H.J.)
2 State Key Laboratory of Desert and Oasis Ecology, Xinjiang Institute of Ecology and Geography, Chinese Academy of Sciences, Urumqi 830011, China; huzengyun@ms.xjb.ac.cn
3 Shenzhen Institutes of Advanced Technology, Chinese Academy of Sciences, Shenzhen 518000, China
4 Department of Medical Engineering and Technology, Xinjiang Medical University, Urumqi 830017, China
* Correspondence: wlei81@126.com
† These authors contributed equally to this work.

Abstract: Pulmonary tuberculosis (PTB) has been a major threat to global public health. The association between meteorological factors and the incidence of PTB has been widely investigated by the generalized additive model, auto-regressive integrated moving average model and the distributed lag model, etc. However, these models could not address a non-linear or lag correlation between them. In this paper, a penalized distributed lag non-linear model, as a generalized and improved one, was applied to explore the influence of meteorological factors (such as air temperature, relative humidity and wind speed) on the PTB incidence in Xinjiang from 2004 to 2019. Moreover, we firstly use a comprehensive index (apparent temperature, AT) to access the impact of multiple meteorological factors on the incidence of PTB. It was found that the relationships between air temperature, relative humidity, wind speed, AT and PTB incidence were nonlinear (showed "wave-type ", "invested U-type", "U-type" and "wave-type", respectively). When air temperature at the lowest value (−16.1 °C) could increase the risk of PTB incidence with the highest relative risk (RR = 1.63, 95% CI: 1.21–2.20). An assessment of relative humidity demonstrated an increased risk of PTB incidence between 44.5% and 71.8% with the largest relative risk (RR = 1.49, 95% CI: 1.32–1.67) occurring at 59.2%. Both high and low wind speeds increased the risk of PTB incidence, especially at the lowest wind speed 1.4 m/s (RR = 2.20, 95% CI: 1.95–2.51). In particular, the lag effects of low and high AT on PTB incidence were nonlinear. The lag effects of extreme cold AT (−18.5 °C, 1st percentile) on PTB incidence reached a relative risk peak (RR = 2.18, 95% CI: 2.06–2.31) at lag 1 month. Overall, it was indicated that the environment with low air temperature, suitable relative humidity and wind speed is more conducive to the transmission of PTB, and low AT is associated significantly with increased risk of PTB in Xinjiang.

Keywords: pulmonary tuberculosis; penalized distributed lag non-linear model; meteorological factors; apparent temperature; cumulative risk

Citation: Gao, C.; Wang, Y.; Hu, Z.; Jiao, H.; Wang, L. Study on the Associations between Meteorological Factors and the Incidence of Pulmonary Tuberculosis in Xinjiang, China. *Atmosphere* **2022**, *13*, 533. https://doi.org/10.3390/atmos13040533

Academic Editor: Kai-Jen Chuang

Received: 24 February 2022
Accepted: 24 March 2022
Published: 28 March 2022

1. Introduction

As one of the widely distributing and potentially fatal infectious diseases, tuberculosis remains one of the top ten causes of mortality worldwide. According to the Global Tuberculosis Report of the World Health Organization in 2020 [1], there were approximately 9.96 million cases and 1.42 million tuberculosis-related deaths worldwide in 2019, of which about 5.94 million were new PTB cases. China accounted for 8.6% of the world's new tuberculosis cases in 2019, ranking third among 30 countries with a high burden of tuberculosis. The incidence of PTB in Xinjiang is relatively serious, which has always been one of the highest in China, and the reported incidence of PTB in 2019 is about three times the national level.

PTB is a chronic respiratory disease caused by *Mycobacterium tuberculosis* (*M. tuberculosis*). A large number of lipids in *M.tb* resist the multiplication of lysozymes and damage macrophages. Hence, it can inhibit the intracellular bactericidal mechanism and cause an inflammatory response in the lungs [2]. Susceptible people can be infected through breathing in dust with *M.tb* and inhaling the droplet nucleus released into the air by actively infectious PTB coughing or sneezing [3]. It is revealed that meteorological factors (such as air temperature, relative humidity and wind speed) may indirectly affect *M.tb* transmission in the environment [4–7]. For instance, within a certain range of air temperature and relative humidity, droplets containing *M.tb* are more likely to be evaporated in the air to form into certain diameters that can be suspended in the air for a longer time (especially, droplets containing *M.tb* can grow and reproduce at 35~37 °C, and it can survive for 4~5 years in −8~−6 °C), to be easily inhaled into the body by susceptible persons [5]. Moreover, changes in air temperature can also affect the human body's physiological response to toxic agents and retard the clearance rate of *M.tb* [6], and activate or inhibit the development of PTB by influencing blood pressure [7]. Suitable humidity can promote the growth and reproduction of *M.tb* and increase the time for *M.tb* to float in the air [8]. When *M.tb* adheres to dust, it is easily affected by the wind speed, i.e., the higher the speed of the wind, the wider spread of *M.tb* [9].

The annual PTB incidence in Xinjiang is still higher, which may be closely related to its unique geographical location and climatic conditions [10]. Xinjiang, located at the northwest border of China, has a temperate continental climate, with long winters, sparse precipitation, dry climate, frequent wind and sand activities, and frequent snowfall [11]. With an annual average temperature of 11 °C, Xinjiang is one of the regions with a lower annual average temperature in China. The heating period in Xinjiang is from October to March of the next year, and the air pollution is serious in this period. *M.tb* is so small that normal air currents can keep the pollution particulates containing *M.tb* airborne and transport them through rooms or some buildings, which may increase the risk of PTB development [3,5,12]. In addition, the snowfall in winter increases the humidity and makes the droplet nucleus ejected by actively infectious PTB stay in the air longer [13], which may create favorable conditions for the transmission of PTB. Therefore, it is of great significance to study the relationship between meteorological factors and the PTB incidence in Xinjiang.

There have been various statistical models to characterize the relationship between meteorological factors and PTB incidence. For instance, a generalized additive model [14] was used to quantitatively evaluate the effects of meteorological factors on the risk of pulmonary tuberculosis in Jiangsu Province. The results illustrated that an environment with low temperature, relatively high wind speed, and low relative humidity is conducive to the transmission of PTB. Li et al. [15] employed an auto-regressive integrated moving average model to exhibit the best predicting performance of PTB incidence by incorporating meteorological factors. However, some results [8,9,15,16] demonstrated there is a non-linear and lag correlation between meteorological factors and the incidence of PTB. The generalized additive model somewhat ignores the collinearity among different lag days [4,17], and the auto-regressive integrated moving average model fails to solve the collinearity problem and ignores the lag effects [12,18]. Fortunately, the distributed lag non-linear model (DLNM) proposed by Armstrong [19] can effectively evaluate the nonlinear and lag relationship between them. For example, Wu et al. [20] investigated the cold and hot effects on mortality at different lags in four subtropical cities. Yang et al. [21] evaluated the association between meteorological factors and the mumps incidence in Guangzhou. Subsequently, a penalized framework DLNM (P-DLNM) was put forward in [22], which introduces a penalized framework to the exposure and lag dimensions in the DLNM, to overcome the complex fitting and the lack of general standards in the selection of basic functions, the number of nodes, the maximum number of lag days and the optimal model in the DLNM. It can control the smoothness of the basis function, and effectively reduce the total number of degrees of freedom to make itself more robust.

Another concerning issue is how the combination of some meteorological factors influenced the incidence of PTB. AT, firstly proposed by Stead-man [23], is a comprehensive index to measure the interaction of air temperature, wind speed and relative humidity, which more objectively represents the actual perception of air temperature. Many research works [15,16,24] evaluated the relationship between meteorological factors and human health by applying AT, such as acute coronary syndromes [15], stroke [16] and acute excessive drinking [24]. However, there are few studies that quantified the impact of meteorological factors and AT on PTB incidence in Xinjiang. Therefore, in this paper, a penalized framework DLNM model is used to analyze the impact of meteorological factors (such as air temperature, wind speed, and relative humidity) on PTB incidence, and AT, is applied to quantify the comprehensive impact of multiple meteorological factors on PTB incidence in Xinjiang from 2004 to 2019, which provides a theoretical basis for the prevention and control of PTB in Xinjiang.

2. Materials and Methods

2.1. Study Area

Xinjiang, located in Northwest China (73°40′–96°18′ E, 34°25′–48°10′ N), has a special topography named as two basins lie in between three mountains, with an average altitude of about 1000 km. From 2004 to 2019, the average annual temperature in Xinjiang was 10.1 °C, the average annual precipitation was 110.0 mm [11], the average annual wind speed was 2.3 m/s, the average annual air pressure was 898 hPa and the average annual relative humidity was 48.3% [25].

2.2. Meteorological and PTB Data

The monthly PTB cases in Xinjiang from 2004 to 2019 were obtained from Public Health Scientific Data Sharing Center (http://www.phsciencedata.cn/, accessed 30 December 2021)and Health Commission of Xinjiang Uygur Autonomous Region (http://wjw.xinjiang.gov.cn/, accessed 8 January 2022).

The monthly average values of meteorological indicators from 2004 to 2019 were derived from 54 stations in Xinjiang (Supplementary Figure S1) from the China Meteorological Data Sharing Center (http://data.cma.cn, accessed 21 November 2021), including average air temperature (°C), average precipitation (mm), average wind speed (m/s), average air pressure (hPa), average relative humidity (%), average sunshine duration (hours/month). AT, is calculated as follows [24]:

$$AT = T + 0.33 \times e - 0.70 \times W - 4.00, \tag{1}$$

$$e = \frac{RH}{100} \times 6.105 \times exp\left(17.27 \times \frac{T}{237.7 + T}\right), \tag{2}$$

where T represents air temperature (°C); e represents water vapor pressure (hPa); RH is relative humidity (%); and W represents wind speed (m/s).

In this paper, based on percentile range the AT was divided into seven categories for analysis of the associations between low AT, high AT and PTB incidence: extreme cold (≤1st percentile), cold (1st–5th percentiles), mild cold (5th–25th percentiles), comfortable (25th–75th percentiles), mild heat (75th–95th percentiles), heat (95th–99th percentiles) and extreme heat (≥99th percentile), respectively. In addition, the influences of low AT (including extreme cold, cold and mild cold) and high AT (including extreme heat, heat, and mild heat) on PTB incidence were investigated [24].

2.3. Spearman's Rank Correlation

In order to measure the nonlinear relationship between meteorological variables and PTB incidence, Spearman's rank correlation analysis was used [26]. The correlation coefficient r_s denotes the strength of an association between two variables, which can be calculated as follows [27]:

$$r_s = 1 - \frac{6\sum d_i^2}{n(n^2-1)}$$

where n is the count of variables and d is the rank difference between each pair of variables. The heatmap can be used to represent the values of the correlation coefficient between multiple variables. In the heatmap, the darker the color grid, the greater the Spearman's rank correlation coefficient, and the blank space is usually used to represent two variables without significant correlation.

2.4. Statistical Method

DLNM, proposed by Gasparrini and Armstrong in 2010 [19], is widely used to evaluate the relationship between meteorological factors and health effects, which is a nonlinear model reflecting the exposure-lag-response relationship by selecting basis functions for exposure-response and lag-response and to form a cross-basis function [28].

Define, $f(x_t;\beta) = {Z_t}^T\beta$ as the exposure-response function, where Z_t is a matrix with dimensions $n \times v_x$ (n represents the length of the time series, v_x represents the degree of freedom), and β is the shape parameter of ${Z_t}^T$. Define, $q_{x,t} = \left[x_{t-\ell_0}, \ldots, x_{t-\ell}, \ldots, x_{t-L}\right]^T$ is lagged effects of the exposure x_t, where ℓ_0 and L are the minimum and maximum lags, respectively, i.e., the lag structure can be formulated as [28] $\ell = [\ell_0, \ldots, L]^T$. The cross-basis function $s(x_t, \eta)$ is constructed by:

$$s(x_t, \eta) = \left(\mathbf{1}_{L-\ell_0+1}^T A_{x,t}\right)\eta = w_{x,t}^T\eta,$$

where $w_{x,t}^T$ is a cross-basis matrix transformed from $A_{x,t}$ (with dimensions $v_x \times v_\ell$), and η is the coefficient of $w_{x,t}^T$. $A_{x,t}$ is computed by a row-wise Kronecker product between the two basis matrices R (with dimensions $(L-\ell_0+1) \times v_x$) and C (with dimensions $(L-\ell_0+1) \times v_\ell$) that obtained by applying basis transformations to vectors $q_{x,t}$ and ℓ, thus $A_{x,t}$ can be expressed as:

$$A_{x,t} = \left(\mathbf{1}_{v_\ell}^T \otimes R_{x,t}\right) \odot \left(C \otimes \mathbf{1}_{v_x}^T\right),$$

where $\mathbf{1}_{v_\ell}^T$ is a unit vector with length v_ℓ, $\otimes$ and $\odot$ denote the Kronecker and Hadamard products, respectively [29].

The distributed lag non-linear model with a penalized framework (P-DLNM) applies varying degrees of penalties in exposure and lag dimensions to smooth the surface of exposure-lag-response, the log-likelihood function $l_p(\eta, \gamma, \lambda)$ of P-DLNM is:

$$l_p(\eta, \gamma, \lambda) = l(\eta, \gamma) - \frac{1}{2}\eta^T\left(\lambda_x\left(S_x \otimes \mathbf{1}_{v_\ell}^T\right) + \lambda_\ell\left(\mathbf{1}_{v_\ell}^T \otimes S_\ell\right)\right)\eta,$$

where the penalty matrices S_x and S_ℓ are separately obtained by introducing the penalties term to the base matrices R and C, λ_x and λ_ℓ represent the penalty parameters of S_x and S_ℓ, respectively. η is the coefficient for the cross-basis, restricted by S_x, S_ℓ and the penalty parameter $\lambda = [\lambda_x, \lambda_l]^T$. In this paper, the ridge penalty is used to add the penalty term for the lag dimension and control the coefficient of S_ℓ to shrink towards the zero value at different lag times [29]. The penalties of lag-response dimension are based on cubic regression splines, i.e., CR smoother, which can be expressed as $S_\ell = P_{v_\ell}$, here P_{v_ℓ} is a pre-specified diagonal matrix with the weight p.

Based on the generalized additive model and combined the cross-basis functions with penalties, meanwhile, controlled the long-term trend of time and other confounding factors, the association between meteorological factors and PTB incidence was investigated by the following model in this paper:

$$\ln(\lambda_t) = \alpha + \sum_{i=1}^{4} s_i(x_i, \eta_i) + ns(MF, df) + ns(time, df) + \gamma S + \varepsilon AR + \theta SF + \tau PC,$$

where λ_t is the case number of PTB on month t_{th}; α is the intercept; are the cross-basis functions with coefficients η_i $(i = 1, \ldots, 4)$, x_i are meteorological variables (such as air temperature, wind speed, relative humidity and AT); $ns(\cdot)$ is a natural cubic spline function, which can be used to control the long-term trends (time with the degree of freedom 10, $df = 10$) and other meteorological factors (air pressure, precipitation and sunshine duration with the degrees of freedom 3, 3 and 1, respectively). S is a categorical variable with coefficient γ to control for the seasonality (spring: Mar.-May., summer: Jun.-Aug., autumn: Sep.-Nov. and winter: Dec.-Feb.); AR is the autoregressive term with coefficient ε to correct for autocorrelation in the residuals [30]; is a binary variable with coefficient θ to control for the potential influences on delayed medical treatment of PTB patients caused by the Spring Festival [12]; PC is a binary variable with coefficient τ to avoid the impact of PTB transmission caused by immigrants entering Xinjiang to pick cotton from August to October every year.

In this paper, taking the median of air temperature, wind speed, relative humidity and AT as references, four cross-basis functions (defined by an exposure-response relationship with four equally spaced intervals knots and a lag-response relationship with four equal space log-knots) were built to more flexibly reflect the 3-D relationship between meteorological factors and the incidence of PTB in Xinjiang.

2.5. Parameter Estimation

To quantify the risk between meteorological factors and PTB incidence, the relative risks (RR) of specific exposure and lag time were used [22]. The values of RR and standard error (SE) of specific exposure level x_p are as follows:

$$RR_{x_p} = exp\left(\hat{\beta}_{x_p}\right) = exp\left(\boldsymbol{A}_{x_p}\hat{\boldsymbol{\eta}}\right)$$

$$SE_{x_p} = exp\left(\sqrt{V\left(\hat{\beta}_{x_p}\right)}\right) = exp\left(\sqrt{\boldsymbol{A}_{x_p}V(\hat{\boldsymbol{\eta}})\boldsymbol{A}_{x_p}{}^{T}}\right)$$

where $\boldsymbol{A}_{x_p}$ denotes the cross-basis matrix with $x = x_p$; $V(\cdot)$ denotes the variance of a random variable. Similarly, the lag-specific risk RR_{ℓ_p} and SE_{ℓ_p} can be estimated. In addition, the overall cumulative effect (RR_c) of such exposure history, with associated SE_c could be computed with:

$$RR_c = exp\left(\hat{\beta}_c\right) = exp(\mathbf{W}^p\hat{\boldsymbol{\eta}}),$$

$$SE_c = exp\left(\sqrt{V\left(\hat{\beta}_c\right)}\right) = exp\left(\sqrt{\mathbf{W}^p V(\hat{\boldsymbol{\eta}})\mathbf{W}^{pT}}\right)$$

where $\mathbf{W}^p$ denotes the cross-basis matrix.

2.6. Sensitivity Analysis

To assess the robustness of the model, a sensitivity analysis was conducted to examine the influence of df of parameters and maximum lag time on the overall cumulative effect of AT. In this paper, df of time in the model was varied from 10, 11 to 12, respectively. The df of air pressure and precipitation in the model were changed between 3 and 5, respectively. The df of sunshine duration was altered between 1 and 3, respectively. The maximum lag time of AT was also set to 8, 10 and 12, respectively.

All statistical analyses were performed by carrying out R software version 4.0.5.

3. Results

3.1. Descriptive Statistics of PTB Cases and Meteorological Factors

During the study period from 2004 to 2019 (192 months), a total of 668,753 PTB cases were included, with an average of 3427 cases per year and a maximum of 74,549 cases in 2018. The case number of PTB in Xinjiang had an obvious seasonal pattern and showed an increasing trend, with a peak from January to April (see Figure 1 and Supplementary Figure S2).

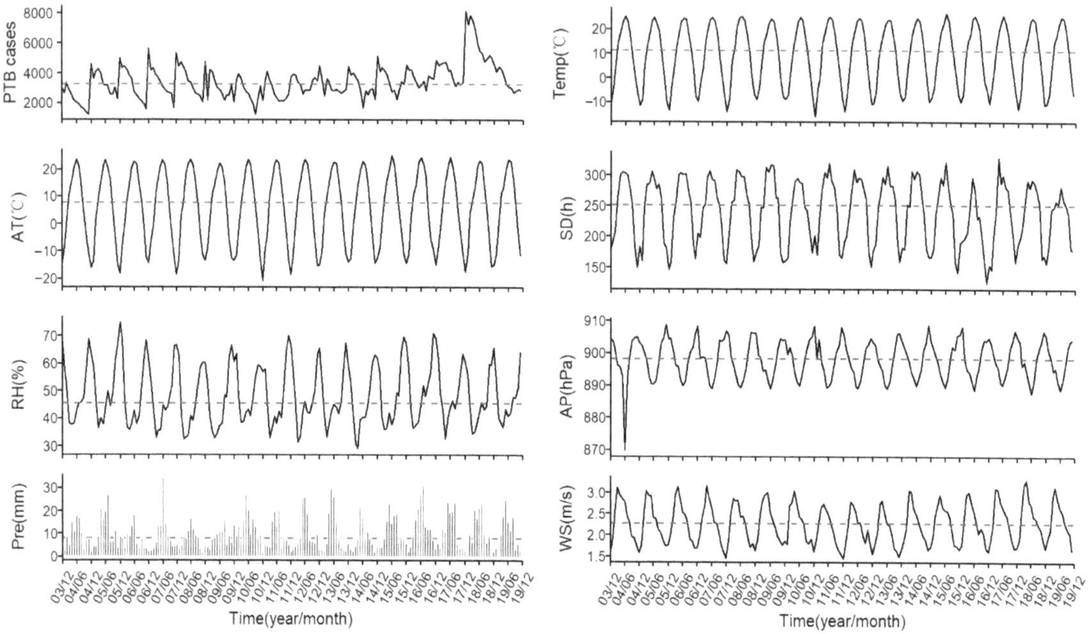

Figure 1. Time series of monthly PTB cases and meteorological factors in Xinjiang from 2004 to 2019. Abbreviations: Temp, air temperature (°C); AT, apparent temperature (°C); SD, sunshine duration (h); RH, relative humidity (%); AP, air pressure (hPa); Pre, precipitation (mm); WS, wind speed (m/s). Note: The blue dotted line represents the median.

There were seasonal fluctuations and periodic trends of meteorological factors in Xinjiang, roughly showing the variation of single peak and single valley (see Figure 1 and Supplementary Figure S3). Air temperature, AT and sunshine duration had similar seasonal patterns, with higher values occurring from April to October (Seasonal index >1 indicates obvious seasonal variation). The peak of relative humidity and air pressure occurred from October to February of the next year, whereas the trough appeared from March to September. A seasonal peak of precipitation emerged from May to September and a trough occurred from October to March of the next year. Wind speed had a seasonal trend with a peak from March to August and a valley from September to February of the next year. In addition, the median of AT, air temperature, relative humidity, air pressure, precipitation, wind speed and sunshine duration were 7.8 °C, 11.3 °C, 45.6%, 898.2 hPa, 8.0 mm, 2.3 m/s and 251.0 h, respectively (see Figure 1).

The descriptive statistics of meteorological factors and the case number of PTB were shown in Table 1. The monthly average of AT was 5.7 °C, which was 3.5 °C lower than the air temperature. The distributions of AT in the1st, 5th, 25th, 75th, 95th and 99th percentiles were −18.5 °C, −15.1 °C, −6.3 °C, 7.8 °C, 18.0 °C, 23.3 °C and 24.3 °C, respectively.

Table 1. Descriptive statistics for monthly PTB cases and meteorological factors in Xinjiang from 2004 to 2019.

Group	Range	$\bar{x} \pm s$	Percentiles						
			P_1	P_5	P_{25}	P_{50}	P_{75}	P_{95}	P_{99}
Pre	0.4~34.2	10.1 ± 7.0	1.7	2.5	4.5	8.0	14.4	24.4	30.0
AP	869.8~908.8	898 ± 6.0	888.5	889.3	893.0	898.2	903.5	906.3	908.3
WS	1.4~3.3	2.3 ± 0.5	1.4	1.6	1.9	2.3	2.6	3.0	3.2
Temp	−16.1~26.5	9.2 ± 12.5	−14.0	−10.6	−1.8	11.3	21.1	24.6	25.4
RH	29.1~74.7	48.3 ± 10.8	31.4	33.4	40.2	45.6	57.9	67.4	70.1
SD	124.8~328.6	239.6 ± 52.7	145.2	157.6	188.1	251.0	286.3	306.9	319.8
PTB cases	1194~8151	3427 ± 1152	1346	1937	2661	3271	4100	5279	7555
AT	−20.9~25.0	5.7 ± 13.5	−18.5	−15.1	−6.3	7.8	18.0	23.3	24.3

Abbreviations: Temp, air temperature (°C); AT, apparent temperature (°C); SD, sunshine duration (h); RH, relative humidity (%); AP, air pressure (hPa); Pre, precipitation (mm); WS, wind speed (m/s).

3.2. Spearman's Rank Correlation Analysis

The Spearman's rank correlation results between monthly PTB cases and meteorological factors in Xinjiang from 2004 to 2019 were provided in Figure 2. AT ($r_s = -0.25$), air temperature ($r_s = -0.23$), sunshine duration ($r_s = -0.17$) and precipitation ($r_s = -0.14$) were negatively correlated with the incidence of PTB, and wind speed ($r_s = 0.15$) is positively correlated with it. There was no significant correlation between relative humidity, air pressure and the incidence of PTB. There were high correlations among meteorological factors, especially between air pressure and AT ($r_s = -0.94$). The p-value < 0.05 was considered statistically significant.

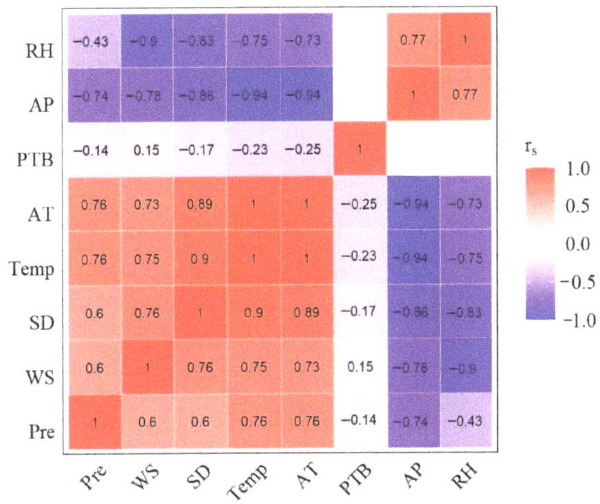

Figure 2. Spearman's rank correlation results between the cases number of PTB and meteorological factors in Xinjiang from 2004 to 2019. Abbreviations: Temp, air temperature (°C); AT, apparent temperature (°C); SD, sunshine duration (h); RH, relative humidity (%); AP, air pressure (hPa); Pre, precipitation (mm); WS, wind speed (m/s); r_s, Spearman's rank correlation coefficient.

3.3. The Influences of Air Temperature on the Incidence of PTB

Taking the overall median of monthly average air temperature (11.3 °C) as a reference, it was obtained that the correlation between average air temperature and PTB incidence was nonlinear at 12 lag months for the general population, and the largest exposed and delayed cumulative effects occurred at −16.1 °C and lag 0 months, respectively (see Figure 3A). The overall exposure-response curve between air temperature and the incidence of PTB took on a U-type when the air temperature below 11.3 °C, with a maximum value, occurred at −16.1 °C (RR = 1.63, 95% CI: 1.21–2.20), whereas an inverted U-type when the air temperature was higher than 11.3 °C with a minimum value appeared at 26 °C (RR = 0.12, 95% CI: 0.06–0.20), as shown in Figure 3B. The association between extreme air temperature and the risk of PTB at specific lag months was shown in Figure 3B and Table S1. The lag 0–1 and lag 11–12 months could significantly increase the risk of PTB at extremely low air temperature −11.9 °C (2.5th percentiles), with a peak at lag 0 (RR = 1.59, 95% CI: 1.53–1.66). The cumulative risk at lag 12 months was largest (RR = 1.19, 95% CI: 1.14–1.24) at the extremely high air temperature 25 °C (97.5th percentiles). The relationship of air temperature-lag-PTB was evaluated by using the heatmap, as shown in Figure 3C.

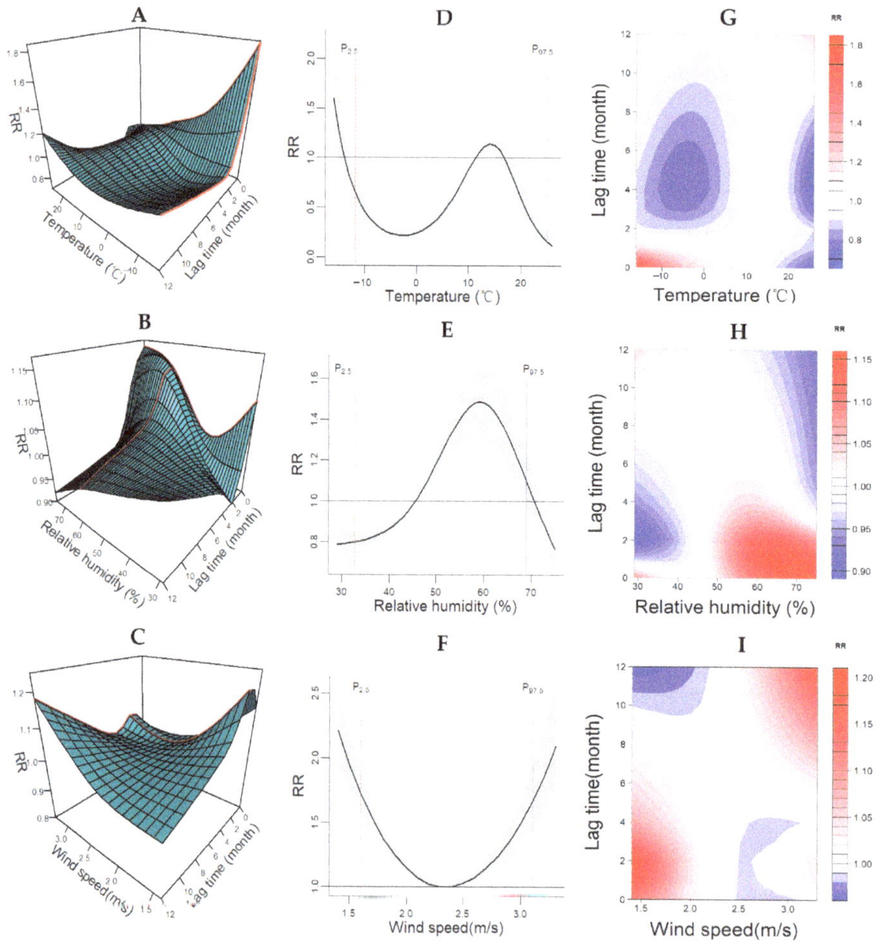

Figure 3. The exposure-lag-response correlations between meteorological factors and PTB incidence in Xinjiang from 2004 to 2019. (**A–C**) represent the 3-D graphs of the relationships between air temperature, relative humidity, wind speed and PTB incidence, respectively. (**D–F**) represent the overall cumulative relative risks of air temperature, relative humidity, wind speed on PTB incidence acrosslag 0–12 months, respectively. (**G–I**) represent the heatmaps of air temperature, relative humidity, wind speed on PTB incidence, respectively.

3.4. The Influences of Relative Humidity on the Incidence of PTB

Setting the overall median of monthly average relative humidity (45.6%) as a reference, it was found that the association between relative humidity and PTB incidence was nonlinear at 12 lag months for the general population, and the largest exposed and delayed cumulative effects occurred at the relative humidity 59.2% and lag 0 months, respectively (see Figure 3D). The overall exposure-response curve between relative humidity and the PTB incidence appeared an inverted U-type, with a peak at a relative humidity of 59.2% (RR = 1.49, 95% CI: 1.32–1.67), as shown in Figure 3E. The relationship between extreme relative humidity and the risk of PTB at specific lag months was revealed in Figure 3E and Table S1. The lag 0, 10, 11 and 12 months could significantly increase the risk of PTB at the extreme low relative humidity of 32.8% (2.5th percentiles), with a top at lag 0 (RR = 1.05, 95% CI: 1.03–1.07). The lag 0–3 months could significantly increase the risk of PTB at the extreme high relative humidity 68.9% (97.5th percentiles), with a top at lag 0 (RR = 1.15, 95% CI: 1.13–1.17). The association of relative humidity-lag-PTB was estimated by using a heatmap, as shown in Figure 3F.

3.5. The Influences of Wind Speed on the Incidence of PTB

Using the overall median of monthly average wind speed of 2 min (2.3 m/s) as a reference (the wind speed corresponding to the lowest risk of PTB), it was found that the influence of wind speed on PTB incidence was nonlinear at 12 lag months for the general population, and the largest exposed and delayed cumulative effects occurred at 1.4 m/s and lag 2 months, respectively (see Figure 3G). The overall exposure-response curve between wind speed and the incidence of PTB showed a U-type with a trough that emerged at 2.3 m/s and a maximum relation arose at 1.4 m/s (RR = 2.20, 95% CI: 1.95–2.51), as shown in Figure 3H. The correlation between extreme wind speed and the risk of PTB at specific lag months was shown in Figure 3H and Table S1; the lag 0–9 months were separately significantly associated with an increased risk of PTB effect at the extreme low wind speed of 1.6 m/s (2.5th percentiles), with a maximum value at lag 2 (RR = 1.15, 95% CI: 1.13–1.16). The lag 6–12 were separately significantly related to an increased risk of PTB at the extreme high wind speed 3.1 m/s (97.5th percentiles), with a maximum value at lag0 (RR = 1.14, 95% CI: 1.12–1.15). The correlation of wind speed-lag-PTB was also assessed by using the heatmap as shown in Figure 3I.

3.6. The Effect of AT on PTB Incidence

Considering the overall median of the monthly AT (7.8 °C) as a reference, it was discovered that the connection between AT and the incidence of PTB was nonlinear at 12 lag months for the total population, and the largest exposed and delayed cumulative effects occurred at −20.9 °C and lag 0 months (see Figure 4A). As shown in Figure 4B, the overall exposure-response curve between AT and the incidence of PTB presented a *U*-type when the AT below 7.8 °C, with a maximum value occurred at −20.9 °C (RR = 2.01, 95% CI: 1.50–2.69), whereas an inverted *U*-type when the AT was higher than 7.8 °C with a maximum risk appeared at 12.6 °C (RR = 1.29, 95% CI: 1.16–1.44). The extreme cold −18.5 °C and mild heat 18 °C had no significantly risky effects on PTB incidence, and the cold −15.1 °C (RR = 0.59, 95% CI: 0.46–0.76), the mild cold −6.3 °C (RR = 0.22, 95% CI: 0.18–0.27), the heat 23.3 °C (RR = 0.58, 95% CI: 0.38–0.90) and the extreme heat 24.3 °C (RR = 0.51, 95% CI: 0.32–0.82) had significantly protective effects on PTB incidence. The connection of AT-lag-PTB was also observed by using the heatmap, as shown in Figure 4C.

It is noteworthy that the lag effects of low AT presented a *V*-shape with the highest risk at lag 0, whereas the lag effects of high AT presented an *N*-shape with the highest risk at lag 12 (see Figure 5). As shown in Figure 5, lag 0 and lag 1 could significantly increase the incidence of PTB at the extreme cold, with the highest risk at lag 0 (RR = 1.81, 95% CI: 1.73–1.90). The cold had a significantly risky effect on increasing PTB incidence at lag 0–1, with the highest risk at lag 0 (RR = 1.60, 95% CI: 1.54–1.67), which was a protective factor for the risk of PTB at lag 2–11. The mild cold had a significantly risky effect on increasing the incidence of PTB at lag 0 and lag 1, with the highest risk at lag 0 (RR = 1.22, 95% CI:1.19–1.26), which was a protective factor for the risk of PTB at lag 2–12. The mild heat had a significantly risky effect on increasing the incidence of PTB at lag 2 and lag 9–12, with the highest risk at lag 12 (RR = 1.11, 95% CI: 1.09–1.14), which did not significantly influence the incidence of PTB at lag 3. The heat and extreme heat had a significantly risky effect on increasing the incidence of PTB at lag10–12, with the highest risk at lag 12 (RR = 1.18, 95% CI: 1.14–1.22 and RR = 1.20, 95% CI: 1.15–1.25), which did not significantly influence PTB incidence at lag 2 and lag 8–9.

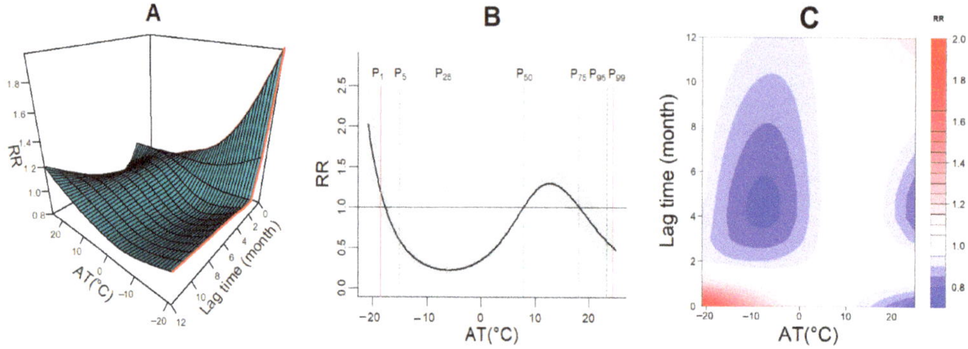

Figure 4. The exposure-lag-response correlations between meteorological factors and the incidence of PTB in Xinjiang from 2004 to 2019. (**A**) represents the 3–D graph of the relationship between AT and PTB incidence, (**B**) represents the overall cumulative relative risk of AT on PTB incidence across lag 0–12 months, (**C**) represents the heatmap of AT on PTB incidence.

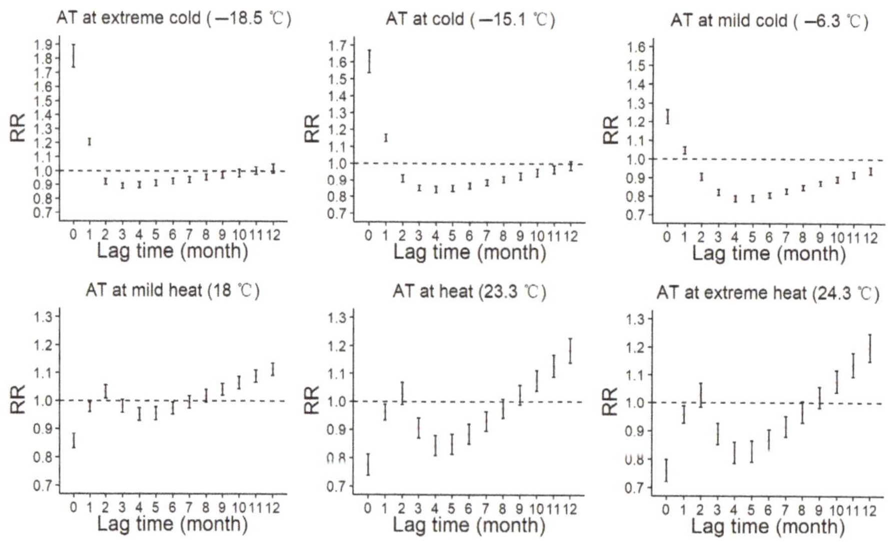

Figure 5. The distribution of RR at various lag months with a reference of medium value 7.8 °C in Xinjiang from 2004 to 2019.

In addition, the cumulative effects of low and high AT on PTB incidence at different lag months were compared (see Table 2). For the extreme cold of low AT, the cumulative effects were risky from lag 0 to lag 0–8, with a maximum value at lag 0–1 (RR = 2.18, 95% CI: 2.06–2.31). For the cold, the cumulative effects were risky effects from lag 0 to lag 0–4, with a maximum value at lag 0–1 (RR = 1.84, 95% CI: 1.74–1.94). For the mild cold the cumulative effects were risky from lag 0 to lag 0–2, with a maximum value at lag 0–1 (RR = 1.28, 95% CI: 1.22–1.34). For the high AT, there was no significant risk of the cumulative effects of the mild heat, heat and extreme heat, in particular, the cumulative effects of mild heat had no significant influences from lag 0–10 to lag 0–12.

Table 2. The cumulative effects of low and high AT on the incidence of PTB by different lag period in Xinjiang from 2004 to 2019.

Lag	P_1	P_5	P_{25}	P_{75}	P_{95}	P_{99}
Lag 0	1.81 (1.73, 1.90)	1.60 (1.54, 1.67)	1.22 (1.19, 1.26)	0.86 (0.83, 0.88)	0.77 (0.74, 0.81)	0.76 (0.72, 0.80)
Lag 0–1	2.18 (2.06, 2.31)	1.84 (1.74, 1.94)	1.28 (1.22, 1.34)	0.84 (0.80, 0.87)	0.74 (0.70, 0.80)	0.73 (0.68, 0.78)
Lag 0–2	2.01 (1.89, 2.14)	1.67 (1.58, 1.77)	1.16 (1.09, 1.22)	0.86 (0.82, 0.91)	0.76 (0.70, 0.84)	0.75 (0.68, 0.82)
Lag 0–3	1.79 (1.65, 1.94)	1.42 (1.32, 1.54)	0.95 (0.88, 1.02)	0.85 (0.78, 0.91)	0.69 (0.61, 0.78)	0.66 (0.58, 0.76)
Lag 0–4	1.61 (1.45, 1.78)	1.20 (1.09, 1.32)	0.74 (0.68, 0.81)	0.80 (0.73, 0.89)	0.58 (0.49, 0.69)	0.54 (0.46, 0.65)
Lag 0–5	1.46 (1.29, 1.66)	1.02 (0.90, 1.14)	0.58 (0.53, 0.65)	0.77 (0.68, 0.87)	0.49 (0.40, 0.61)	0.45 (0.36, 0.56)
Lag 0–6	1.35 (1.17, 1.57)	0.88 (0.77, 1.01)	0.47 (0.42, 0.53)	0.75 (0.64, 0.87)	0.44 (0.34, 0.56)	0.39 (0.30, 0.51)
Lag 0–7	1.27 (1.07, 1.51)	0.78 (0.67, 0.91)	0.39 (0.34, 0.45)	0.74 (0.63, 0.88)	0.40 (0.31, 0.54)	0.36 (0.26, 0.48)
Lag 0–8	1.21 (1.01, 1.47)	0.70 (0.59, 0.84)	0.33 (0.28, 0.38)	0.76 (0.63, 0.91)	0.39 (0.29, 0.54)	0.34 (0.24, 0.48)
Lag 0–9	1.18 (0.95, 1.45)	0.65 (0.54, 0.79)	0.29 (0.24, 0.34)	0.79 (0.64, 0.97)	0.40 (0.28, 0.57)	0.35 (0.24, 0.51)
Lag 0–10	1.16 (0.92, 1.46)	0.62 (0.50, 0.76)	0.25 (0.21, 0.31)	0.84 (0.67, 1.05)	0.43 (0.30, 0.63)	0.38 (0.25, 0.57)
Lag 0–11	1.16 (0.90, 1.49)	0.59 (0.47, 0.75)	0.23 (0.19, 0.28)	0.91 (0.71, 1.16)	0.49 (0.32, 0.73)	0.43 (0.27, 0.66)
Lag 0–12	1.18 (0.90, 1.55)	0.59 (0.46, 0.75)	0.22 (0.18, 0.27)	1.01 (0.78, 1.31)	0.58 (0.37, 0.89)	0.51 (0.32, 0.82)

3.7. Sensitivity Analysis

The result of sensitivity analysis showed the model was robust when the *df* were altered for the time trend (*df* = 10–12), air pressure (*df* = 3–5), precipitation (*df* = 3–5) and sunshine duration (*df* = 1–3) (see Figures S4–S6). Changing the maximum lag day into 8, 10 and 12 in the model did not show an obvious difference for fitting the overall effect curve of AT in the model (see Figure S7).

4. Discussion and Conclusions

In this paper, the effects of meteorological factors (air temperature, wind speed and relative humidity) on PTB incidence in Xinjiang were investigated by using a P-DLNM model, and the relationship between AT and PTB incidence was evaluated. It was found that the correlations between air temperature, relative humidity, wind speed, AT and PTB incidence were nonlinear and lagged.

The overall effect between air temperature and PTB incidence showed a curve fluctuation, and the low air temperature could have a significantly risky effect on PTB incidence, which was consistent with previous studies [4,31,32]. It was reasonable that Xinjiang has a longer winter and thus, people are more susceptible to being infected with *M.tb* because the time of outdoor activities for humans was shorted [31]. Another reason is that the air pollution is serious during the heating period in Xinjiang (from October to March of the next year), then pollution particulates may attach more pathogenic bacteria including *M.tb* [3,12]. Moreover, vitamin D was demonstrated as an important factor that affected people's immune response to resist and remove *M.tb* when someone was during incubation, and with a long winter and low levels of vitamin D, latent infected PTB peoples are sensitive to *M.tb* [32]. The association between high air temperature with PTB incidence is negative, which may be the result that the recombinant strain of *M.tb* may stop growing or even be destroyed when the temperature exceeds 37 °C. In addition, sunshine is the main route of vitamin D synthesis. More sunbathing could help humans to enhance immunity so that the risk of transmission of PTB is relatively lower [33]. As a reference of the overall median of monthly average relative humidity (45.6%), it was obtained that the relative humidity was possible to increase the risk of PTB when it was between 44.5% and 71.8%, while it decreased the risk if it $\geq$ 72%, which is consistent with the results in reference [4]. One explanatory hypothesis for the high relative humidity is a lower chance of infection with

PTB because Xinjiang is an arid/semi-arid region with a dry climate, scarce precipitation and infrequent high relative humidity environment [11]. Another explanatory hypothesis is high relative humidity promotes the production of protective mucus on the surface of the respiratory tract thereby resisting the invasion of *M.tb* [34]. The overall effect of wind speed on the incidence of PTB was a *U*-shape, low and high wind speed are significantly positively correlated with the incidence of PTB, which is different from previous results that high wind speed was positively correlated with PTB [31,35]. It can be seen in the results that high wind speed may affect the transmission of *M.tb*, which may be because the frequent dust weather in Xinjiang makes *M.tb* float in the air for a longer time [9]. Moreover, the lower wind speed may also increase the chance of *M.tb* floating in the air in Xinjiang because the wind speed in winter is lower but the frequency is higher [36].

One of the highlights of this paper is to illustrate that there is a significant association between low AT and PTB incidence in Xinjiang. Generally, people are exposed to multiple meteorological factors in the environment, with low air temperature, wind and high relative humidity the sensations experienced by the human body are usually lower than the air temperature. AT is a comprehensive index combining ambient temperature, wind speed and humidity, which characterizes the physiological experience better than just air temperature alone, then it may be more realistic and objective. In this paper, it was shown that there was a nonlinear and lagged relationship between AT and PTB incidence in Xinjiang, and low AT can significantly increase the risk of PTB. Using the overall median of monthly AT (7.8 °C) as a reference, it was discovered that the lag effects of low AT presented a *V*-shape with the highest risk at lag 0, while the lag effects of high AT emerged as an *N*-shape with the highest risk at lag 12. Some specific biological mechanisms can account for why low AT can significantly increase the risk of PTB [37]. In a low air temperature, the human body is more sensitive to cold stimulation than heat stimulation, and the sympathetic nervous system is stimulated if the human body is invaded by cold air, which leads to the contraction of bronchial smooth muscle and excessive response, thus affecting the pulmonary ventilation and the decline of human lung function [38]. Furthermore, the human body's perception of cold and heat is mainly controlled by the thermoregulation system, nervous system and endocrine system [39], *M.tb* is more likely to invade the body in extreme cold weather, thus the human body's autonomous regulation is limited, the defense ability of the immune system is weakened. AT, as a comprehensive index combining air temperature, wind speed and relative humidity can be used to evaluate the impact of meteorological factors on PTB incidence, thus it can better reflect the relationship among human body, environment and disease.

The adjustment of energy structure has improved air quality in Xinjiang, and the incidence of PTB decreased slowly in recent years [25]. However, there is still a long way to go to mitigate the adverse effect of meteorological factors on the incidence of PTB in Xinjiang. Some measures should be implemented to improve the air quality. For example, the government should encourage the use of clean energy during the heating period. Moreover, more trees should be planted to alleviate the harm of dust storms. People should be appealed to wear masks in public places (especially in dusty weather) and enhance their immunity by getting more sunlight (at a favorable temperature) and exercising.

This paper also has some limitations. Firstly, the incidence of PTB is also affected by many variables (such as age, gender and air pollutant concentration) [40]. If these confounding variables are included in the model, the impact of meteorological factors and AT on PTB incidence can be more accurately evaluated. Secondly, other periodic dynamic models also can be applied to investigate the relationship between meteorological factors and PTB incidence.

Supplementary Materials: The following supporting information can be downloaded at: https://www.mdpi.com/article/10.3390/atmos13040533/s1, Figure S1: The study area and locations of weather stations; Figure S2. The plot of partial auto-correlation function in the distributed lag non-linear model; Figure S3. Seasonal decomposition of the time series of meteorological factors and the case in-cidence of PTB in Xinjiang from 2004 to 2019. Abbreviations: PTB, pulmonary tuberculosis;

Temp, temperature; AT, apparent temperature; SD, sunshine duration; RH, relative humidity; AP, air pressure; Pre, precipitation; WS, wind speed; Figure S4. Sensitivity analysis when altering the degrees of freedom (*df* = 10–12) of time for con-trolling for the long-term trend in the model in Xinjiang from 2004 to 2019; Figure S5. Sensitivity analysis when altering the degrees of freedom (*df* = 3–5) of air pressure and precipitation for controlling for the effect of confounding factors in the model in Xinjiang from 2004 to 2019; Figure S6. Sensitivity analysis when altering the degrees of freedom (1–3) of sunshine duration for controlling for the effect of confounding factors in the model in Xinjiang from 2004 to 2019; Figure S7. Sensitivity analysis when altering the maximum lag periods for 8, 10 and 12 months in the model in Xinjiang from 2004 to 2019; Table S1. Estimated relative risks (95% CI) of pulmonary tuberculosis cases with extremely low temperature (2.5th percentile, −11.9 °C) and ex-tremely high temperature (97.5th percentile, 25 °C) and extremely low relative humidity (2.5th percentile, 32.8%) and extremely high relative humidity (97.5th percentile, −68.9%) and extremely low wind speed (2.5th percentile, 1.6 m/s) and extremely high wind speed (97.5th percentile, 3.1 m/s) at lagged months in Xinjiang from 2004 to 2019.

Author Contributions: Conceptualization, C.G. and L.W.; Data curation, H.J.; Formal analysis, L.W.; Funding acquisition, L.W.; Methodology, C.G. and Y.W.; Software, Y.W. and H.J.; Writing—original draft, C.G., Y.W. and L.W.; Writing—review & editing, C.G., Z.H. and L.W. All authors have read and agreed to the published version of the manuscript.

Funding: The work is supported by the Natural Science Foundation of Xinjiang (Grant No. 2019D01C20), the National Natural Science Foundation of China (Grant Nos. 12061079, 12026236, 12026222), the Shenzhen science and technology innovations Committee (JCYJ20210324101406019) and the Alliance of International Science Organizations (Grant No. ANSO-CR-KP-2021-02).

Institutional Review Board Statement: Not applicable.

Informed Consent Statement: Not applicable.

Data Availability Statement: The data used and/or analyzed in the current study are publicly available in [the Public Health Scientific Data Sharing Center] at [http://www.phsciencedata.cn/. Accessed 30 December 2021], [the Health Commission of Xinjiang Uygur Autonomous Region] at [http://wjw.xinjiang.gov.cn/. Accessed 8 January 2022] and [the China Meteorological Data Sharing Center] at [http://data.cma.cn. Accessed 21 November 2021].

Conflicts of Interest: The authors declare no competing interests.

References

1. World Health Organization. *Global Tuberculosis Report*; World Health Organization: Switzerland, Geneva, 2020.
2. Turner, R.D. Cough in pulmonary tuberculosis: Existing knowledge and general insights. *Pulm. Pharmacol. Ther.* **2019**, *55*, 89–94. [CrossRef] [PubMed]
3. Cai, Y.L.; Zhao, S.; Niu, Y.; Peng, Z.; Wang, K.; He, D.H.; Wang, W.M. Modelling the effects of the contaminated en-vironments on tuberculosis in Jiangsu, China. *J. Theor. Biol.* **2021**, *508*, 110453. [CrossRef] [PubMed]
4. Li, Z.Q.; Liu, Q.; Zhan, M.Y.; Tao, B.L.; Wang, J.M.; Lu, W. Meteorological factors contribute to the risk of pulmonary tu-berculosis: A multicenter study in eastern China. *Sci. Total Environ.* **2021**, *793*, 148621. [CrossRef]
5. Rivas-Santiago, C.E.; Sarkar, S.; Cantarella, P., IV; Osornio-Vargas, Á.; Quintana-Belmares, R.; Meng, Q.; Kirn, T.J.; Strickland, P.O.; Chow, J.C.; Watson, J.G.; et al. Air pollution particulate matter alters antimycobacterial respiratory epithelium innate immunity. *Infect. Immun.* **2015**, *83*, 2507–2517. [CrossRef] [PubMed]
6. Niu, Z.C.; Qi, Y.J.; Zhao, P.Q.; Li, Y.D.; Tao, Y.; Peng, L.; Qiao, M.L. Short-term effects of ambient air pollution and meteorological factors on tuberculosis in semi-arid area, northwest China: A case study in Lanzhou. *Environ. Sci. Pollut. Res. Int.* **2021**, *28*, 69190–69199. [CrossRef]
7. Petersen, W.F. Tuberculosis: Weather and resistance. In Proceedings of theAnnual Meeting of the American College of Chest Physicians, Atlantic City, NJ, USA, 6 June 1942.
8. Guo, C.; DU, Y.; Shen, S.Q.; Lao, X.Q.; Qian, J.; Ou, C.Q. Spatiotemporal analysis of tuberculosis incidence and its associated factors in mainland China. *Epidemiol. Infect.* **2017**, *145*, 2510–2519. [CrossRef]
9. Zhang, C.Y.; Zhang, A. Climate and air pollution alter incidence of tuberculosis in Beijing, China. *Ann. Epidemiol.* **2019**, *37*, 71–76. [CrossRef]
10. Yang, J.; Zhang, M.; Chen, Y.; Ma, L.; Yadikaer, R.; Lu, Y.; Lou, P.; Pu, Y.; Xiang, R.; Rui, B. A study on the relationship between air pollution and pulmonary tuberculosis based on the general additive model in Wulumuqi, China. *Int. J. Infect. Dis.* **2020**, *96*, 42–47. [CrossRef]

11. An, Q.; He, H.; Gao, J.; Nie, Q.; Cui, Y.; Wei, C.; Xie, X. Analysis of Temporal-Spatial Variation Characteristics of Drought: A Case Study from Xinjiang, China. *Water* **2020**, *12*, 741. [CrossRef]
12. Xu, M.; Li, Y.; Liu, B.; Chen, R.; Sheng, L.; Yan, S.; Chen, H.; Hou, J.; Yuan, L.; Ke, L.; et al. Temperature and humidity associated with increases in tuberculosis notifications: A time-series study in Hong Kong. *Epidemiol. Infect.* **2021**, *149*, e8. [CrossRef]
13. Ghadimi-Moghadam, A.; Salahi, M.; Ghatee, M.A.; Kanannejad, Z.; Mosavi, A.; Ramshk, O.; Khoramrooz, S. Environmental and climatic factors influencing the occurrence and distribution of tuberculosis in southwest Iran: A GIS-based study. *Acta Med. Mediterr.* **2020**, *36*, 557–563.
14. Li, Z.-Q.; Pan, H.-Q.; Liu, Q.; Song, H.; Wang, J.-M. Comparing the performance of time series models with or without meteorological factors in predicting incident pulmonary tuberculosis in eastern China. *Infect. Dis. Poverty* **2020**, *9*, 151. [CrossRef] [PubMed]
15. Li, N.; Ma, J.; Liu, F.; Zhang, Y.; Ma, P.; Jin, Y.; Zheng, Z.-J. Associations of apparent temperature with acute cardiac events and subtypes of acute coronary syndromes in Beijing, China. *Sci. Rep.* **2021**, *11*, 15229. [CrossRef]
16. Royé, D.; Zarrabeitia, M.T.; Riancho, J.; Santurtún, A. A time series analysis of the relationship between apparent temperature, air pollutants and ischemic stroke in Madrid, Spain. *Environ. Res.* **2019**, *173*, 349–358. [CrossRef] [PubMed]
17. Yang, L.; Yang, J.; Liu, M.; Sun, X.; Li, T.; Guo, Y.; Hu, K.; Bell, M.L.; Cheng, Q.; Kan, H.; et al. Nonlinear effect of air pollution on adult pneumonia hospital visits in the coastal city of Qingdao, China: A time-series analysis. *Environ. Res.* **2022**, *209*, 112754. [CrossRef] [PubMed]
18. Chaudhuri, S.; Dutta, D. Mann–Kendall trend of pollutants, temperature and humidity over an urban station of India with forecast verification using different ARIMA models. *Environ. Monit. Assess.* **2014**, *186*, 4719–4742. [CrossRef] [PubMed]
19. Gasparrini, A.; Armstrong, B.; Kenward, M.G. Distributed lag non-linear models. *Stat. Med.* **2010**, *29*, 2224–2234. [CrossRef]
20. Wu, W.; Xiao, Y.Z.; Li, G.C.; Zeng, W.; Lin, H.; Rutherford, S.; Xu, Y.; Luo, Y.; Xu, X.; Chu, C.; et al. Temperature-mortality relationship in four subtropical Chinese cities: A time-series study using a distributed lag non-linear model. *Sci. Total Environ.* **2013**, *449*, 355–362. [CrossRef]
21. Yang, Q.Y.; Yang, Z.C.; Ding, H.Y.; Zhang, X.; Dong, Z.; Hu, W.; Liu, X.; Wang, M.; Hu, G.; Fu, C. The relationship between meteorological factors and mumps incidence in Guangzhou, China, 2005–2012. *Hum. Vaccines Immunother.* **2014**, *10*, 2421–2432. [CrossRef]
22. Gasparrini, A.; Scheipl, F.; Armstrong, B.; Kenward, M.G. A penalized framework for distributed lag non-linear models. *Bio-metrics* **2017**, *73*, 938–948. [CrossRef]
23. Steadman, R.G. A universal scale of apparent temperature. *J. Appl. Meteorol. Climatol.* **1984**, *23*, 1674–1687. [CrossRef]
24. Liu, X.; Wen, Y.; Zhang, K.; Duan, Y.; Li, H.; Yan, S.; Yin, P.; Cheng, J.; Jiang, H. Examining the association between apparent temperature and incidence of acute excessive drinking in Shenzhen, China. *Sci. Total Environ.* **2020**, *741*, 140302. [CrossRef] [PubMed]
25. Statistic Bureau of Xinjiang Uygur Autonomous Region. Available online: http://tjj.xinjiang.gov.cn/tjj/zhhvgh/list_nj1.shtml (accessed on 22 March 2022).
26. Hauke, J.; Kossowski, T. Comparison of Values of Pearson's and Spearman's Correlation Coefficients on the Same Sets of Data. *Quaest. Geogr.* **2011**, *30*, 87–93. [CrossRef]
27. Ha, Y.S.; Seongjin, P. An Analysis of Correlation between Personality and Visiting Place using Spearman's Rank Correlation Coefficient. *KSII Trans. Internet Inf. Syst. (TIIS)* **2020**, *14*, 1951–1966.
28. Gasparrini, A.; Armstrong, B. Reducing and meta-analysing estimates from distributed lag non-linear models. *BMC Med. Res. Methodol.* **2013**, *13*, 1. [CrossRef]
29. Gasparrini, A. Modeling exposure-lag-response associations with distributed lag non-linear models. *Stat. Med.* **2014**, *33*, 881–899. [CrossRef] [PubMed]
30. Kirolos, A.; Thindwa, D.; Khundi, M.; Burke, R.M.; Henrion, M.Y.R.; Nakamura, I.; Divala, T.H.; Nliwasa, M.; Corbett, E.L.; MacPherson, P.; et al. Tuberculosis case notifications in Malawi have strong seasonal and weather-related trends. *Sci. Rep.* **2021**, *11*, 4621. [CrossRef]
31. Xiao, Y.; He, L.; Chen, Y.; Wang, Q.; Meng, Q.; Chang, W.; Xiong, L.; Yu, Z. The influence of meteorological factors on tuberculosis incidence in Southwest China from 2006 to 2015. *Sci. Rep.* **2018**, *8*, 10053. [CrossRef]
32. Jaimni, V.; Shasty, B.A.; Madhyastha, S.P.; Shetty, G.V.; Acharya, R.V.; Bekur, R.; Doddamani, A. Association of Vitamin D Deficiency and Newly Diagnosed Pulmonary Tuberculosis. *Pulm. Med.* **2021**, *2021*, 5285841. [CrossRef]
33. Pinto, C.T.; Nano, F.E. Stable, temperature-sensitive recombinant strain of Mycobacterium smegmatis generated through the substitution of a psychrophilic ligA gene. *FEMS Microbiol. Lett.* **2015**, *362*, fnv152. [CrossRef]
34. Fahy, J.V.; Dickey, B.F. Airway Mucus Function and Dysfunction. *N. Engl. J. Med.* **2010**, *363*, 2233–2247. [CrossRef] [PubMed]
35. Li, Q.; Liu, M.; Zhang, Y.; Wu, S.; Yang, Y.; Liu, Y.; Amsalu, E.; Tao, L.; Liu, X.; Zhang, F.; et al. The spatio-temporal analysis of the incidence of tuberculosis and the associated factors in mainland China, 2009–2015. *Infect. Genet. Evol.* **2019**, *75*, 103949. [CrossRef] [PubMed]
36. Wang, J.; Qin, S.; Zhou, Q.; Jiang, H. Medium-term wind speeds forecasting utilizing hybrid models for three different sites in Xinjiang, China. *Renew. Energy* **2015**, *76*, 91–101. [CrossRef]
37. Halonen, J.I.; Zanobetti, A.; Sparrow, D.; Vokonas, P.S.; Schwartz, J. Relationship between outdoor temperature and blood pressure. *Occup. Environ. Med.* **2010**, *68*, 296–301. [CrossRef]

38. Khadadah, M.; Mustafa, S.; Elgazzar, A. Effect of acute cold exposure on lung perfusion and tracheal smooth muscle contraction in rabbit. *Eur. J. Appl. Physiol.* **2011**, *111*, 77–81. [CrossRef]
39. Hensel, H.; Schafer, K. Thermoreception and Temperature Regulation in Man. *Recent Adv. Med. Thermol.* **1984**, *38*, 51–64. [CrossRef]
40. Smith, G.S.; Schoenbach, V.J.; Richardson, D.B.; Gammon, M.D. Particulate air pollution and susceptibility to the development of pulmonary tuberculosis disease in North Carolina: An ecological study. *Int. J. Environ. Health Res.* **2013**, *24*, 103–112. [CrossRef]

atmosphere

Article

Impacts of Climate and Environmental Change on Bean Cultivation in China

Sidan Li [1,*], Shibing You [1], Ze Song [1], Li Zhang [1] and Yixuan Liu [2]

1 School of Economics and Management, Wuhan University, Wuhan 430072, China; sbyou@whu.edu.cn (S.Y.); songze@whu.edu.cn (Z.S.); lizhang_justin@whu.edu.cn (L.Z.)
2 The School of Economics, Shanghai University of Finance and Economics, Shanghai 200433, China; 2019310031@live.sufe.edu.cn
* Correspondence: sidanli0617@whu.edu.cn; Tel.: +86-18971564571

Abstract: The impact of human-caused environmental pollution and global climate change on the economy and society can no longer be underestimated. Agriculture is the most directly and vulnerably affected sector by climate change. This study used beans, the food crop with the largest supply and demand gap in China, as the research object and established a panel spatial error model consisting of multiple indicators of four factors: climate environment, economic market, human planting behavior and technical development level of 25 provinces in China from 2005 to 2019 to explore the impact of climate environmental changes on the yields of beans. The study shows that: (1) The increase in precipitation has a significant positive effect on bean yields; however, the increase in temperature year by year has a significant negative effect on bean yields; (2) carbon emissions do not directly affect bean production at present but may have an indirect impact on bean production; (3) artificial irrigation and fertilization behavior on bean production has basically reached saturation, making it difficult to continue to increase bean yields and (4) the development of technology and human activity is a mixed blessing, and the consequent inhibiting effects on bean production are currently unable to offset their promoting effects. Thus, when it comes to bean cultivation, China should focus mainly on the overall impact of environmental changes on its production, rather than technical enhancements such as irrigation and fertilization.

Keywords: yield per unit area of beans; climate change; panel spatial error model

Citation: Li, S.; You, S.; Song, Z.; Zhang, L.; Liu, Y. Impacts of Climate and Environmental Change on Bean Cultivation in China. *Atmosphere* **2021**, *12*, 1591. https://doi.org/10.3390/atmos12121591

Academic Editors: Zengyun Hu, Xuguang Tang and Qinchuan Xin

Received: 14 October 2021
Accepted: 26 November 2021
Published: 29 November 2021

1. Introduction

In recent decades, the global climate has undergone significant changes due to natural environmental changes and human activities, and this change is mainly characterized by global warming. The main reason for the warming is the emission of greenhouse gases such as carbon dioxide caused by human activities [1]. Climate change research intends to assess the effects of climate change on ecosystems, economies and societies by predicting possible future changes of climate [2]. The influence of climate change on human beings is comprehensive, multi-level and multi-scale. Agriculture is the sector most directly affected by and vulnerable to climate change. It is also the sector on which human society depends for the output of basic means of living, and the sustainable development of agriculture is directly related to the survival and development of human society [3,4]. As grain is a crucial agricultural product, grain security, grain varieties and grain production have always been the top priority of the global economy. With the advancement of agricultural technologies and the improvement of life quality, the yields of crops and productivity have been boosted over time. At the same time, climate warming, precipitation instability, extreme weather and disasters have brought negative impacts on food production [5]. Studies have shown that the determinants of crop production are climatic (e.g., rainfall, high temperatures) and weather extremes (e.g., floods, droughts and storms) [6]. Changes in rainfall patterns and increased temperature could have a marked impact on food production [7]. According to

IPCC (2012), crop production is sensitive to climate change associated with temperature increases, changes in rainfall patterns and extreme weather events. In addition to the changes of climate and temperature, carbon dioxide also affects the food production, and it has a higher impact on C3 species (including wheat, rice and bean) than C4 species (including maize and sorghum) [8]. Food production in countries around the world is currently under serious threat from climate change, which leads to challenges to sustainable development as well [9]. Climate change is also an important challenge for food production and security in China. In the context of global warming, changes in temperature and precipitation patterns could lead to a 20–36% decline in maize, wheat, and rice yields in China over the next 20–80 years [10]. China's grain production with the background of climate change has the following three main problems: Firstly, grain production becomes more volatile; secondly, the structure and layout of food production may change, and crop cultivation systems need to change accordingly and thirdly, climate change will substantially increase the cost of grain production [11].

China is a global grain-producing country. For example, in 2020, the total grain production in China reached 669 million tons, an increase of 0.85% year-on-year, accounting for about 24% of the world's grain production, and the grain supply was more than adequate. However, the production of bean products accounted for only 3.42%, or about 22.87 million tons. In contrast to the reality of limited bean supply, China is a major consumer of beans in the world, and bean consumption is rising annually, with bean consumption demand exceeding 120 million tons in 2018 and 2019, implying a huge domestic bean supply and demand gap in China and high dependence of the bean industry on imports [12,13]. Therefore, this study turned the research perspective to the Chinese bean industry, which is severely limited in self-sufficiency, and investigated the impact of climate change on bean production so as to provide some practical suggestions for the bean industry in China regarding future changes in the structure, layout of food production and the long-term development.

The impact of climate change on agricultural production is an interdisciplinary subject between climatology, agronomy and economics, and different methods and emphases have been adopted by different disciplines in the study of this subject. In existing studies, the impact of agronomic climate change on agricultural production has been mainly used in the crop growth simulation model (CGSM), which examines the changes in crop growth, development and yield due to changes in climatic factors by dynamically simulating crop growth and development processes and their relationships with climatic factors, soil properties and management techniques and provides a quantitative tool for predicting early warning and assessing the effects of crop productivity under different conditions [14–16]. The agronomic approach focuses on the process of crop growth and development cycles without considering economic factors, while the economic approach focuses more on the economic value dimension. The representative studies in economics have measured agricultural production by constructing land value indicators and with the help of crop yields, progress in controlling technology and economic factors using a large quantity of statistical data and relevant econometric models to examine the causal relationship between climate factors and agricultural production [17,18]. A review of representative literature reveals that the current research trend gradually shifts to examine specific crop yields, and the empirical method of causality shifts from traditional cross-sectional regression to panel spatial econometric empirical analysis [19–21]. Spatial measurement is a common econometric method. When spatial correlations and spatial differences exist among individual research objects, spatial measurement can be used to consider the variability under multiple influencing factors in different spaces at the same time, removing these differential influences and focusing on the main influencing factors of the research object. In the present studies on the relationship between grain production and climate in China, most of the research objects are cereal grains such as wheat and rice, and the influencing factors are covered comprehensively with various analysis methods; however, the research on beans is relatively insufficient and covered with fewer dimensions of influencing factors. This study used bean productivity as the research object and analyzed the influence of four influencing factors on bean production: climate environment, economic market, human planting behavior and technology development level. The spatial error model was selected in this study for empirical analysis, which focuses on climatic environmental factors as the variables of interest, while economic market, human planting behavior and technology development level are the variables that need to be controlled. Moreover, the inter-regional factors such as geographic rationality and policy institutional differences should be excluded.

2. Materials and Methods

2.1. Variable Selection

In this study, the unit area yield of bean crops was selected as the explanatory variable, and the four factors affecting the unit area yield of bean crops were climate environment, economic market, human cultivation behavior and technology development level. Economic market factors mainly include economic factors such as market price and cost of food. Human cultivation behavior refers to the human intervention in food crops, including both fertilization and irrigation. The level of technological development refers to the level of technology in growing food, which is reflected in the paper using the degree of modernization of the cultivation area. Therefore, the specific indicators including the explanatory variables and the four aspects of explanatory variables selected in this study are shown in Table 1 below.

Table 1. Table for selection of variable indicators.

Explained and Explanatory Variables	Name of Indicator
Production of beans	Bean yields (per unit)
Climatic environment	Effective cumulative temperature Precipitation (meteorology) Extent of disaster Carbon Emissions Sulfur dioxide emissions from exhaust gases Total wastewater discharge
Economic market	Food benefit–cost ratio
Human cultivation behavior	Fertilizer use per unit area Effective irrigated area ratio
Level of technological development	Rural electricity consumption

2.2. Modeling

In this study, a panel spatial error model was constructed to analyze the effect of climatic environmental changes on the yield per unit area of beans. The model was constructed as shown below [22].

$$Y_{i,t} = H_{i,t}\beta_0 + P_{i,t}\beta_1 + X_{i,t}\beta_2 + E_{i,t}\beta_3 + A_{i,t}\beta_4 + T_{i,t}\beta_5 + \lambda_t + \varepsilon_{i,t} \quad (1)$$

$$\varepsilon_{i,t} = \rho\sum_{i'} W_{i,i'}\varepsilon_{i',t} + \eta_{i,t} \quad (2)$$

(1) In the equation, i and t denote province and year, respectively; $Y_{i,t}$ denotes grain yield; $H_{i,t}$, an indicator of regional heat resources, denotes effective cumulative temperature, the sum of effective temperature of crops during the reproductive period, reflecting the heat demand of biological growth and development; $p_{i,t}$ denotes total annual precipitation of each province and $X_{i,t}$ denotes other climate variables, including the degree of disaster, carbon emission, SO2 emission in exhaust gas and total wastewater discharge. Considering that coal is the main source of carbon emissions, coal usage was chosen to represent carbon emissions, while the degree of disaster was calculated by dividing the number of disasters

by the total sown area in each province. $E_{i,t}$ represents economic market factors, expressed as the benefit–cost ratio of grain cultivation. Due to the interplay between economic market factors and bean yields, there is a significant endogeneity problem between the two variables. A one-period lagged price variable was used in this study as a proxy for current period prices. The rationale for this treatment is that climatic conditions and market factors in the previous year have a strong influence on farmers' decisions to plant in the following year, which in turn changes farmers' expectations of the returns and costs of planting in the current year, but there is no direct link between climatic conditions and economic factors in the previous year and crop yields in the current year (i.e., the residuals of the model). Thus, the price index of agricultural production with a lag of one period can be considered as the expected return to planting in the current period, while the price index of agricultural production materials in the current period represents the cost of planting faced in the current period, and the ratio of these two price indices, i.e., expected return/cost of planting, characterizes economic market factors, with higher ratios indicating greater returns to planting [23]. $A_{i,t}$ characterizes human behavioral interventions, including fertilizer use per unit area and effective irrigated area. The ratio of effective irrigated area is calculated by dividing the effective irrigated area by the total sown area [24]. The factor $T_{i,t}$ characterizes technology development, i.e., the level of modernization of the growing area, as reflected by rural electricity consumption, and the time fixed effect λ_t captures factors that do not change with the area, such as policy regimes.

In addition, other influences not captured by the independent variables in Equation (1) were all included in the residual term ($\varepsilon_{i,t}$), including regional farming systems, regional soil environments, and regional cropping habits that are both highly correlated with local climate and affect bean yields and therefore also affect the consistency of the estimation results. To compensate for potential omitted variable bias, this study allows for spatial correlation between samples and uses residuals from neighboring provinces ($\varepsilon_{i',t}$) to explain province-specific residuals ($\varepsilon_{i,t}$) to capture all omitted variables with regional characteristics. The basic idea behind the ability of the spatial error models to correct for omitted variable bias is that anything that cannot be captured by the independent variables in the regression model will enter the model residual term ($\varepsilon_{i,t}$); however, once the unobservable omitted variables share common spatial regional characteristics (i.e., regional characteristics common to a particular province and adjacent provinces, such as regional farming systems, regional natural hazards, regional varieties and soil types), using the residuals $\varepsilon_{i',t}$ as an additional independent variable to explain crop yields ($Y_{i,t}$) in a given province from adjacent provinces can reflect all common regional characteristics as shown in Equation (2). In Equation (2), $W_{i,i'}$ is the spatial weighting matrix that specifies the range of spatially correlated influences. The spatial adjacency matrix was used in this study to reflect the spatial correlation of the samples, with a priori assumption that there are spatially correlated influences between neighboring sample provinces but no spatial correlation between non-adjacent sample provinces. The spatial adjacency matrix is a square matrix whose elements take the values (0, 1). It takes the value 1 if the sample county domains are adjacent and 0 if they are not. The spatial adjacency matrix was chosen as the baseline case in this study for three reasons: first, to be consistent with the existing literature [17,22]; second, because existing studies [17,25] confirm that while the choice of different spatial weighting matrices affects the magnitude of the spatially relevant degree, indicator ρ does not significantly change the coefficient estimates of the climate variables of interest to the empirical analysis model. Furthermore, in the panel data case, it is often assumed that the extent and magnitude of spatially correlated impacts (i.e., spatial weighting matrices $W_{i,i'}$) do not change over time [26]. The degree of spatial correlation between neighboring sample provinces is reflected by the estimated coefficients. Equations (1) and (2) form the panel spatial error model in this study, after stripping out $\varepsilon_{i,t}$ the spatial correlation, where $\eta_{i,t}$ is the true residual term of the empirical analysis model.

Thus, the economic significance of the climate factor coefficients (β_0, β_1, β_2) in this paper is the marginal impact per unit ($H_{i,t}$, $P_{i,t}$, $X_{i,t}$) change in climate factors on crop yields, with all else remaining the same($E_{i,t}$, $A_{i,t}$, $T_{i,t}$ and λ_t).

2.3. Data Sources and Processing

The sample data used in this study contained the following four aspects: planting data, basic climate data, environmental data and socioeconomic data. The time interval was 2005–2019, and the research sample area was initially selected from 31 provinces in China. Among them, considering the special geographical environment of Hainan Province and Tibet Autonomous Region, their low bean yield and missing data, these regions were excluded. At the same time, the four municipalities directly under the central government of Beijing, Tianjin, Shanghai and Chongqing have less arable land compared with other provinces and have a higher level of modernization, which makes them unsuitable for the cultivation of beans and other agricultural products and not representative for the analysis of factors influencing bean yields in this study, so they were also excluded. The specific sources of data for the four areas are as follows.

Planting data: Among the 25 provinces, planting data were selected to include bean planting area, yield per unit area, effective irrigated area and fertilizer use per unit area for each province. The data were obtained from the Yearbook of the National Bureau of Statistics of China and the CSMAR database, respectively, and all 375 complete sample data were collected.

Basic climate data: The basic climate data were obtained from 820 meteorological observation sites across the country, and the meteorological data of daily values from 25 urban meteorological stations with the highest bean production in each province were selected as the representative of the basic climate data in the province. The growing temperature of bean crops is above 10 °C, so the average daily temperature greater than 10 °C in each station in each year was accumulated in this study as the annual effective cumulative temperature value in the province, which reflects the influence of temperature in bean cultivation. Precipitation data for each province were obtained from the China National Meteorological Science Data Sharing Service Platform—China Terrestrial Climate Information Daily Value Dataset v3.0. These data were complete and free of defects.

Environmental data: Environmental data mainly include total carbon emissions, total SO_2 emissions in exhaust gas, total wastewater emissions and severity of disaster, where severity of disaster is expressed as the number of disasters per year in the province divided by the area of bean cultivation. The data on total carbon emissions and the number of disasters were obtained from the yearbook of the National Bureau of Statistics of China, in which there were 25 missing values for the indicator of total carbon emissions; the data on total SO_2 emissions in exhaust gas and total wastewater emissions were obtained from the statistical database of the China Economic Network, in which there were 25 missing values for total wastewater emissions, and the missing values above were supplemented by the interpolation method.

Socioeconomic data: The socioeconomic data include the benefit–cost ratio of food and rural electricity consumption; the benefit–cost ratio is calculated by dividing the relative price index of agricultural production in the previous period by the relative price index of agricultural production materials in the current period. The data on the relative price index of agricultural production and the relative price index of agricultural production materials were obtained from the CSMAR database, and the data were complete and free of missing values; the data on rural electricity consumption were obtained from the CSMAR database, and the data had 25 missing values, which were completed by interpolation.

Numerous studies have shown that precipitation has an important influence on the production of crops. The national precipitation is plotted for the years 2005–2009, 2010–2014 and 2015–2019 as shown in Figures 1–3, respectively. The major growing provinces of beans in China are in Xinjiang Uyghur Autonomous Region, Fujian Province, Shandong Province and Guangdong Province, and as can be seen from the figure, the distribution of precipitation in these provinces varies widely, indicating that beans are not very sensitive to precipitation. The result is not quite consistent with the conclusion that the yields of many cereal crops are more sensitive to changing patterns of precipitation.

For the 11 independent variables selected in this study, the same descriptive statistics are shown in Table 2 for each of the 5-year periods 2005–2009, 2010–2014 and 2015–2019.

As can be seen in Table 2, the average bean yields in China increased steadily; there was also a gradual increase in temperature levels and average precipitation across the country. The level of disasters decreased year by year, although this was not because the number of disasters became less frequent. In contrast, the frequency of disasters increased gradually, but the area of bean cultivation in China decreased year by year, and the area of efficient, bean-adapted arable land also decreased year by year. Emissions of SO_2 from exhaust gases decreased significantly from 2015 to 2019. As a positive effect of policy interventions, the control of emissions from wastewater increased year by year. There is a more pronounced downward trend in the benefit–cost ratio of food, which implies that the willingness of Chinese farmers to grow on their own may have declined to some extent, and the use of fertilizer per unit area remains largely stable or even slightly declines as an indication of this phenomenon. Rural electricity consumption, on the other hand, has shown a very significant upward trend, indicating that the modernization and scientific and technological development of China's rural areas over the past 15 years have had a remarkable effect.

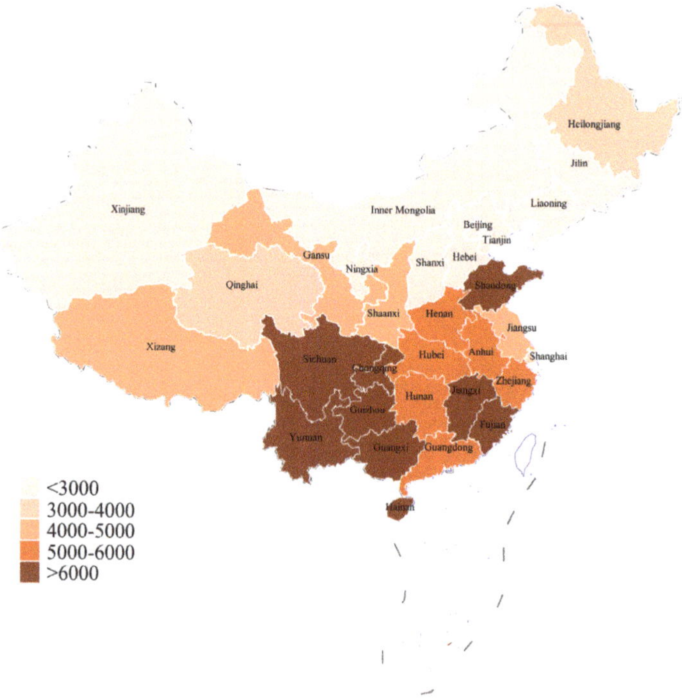

Figure 1. Distribution of precipitation in China, 2005–2009 (unit: millimeter, aggregated data from 2005 to 2009).

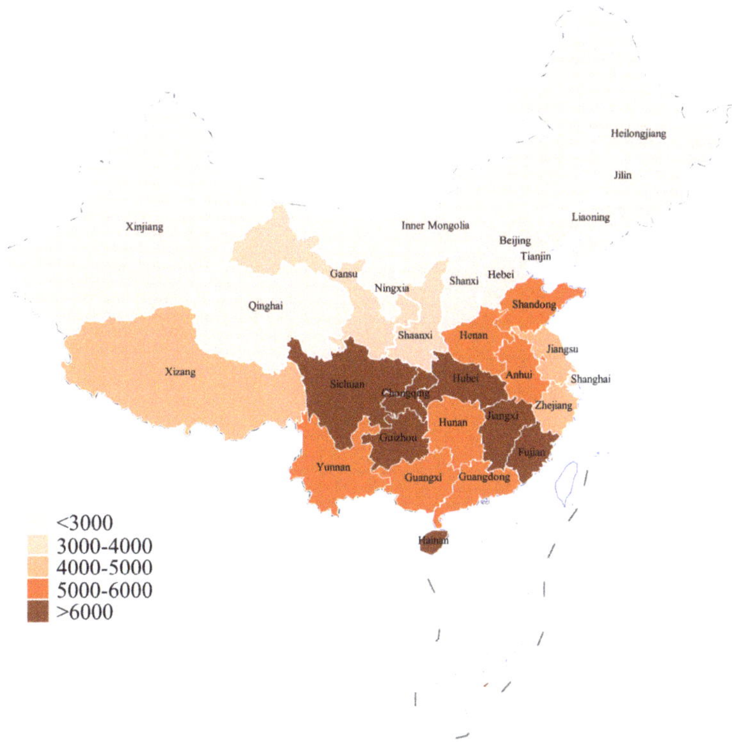

Figure 2. Precipitation distribution in China, 2010–2014 (unit: millimeter, aggregated data from 2010 to 2014).

Figure 3. Precipitation distribution in China, 2015–2019 (unit: millimeter, aggregated data from 2015 to 2019).

Table 2. Descriptive statistics of the sample.

Variable Name	Variable Identification	Unit	2005–2009		2010–2014		2015–2019	
			Average Value	Standard Deviation	Average Value	Standard Deviation	Average Value	Standard Deviation
Bean Yields	Yield	kg/ha	1877.857	565.6416	1921.798	566.7584	2024.988	581.6842
Effective Cumulative Temperature	EAT	degrees centigrade	2566.77	972.2334	2525.814	948.9907	2598.245	969.6687
Precipitation (Meteorology)	PRE	mm	894.4729	416.7491	933.9605	436.7277	990.1309	478.927
Extent of Disaster	DISASTER	times/ha	0.3049891	0.156439	0.2023446	0.1127987	0.1386728	0.1045211
Carbon Footprint	COAL	million tons	256.1646	166.7895	396.4381	262.6799	446.6218	330.7543
Sulfur Dioxide Emissions from Exhaust Gases	SO_2	million tons	90.30312	44.34946	79.5552	37.38318	33.45112	26.404
Total Wastewater Discharge	WASTE	million tons	198,086.6	153076.4	245,094.4	183,915.9	265,363.5	203,324
Food Benefit–cost Ratio	RATIO		1.0303	0.0919729	1.024677	0.0604204	0.9843792	0.0406275
Fertilizer Use Per Unit Area	FER	kg/ha	326.8883	108.8031	366.5041	126.2272	360.5308	134.1082
Effective Irrigated Area Ratio	IRRI		0.3855871	0.1445012	0.4070513	0.1502828	0.423296	0.1480573
Rural Electricity Consumption	POWER	billion kWh	201.9817	286.5458	282.9348	402.312	323.7136	450.8923

3. Model Results and Conclusions

3.1. Testing for Spatial Effects

Moran's index and Geary's c coefficient, which measure the existence of spatial autocorrelation between global areas, and the spatial adjacency matrix (W) was used as the baseline matrix reflecting the spatial correlation of the sample, which is defined in the same way as in Section 2.2. If both Moran's index and Geary's c coefficient values are significant, this indicates that there is correlation between bean unit yield samples. The positive or negative sign of the coefficient reflects the corresponding spatial correlation between samples. Table 3 shows the results of the test for spatial correlation for this indicator of bean yields for the sample studied in this work.

Table 3. Results of spatial autocorrelation test.

Test Coefficient	Bean Yields
Moran's I	0.201 **
Geary's c	0.733 **

** denotes 5% level of significance.

The results of the above tests indicate that there is a significant positive spatial correlation effect between the bean unit yield samples. The traditional panel model is inadequate to deal with the spatial correlation problem since its estimation results are not only biased but could greatly overestimate the significance level of the variables. The bias essentially stems from the omission of spatial factors, and therefore, a spatial econometric model was used in this study for the empirical analysis.

3.2. Model Selection and Robustness Tests

The stability test is a necessary step before using the model for panel data. It is necessary to ensure that each variable is a stable series and that a cointegration between the variables can be modeled. The test revealed that bean yields were not stable, so they were logarithmically processed. The results of the stability test of the processed bean yields with the rest of the variables are shown in Table 4.

In this study, the Levin–Lin–Chu test, Hadri (2000) panel unit root test and Im–Pesaran–Shin test were selected for the above 11 variables. In the test results of these three coefficients, more than two coefficients needed to be significant to consider the variable as a stable series. The results in Table 4 show that the panel data of the above variables are all stable series.

Section 3.1 validates the findings of this study using a spatial econometric model from the perspective of a single variable, grain yields. The LM test, on the other hand, tests whether a spatial regression method or the OLS regression method should be used for ordinary panel data from the perspective of the panel data containing all variables. The LM test can test whether there are spatial error effects and spatial lag effects in the overall panel data. Table 5 shows the results of LM test for the overall model. The three test coefficients of spatial error effect are significant, indicating that the panel data has spatial error effect, while the robust Lagrange multiplier in spatial lag effect is not significant, so it is not fully confirmed that the panel data has a spatial lag effect.

Table 4. Table of results of stability tests.

	Inspection Method	Test Results
Ln(Yield)	LLC	−9.013 ***
	HARDI	15.238 ***
	IPS	−1.781 *
EAT	LLC	−11.331 ***
	HARDI	7.187 ***
	IPS	−2.304 ***
PRE	LLC	−12.314 ***
	HARDI	1.216
	IPS	−2.236 ***
DISASTER	LLC	−14.212 ***
	HARDI	12.848
	IPS	−2.550 ***
COAL	LLC	−4.327 **
	HARDI	37.559 ***
	IPS	−1.175
SO_2	LLC	−3.725
	HARDI	33.522 ***
	IPS	−0.717
WASTE	LLC	−5.141 *
	HARDI	34.790 ***
	IPS	−0.988
RATIO	LLC	−19.627 ***
	HARDI	−1.484
	IPS	−3.602 ***
FER	LLC	−8.599 ***
	HARDI	27.550 ***
	IPS	−1.588
IRRI	LLC	−12.055 ***
	HARDI	28.060 ***
	IPS	−2.329 ***
POWER	LLC	−7.422 ***
	HARDI	37.509 ***
	IPS	−1.374

*** denotes 1% level of significance. ** denotes 5% level of significance. * denotes 10% level of significance.

Table 5. LM test results.

Spatial Error	
Moran's I	15.347 ***
Lagrange multiplier	204.345 ***
Robust Lagrange multiplier	13.433 ***

*** denotes 1% level of significance.

The test provides strong support for the choice of the spatial error model, while for panel data, whether to use the fixed effect or random effect model depends on the Hausman test. Table 6 shows the results of the Hausman test.

The original hypothesis of the Hausman test is that there is a random effect in these panel data. As can be seen from Table 6, the p-value is 0.0008, so the original hypothesis can be rejected at the 95% confidence level. That is, the panel data do not have a random effect but a fixed effect, and the fixed effect model should be selected. The four tests in Tables 3–6 provide support for model Equations (1) and (2) outlined in Section 2.2 of this paper.

Table 6. Hausman test results.

	Coef.	Std. Err.	p-Value
_Cons	7.200671 ***	0.2130221	0.000
EAT	0.000091 *	0.0000525	0.083
PRE	0.0000683	0.0000633	0.280
DISASTER	−0.1533701 **	0.0771955	0.047
COAL	2.58929 ***	0.7248598	0.000
SO_2	−0.4844776	3.493605	0.890
WASTE	−0.0051428 **	0.0022797	0.024
RATIO	−0.1334411	0.1251934	0.286
FER	−0.0001154	0.0002452	0.638
IRRI	0.6176916 ***	0.2025145	0.002
POWER	0.0000286	0.0001004	0.775

*** denotes 1% level of significance. ** denotes 5% level of significance. * denotes 10% level of significance. H_0: Difference in coefficients not systematic chi2(11) = 31.74 Prob > = chi2 = 0.0008.

3.3. Model Results

According to the settings of panel spatial error model (1) and (2), the goodness of fit of the three models of time fixed effect, individual fixed effect and double fixed effect was 0.27, 0.06 and 0.03, respectively, so the time fixed effect model was chosen to examine the effects of the four types of factors on bean yields in China, and the regression results are shown in Table 7.

Table 7. Regression results of panel spatial error model.

	Coef.	Std. Err.	p-Value
EAT	−0.0002578 ***	0.0000468	0.000
PRE	0.0004895 ***	0.0000757	0.000
DISASTER	−0.6823345 ***	0.1305011	0.000
COAL	−0.00009	0.000079	0.254
SO_2	0.0015519 ***	0.0006016	0.010
WASTE	7.67×10^{-7} ***	2.38×10^{-7}	0.001
RATIO	0.227392	0.3739179	0.543
FER	0.0002584	0.0001873	0.168
IRRI	1.03056 ***	0.128758	0.000
POWER	−0.000172 **	0.0000844	0.041
	within	**between**	**overall**
R-sq	0.0094	0.5439	0.2793

*** denotes 1% level of significance. ** denotes 5% level of significance.

Table 7 reflects the following findings. The overall R squared is 0.2793, which indicates that the overall fit of the model is good. The within R squared is 0.0094, which shows that the within fit of the model is poor. The between R squared is 0.5439, indicating that the between fit of the model is good.

(1) Biologically, similar to cereals such as wheat and rice [27], there is an "inverted U-shaped" relationship between temperature and precipitation and unit yield of beans, i.e., the effect of temperature and precipitation on bean production is a non-linear relationship that increases first and then decreases. The model results showed that the current effective cumulative temperature had a negative effect on bean yields (Coef. = −0.0002578, $p < 0.01$), i.e., for every 1 °C increase in effective cumulative temperature, bean yields decreased by 0.02578%. The result not only implies that the effect of extreme weather on bean yields is significant, with either lower or higher cumulative temperatures causing a reduction in bean yields, but also, combined with the descriptive statistics in Table 2, we can conclude that the warming phenomenon occurring in China significantly and negatively affects bean production. In the long run, elevated temperatures can adversely affect crop production [28,29], and the results of this paper suggest that the relationship between legumes and temperature factors is consistent with this conclusion. There is also a

positive relationship between precipitation and bean yields (Coef. = 0.0.0015519, $p < 0.01$), i.e., grain yields increase with increasing rainfall, with each 100 mm increase in precipitation increasing bean yields by 4.895%. In countries such as China, where agriculture is more dependent on rainfall, climate change is more significant for crop production, and changes in rainfall may adversely affect crop production [30]. There is a negative relationship between rice production and precipitation in China, which is contrary to the findings of the relationship between legumes and precipitation studied in this research. The current precipitation situation in China remains in an interval of positive effects on bean production and growth.

(2) Among the climatic variables in the environment category, there are significant relationships between SO_2 emissions from exhaust gases (Coef. = 0.0015519, $p < 0.05$), wastewater emissions (Coef. = 7.67×10^{-7}, $p < 0.05$) and severity of damage (Coef. = -0.68233, $p < 0.01$) and bean yields. Among them, the effect of SO_2 emissions from waste gas on bean yields is the largest, with an increase of 0.155194% in bean yields for every 10,000 tons of SO_2 emissions from waste gas. The effect of wastewater emissions on bean yields is the smallest and almost negligible, with an increase of 0.0000766% in bean yields for every 10,000 tons of increase in wastewater emissions. The results show that there is a positive relationship between SO_2 emissions from exhaust gas and wastewater emissions and bean yields, which is different from the hypothesis.

With the rapid development of China's economy, the domestic energy situation is becoming increasingly scarce, and the demand for natural gas as a clean gas fuel and an important chemical raw material is increasing. Therefore, some domestic scholars have studied the short-term and long-term effects of SO_2 emissions from waste gas on the environment, soil and crops. Some studies have shown that although SO_2 is higher than the requirements of the Limits of Concentration of Air Pollutants for the Protection of Crops, it does not cause acute harm to wheat, rice, beans, corn or rapeseed in the predicted impact area, let alone to their long-term growth [31]; some studies have also shown that high sulfur gas field SO_2 emissions from high sulfur gas fields can significantly reduce the thousand-grain-weight of rice seeds, thereby affecting rice yield and reducing rice quality [32]. There are no studies that support a positive effect of SO_2 emissions on bean yield, which is contrary to the results of the study, probably due to the inaccurate location of the selected emissions data collection, which does not fully reflect the effect on bean cultivation, and the fact that emissions are related to human activities and industrial production, which may have omitted variables and endogeneity problems. Similarly, the same problem exists in the relationship between wastewater emissions and bean yields. It has been suggested that each percentage point increase in CO_2 emissions decreases cereal production by 0.29%, showing a negative correlation [30,33–35], and it has also been suggested that there is no significant relationship between CO_2 emissions and cereal production [36]. The effect of bean yields and CO_2 emissions studied in this work is similar to the findings of Sossou et al., who studied cereal yields, where the effect of carbon emissions (Coef. = -0.00009, $p > 0.05$) on bean yields was not significant. The reason may be that on the one hand, the increase in carbon emissions brings negative climate effects such as extreme weather and global warming, which have a negative effect on the growth of beans, and on the other hand, CO_2 is a condition for photosynthesis in beans, which has a positive effect on their growth, and the two effects are currently in a state of offset; carbon emissions have not yet had a significant effect on bean yields. There is a negative relationship between the degree of damage and bean yields. In bean growing areas, each increase in the degree of damage per hectare reduces bean yields by 68.233%, indicating that the degree of damage is the most serious factor affecting bean yields per unit area. This is in agreement with Chijioke et al. who concluded that extreme climatic events such as floods and droughts significantly affect crop yields [37], suggesting that beans as a crop are also sensitive to the degree of damage.

(3) The benefit–cost ratio of grain does not have an effect on bean yields (Coef. = 0.227392, $p > 0.05$), and farmers' expected returns and costs of growing beans are not significantly related to bean yields.

(4) Among the human behavioral intervention factors, there is a positive relationship between effective irrigated area ratio (Coef. = 1.030, $p < 0.001$) and bean yields, i.e., an increase in effective irrigation ratio boosted grain yields, whereas the effect of fertilizer being used per unit area (Coef. = 0.0002584, $p > 0.005$) on grain yield is insignificant. This may be due to the good soil environment of the farmland, where fertilizer self-supply and micro-ecological cycles are at a better level and the farmers' fertilizer application for bean tillage is at a more desirable level, which has a weaker effect on bean growth.

(5) There is a negative relationship between rural electricity consumption (Coef. = −0.0001, $p < 0.05$), a proxy for the technology level factor, and bean yields, with an increase in rural electricity consumption of 100 million kWh reducing bean yields by 0.01%. An increase in electricity consumption in turn reduces bean yields. The development of modern technology has effectively improved agricultural cultivation techniques, which will increase the production capacity of beans to a certain extent, but from another perspective, the modernization of the countryside means an increase in human activities, overexploitation of the ecological environment and a reduction in the area of high-quality arable land, all of which have a negative impact on the production capacity per unit area of beans, and the conclusion shows that the negative impact under this factor is slightly greater than the positive effects.

4. Discussion

A large number of observations and research results show that climate change has different impacts on crop growth and development, cropping systems and yield quality, with both advantages and disadvantages, but the negative impacts outweigh positive ones. The relationship between climate change and grain production in China is complex. From an ecosystem perspective, not only are there two major agroecosystems, the southern paddy field and the northern dryland, which play an important role in crop growth, but grain production is also affected differently by climate change due to the vast geographical area, the diversity of crops, the many varieties, the differences in cultivation practices, cropping systems and production structures [38]. In this study, based on panel data of 25 Chinese provinces from 2005 to 2019, we analyzed the effects of the above factors on the unit yield of beans in China in four dimensions: climatic environment, economic market, human cultivation behavior and technological development level, where the focus is on the climatic environment factor and other factors are considered as control variables in the model.

The results of this study show that in recent years, all aspects of bean production in China have been characterized by progressive changes in the context of climate change. During 1979–2002, climate warming was more favorable for bean growth in China, especially in the northeast, where warmer temperatures could extend the growing period of the crop and reduce the impact of frost damage on beans [39]. However, the results of this study of 2005–2019 show that as global warming intensifies and the growing season accumulation temperature exceeds the inflection point optimum accumulation level, the increase in temperature begins to show a suppressive effect on bean yields, which is different from the relationship between temperature change and bean yields in 1979–2002. In addition, carbon emissions increase each year. Kumar, P. et al. indicated a positive relationship between carbon emissions and cereal production [40]; Pickson, R. B et al. stated that CO_2 emissions have a significant negative impact on cereal production in the long run [30] and show a unidirectional causality and Ahsan F et al. had the same conclusion as Pickson, R. B et al. on the relationship between cereal production and CO_2 emissions, but their study showed a bidirectional causal relationship between CO_2 emissions and cereal production [35]. These findings are not the same as the relationship between bean yields and CO_2 emissions in this paper, but this does not mean that it has no effect on the growth of bean crops in China. Excessive carbon emissions are an important cause of large changes and significant fluctuations in climate, and the most direct result is the increase in the number of disasters each year, which has a significant effect on the reduction in bean yields. It can be seen that the increasing temperature will cause changes in the spatial and temporal distribution of light, temperature, water and other climatic resources in the main grain-producing regions of China, which will lead to changes in soil organic matter, soil microorganisms and soil fertility and intensify the outbreak of agricultural pests and meteorological disasters in local areas.

At the same time, climate change will also lead to changes in the physiological and ecological characteristics of grain crop varieties by changing environmental factors, thus affecting the grain yield, cropping system, production method and structure in China. Climate change also leads to changes in the physiological and ecological characteristics of food crop varieties through changes in environmental factors, thus having far-reaching effects on China's grain production, cropping systems, production methods, structural layout and variety quality. In addition, the high level of rural modernization has started to have a slight negative impact on the growing environment of beans, which is also a conclusion that deserves our attention. The growing industry is one of the most sensitive areas to climate change, which has caused changes in crop fertility, farming systems, etc., and increased frequency and intensity of disasters, posing risks to and increasing pressure on global food production systems and food security. Ensuring sustainable agricultural development and food security is one of the important objectives of addressing climate change.

In international studies, many other countries have studied the relationship between local bean production and climate, for example, Wurr, D. et al. conducted a detailed analysis of French bean production in relation to CO_2 emissions and temperature variation. They concluded that temperature has a large positive effect on French bean production, while CO_2 has a negative or even no significant effect on French bean growth [41]. The current studies on bean crop yield in China are insufficient, and the factors involved are not comprehensive. The panel spatial error model used in this study explains the impact of current climatic and environmental factors on the yield per unit area of beans in China in four aspects, and the conclusions of the paper provide a reference for Chinese bean cultivation in response to climate change, while showing a positive correlation between CO_2 and SO_2 emissions, wastewater emissions and bean yield and the negative correlation between technological level development and bean yield. The two findings, which are different from cereal yields, are worthy of our attention and consideration. However, the limitations of the index still hindered the study somewhat. When considering the impact of air pollution and water environment pollution on bean production, SO_2 emissions from exhaust gas and total wastewater discharge essentially reflect the emission behavior of human life, which can only reflect the actual situation of air pollution and water pollution to a certain extent, while the distinction between planting and non-planting areas in these two indicators appears difficult to achieve, which is due to the low selectivity of statistical indicators for planting areas. Therefore, only the overall indicator can be chosen to replace the indicator for planted areas. The errors introduced by this indicator choice may lead to problems such as lower overall model fit superiority and insignificant estimated coefficients. In later studies, other alternative variables or instrumental variables that are highly accurate and can be collected to respond to pollutant concentrations in air and water pollution need to be considered.

Author Contributions: S.L. conducted the analysis of conditions, numerical simulation, data interpretation and drafted the manuscript; S.Y. initiated the study; Z.S. coordinated the data collection; L.Z. contributes to checking and translating; Y.L. contributed to checking and proofreading. All authors have read and agreed to the published version of the manuscript.

Funding: This research received no external funding.

Institutional Review Board Statement: Not applicable.

Informed Consent Statement: Not applicable.

Data Availability Statement: The data used in the study comes from the National Bureau of Statistics of China, China's National Meteorological Science Data Sharing Service Platform-China Surface Climate Data Daily Value Dataset V3.0, and the China Economic Information Network database, above data can be accessed online. The CSMAR database can be accessed only after subscription.

Acknowledgments: The authors would like to thank the editor and anonymous reviewers for their valuable comments and suggestions to this paper.

Conflicts of Interest: The authors declare no conflict of interest.

References

1. Lovejoy, E.T.; Hannah, L. *Climate Change and Biodiversity*; Yale University Press: New Heaven, CT, USA; London, UK; Sheridan Books: Ann Arbor, MI, USA, 2005.
2. Li, X.; Qin, D.; Li, J. *National Assessment Report on Climate Change*; Science Press: Beijing, China, 2007.
3. Aryal, J.P.; Sapkota, T.B.; Khurana, R.; Khatri-Chhetri, A.; Rahut, D.B.; Jat, M.L. Climate change and agriculture in South Asia: Adaptation options in smallholder production systems. *Env. Dev. Sustain.* **2019**, *22*, 5045–5075. [CrossRef]
4. Banerjee, C.; Adenaeuer, L. Up, up and away! The economics of vertical farming. *J. Agric. Stud.* **2014**, *2*, 40. [CrossRef]
5. Ali, T.; Huang, J.; Wang, J.; Xie, W. Global footprints of water and land resources through China's food trade. *Glob. Food Secur.* **2017**, *12*, 139–145. [CrossRef]
6. Adams, R.M.; Hurd, B.H.; Lenhart, S.; Leary, N. Effects of global climate change on agriculture: An interpretative review. *Clim. Res.* **1998**, *11*, 19–30. [CrossRef]
7. Zhao, C.; Liu, B.; Piao, S.; Wang, X.; Lobell, D.B.; Huang, Y.; Asseng, S. Temperature increase reduces global yields of major crops in four independent estimates. *Proc. Natl. Acad. Sci. USA* **2017**, *114*, 9326–9331. [CrossRef] [PubMed]
8. Shankar, S. Impacts of Climate Change on Agriculture and Food Security. In *Biotechnology for Sustainable Agriculture: Emerging Approaches and Strategies*; Lakahan, R., Mondal, S., Eds.; Elsevier: Amsterdam, The Netherlands, 2017.
9. Howden, S.M.; Soussana, J.-F.; Tubiello, F.N.; Chhetri, N.; Dunlop, M.; Meinke, H. Adapting agriculture to climate change. *Proc. Natl. Acad. Sci. Sci. USA* **2007**, *104*, 19691–19696. [CrossRef]
10. Xiong, W.; Lin, E.; Ju, H.; Xu, Y. Climate change and critical thresholds in China's food security. *Clim. Chang.* **2007**, *81*, 205–221. [CrossRef]
11. Zhou, W. *Impact of Climate Change on Grain Production in China and Coping Strategies*; Nanjing Agricultural University: Nanjing, China, 2012.
12. Jiang, M. *Research on Bean Improvement, Production and Utilization in China in the Twentieth Century*; Nanjing Agricultural University: Nanjing, China, 2006.
13. Meng, L. Review of domestic and international bean markets in 2011 and outlook for 2012. *Agric. Outlook* **2011**, *11*, 10–13.
14. Lin, E. *Simulation of the Impact of Global Climate Change on Chinese Agriculture*; China Agricultural Science and Technology Press: Beijing, China, 1997.
15. Cline, W. *Global Warming and Agriculture: Impact Estimates by Country*; Peterson Institute: Washington, DC, USA, 2007.
16. Pan, G.-X.; Gao, M.; Hu, G.-H.; Wei, Q.-P.; Yang, X.-G.; Zhang, W.-Z.; Zhou, G.-S.; Zou, J.-W. Impacts of Climate Change on Agricultural Production in China. *J. Agric. Environ. Sci.* **2011**, *30*, 1698–1706.
17. Schlenker, W.; Hanemann, W.; Fisher, A. The Impact of Global Warming on US Agriculture: An Econometric Analysis of Optimal Growing Conditions. *Rev. Econ. Stat.* **2006**, *88*, 113–125. [CrossRef]
18. Lobell, D.; Bänziger, M.; Magorokosho, C.; Vivek, B. Nonlinear Heat Effects on African Maize as Evidenced by Historical Yield Trials. *Nat. Clim. Chang.* **2011**, *1*, 42–45. [CrossRef]
19. Sarker, M.A.R.; Alam, K.; Gow, J. Assessing the effects of climate change on rice yields: An econometric investigation using Bangladeshi panel data. *Econ. Anal. Policy* **2014**, *10*, 405–416. [CrossRef]
20. Loum, A.; Fogarassy, C. The effects of climate change on cereals yield of production and food security in Gambia. *Appl. Stud. Agribus. Commer.* **2015**, *9*, 83–92. [CrossRef]
21. Susanto, J.; Zheng, X.; Liu, Y.; Wang, C. The impacts of climate variables and climate-related extreme events on island country's tourism: Evidence from Indonesia. *J. Clean. Prod.* **2020**, *276*, 124204. [CrossRef]
22. Schlenker, W.; Roberts, M. Nonlinear Temperature Effects Indicate Severe Damages to US Crop Yields under Climate Change. *Proc. Natl. Acad. Sci. USA* **2009**, *106*, 15594–15598. [CrossRef]
23. Welch, J.; Vincent, J.; Auffhammer, M.; Moya, P.; Dobermann, A.; Dawe, D. Rice Yields in Tropical/subtropical Asia Exhibit Large but Opposing Sensitivities to Minimum and Maximum Temperatures. *Proc. Natl. Acad. Sci.USA* **2010**, *107*, 14562–14567. [CrossRef]
24. Deschênes, O.; Greenstone, M. The Economic Impacts of Climate Change: Evidence from Agricultural Output and Random Fluctuations in Weather. *Am. Econ. Rev.* **2007**, *97*, 354–385. [CrossRef]
25. Chen, S.; Chen, X.; Xu, J. Impacts of Climate Change on Agriculture: Evidence from China. *J. Environ. Econ. Manag.* **2016**, *76*, 105–124. [CrossRef]

26. Elhorst, J. Spatial Panel Data Models. In *Spatial Econometrics*; Elhorst, J., Ed.; Springer: Berlin/Heidelberg, Germany, 2014.
27. Chen, S.; Xu, J.; Zhang, H. Impact of climate change on grain production in China—An empirical analysis based on county-level panel data. *China Rural. Econ.* **2016**, *5*, 2–15.
28. Peiris, D.R.; Crawford, J.W.; Grashoff, C.; Jefferies, R.A.; Porter, J.R.; Marshall, B. A simulation study of crop growth and development under climate change. *Agric. Meteorol.* **1996**, *79*, 271–287. [CrossRef]
29. Joshi, N.P.; Maharjan, K.L.; Piya, L. *Effect of Climate Variables on Yield of Major Food Crops in Nepal-A Time-Series Analysis*; MPRA Paper No. 35379; Springer: Tokyo, Japan, 2013.
30. Pickson, R.B.; He, G.; Ntiamoah, E.B.; Li, C. Cereal production in the presence of climate change in China. *Environ. Sci. Pollut. Res.* **2020**, *27*, 45802–45813. [CrossRef]
31. Liu, Z. *Prediction Study on the Impact of SO_2 Emissions from a Large Natural Gas Purification Plant on Regional Crops*; Southwest Jiaotong University: Chengdu, China, 2012.
32. Yang, G.; Zhong, G.; Wu, J. Deng Shihuai. Effects of SO_2 emissions from high sulfur gas fields on rice growth and yield. *Chin. J. Ecol. Agric.* **2010**, *18*, 827–830. [CrossRef]
33. Janjua, P.Z.; Samad, G.; Khan, N. Climate change and wheat production in Pakistan: An autoregressive distributed lag approach. *NJAS Wagening J. Life Sci.* **2014**, *68*, 13–19. [CrossRef]
34. Casemir, B.H.; Diaw, A. Analysis of climate change effect on agricultural production in Benin. *Asian J. Agric. Ext. Econ. Soc.* **2018**, *24*, 1–12.
35. Ahsan, F.; Chandio, A.A.; Fang, W. Climate change impacts on cereal crops production in Pakistan. *Int. J. Clim. Chang. Strateg. Manag.* **2020**, *12*, 257–269. [CrossRef]
36. Sossou, S.; Igue, C.B.; Diallo, M. Impact of climate change on cereal yield and production in the Sahel: Case of Burkina Faso. *Asian J. Agric. Ext. Econ. Soc.* **2019**, *37*, 1–11. [CrossRef]
37. Chijioke, O.B.; Haile, M.; Waschkeit, C. Implications of climate change on crop yield and food accessibility in sub-saharan Africa. In *Interdisciplinary Term Paper*; Center for Development Research, University of Bonn: Bonn, Germany, 2011; Available online: http://www.zef.de/fileadmin/downloads/forum/docprog/Termpapers/2011_1_Oyiga__Haile_Waschkeit.pdf (accessed on 13 September 2013).
38. Qin, Z.; Tang, H.; Li, W. Frontiers of research on the impact of climate change on food production systems in China. *China Agric. Resour. Zoning* **2015**, *36*, 1–8.
39. Tao, F.; Yokozawa, M.; Liu, J.; Zhang, Z. Climate-crop yield relationships at provincial scales in China and the impacts of recent climate trends. *Clim. Res.* **2008**, *38*, 83–94. [CrossRef]
40. Kumar, P.; Sahu, N.C.; Kumar, S.; Ansari, M.A. Impact of climate change on cereal production: Evidence from lower-middle-income countries. *Environ. Sci. Pollut. Res.* **2021**, *28*, 51597–51611. [CrossRef] [PubMed]
41. Wurr, D.; Edmondson, R.; Fellows, J. Climate change: A response surface study of the effects of CO_2 and temperature on the growth of French beans. *J. Agric. Sci.* **2000**, *135*, 379–387. [CrossRef]

atmosphere

MDPI

Article

The Economic Loss Prediction of Flooding Based on Machine Learning and the Input-Output Model

Anqi Chen [1,*], Shibing You [1], Jiahao Li [2] and Huan Liu [1]

[1] Economics and Management School, Wuhan University, Wuhan 430072, China; sbyou@whu.edu.cn (S.Y.); liuhuan@whu.edu.cn (H.L.)
[2] Faculty of Engineering, The University of Sydney, Camperdown, NSW 2006, Australia; jili9122@uni.sydney.edu.au
* Correspondence: anqichen@whu.edu.cn

Abstract: As climate change becomes increasingly widespread, rapid, and intense, the frequency of heavy rainfall and floods continues to increase. This article establishes a prediction system using feature sets with multiple data dimensions, including meteorological data and socio-economic data. Based on data of historical floods in 31 provinces and municipalities in China from 2006 to 2018, five machine learning methods are compared to predict the direct economic losses. Among them, GBR performs the best with a goodness-of-fit of 90%. Combined with the input-output (IO) model, the indirect economic losses of agriculture to other sectors are calculated, and the total economic losses caused by floods can be predicted effectively by using the GBR-IO model. The model has a strong generalization ability with a minimum requirement of 80 pieces of data. The results of the data show that in China, provinces heavily reliant on agriculture suffered the most with the proportion of direct economic losses to provincial GDP exceeding 1‰. Therefore, some policy implications are provided to assist the government to take timely pre-disaster preventive measures and conduct post-disaster risk management, thereby reducing the economic losses caused by floods.

Keywords: economic loss prediction; machine learning; input-output model; flooding

Citation: Chen, A.; You, S.; Li, J.; Liu, H. The Economic Loss Prediction of Flooding Based on Machine Learning and the Input-Output Model. *Atmosphere* **2021**, *12*, 1448. https://doi.org/10.3390/atmos12111448

Academic Editors: Zengyun Hu, Xuguang Tang and Qinchuan Xin

Received: 8 October 2021
Accepted: 30 October 2021
Published: 2 November 2021

1. Introduction

The acceleration of climate change has intensified the water cycle, affected rainfall patterns, and caused rising sea levels, triggering more frequent heavy rainfall and worsening flood situations [1]. Due to global warming, research shows that the death toll and economic damage of floods have risen by around 75% and 200%, respectively [2]. According to a comprehensive analysis published in Geneva, 23 July 2021 by the World Meteorological Organization (WMO), of the top 10 meteorological disasters, floods have inflicted the top three largest economic tolls and human losses around the world over the past half-century. Fatal floods, incurred by heavy rains, occur with a low frequency, a wide range of influence, and can cause serious economic losses [3]. The economic loss from such natural disasters should be evaluated as soon as possible and effective countermeasures should be taken to manage natural hazards [4]. Hence, if potential economic losses of floods can be accurately predicted prior to or just after the beginning of heavy rainfall, more time can be provided for people to take precautions and put in place measures to avoid and alleviate unnecessary losses and secondary hazards.

In response, relevant studies of economic loss assessment have been conducted. Generally, economic losses, which can be measured with monetary value, are classified into two sorts: direct economic losses and indirect economic losses [2]. Direct economic losses refer to property losses of residences, asset losses of enterprises, loss of infrastructure, loss of natural resources, etc., while indirect economic losses arise from the suspension and reduction of production, investment premiums, material shortages, overstock, and losses from related industries [5].

For direct economic losses, assessments are normally conducted by using a flood stage-damage calculation [6], loss rate approach, comprehensive loss value method, Category-Unit Loss Functions [7], approach based on RS and GIS [8], and Regression Analysis Rapid Evaluation Model [9], etc. By far, most of the direct economic assessments are conducted post-disaster, while pre-disaster assessment is currently at an initial stage. Yet with the boom in artificial intelligence, advanced approaches, such as backpropagation neural networks (BPNN) [10], random forest, support vector regression (SVR) [11], and gradient boosted regression tree (GBR) [12], etc., are implemented to predict the direct economic losses of natural disasters. The most popular approach among these is BPNN. Although BPNN better fits the complex nonlinear function relationship and has higher assessment accuracy, it requires a huge amount of data, which is lacking in extreme weather data as the number of flood occurrences across history are limited [10]. Another commonly used method, SVR—the extension of support vector machine (SVM)—is suitable for solving high-dimensional regression problems using a small number of samples, which is in accordance with the features of natural disasters [13]. Additionally, ensemble learning—a commonly used algorithm—integrates several machine learning methods to reduce deviation and improve prediction accuracy [14]. Extreme Gradient Boosting Regressor (XGB) has been demonstrated to be a good fit for evaluating economic losses of natural disasters, but the generalization ability is unsatisfactory [12]. Sun et al. used GBR in the field of direct economic loss evaluation caused by storm surge disasters, improving the prediction accuracy and reducing model overfitting caused by small datasets [11]. However, there has not yet been any report of research applying GBR to the disaster assessment of floods.

For indirect economic loss assessment, empirical analysis methods, Computable General Equilibrium (CGE) models, and Input-Output (IO) models are commonly used. The empirical analysis method assumes a certain proportional relationship between the indirect economic losses caused by a flood to different sectors and industries and its direct economic losses in the inundated area, but the result is relatively ineffective [15]. The dynamic CGE model can realize a comprehensive assessment of the indirect economic impact of flood disasters through parameter simulation, flow restriction simulation, and variable shock simulation [16]. However, CGE models are based on more restrictive assumptions, which typically assume optimizing behavior and equilibrium economy. Such assumptions are easily violated under real-world economic conditions [17]. In contrast, the input-output approach, an improvement in methodologies, is more suitable for assessing economic losses caused by exogenous shocks [18]. Zhang et al. implemented the IO model to analyze the indirect economic losses of floods in Hunan Province, China, in 1998 [2]. It is proven that the model is an effective and flexible approach to assess indirect economic losses as it can select the number of sectors according to the retrieved data.

Therefore, in this paper, we establish a comprehensive prediction system to assess the direct and indirect economic losses caused by floods in 31 regions of China. There are two main contributions in this paper. Firstly, we designed a timely and effective pre-disaster prediction system with the GBR-IO model by using indicators which can be collected pre-disaster or at the beginning of a flooding episode and found the minimum amount of data required for the prediction system to be able to provide results with high effectiveness. Secondly, based on the meteorological features of natural disasters, we combined machine learning methods, using advanced regression prediction approaches, with the input-output model, a traditional economic method, resulting in what can have a profound impact on inter-disciplinary research. Thirdly, the method we used has a high generalization ability, meaning that it can be applied to other countries and regions which experience flooding, especially those with small datasets.

2. Materials and Methods

2.1. Study Area

Statistics show that Asia is the continent where floods are, by far, the most frequent and devastating natural disasters around the world [19]. Among Asian countries, China—with a monsoon climate and major rivers of the world—is the most frequently affected country. Around the world, the frequency of floods and economic losses caused by floods in China ranked first, and the casualties and death toll ranked second during the last decade [20]. Recently, in mid to late July 2021, rainstorms and devastating floods battered large portions of North China and the Huanghuai Region, especially Zhengzhou, the state capital of Henan province, causing the deaths of 302 people, 50 people to be missing, and 114.3 billion RMB yuan (US $17.7 billion) in direct economic losses [21]. For studying floods or the economic losses caused by floods, China is ideal in terms of sample size and practical significance. Thus, in this research we focused on 31 provinces in China (excluding the Hong Kong, Macao, and Taiwan regions) with the aim having the ability to extend the prediction system to other countries affected by flood threats.

2.2. Data

The process of establishing the pre-disaster prediction system is data-driven. We primarily used two sorts of data: first, meteorological data; and second, socio-economic data. To predict the direct economic losses in terms of geographical information and timescale, while considering the integrity, continuity, and variety of data, we selected a time-series dataset of 31 provinces and municipalities in China composed of a period of 13 years, from 2006 to 2018. With the aim of ensuring the timeliness of the prediction system, we chose variables based on significance and whether they can be retrieved pre-disaster or at the beginning of a flooding episode. Therefore, 23 independent variables and a dependent variable, the latter being direct economic loss, were used in the study, as shown in Table 1. There were 403 pieces of data in total showing the meteorological and socio-economic conditions. To predict the indirect economic losses, which are based on the direct economic losses of the agricultural sector, we further utilized the direct economic loss data as the input of the IO model. With the combination of other socio-economic data, we generated the indirect economic losses caused by the demand reduction in the agricultural sector.

Table 1. Floods disasters direct economic loss prediction index system.

Criteria	Indicators	Variables
Disaster-inducing factors	Year	X_1
	Daily Maximum Precipitation	X_2
	Precipitation Anomaly Percentage	X_3
	Precipitation Anomaly Percentage in Spring	X_4
	Precipitation Anomaly Percentage in Summer	X_5
	Precipitation Anomaly Percentage in Autumn	X_6
	Precipitation Days	X_7
	Moderate Rainy Days	X_8
	Heavy Rainy Days	X_9
	Torrential Rainy Days	X_{10}
	Maximum Continuous Precipitation	X_{11}
	Maximum Annual Rainfall	X_{12}
	Maximum Annual Continuous Rainy Days	X_{13}

Table 1. *Cont.*

Criteria	Indicators	Variables
Disaster-affected bodies	Casualties	X_{14}
	Death Toll	X_{15}
	Sown Area with 10% Reduced Production	X_{16}
	Sown Area with 30% reduced production	X_{17}
	Sown Area with 80% reduced production	X_{18}
	Railway Disruption	X_{19}
	Road Disruption	X_{20}
	Reservoir Loss	X_{21}
	Province	X_{22}
Disaster Prevention Capabilities	Number of Reservoirs	X_{19}
	Capacity of Reservoirs	X_{20}
	Area with Flood Prevention Measures	X_{21}
	Areas with Soil Erosion under Control	X_{22}
	City Sewage Pipes Length	X_{23}

2.2.1. Meteorological Data

Meteorological data were obtained from "Daily meteorological dataset of basic meteorological elements of the China National Surface Weather Station (V3.0)" including China's national basic weather stations, reference climatological station, and general weather stations, with a total of 2474 stations of the China Meteorological Administration. All of the research data were retrieved at an annual provincial level. The data included daily maximum precipitation, precipitation anomaly percentage, precipitation anomaly percentage in spring, precipitation anomaly percentage in summer, precipitation anomaly percentage in autumn, precipitation days, moderate rainy days, heavy rainy days, torrential rainy days, and maximum continuous precipitation days.

2.2.2. Socio-Economic Data

To predict the direct losses, the socio-economic data included: casualties, death toll, direct economic losses, sown area with 10% reduced production, sown area with 30% reduced production, sown area with 80% reduced production, railway disruption, road disruption, number of reservoirs, capacity of reservoirs, area with flood prevention measures, area of soil erosion under control, and length of city sewage pipes. The above data were retrieved from the Bulletin of flood and drought disasters in China (2006–2018), published by the Ministry of Water Resources of the People's Republic of China, China statistical Yearbook (2006–2018). To predict the indirect economic losses, the Input and Output table with 42 sectors of 31 provinces from the National Bureau of Statistics (2017) was used to calculate the industry linkage [22].

2.2.3. Data Processing

Step 1 processing the missing data: As mentioned in the previous section, all data used in the study were collected online and offline from official datasets, official reports, and reference books. For some missing data, if the data were proved to be below the statistical standard, we substituted it with 0. For the data where the true value was missing, the data were substituted with the average figure in order to mitigate the impact on the prediction model.

Step 2 normalization: Each sample of the original dataset had 23 features (independent variables) to reflect the flood disasters from a specific aspect or information related to the local province. If components are with different magnitudes, they will not make equal contributions to fit the model and are likely to cause a bias. Due to this, we used the normalization process of Min-Max normalization to make the indicators comparable. By doing so, all features were transformed into the range [0, 1], meaning that the minimum value of a feature becomes 0, while the maximum value of that becomes 1. The formula is as follows:

$$x_{normalised} = \frac{x - \min(x)}{\max(x) - \min(x)} \quad (1)$$

where x denotes the input of independent variables.

2.3. Methods

2.3.1. Gradient Boosting Regression Trees (GBR)

A new machine learning method, Gradient Boosting Regression Trees (GBRT or GBR) was adopted in our research to efficiently predict the direct economic losses of floods. It is a modification of the gradient boosting (GB) algorithm and Classification and Regression Trees (CARTs) by using a regression tree of fixed size as weak learners [23].

The gradient boosting algorithm, proposed by Friedman in 1999 [24], integrates various machine learning and statistical methods, such as gradient algorithm, boosting algorithm and tree algorithm. The rationale of the gradient boosting algorithm is to create new base learners which are maximally correlated with a negative gradient of the loss function. Compared with traditional boosting algorithms, along the direction of the gradient, every new model is built with the aim of reducing the loss function of the previous model. By using the gradient boosting algorithm during the training process, regression was achieved by continuously reducing the residuals [25]. The process of gradient boosting with multiple iterations is shown in Figure 1. To elaborate on the training process, in each iteration, a weak classifier is generated by the model, which goes through further training according to the residuals of the previous classifier. Based on the performance of a classifier, the weight is generated. The poorer the performance, the more weight will be given. Finally, the ensemble model is achieved by the weighted summarization of all weak classifiers. With a variance reduction, the prediction accuracy of a classifier is improved. Both continuous and discrete values can be dealt with gradient boosting, and this algorithm can mitigate the drawbacks of overfitting. In the process, a base classifier, also known as a weak classifier, is generally CARTs, a series of decision tree regression models. The weight is generated based on the performance of weak classifiers, and the process of learning consists of inputting the residuals from the previous iteration as the object function of the next base classifier.

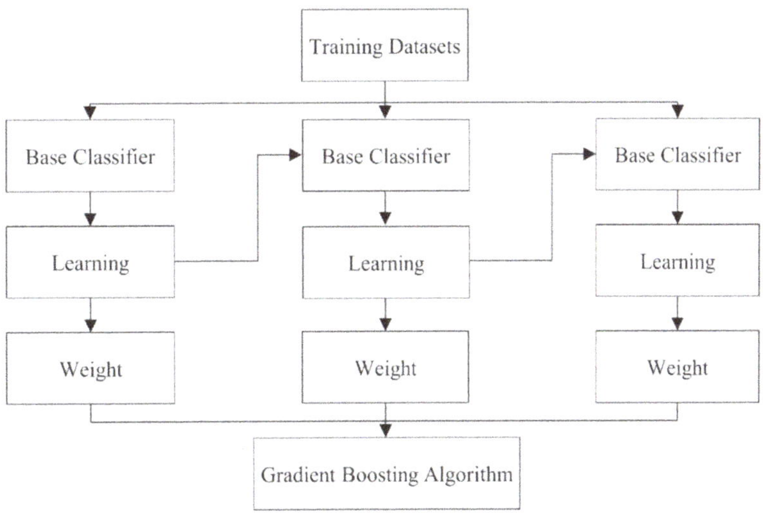

Figure 1. Training Process of Gradient Boosting.

Classification and regression trees (CARTs), proposed by Breiman et al. in 1984 [26], can be used for both classification and regression models [27–29]. The two types of trees used in these two models are decision trees and regression trees. Compared with other artificial intelligence models, CARTs have better performance in terms of prediction as they can obtain nonlinear relationships without requiring prior information about the probability distribution of variables.

The GBR algorithm combines weak learners by iteratively concentrating on the errors resulting at each step until a sum of the successive weak learners can a create suitable strong learner.

Given the Dataset $\{X_i, Y\}_{i=1}^{n}$ (i.e., historical flooding dataset), the loss function is used to evaluate a set of weights to seek a minimized error with the aim of optimizing the model. Let X_i denote a set of explanatory variables (i.e., disaster-inducing factors, disaster-affected bodies and the disaster-prevention capabilities) and Y be dependent variables (i.e., direct economic losses). There are six main steps of the GBR, which can be expressed as follows:

1. Initialize the parameters in the learning machine with the following equation:

$$F_0(x) = \text{argmin}_\rho \sum_{i=1}^{N} L(y_i, \rho) \tag{2}$$

where $F_0(x)$ is the parameter set. ρ is the parametric variable that minimizes the loss function, and $L(y_i, \rho)$ is the square error loss function in our model. The negative gradient of the loss function is used in the current model as an approximation of the residual. The calculation process of the residual is shown as follows:

$$\overline{y_i} = -\left[\frac{\partial L(y, F(x_i))}{\partial F(x_i)}\right]_{F(x)=F_{m-1}(x)}, i = 1, 2, \cdots, N \tag{3}$$

For iterations m = 1 to M:
where $F(x_i)$ refers to the objective function.

2. A regression tree is generated with J leaf nodes, described as follows:

$$\{R_{jm}\}_1^J = J - TNT\left(\{\overline{y_i}, x_i\}_i^N\right) \tag{4}$$

where R_{jm} refers to a regression tree with J leaf nodes, TNT refers to terminal node tree.

3. Estimating the value of the leaf nodes in the regression tree. The value can be estimated by the following equation:

$$\gamma_{jm} = \text{argmin}_\gamma \sum_{x_i \in R_{jm}} L(y_i, F_{m-1}(x_i) + \gamma) \tag{5}$$

where γ refers to the value of the leaf nodes in the regression tree.

4. The learning machine of this iteration can be obtained, as shown in the following:

$$F_m(x) = F_{m-1}(x) + \sum_{j=1}^{J} \gamma_{jm} I(x \in R_{jm}) \tag{6}$$

5. After iterations, the final regression model can be shown as follows:

$$F(x) = F_m(x) = \gamma + \sum_{m=1}^{M} \sum_{j=1}^{J} \gamma_{jm} I(x \in R_{jm}), \text{ where } I(x \in R_{jm}) = \begin{cases} 1, x \in R_{jm} \\ 0, x \notin R_{jm} \end{cases} \tag{7}$$

Numerous researchers have developed an index system-based assessment method to gain a better understanding of the relationship between factors and natural disaster loss. Sun et al. classified the indicators into several different categories, including disaster-inducing factors, disaster-affected factors and disaster-prevention capabilities [30]. In our research, we also designed an index system for independent variables as the input of GBR, as shown in Table 1.

To predict the direct economic losses of floods pre-disaster or at the beginning of a flooding episode is to establish a regression model. Since the dataset of flooding, a disaster occurring with a low frequency, is discrete with a small scale and large time span, five machine learning methods are generally used, including Bayesian Ridge, Line Linear, Elastic Net, XGB, and GBR [31]. Despite the advantages of the other four methods, we compared the five different models to verify the effectiveness of GBR. The comparative experiment is designed as follows in Figure 2:

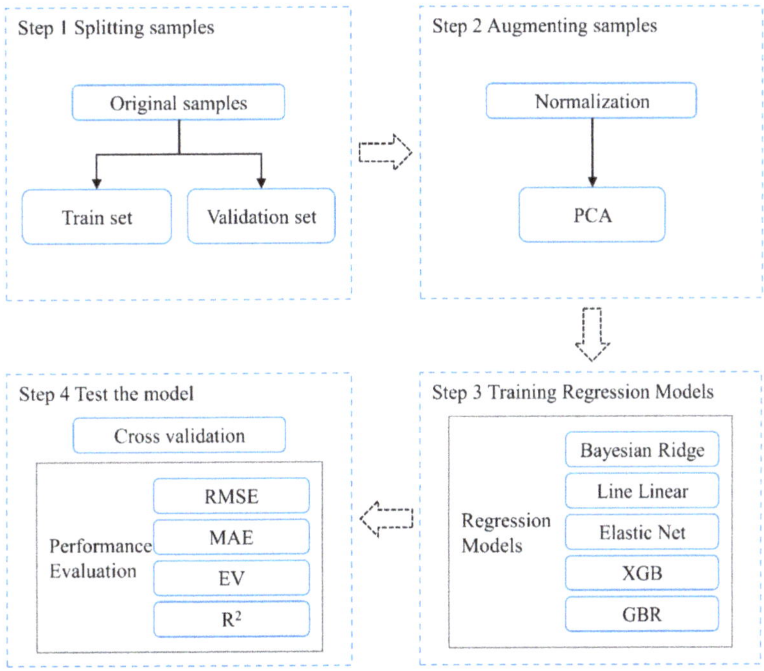

Figure 2. The flowchart of model comparison.

To further reduce the autocorrelation between the variables, principal component analysis (PCA) is implemented. In this study, we used a randomized truncated singular value decomposition (SVD) by the method Halko et al. 2009 [32].

To compare the validation of the performance of the five models, four indicators are used, including Root-Mean-Square Error (RMSE), Mean Absolute Error (MAE), Explained Variance (EV), and R-squared score(R^2). The formulas are as follows:

$$RMSE = \sqrt{\frac{\sum_{i=1}^{n} (y_i - \hat{y}_i)^2}{n}} \quad (8)$$

$$MAE = \frac{\sum_{i=1}^{n} |y_i - \hat{y}_i|}{n} \quad (9)$$

$$EV(y_i, \hat{y}_i) = 1 - \frac{Var(y_i - \hat{y}_i)}{Var(y_i)} \quad (10)$$

$$R^2(y_i, \hat{y}_i) = 1 - \frac{\sum_{i=1}^{n} (y_i - \hat{y}_i)^2}{\sum_{i=1}^{n} (y_i - \overline{y})^2} = 1 - \frac{RMSE}{Var(y_i)} \tag{11}$$

where y_i is the actual measurement, $\hat{y}_i$ is the predicted value $\overline{y} = \frac{1}{n}\sum_{i=1}^{n} y_i$, n is the number of measurements.

In this article, Anaconda Navigator 2.0.3, python 3.8.8, Jupyter notebook 6.3.0 were implemented.

2.3.2. Input-Output (IO) Model

Natural disasters, especially floods, have detrimental impacts on the agricultural industry and disrupt public transportation. The agricultural industry and the transportation industry are closely related to other industries, such as real estate, construction, warehousing and retail, accommodation, and catering, etc. Floods also cause indirect losses in upstream and downstream industries that are not directly related, such as the financial, mining, and other industries. Therefore, losses of directly affected industries were predicted by using the machine-learning prediction system, and then as the input, the predicted direct losses were used to evaluate the indirect losses of other industries using the Input-Output (IO) model.

The IO model has been implemented to assess the economic effect of natural disasters since the 1970s. Results have shown that the model can assess related economic losses effectively. In this article, indirect economic losses among the industries are evaluated by using a static IO model [18].

The correlations among the industries in the IO table can be expressed as:

$$AX + Y = X \tag{12}$$

That is,

$$\sum_{i,j=1}^{n} a_{ij}X_j + Y_i = X_i \ (i,j = 1,2,\ldots,n), \tag{13}$$

where a_{ij} is the direct consumption coefficient, X_i is the total output of sector i, and Y_i is the final demand for sector i. The above formula can then be transformed as:

$$X = (I - A)^{-1}Y \tag{14}$$

where I is the identity matrix and $(I - A)^{-1}$ is the inverse matrix of Leontief.

Taking the sectional direct economic losses as losses in final products, $\Delta Y = (\Delta Y_1, \Delta Y_2, \ldots, \Delta Y_n)^T$. The total product loss is then:

$$\Delta X = (I - A)^{-1}\Delta Y. \tag{15}$$

where ΔX denotes the total economic losses, and ΔY denotes the direct economic losses Thus, the loss of indirect input is expressed by the reduction of intermediate input, as $\Delta X - \Delta Y$.

To improve the accuracy of the indirect loss assessment of various departments, this paper uses the complete consumption coefficient for analysis. Let B be a complete consumption coefficient matrix obtained by transforming the direct consumption coefficient matrix A, then $B = (I - A)^{-1} - I$. Therefore, the total loss of the product can be further expressed as:

$$\Delta X = (B + I)\Delta Y. \tag{16}$$

Assume that ΔY_i is the economic loss in the sector i caused by floods and the final use of other sectors has no change. The total output of the entire economic system then becomes:

$$\begin{pmatrix} \Delta X_1 \\ \Delta X_2 \\ \cdot \\ \cdot \\ \cdot \\ \Delta Xn \end{pmatrix} = \begin{pmatrix} b_{1i}\Delta Y_i \\ b_{2i}\Delta Y_i \\ \cdot \\ \cdot \\ \cdot \\ b_{ni}\Delta Y_i \end{pmatrix} + \begin{pmatrix} 0 \\ 0 \\ \cdot \\ \Delta Y_i \\ \cdot \\ 0 \end{pmatrix}, \tag{17}$$

where $b_{ij}(i,j = 1,2,\ldots,n)$ are the complete consumption coefficients. The total output loss of sector i is then:

$$\Delta X_i = b_{ii}\Delta Y_i + \Delta Y_i, \tag{18}$$

where ΔY_i is the direct economic loss of sector i and $b_{ii}\Delta Y_i$ is the indirect economic loss of sector i. The total product losses of other sectors are:

$$\Delta X_n = b_{ni}\Delta Y_i,\ n \neq i. \tag{19}$$

2.3.3. The Pre-Disaster Prediction System

The overall prediction system process of the GBR-IO model is shown, as follows, in Figure 3.

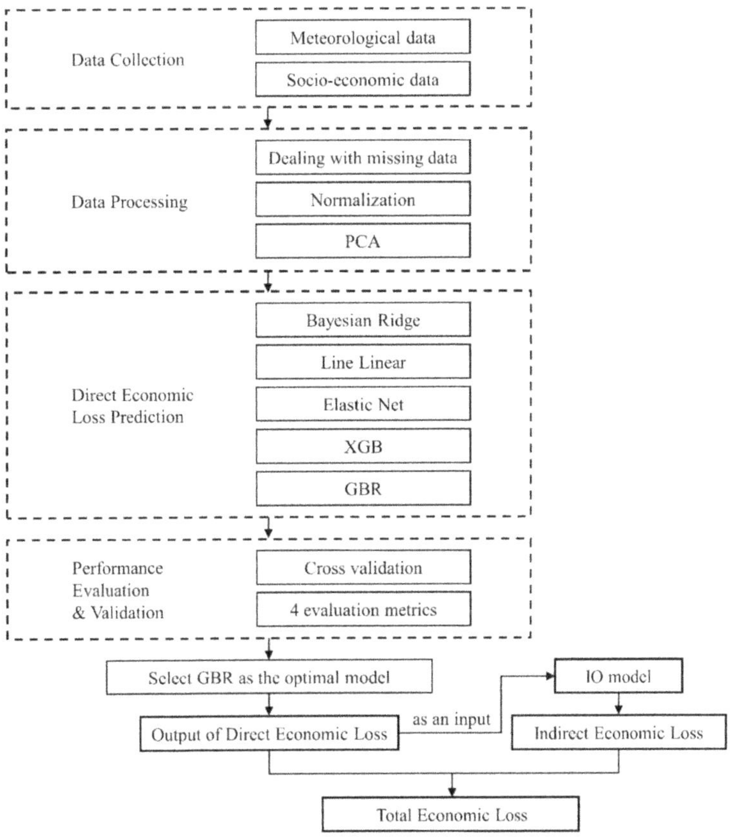

Figure 3. Prediction system process.

3. Results & Discussion

3.1. Direct Economic Loss Prediction

To analyze which features make the most contribution in causing direct economic losses, we firstly retrieved feature importance indicating the usefulness and value of each feature in the construction of the boosted regression trees within the model. The more an attribute is used to make key decisions with decision trees, the higher the F score is. From Figure 4, below, it can be seen that reservoir loss, sown area with 10% reduced production, road disruption, annual anomaly percentage and daily maximum precipitation are the top five features of importance.

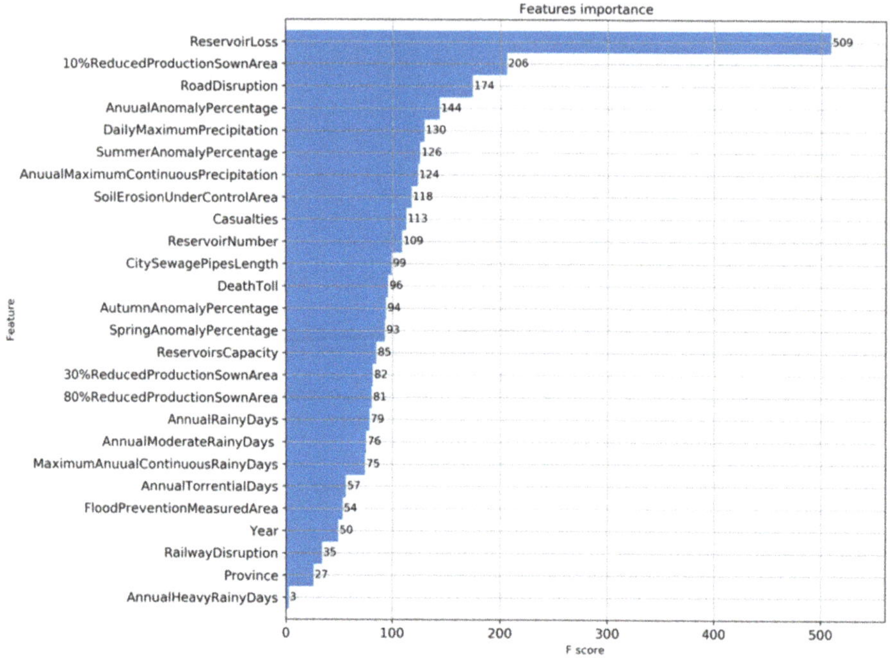

Figure 4. Features of importance, measured with the F score.

Further, we developed a heat map of the correlation coefficient between the top 10 features and the model. The heat map is used to directly illustrate the correlation between the indicators and economic losses of flooding. The color shade of the heat map shows the degree of the correlation; that is, the greater the correlation, the lighter the color. The correlation between the indicators and direct economic losses, ranging between −1 to 1, shows whether they are positively correlated or negatively correlated. From the heat map, shown in Figure 5, it is notable that reservoir loss (with the correlation of above 0.8), sown area with 10% reduced production (correlation of 0.6–0.8), casualties (correlation of 0.6), road disruption, number of reservoirs, annual anomaly percentage, summer anomaly percentage and daily maximum precipitation are positively correlated with direct economic losses. In contrast, length of city sewage pipes (correlation of −0.2–0), reservoir capacity (correlation of −0.3) and annual rainy days are negatively correlated (correlation of below −0.2). From the results, it can be inferred that: (1) the agricultural industry is the most directly affected sector; (2) there is a time lag in constructing reservoirs for the mitigation of economic impacts caused by floods; (3) since a region with high annual rainy days is less likely to suffer from economic losses of floods, a possible explanation could be that such regions have a balanced precipitation trend and are unlikely to suffer from floods caused by short-term torrential or heavy rains; (4) the improvement of disaster prevention capabilities is effective in reducing economic impacts; and (5) reservoir loss contributes the most to direct economic losses, among other factors.

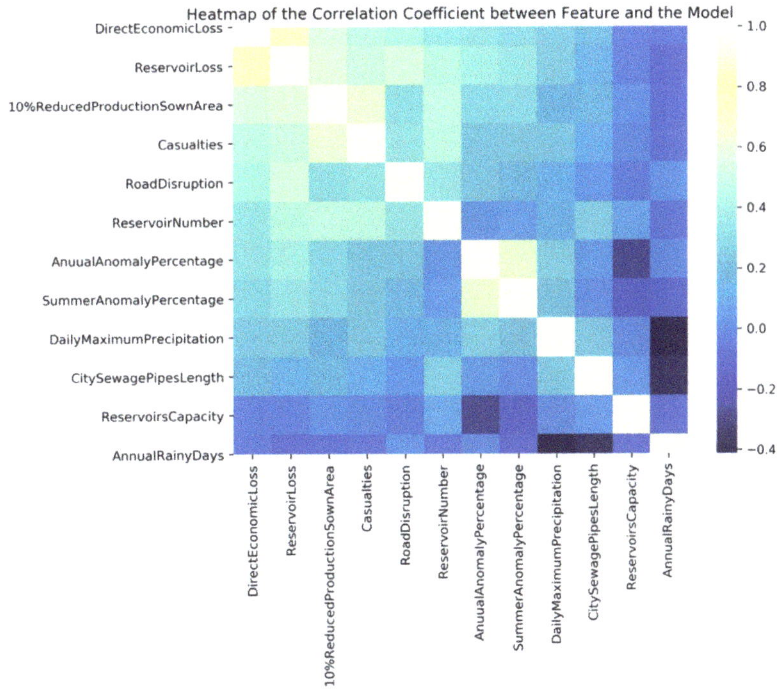

Figure 5. Heat map of the correlation coefficient between features and the model.

To further reduce the autocorrelation between variables, principal component analysis (PCA) was implemented. PCA is used to decompose a multivariate dataset in a set of successive orthogonal components that explain a maximum amount of variance. The PCA screen plot in Figure 6 shows the explained variance of all of the variables used in the model. Based on the ranking of variance that is explained, we selected the top 16 principal components as new features, i.e., F1 to F16 in Figure 6. Such new features retained 90.75% of the original information. Due to the characteristics of decision tree algorithms, they do not address the value of variables. Therefore, there is no need for feature normalization and PCA. The tree models (GBR and XGB) in the study were not processed with PCA.

The above new features are input into the prediction system to be trained with five different machine learning algorithms. During the training process, K-fold cross-validation (k = 5 in this prediction system) is performed. Firstly, the original data set is split into five sets. We then use the four of the five sets as the training set to train the model without repetition. Such an approach can obtain a substantial amount of information from limited data. The cross-validation results are shown in Figure 7. Among the five machine learning methods, Bayesian Ridges and GBR performed better than the other three methods, with a cross score of around 0.7. To further determine which method is better, four regression model evaluation metrics, including RMSE, MAE, EV, and R^2, were used to compare the prediction effect of the model.

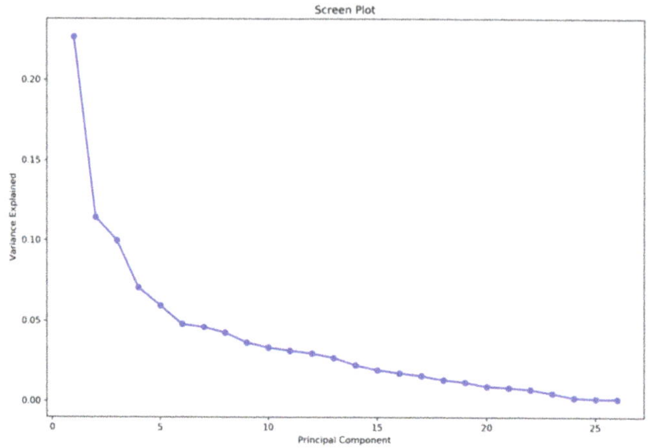

Figure 6. PCA screen plot.

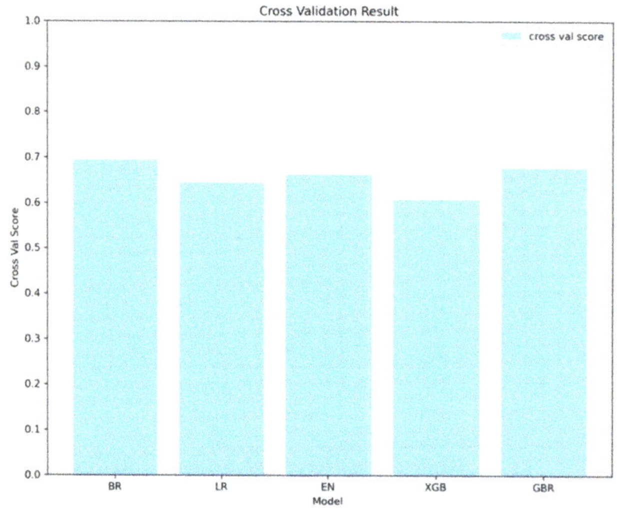

Figure 7. Cross validation of machine learning models.

RMSE, MAE, EV, and R^2 were calculated for the gradient boosting regression technique, as well as the other four machine learning methods. The lower value of RMSE and MAE and the higher value of EV indicate the more accurate prediction result, and the higher value of R^2 indicates a heightened match between the analytical and predicted values. The advancement of gradient boosting regression, compared to other approaches, is shown in Table 2. Observing the regression metrics, it is notable that, although Bayesian Ridge, Linear Regression, and Elastic Net perform better than GBR in some error metrics, the prediction of a specific dataset can substantially deviate from the true value, indicating overfitting. GBR has the highest R^2 and low RMSE, MAE and high EV in comparison to the predicted and the actual value, indicating that GBR has the best fit of values and a better fit than the other four methods. Therefore, combining the cross-validation results and regression metrics, we obtained GBR as the best machine learning regression model in predicting the direct economic losses of floods.

Table 2. Regression metrics.

	RMSE	MAE	EV	R^2
Bayesian Ridge	32.80	24.61	0.86	0.84
Linear Regression	36.66	28.97	0.84	0.80
Elastic Net	34.03	26.06	0.85	0.82
XGB	31.76	18.64	0.85	0.85
GBR	25.57	16.49	0.90	0.90

The learning curve is generally used in machine learning to evaluate both the performance of the training and the validation of datasets to diagnose whether the model is underfit, overfit, or well-fit [33]. Training samples on the horizontal axis show the size of datasets used in the learning process, while the score on the vertical axis represents R^2, which is used to evaluate the overall performance. The learning curve of the GBR model is shown in Figure 8. With the expansion of the training set, the cross-validation score approaches a desirable level, and it converges with the training score. It can be seen from the figure that when the training data size reaches about 80 pieces, the over-fitting and under-fitting of the model have been significantly reduced, proving that the GBR model is well-fit. Adding more training data can decrease the variance and bias of the GBR model. When the data size reaches about 150 pieces, the variance and deviation of the prediction results can be further mitigated. Therefore, when the model is generalized to analyze similar problems in other countries or regions, the required data size must reach at least 80 pieces, and the prediction accuracy will be improved with the expansion of data size.

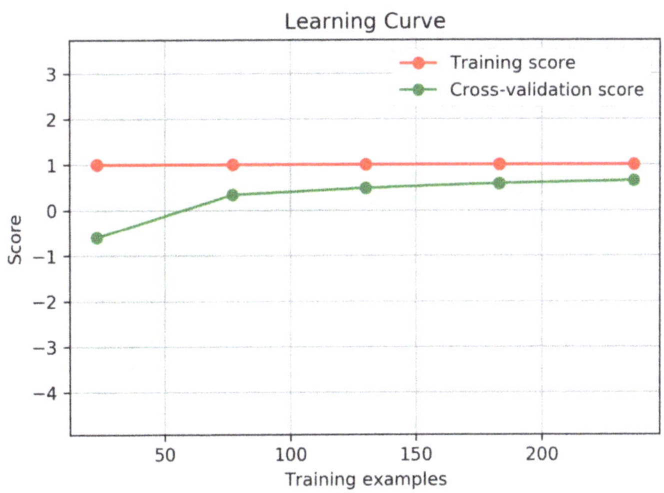

Figure 8. Learning curve of GBR model.

Integrated learning is considered to have high predictive precision, especially in terms of those algorithms which use a decision tree as the base learner. To validate GBR as the optimal combined model, we compared the regression result with Bayesian Ridge, Linear Regression, Elastic Net, XGB, and GBR, as is shown in Figure 9. It can be seen that GBR is the best fit for the true value.

To verify the effectiveness of the prediction model, we did descriptive analysis and empirical analysis based on the dataset in China from 2006 to 2018.

According to the descriptive statistical analysis for the years 2006 to 2018, the direct economic losses caused by floods in China were overwhelmingly higher in the years 2010 and 2016 than of that in other years and showed an overall increasing trend, as shown in Figure 10. Among the floods in China over this period, floods in 2010 caused by torrential rainfall in southern China were the most devastating to occur since 1998, leading to the worst economic conditions in the past half century [34]. Global warming had caused a surge of moisture in the atmosphere. With the combination of a strong El Niño effect and climate change, severe flooding occurred along the Yangtze River in the summer of 2016 [35]. The floods that broke out in these two years were the worst in the recent decades resulting in the direct economic losses being much more severe than in the other years analyzed.

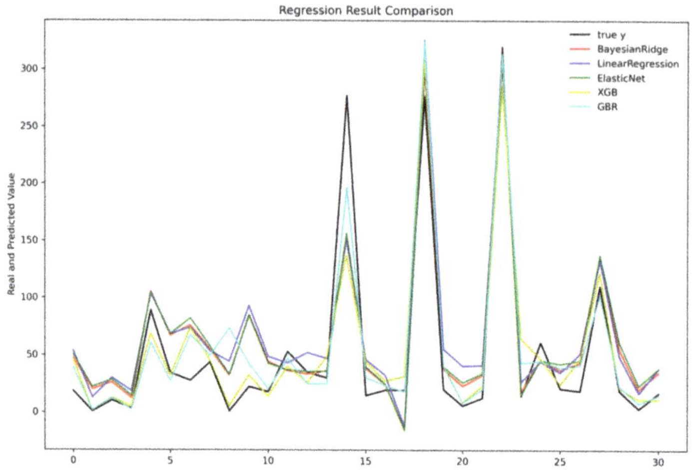

Figure 9. The fitting and prediction result of compared models.

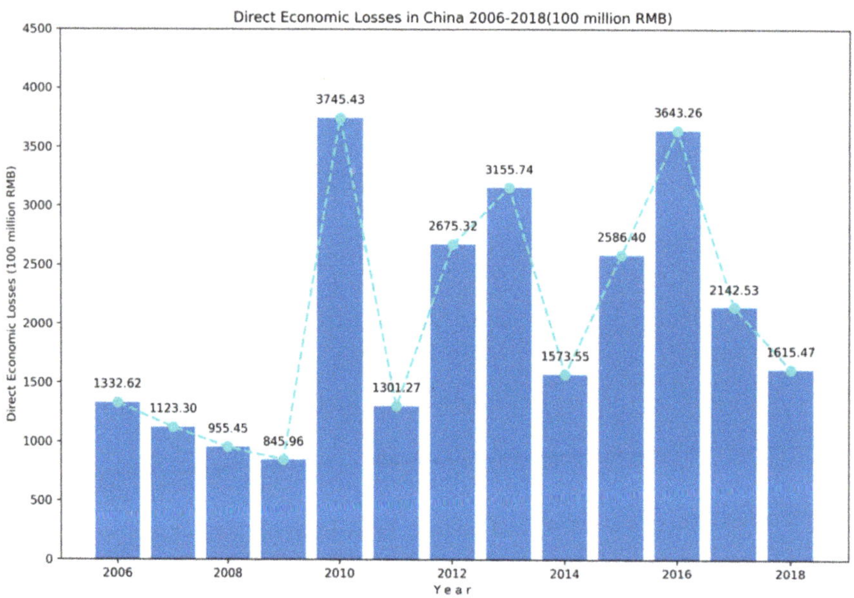

Figure 10. Direct economic losses (billion RMB).

As for the empirical analysis, by splitting the samples into training sets (data from 2006 to 2017) and the validation set (data in 2018), we gained the regression result of direct economic loss with GBR, as shown in Table 3. The comparison between the true value and predicted value is shown in Figure 11. It is notable that the deviation between the predicted direct economic losses and the true economic losses is relatively negligible. From Table 3, we can see that among 31 provinces in China, Heilongjiang, Shandong, Inner Mongolia, and Gansu provinces, which are heavily reliant on agriculture, suffered the most, with the proportions of the direct economic losses to provincial GDP at 1.73‰, 0.94‰, 0.86‰, and 0.75‰ respectively.

Table 3. Predicted direct economic losses.

Province	Predicted Direct Economic Losses (PDEL, billion RMB)	PDEL/ Provincial GDP (‰)	Province	Predicted Direct Economic Losses (PDEL, billion RMB)	PDEL/ Provincial GDP (‰)
Beijing	3.84	0.02	Hubei	2.23	0.15
Tianjin	0.2	0.05	Hunan	0.64	0.05
Hebei	1.28	0.14	Guangdong	32.55	0.11
Shanxi	0.16	0.04	Guangxi	3.92	0.32
Inner Mongolia	5.77	0.86	Hainan	0.64	0.23
Liaoning	2.71	0.42	Chongqing	3.79	0.03
Jilin	6.77	0.3	Sichuan	32.07	0.22
Heilongjiang	5.08	1.73	Guizhou	4.14	0.08
Shanghai	7.08	0.02	Yunnan	4.42	0.11
Jiangsu	3.98	0.08	Xizang	3.68	0.16
Zhejiang	2.08	0.04	Shaanxi	3.7	0.06
Anhui	4.63	0.39	Gansu	10.15	0.75
Fujian	2.32	0.07	Qinghai	2.25	0.17
Jiangxi	3.81	0.42	Ningxia	0.88	0.12
Shandong	20.04	0.94	Xinjiang	1.15	0.06
Henan	3.1	0.35			

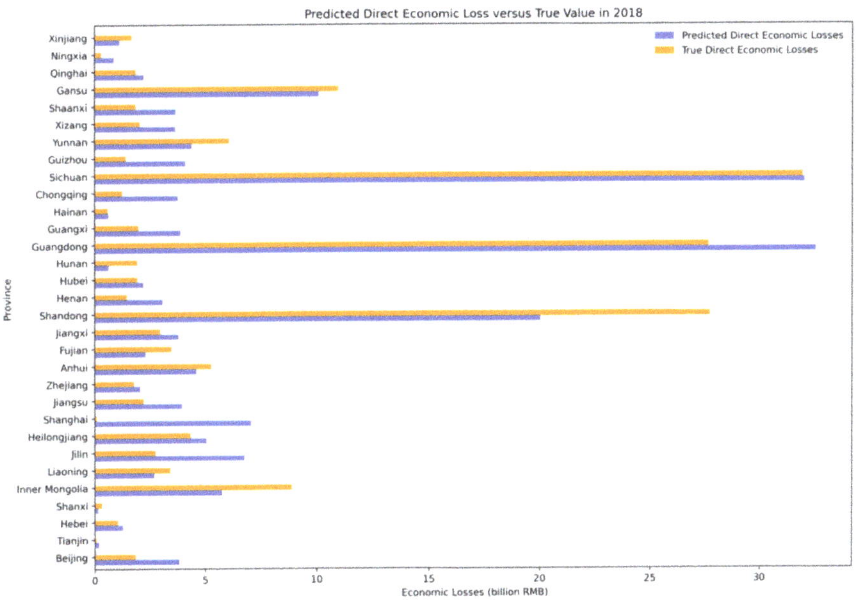

Figure 11. Predicted direct economic loss versus true value in 2018.

From the charts above, it can be seen that, for provinces suffering mild losses from natural disasters, the GBR model can provide relatively accurate prediction results in terms of the direct economic losses, while for the losses caused by extreme weather, despite the suddenness and uncertainty, the prediction results can still partially reflect the characteristics; reflecting the robustness and effectiveness of the model. It is also notable that the provincial direct economic losses caused by floods are geographically unevenly distributed. Gansu, Guangdong, Shandong, Inner Mongolia, and Yunnan suffered more direct economic losses, while Ningxia, Shanxi, and Tianjin suffered less. Such distribution can be attributed to the following reasons: (1) coastal regions are more susceptible to floods caused by typhoons and monsoon climates; (2) provinces with a high reliance on agriculture, such as Henan, Shandong, and Gansu, are more likely to be economically affected by floods as the agriculture sector is directly related to natural disasters; (3) provinces with more flood prevention facilities and infrastructure, such as Beijing, Tianjin, and Shanghai have strong resistance to floods.

3.2. Indirect Economic Loss Prediction

In this paper, the IO model was introduced to predict the indirect economic losses caused by floods in 42 sectors of 31 provinces in China. The precondition of the IO model is to apply the economic loss data of directly affected sectors. Since agriculture is the most directly affected sector among all of the 42 sectors, according to the Chinese National Bureau of Statistics, we took it as the directly affected sector to calculate the indirect losses of all of the sectors by using the industry linkage generated from the IO table [36].

For the direct economic loss of the agricultural sector, according to the descriptive statistical analysis (as shown in Figure 11), most of the floods in Northern China happened in autumn, while those in Southern China happened in summer. The sown areas in both northern China in autumn and southern China in summer were used for planting rice. According to China's Ministry of Agriculture and Rural Affairs of the People's Republic of China, the price of rice and the average yield per hectare are published in each season and can be denoted with P_i. To verify the model validation, we obtained the official average price in 2018, i.e., 2.56 RMB/kg, and the average yield per hectare, which was 2347.6667 kg/hectare. Thus, the average production per hectare is 2.56 * 2347.6667 = 6010.0267 RMB/hectare. Since the national standard for affected sown areas includes the sown areas with 10%, 30% and 80% reduced production, according to the bulletin of flood and drought disasters in China (2006–2019), we did a scenario analysis to reflect an average scenario of affected production with the assumption that 40% of the production was reduced due to flood disasters [37]. In other extreme situations, related calculations can be also conducted according to the specific situation. The estimated direct economic losses of the agriculture sector in 31 provinces in 2018 is shown, as follows, in Figure 12.

Further, to estimate the indirect economic losses of floods of other industries related to the agricultural industry, we analyzed the IO model. Although the 42 sectors are integrated and related, in the Input-Output table, they are split into different sectors without overlapping. Considering the analysis method of the input-output relationship from direct and indirect economic losses, we chose the direct economic loss of the directly affected sectors as the final product loss, where ΔY and $\Delta X - \Delta Y$ are defined the same as above. The complete consumption coefficient was introduced to obtain an accurate assessment of the indirect input loss of various sectors. The complete consumption coefficient is the quantity of the products in sector i that need to be consumed directly and indirectly to produce final products per unit of sector j. As defined above, B was used to represent the complete consumption coefficient matrix. Accordingly, $B = (I - A)^{-1} - I$ is the relationship between the complete consumption coefficient and the direct consumption coefficient. Thus, B is obtained by the direct consumption coefficient matrix of each one given by the 42 sectors in the IO table of nationwide provinces in 2017. In this paper, we investigated the most directly hit industry—the agricultural sector—which has an indirect effect on all industries. Here we only demonstrated the top six most affected industries' complete consumption coefficients of agriculture out of 42 industries, as shown in Table 4. The figure indicates the correlation between other sectors and the agricultural sector, which can be treated as the input for the following calculations.

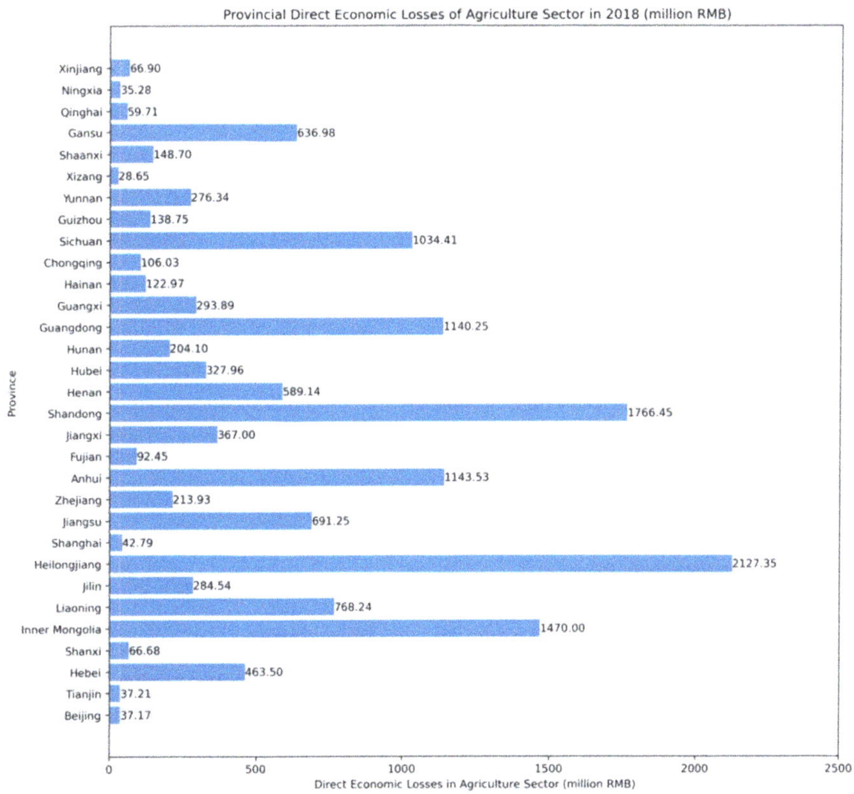

Figure 12. Direct economic losses of agriculture sector in 2018.

Table 4. Matrix of cumulative input coefficients of agriculture, forestry, animal husbandry and fishery sectors to other sectors.

Province \ Sector	Agriculture Forestry Animal Husbandry and Fishery	Food and Tobacco Processing	Manufacture of Chemical Products	Smelting and Processing of Metals	Repair of Metal Products, Machinery and Equipment	Wholesale and Retail Trades	Real Estate
Beijing	0.28	0.28	0.23	0.13	0.18	0.11	0.22
Tianjin	0.24	0.26	0.20	0.06	0.06	0.14	0.07
Hebei	0.19	0.19	0.11	0.02	0.05	0.06	0.02
Shanxi	0.16	0.06	0.20	0.03	0.04	0.04	0.01
Inner Mongolia	0.22	0.12	0.13	0.02	0.04	0.09	0.03
Liaoning	0.33	0.18	0.21	0.04	0.05	0.05	0.02
Jilin	0.30	0.18	0.15	0.05	0.03	0.06	0.04
Heilongjiang	0.32	0.10	0.13	0.01	0.03	0.05	0.02
Shanghai	0.24	0.21	0.25	0.04	0.06	0.14	0.10
Jiangsu	0.20	0.13	0.19	0.04	0.04	0.06	0.03
Zhejiang	0.09	0.11	0.22	0.03	0.08	0.07	0.02
Anhui	0.23	0.16	0.18	0.04	0.04	0.05	0.04
Fujian	0.03	0.02	0.22	0.00	0.03	0.00	0.01
Jiangxi	0.04	0.02	0.35	0.00	0.02	0.02	0.02
Shandong	0.03	0.06	0.21	0.05	0.11	0.18	0.05
Henan	0.03	0.01	0.29	0.03	0.05	0.06	0.02
Hubei	0.02	0.01	0.08	0.07	0.04	0.05	0.02
Hunan	0.22	0.18	0.17	0.02	0.04	0.04	0.02
Guangdong	0.21	0.26	0.13	0.04	0.05	0.04	0.02
Guangxi	0.02	0.00	0.09	0.00	0.01	0.00	0.01
Hainan	0.10	0.12	0.20	0.01	0.02	0.10	0.05
Chongqing	0.09	0.09	0.10	0.02	0.05	0.04	0.03

Table 4. *Cont.*

Province \ Sector	Agriculture Forestry Animal Husbandry and Fishery	Food and Tobacco Processing	Manufacture of Chemical Products	Smelting and Processing of Metals	Repair of Metal Products, Machinery and Equipment	Wholesale and Retail Trades	Real Estate
Sichuan	0.22	0.14	0.20	0.02	0.02	0.04	0.02
Guizhou	0.20	0.04	0.16	0.02	0.05	0.08	0.01
Yunnan	0.21	0.07	0.15	0.02	0.04	0.04	0.03
Xizang	0.25	0.17	0.15	0.01	0.03	0.05	0.02
Shaanxi	0.17	0.11	0.21	0.04	0.03	0.05	0.03
Gansu	0.19	0.06	0.17	0.02	0.06	0.06	0.02
Qinghai	0.16	0.14	0.14	0.02	0.07	0.05	0.01
Ningxia	0.20	0.14	0.19	0.02	0.09	0.09	0.03
Xinjiang	0.27	0.09	0.29	0.01	0.03	0.07	0.01

Based on the above matrix, taking the computation of indirect economic loss of agriculture, forestry, animal husbandry and fishery to other sectors, the total indirect economic loss of the 31 provinces was calculated, as shown in Figure 13. It can be seen that Shandong, Henan, Heilongjiang, and Inner Mongolia, all of which are provinces that are reliant on agriculture, suffered the most indirect economic losses among all provinces, with the predicted indirect economic losses reaching 6235.55 million RMB, 1770.49 million RMB, 2227.45 million RMB, and 1389.13 million RMB, respectively. Thus, to prevent unnecessary economic losses and secondary disasters, these provinces should be given more attention.

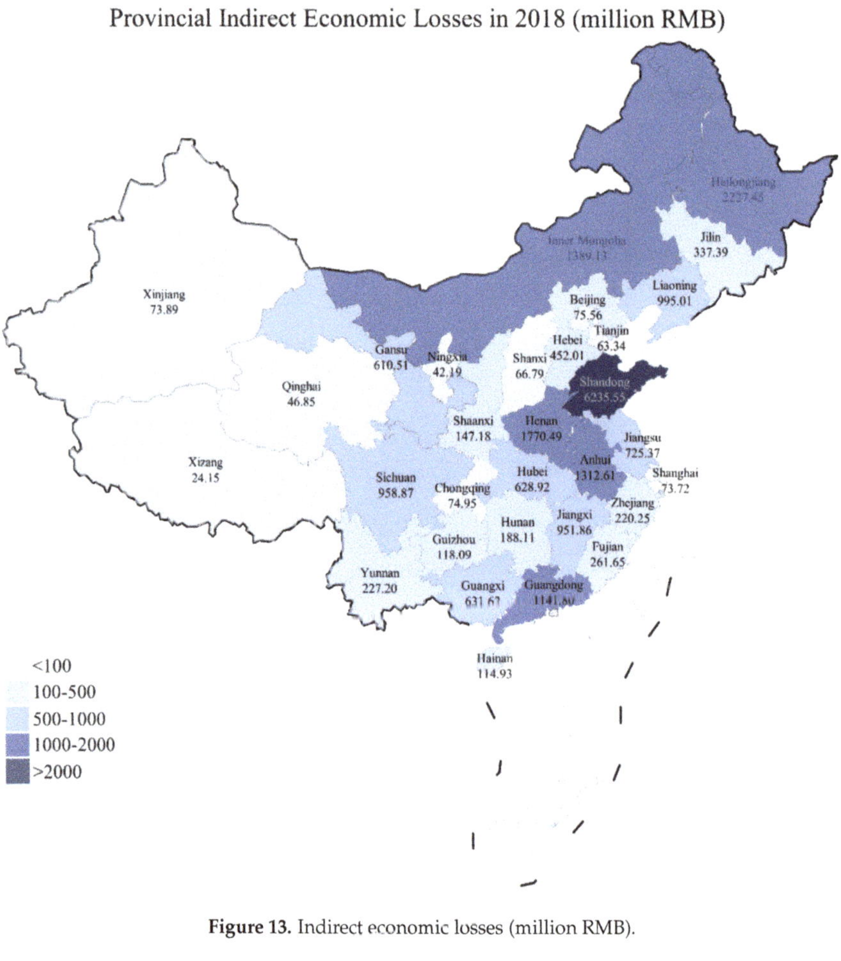

Figure 13. Indirect economic losses (million RMB).

4. Conclusions

This paper proposes an effective prediction model consisting of a cutting-edge machine learning regression model and a traditional economic model, i.e., the GBR-IO model. The performance of the GBR-IO model used in the paper is superior to other models because it improves the prediction effectiveness, reduces the issue of overfitting and combines inter-disciplinary methods. The prediction model obtains outstanding results of direct economic losses on datasets in 31 provinces (excluding Hong Kong, Macau, and Taiwan) in China from 2006 to 2018, with a goodness-of-fit of 90%. Based on the predicted direct economic losses, we obtained the indirect economic losses by using the Input-Output model. Compared with previous studies, the GBR-IO model can predict regional direct economic losses and indirect economic losses pre-disaster or at the beginning of a flooding episode with effectiveness and efficiency. Further, the GBR-IO model has a high generalization ability, which can be applied to other countries, especially to those with small datasets at the minimum requirement of 80 pieces.

However, there is still room for improvement of the pre-disaster prediction model. First, as flood forecasting systems and the numerical simulation technology advance in meteorology, the variables used in the model could be estimated more accurately before a disaster happens. Second, the limitation of the datasets results in a failure to categorize floods. If floods are categorized into river floods, flash floods, and drainage problem floods, the accuracy of the model can be further improved. Third, since geographical conditions are not included in the system, global climate models (GCMs) and regional climate models (RCMs) can be further integrated into the system to boost performance. With the above improvements, the model performance can be further enhanced.

According to the prediction model, it can be concluded that positively correlated indicators include reservoir loss, sown area with 10% reduced production, casualties, road disruption, annual anomaly percentage, and daily maximum precipitation while the length of city sewage pipes, reservoir capacity and annual rainy days are negatively correlated with economic losses. From the empirical case based on datasets in China, we can also conclude that from the proportion of direct economic losses to provincial GDP, it is notable that the impact of flooding on the economy is relatively significant. Among the 31 provinces in China, especially Gansu, Inner Mongolia, Shandong, and Heilongjiang, the direct economic losses even approached or exceeded one-thousandth of the province's GDP.

Thus, some policy implications can be drawn from the results. Since the GBR-IO model has a strong generalization ability, regional flooding databases are encouraged to be established in order to improve the prediction accuracy with more data. Although there is a time lag of reservoir capacity, giving priority to reservoir construction and city drainage capacity can mitigate the economic impact of flooding. Policymakers could also pay more attention to those regions that are heavily reliant on agriculture as they are more vulnerable to flood disasters. Further, if the accuracy of meteorological forecasts can be improved, more effective measures can be taken in advance.

Author Contributions: Conceptualization, A.C. and J.L.; methodology, A.C. and J.L; software, J.L.; validation, A.C. and J.L.; formal analysis, H.L.; investigation, H.L.; resources, A.C. and J.L; data curation, J.L.; writing—original draft preparation, A.C.; writing—review and editing, A.C.; visualization, A.C. and J.L.; supervision, S.Y.; project administration, S.Y. All authors have read and agreed to the published version of the manuscript.

Funding: This research received no external funding.

Institutional Review Board Statement: Not applicable.

Informed Consent Statement: Not applicable.

Data Availability Statement: The Meteorological datasets generated for this study are available from China Meteorological Administration. The socio-economic datasets presented in the study are available at Ministry of Water Resources of the People's Republic of China, China statistical Yearbook

(2006–2018). The Input and Output table of provinces in 2017 is available at the National Bureau of Statistics.

Conflicts of Interest: The authors declare no conflict of interest.

References

1. Climate Change Widespread, Rapid, and Intensifying—IPCC. Available online: https://public.wmo.int/en/media/press-release/climate-change-widespread-rapid-and-intensifying-%E2%80%93-ipcc (accessed on 26 August 2021).
2. Huang, X.; Tan, H.; Zhou, J.; Yang, T.; Benjamin, A.; Wen, S.W.; Li, S.; Liu, A.; Li, X.; Fen, S. Flood hazard in Hunan province of China: An economic loss analysis. *Nat. Hazards* **2008**, *47*, 65–73. [CrossRef]
3. Wu, X.; Guo, J. A new economic loss assessment system for urban severe rainfall and flooding disasters based on big data fusion. In *Economic Impacts and Emergency Management of Disasters in China*; Springer: Berlin/Heidelberg, Germany, 2021; pp. 259–287.
4. Wang, D.; Huang, C.; Mai, B. To Facilitate the Advance of Risk Analysis and Crisis Response in China. *Environ. Res.* **2016**, *148*, 547–549. [CrossRef]
5. Kreimer, A. Social and economic impacts of natural disasters. *Int. Geol. Rev.* **2001**, *43*, 401–405. [CrossRef]
6. Das, S.; Lee, R. A nontraditional methodology for flood stage-damage calculations 1. *JAWRA J. Am. Water Resour. Assoc.* **1988**, *24*, 1263–1272. [CrossRef]
7. Krzysztofowicz, R.; Davis, D.R. Category-unit loss functions for flood forecast-response system evaluation. *Water Resour. Res.* **1983**, *19*, 1476–1480. [CrossRef]
8. Huabin, W.; Gangjun, L.; Weiya, X.; Gonghui, W. GIS-based landslide hazard assessment: An overview. *Prog. Phys. Geogr.* **2005**, *29*, 548–567. [CrossRef]
9. Sanders, B.F.; Schubert, J.E.; Detwiler, R.L. ParBreZo: A parallel, unstructured grid, Godunov-type, shallow-water code for high-resolution flood inundation modeling at the regional scale. *Adv. Water Resour.* **2010**, *33*, 1456–1467. [CrossRef]
10. Koç, G.; Natho, S.; Thieken, A.H. Estimating direct economic impacts of severe flood events in Turkey (2015–2020). *Int. J. Disaster Risk Reduct.* **2021**, *58*, 102222. [CrossRef]
11. Sun, H.; Wang, J.; Ye, W. A data augmentation-based evaluation system for regional direct economic losses of storm surge disasters. *Int. J. Environ. Res. Public Health* **2021**, *18*, 2918. [CrossRef]
12. Chen, T.; Guestrin, C. Xgboost: A scalable tree boosting system. In Proceedings of the 22nd Acm Sigkdd International Conference on Knowledge Discovery and Data Mining, San Francisco, CA, USA, 13–17 August 2016; pp. 785–794.
13. Tian, W.; Wu, J.; Cui, H.; Hu, T. Drought Prediction Based on Feature-Based Transfer Learning and Time Series Imaging. *IEEE Access* **2021**, *9*, 101454–101468. [CrossRef]
14. Sagi, O.; Rokach, L. Ensemble learning: A survey. *Wiley Interdiscip. Rev. Data Min. Knowl. Discov.* **2018**, *8*, e1249. [CrossRef]
15. Li, J.; Crawford-Brown, D.; Syddall, M.; Guan, D. Modeling imbalanced economic recovery following a natural disaster using input-output analysis. *Risk Anal.* **2013**, *33*, 1908–1923. [CrossRef] [PubMed]
16. Narayan, P.K. Macroeconomic impact of natural disasters on a small island economy: Evidence from a CGE model. *Appl. Econ. Lett.* **2003**, *10*, 721–723. [CrossRef]
17. Rose, A. Input-output economics and computable general equilibrium models. *Struct. Chang. Econ. Dyn.* **1995**, *6*, 295–304. [CrossRef]
18. You, S.; Wang, H.; Zhang, M.; Song, H.; Xu, X.; Lai, Y. Assessment of monthly economic losses in Wuhan under the lockdown against COVID-19. *Humanit. Soc. Sci. Commun.* **2020**, *7*, 1–12. [CrossRef]
19. Organisation, W.M. Heavy Rains and Flooding Hit Large Parts of Asia. Available online: https://public.wmo.int/en/media/news/heavy-rains-and-flooding-hit-large-parts-of-asia (accessed on 18 September 2021).
20. Dutta, D.; Herath, S. Trend of floods in Asia and flood risk management with integrated river basin approach. In Proceedings of the 2nd international conference of Asia-Pacific hydrology and water resources Association, Singapore, 5–9 June 2004; pp. 55–63.
21. Death Toll from Floods in China's Henan Province Rises to 302. Available online: https://www.reuters.com/world/china/death-toll-flooding-chinas-henan-province-rises-302-2021-08-02/ (accessed on 30 August 2021).
22. The Input and Output Table with 42 Sectors. 2017. Available online: https://data.stats.gov.cn/ifnormal.htm?u=/files/html/quickSearch/trcc/trcc01.html&h=740 (accessed on 18 September 2021).
23. Elith, J.; Leathwick, J.R.; Hastie, T. A working guide to boosted regression trees. *J. Anim. Ecol.* **2008**, *77*, 802–813. [CrossRef]
24. Friedman, J.H. Greedy function approximation: A gradient boosting machine. *Ann. Stat.* **2001**, *29*, 1189–1232. [CrossRef]
25. Mason, L.; Baxter, J.; Bartlett, P.; Frean, M. Boosting algorithms as gradient descent in function space. *Proc. NIPS* **1999**, *12*, 512–518.
26. Breiman, L.; Friedman, J.H.; Olshen, R.A.; Stone, C.J. *Classification and Regression Trees*; Routledge: Oxfordshire, UK, 2017.
27. Prasad, A.M.; Iverson, L.R.; Liaw, A. Newer classification and regression tree techniques: Bagging and random forests for ecological prediction. *Ecosystems* **2006**, *9*, 181–199. [CrossRef]
28. Ding, C.; Wu, X.; Yu, G.; Wang, Y. A gradient boosting logit model to investigate driver's stop-or-run behavior at signalized intersections using high-resolution traffic data. *Transp. Res. Part C Emerg. Technol.* **2016**, *72*, 225–238. [CrossRef]
29. Li, H.; Sun, J.; Wu, J. Predicting business failure using classification and regression tree: An empirical comparison with popular classical statistical methods and top classification mining methods. *Expert Syst. Appl.* **2010**, *37*, 5895–5904. [CrossRef]
30. Sun, R.; Gong, Z.; Gao, G.; Shah, A.A. Comparative analysis of Multi-Criteria Decision-Making methods for flood disaster risk in the Yangtze River Delta. *Int. J. Disaster Risk Reduct.* **2020**, *51*, 101768. [CrossRef]

31. Doan, T.; Kalita, J. Selecting machine learning algorithms using regression models. In Proceedings of the 2015 IEEE International Conference on Data Mining Workshop (ICDMW), Atlantic City, NJ, USA, 14–17 November 2015; pp. 1498–1505.
32. Halko, N.; Martinsson, P.-G.; Tropp, J.A. Finding structure with randomness: Probabilistic algorithms for constructing approximate matrix decompositions. *SIAM Rev.* **2011**, *53*, 217–288. [CrossRef]
33. Perlich, C.; Provost, F.; Simonoff, J. Tree Induction vs. Logistic Regression: A learning-Curve Analysis. *J. Mach. Learn. Res.* **2003**, *4*, 211–255.
34. China Flooding Causes Worst Death Toll in Decade. Available online: https://www.theguardian.com/world/2010/jul/21/china-flooding-worst-decade (accessed on 18 September 2021).
35. China's Historic Floods Are among the Earth's Most Costly Weather-Related Disasters. Available online: https://www.climatesignals.org/events/china-floods-june-july-2016 (accessed on 19 September 2021).
36. Wang, H.; Wang, Z.; Dong, Y.; Chang, R.; Xu, C.; Yu, X.; Zhang, S.; Tsamlag, L.; Shang, M.; Huang, J. Phase-adjusted estimation of the number of coronavirus disease 2019 cases in Wuhan, China. *Cell Discov.* **2020**, *6*, 1–8. [CrossRef]
37. Bulletin of Flood and Drought Disasters in China (2006–2019). Available online: http://www.mwr.gov.cn/sj/tjgb/zgshzhgb/ (accessed on 19 September 2021).

atmosphere

Article

Study on the Agricultural Air Pollution Aggravated by the Rural Labor Migration

Ying Liu [1], Shibing You [1,*], Nan Li [2], Junsheng Fang [1], Jie Jia [1], Xuesong Li [1,*] and Jingru Ren [3]

1 School of Economics and Management, Wuhan University, Wuhan 430072, China; YingLiuYL@whu.edu.cn (Y.L.); jasonfang@whu.edu.cn (J.F.); jiejia@whu.edu.cn (J.J.)
2 China Bohai Bank Co., Ltd., Tianjin 300012, China; nan05.li@cbhb.com.cn
3 Shenzhen Institute of Information Technology, Shenzhen 518172, China; 2019000218@sziit.edu.cn
* Correspondence: sbyou@whu.edu.cn (S.Y.); xsli@whu.edu.cn (X.L.)

Abstract: In recent years, air pollution has received serious concerns from researchers, media, and the public sectors, but air pollution from agricultural production activities has not received enough attention. This paper focuses on agricultural air pollution in central China, which is aggravated by the ongoing rural labor migration trend. With a set of panel data released from Hubei and Hunan provinces in China, we adopt the mediating effect model to explore the relationship between rural labor migration and air pollution caused by agricultural activity in China. First, we use the inventory analysis method and principal component analysis method to calculate the comprehensive index of the air pollution of agriculture in 152 counties and districts from Hubei and Hunan provinces, and we empirically test the impact of labor migration on air pollution with a mediating effect model as well as carry out regional heterogeneity analysis on the pollution effect of these two provinces mentioned above. The analysis above indicates that the increase of labor migration has intensified the comprehensive index of air pollution caused by agricultural activity by changing the supply of labor force in the agricultural sector, the budget line of rural residents, and the scale of agricultural production and crop planting structure, but there is a difference in the indirect total effect between the two provinces mentioned above according to our regional heterogeneity analysis. This study is a necessary extension to studies on alleviating and controlling air pollution in China.

Keywords: agricultural air pollution; labor migration; mediation effect; income effect; economy of scale

Citation: Liu, Y.; You, S.; Li, N.; Fang, J.; Jia, J.; Li, X.; Ren, J. Study on the Agricultural Air Pollution Aggravated by the Rural Labor Migration. *Atmosphere* **2022**, *13*, 174. https://doi.org/10.3390/atmos13020174

Academic Editors: Zengyun Hu, Xuguang Tang and Qinchuan Xin

Received: 30 December 2021
Accepted: 20 January 2022
Published: 21 January 2022

1. Introduction

The current COVID-19 pandemic interrupted many production and industrial activities, but the agriculture industry seems to have suffered less. Evidence from Lovarelli's research suggests that during the quarantine, the increase in ammonia emissions is mainly due to agricultural production activities, as there was a significant reduction in transportation and industrial activities [1]. Air pollution control in rural areas has fallen into a state of considerable stress and crucial difficulty regarding protecting the environment and ecosystem in China, which has a direct influence on the process of China's sustainable development. There are two sources of air pollution: human activities and natural processes. Agricultural production is one of the typical human activities that brings air pollution. Hellin (2021) has evaluated the impact of agricultural production on the atmosphere in India, which confirmed that agricultural production will pollute the atmosphere [2].

Agricultural activities include cultivating the soil, producing crops, and raising livestock, most of which will pollute the atmosphere. Crop farming pollutes the atmosphere when straw is burned, pesticides are sprayed, and fertilizer is applied. In China, a large amount of straw is burned artificially in the open air after harvest, causing serious air pollution. The emissions from biomass combustion are considered to be the sources of greenhouse gases such as carbon dioxide, methane, and nitric oxide. In addition, aerosols emitted from biomass burning have high mass concentrations of $PM_{2.5}$, organic carbon, elemental carbon, and potassium [3]. In Jiaodong Peninsula, the North China Plain, East China, and other areas, the particulate matter produced by farmland incineration of agricultural residues accounts for more than 35% of the aerosol optical thickness and more than 60% in some areas of Shandong, Henan, and Jiangsu provinces in China [4]. Agriculture is the second largest carbon emission source after power generation and waste gas, accounting for 14% of the total emissions of human activities in the world, according to the Intergovernmental Panel on Climate Change (IPCC) fourth assessment report in 2007. Furthermore, as a traditional agricultural country, China's agriculture has produced nearly 500 million tons of carbon dioxide emissions [5]. Wang (2022) introduced the ecosystem health risk index (EHR) to evaluate the impact of agricultural pollutants on the health of natural ecosystems and found that in the process of agricultural production related to nitrogen fertilizer, air pollutants (including NH_3 and N_2O) cause much more damage to ecosystem health than soil or water pollutants [6]. There is also evidence from Stacy's research (2011) that the livestock industry will also exacerbate the degree of air pollution [7]. Huang (2020) found that livestock and fertilizer application are the two main NH_3 emission sources that will exacerbate air pollution [8].

This paper has great significance regarding balancing urbanization and air pollution control in China. Over the past forty years, there has been a large scale of rural labor migration from the traditional agricultural sector to various types of non-farm employment in China. Guo's study show that they are also one of the most important driving forces to the emerging urbanization and rapid economic growth [9]. However, the ecosystem cannot be neglected in order to maintain the fast pace of the rural economy. Air pollution control in rural areas has become a crucial difficulty to the environment protection and sustainable development for China.

The vast peasantry are the major stakeholders of rural ecological environment status, whose economic and social behavior will also inevitably have a profound impact on agricultural air pollution. The continuous large-scale labor migration provides the internal impetus for the sustainable growth of the regional economy, but in Xu's study, it also affects the household land use arrangements [10], afterwards exerting an influence on agricultural air pollution.

2. Theoretical Mechanism

Agricultural air pollution refers to air pollutants generated during agricultural production activities, such as straw burning, application of volatile fertilizers, rural livestock feces, etc. Accordingly, by what means does the migration of the rural labor force change the mode of agricultural activities, therefore affecting the agricultural air pollution? Summarizing the currently available academic research, there are basically four paths listed as follows:

First, Bhandari et al. (2016) and Tian et al. (2020) suggested a substitution relationship between farming supplies such as chemical fertilizers and labor force in agricultural production [11,12]. Bounded by a limited labor force, farmers tend to increase their usage of farming supplies, especially chemical fertilizers to offset the shortage of labor force and gain the same amount of output. Likewise, Luan (2017) conducted an empirical study on the relationship between the outflow of rural labor force and the application of agricultural chemical fertilizer based on the panel data of major grain-growing provinces in China from 2004 to 2015. The regression results confirmed the substitution relationship between rural labor migration and fertilization [13]. Hence, there is the first hypothesis of this paper.

Hypothesis 1. *Rural labor migration leads to the reduction of labor force supply for agricultural production, which increases the input intensity of farming supplies as its substitution factors and then aggravates the agricultural air pollution level.*

Second, there is the income effect. Changes in income shift the consumer choice, increasing or decreasing the consumption of consumables and other agricultural goods. According to the consumer equilibrium theory, under the condition of unchanged consumer utility preferences, income growth will expand the income constraints, thus affecting the demand for certain consumption. As a matter of fact, with different utility preferences, the impact of income growth on different variety of goods is also different.

Changes in income can also have an impact on labor supply. In the labor market, workers provide different amounts of work according to their income and leisure estimates at different wages, and the results are shown in two opposite dimensions. On one hand, a growth of wage rate increases the opportunity cost of leisure, which can result in a rise of working hours; on the other hand, a growth of wage rate increases the savings and consumption level of the workers, which then leads to an increase of the utility demand preference for leisure, thus bringing a reduction of the working time.

As for the influence of non-farm income on agricultural production, some scholars believed that the increase of non-farm income expanded the budget for their family members who were left behind in rural areas, providing them with the economic capacity of purchasing more farming supplies, resulting in applying more fertilizers (Goodwin and Mishra, 2004; Luan et al., 2016) [14,15]. On the contrary, some other scholars believe that the family's increased income is actually not for farming supplies but for household fixed assets investment, education, and life consumption. So, the increase in income does not affect the chemical fertilizers usage and does not exacerbate air pollution (Cheng et al., 2015) [16]. Meanwhile, the non-farm income increased opportunity costs for farming, leading to extensively rough agricultural management and even abandoned farmland, resulting in a decrease of various farming supplies including chemical fertilizers (Carletto et al., 2013; Qiu, 2018) [17,18].

Hypothesis 2. *The migration of rural labor force leads to an increase of non-farm income. On one hand, it can increase the purchase and application of farming supplies because of the alleviating financial constraints, aggravating agricultural air pollution; on the other hand, it also increases the opportunity cost for farming, encouraging discarded farmland, which alleviates agricultural air pollution. The overall effect is uncertain.*

Third, there is economy of scale. The transfer of land use rights can affect agricultural production efficiency by reducing the degree of land division and expanding the business scale, and it can also produce a "leveling effect", that is, the transfer of land use right from farmers with less land to farmers with more land. Based on the survey data of Southern Henan, a major grain-producing area in China, Gao (2020) conducted an empirical study on the impact of farmers' migration to rural land transfer using the threshold model and CHARLS 2015 data. This study empirically analyzed the impact of the labor migration scale on the ratio of rural transferred land to total land area and concluded that the migration of rural labor force has affected the turnover speed of farmland circulation and the concentration of agricultural operation [19]. The migration of labor force and the increase of non-farm income reduce the dependence of rural families on farmland, promoting the transfer of use rights of agricultural land, making extensive running possible and bringing more air pollution. Therefore, we have the third hypothesis of this paper:

Hypothesis 3. *The migration of rural labor force accelerates the transfer of use rights of land and expands the production scale, thus improving the productivity efficiency of farming supplies, reducing the average application intensity, and then contributing to aggregating the agricultural air pollution.*

Fourth, there is the effect of crop planting structure. Two different views appeared related to the impact of rural labor migration on the crop planting structure (proportion of cash crop planting area). From the perspective of the farmers' own families, Zhong et al. (2016) believed that the migration of the labor force reduces the core labor force of rural families, leading to the trend of farmers' aging and part-time agriculture. They prefer mechanized agriculture because mechanized agriculture is very convenient [20]. However, from the perspective of transfer of use rights of land, Wan (2014) believed that under the stimulation of maximizing short-term benefits, land contractors were more willing to shift from planting "safety first" food crops to cash crops with higher planting density, more complex compounds, and basically higher yield for production. It makes it difficult to control the application amount of pesticides, chemical fertilizers, and various additives, thus aggravating agricultural air pollution [21]. Consequently, there we have the fourth hypothesis of our research.

Hypothesis 4. *There are two opposite dependent paths of how rural labor migration affects the agricultural crop planting structure, and therefore, the overall impact on agricultural air pollution is temporarily uncertain.*

In summary of all the assumptions above, the substitution effect and economy of scale effect aggravate the pollution level of agricultural air; meanwhile, the influence of income effect and crop planting structure effect is temporarily uncertain. Therefore, the overall impact of rural labor force migration on agricultural air pollution needs to be tested.

3. Data Description and Research Methodology

3.1. Data Description

This paper selected the rural agricultural data from every county and district of Hubei and Hunan provinces from 2007 to 2017 in China as the data sample. All the underlying data involved are from the Statistical Yearbook, Rural Statistical Yearbook, and China County Statistical Yearbook of each province and city. Based on the availability of data, we remove outliers and deal with the missing values of some data with the method of line deduction in data from 69 counties and districts in Hubei Province and 83 counties and districts in Hunan Province. Finally, 152 counties and districts were retained as valid samples. To eliminate the possible heteroscedasticity of time-series data, some variables in this paper use natural logarithms in empirical analysis.

According to the research purpose, the comprehensive index of agricultural air pollution (AAPI) is taken as the prediction variable. The main explanatory variables are the scale of labor migration, per-capita non-farm income, farmland scale, and the cash crop ratio. In addition, other factors are considered as control variables, including the size of the rural population, the output of agricultural production, agricultural irrigation area, and the mechanical power for agricultural use.

Table 1 below clearly demonstrates the setting and statistical description of the variables. Among them, the minimum and maximum values of the core variable labor migration are 3.438 and 87.698, respectively, with a standard deviation of 14.186. It reflects that the data samples of rural labor migration have great differences and changes in the region. The comprehensive index of agricultural air pollution was established based on principal component analysis (PCA). The minimum and maximum value of the comprehensive agricultural air pollution index are 0.445 and 34.686, respectively, with a standard deviation of 6.638. Large variations in variables can also meet research needs.

Figure 1 demonstrates a fitting diagram between the comprehensive index of agricultural air pollution (AAPI) and rural labor force migration in Hubei and Hunan provinces. It is obvious that there is a positive correlation between labor migration and agricultural air pollution, indicating that the higher the migration scale of rural labor force, the more serious the air pollution, providing an intuition to the preliminary research.

3.2. Research Methodology

3.2.1. Inventory Analysis Method

The inventory analysis method is commonly used in environmental pollution research. First, we determine the pollutant generating unit, and then, we use the pollutant producing coefficients and accounting formula of the pollutant producing unit to indirectly calculate the amount of agricultural pollution. The pollution production unit, pollution production coefficient, and accounting model constitute the core considerations of this method. In this paper, chemical fertilizers, straw, and mammal livestock are selected as the pollution production unit, referring to the study of Chen et al. (2006) and Fan (2014) [22,23].

Table 1. Variables setting and statistical description.

Variable	Indicator	Unit	Mean	Standard Deviation	Min	Max
Rural labor migration (LM)	Rural non-farm payroll employments/total rural labor force	%	48.340	14.186	3.438	87.698
Non-farm income (NI)	Per-capita non-farm income of rural residents	hundred US dollars	12.689	7.351	1.413	46.880
Farmland scale (FS)	Total cultivated farmland/total number of agricultural employments	0.1 hectare per person	7.112	3.435	2.224	24.697
Cash crop ratio (CCR)	Cash crop sown area/total sown area	%	41.724	9.819	7.468	69.393
Rural population (Rpop)	Rural resident population	ten thousand	40.799	22.518	2.330	129.250
Agricultural output (Aoutput)	Gross output of agricultural production at constant price	100 million	47.687	33.875	1.274	208.257
Total mechanical power (Mpower)	Total energy consumption of agricultural machinery	10 megawatts	47.819	34.165	0.679	178.249
Agricultural irrigation area (Irrigation)	Total area of cultivated farmland for agricultural irrigation	thousands of hectares	32.852	26.097	0.770	206.390

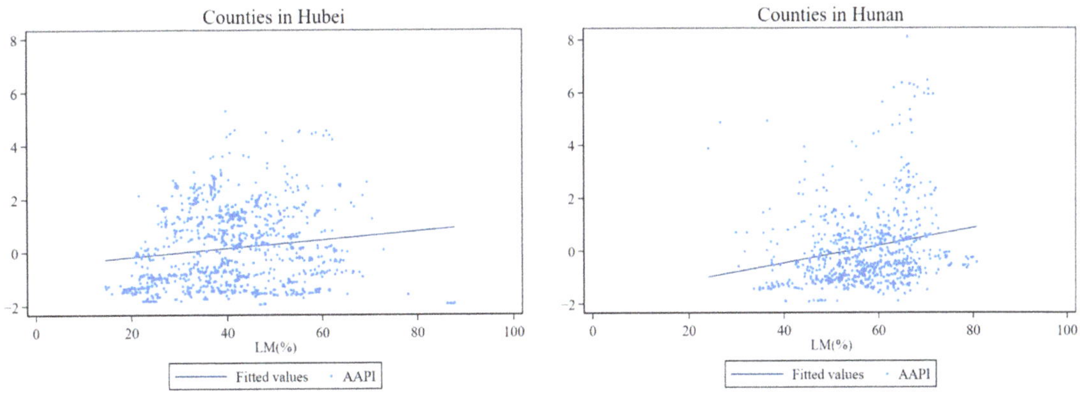

Figure 1. AAPI and rural labor migration (LM, %) of counties in Hubei and Hunan.

According to the above text, we know that agricultural activity affects air pollutants mainly by crop farming and livestock breeding. Hence, it is reasonable to choose an index related to crop farming and livestock breeding. We use nitrogenous fertilizer and application to describe the chemical fertilizer part because nitrogenous fertilizer is one of the main sources of air pollutants such as N_2O and NH_3. Yichao Wang (2020) calculated cropland N_2O and NH_3 emission factors from fertilized farmland, and we adapt this index [6]. Straw residue burning is one of the most important polluting agricultural activities that affects the atmosphere, and we choose the total output of rice, maize, wheat, beans, and tubers and other air pollutants emitted via straw burning to calculate air pollutants emitted via straw burning. The grain–straw ratio and straw open burning ratio in Hubei and Hunan are from Cao's research in 2005 [24]. Livestock breeding also contaminates air quality, and ammonia (NH_3) emissions, the majority of which arise from livestock production, are linked to the high concentration of $PM_{2.5}$ and degraded air quality in China, according to Haodan Wang's study in 2021. The NH_3 emitted ratio from livestock breeding is from Wang's paper and livestock breeding total nitrogen are selected from the announcements published by the Ministry of Ecology and Environment of China [25,26]. Other data are collected from China rural statistical yearbook, and China County Statistical Yearbook. The specific accounting method is shown in Table 2.

Table 2. Accounting method of pollution production unit.

Agricultural Activity	Category		Accounting Method	Result
Crop farming	chemical fertilizer	rice maize wheat	The amount of fertilizer air pollutants = crop planting area(ha) × cropland emission factors (N_2O and NH_3; kg/ha)	Total agricultural emission of multiple air pollutants (kg)
	straw	rice maize wheat beans tubers	The amount of air pollutants emitted via straw burning = total crop output × grain–straw ratio^{-1} × straw open burning ratio (of Hubei and Hunan, estimated) × emission factors (CH_4,NH_3,SO_2,NO_2,$PM_{2.5}$; g/kg)/1000	
Livestock breeding	livestock waste	hog cattle fowl	The amount of livestock air pollutants = livestock breeding numbers × livestock breeding total nitrogen (TN, g/day per animal) coefficient × livestock output average duration (day) × NH_3 emitted ratio/1000	

3.2.2. Principal Component Analysis (PCA) on the Agricultural Air Pollution Comprehensive Index (AAPI) as the Dependent Variable in Subsequent Regression Model

Principal component analysis is a statistical method. Through the orthogonal transform, a set of possible relevant variables was transformed into a set of uncorrelated variables. The transformed group of variables is called the principal component.

In order to simplify the steps of regression, the agricultural air pollutants comprehensive index (AAPI) is calculated by synthesizing the air pollutants released by chemical fertilizer (CF), air pollutants emitted via straw burning (ST), and air pollutants released by livestock breeding (LS) using principal component analysis (PCA). Based on the principle of a feature root greater than 1, a principal component, indicated as "comp" reflecting the comprehensive level of agricultural air pollutants of the total sample, is extracted, with a feature root of 2.595 and a variance interpretation of 86.51%, and the calculation function is shown as:

$$\text{comp} = 0.602 \times \text{CF} + 0.583 \times \text{ST} + 0.546 \times \text{LS}.$$

3.2.3. Causal Steps Approach in Mediating Effect Model

Mediating effect models are often used to study the complex process and mechanism of the explanatory variables on the predicted variables. Mediation effects are indirect effects and can be modeled using structural equation models. The main methods to test the mediation effect include the product of coefficient approach, causal steps regression test, and significance test of difference (Wen et al., 2004) [27]. To examine the mechanism by which the rural labor migration affects agricultural production activities, thereby affecting agricultural air pollutants, this paper selected the causal steps approach, a more commonly used method by domestic scholars, to test the mediation effect in this scenario. Under the assumptions of the hypothesis mentioned previously, the following five regression equations are set to constitute the mediating effect model as Function (1) to (5):

$$lnAAPI_{it} = \alpha_i + \beta_0\, LM_{it} + \sum_j \beta_j X_{it} + v_t + \xi_{it} \quad (1)$$

$$lnNI_{it} = \alpha_i + \lambda_1\, LM_{it} + \sum_j \lambda_j X_{it} + v_t + \xi_{it} \quad (2)$$

$$lnFS_{it} = \alpha_i + \pi_1\, LM_{it} + \sum_j \pi_j X_{it} + v_t + \xi_{it} \quad (3)$$

$$CCR = \alpha_i + \theta_1\, LM_{it} + \sum_j \theta_j X_{it} + v_t + \xi_{it} \quad (4)$$

$$lnAAPI_{it} = \alpha_i + \beta_1\, LM_{it} + \rho_1 lnNI_{it} + \rho_2 lnFS_{it} + \rho_3 CCR_{it} + \sum_j \beta_j X_{it} + \nu_t + \xi_{it} \quad (5)$$

X are control variables, including the rural population size, agricultural production, agricultural irrigation, and agricultural machinery power; these control variables affect the degree of agricultural air pollution. α is the constant of the regression, v is the time variable, and ξ represents the regression residual in the equation. Among them, Function (1) is the benchmark equation for the agricultural air pollutants comprehensive index; Function (2) to (4) test the mediation effect of the income effect, the economy of scale effect, and the effect of the crop-planting structure, respectively; Function (5) combines the direct effect variable labor migration (LM) with the mediation effect variables of non-farm income (NI), farmland scale (FS), and cash crop ratio (CCR) to measure the total effect on agricultural air pollution.

4. Analysis Process

In order to exclude heteroscedasticity, some variables were converted to their logarithmic form and estimated using the fixed effect model (FE) based on the p value of the Hausman test being zero.

4.1. Total Sample Analysis: The Effect of Labor Migration on Agricultural Air Pollution

Table 3 shows the impact results of labor migration, LM, on the comprehensive index of agricultural air pollution, AAPI.

Based on the research design of Model (1-1) to (1-5) presented in the previous section, the calculation of the pollution effect of rural labor migration is divided into the following three steps:

4.1.1. Benchmark Model Test

The explanatory variables in Model (1-1) include only labor migration variables and four related controlled variables, which is the benchmark model to test the significance of rural labor migration on the AAPI. In the results, the coefficient of labor migration is significantly positive (0.0023), which suggests that the total effect of labor migration on AAPI is positive. This is consistent with the previous theoretical hypothesis.

4.1.2. Mediation Test of Indirect Effect of Labor Migration

In the second step, we tested the mediation effect variables (non-farm income, farmland scale, and crop-planting structure) with Model (1-2), (1-3), and (1-4) separately. The test results show that the coefficient of rural labor migration on the non-farm income variable and farmland scale variable are significantly positive (0.1248 and 0.159), which means that the migration of rural labor force has improved the per-capita income of rural resident households, promoting farmland concentration and increasing the per-capita farmland for rural residents. Whereas, the coefficient of rural labor migration on the crops planting structure variable is not significant, suggesting that the migration of labor force in Hubei and Hunan provinces has not significantly increased the proportion of cash crops, also confirming the research conclusion from Zhong (2016) [20].

Table 3. Results of mediating effect model analysis.

Model	1-1	1-2	1-3	1-4	1-5
	Benchmark Model	Mediation Test of the Income, Scale and Crop-Planting Structure Effect of Rural Labor Force Migration			Direct and Indirect Effect of Labor Migration on AAPI
Variable	lnAAPI [1]	lnNI	lnFS	CCR	lnAAPI
lnNI					0.0564 * [2] (0.0294) [3]
lnFS					0.1970 *** (0.0467)
CCR					−0.0098 *** (0.0013)
LM	0.0023 *** (0.0006)	0.1248 *** (0.0474)	0.1590 *** (0.0229)	−0.0396 (0.0990)	−0.0030 ** (0.0013)
lnRpop	0.0869 * (0.0453)	−0.0452 (0.0471)	−0.0347 ** (0.0227)	−0.0041 (0.0169)	0.0899 * (0.0455)
lnAoutput	0.227 *** (0.0679)	1.1279 *** (0.0402)	0.3005 *** (0.0194)	0.0622 *** (0.0143)	0.2080 *** (0.0693)
lnMpower	0.0724 ** (0.0330)	0.2557 *** (0.0532)	0.0227 (0.0256)	0.0082 (0.0190)	0.0724 ** (0.0307)
lnIrrigation	0.0443 (0.0332)	0.1008 *** (0.0303)	−0.0745 ** (0.0146)	0.0621 * (0.0108)	0.0334 (0.0351)
Constant	−0.966 ** (0.381)	−1.399 *** (0.2561)	0.4488 *** (0.1234)	3.6953 *** (0.0916)	−0.802 ** (0.390)
Fixed location	controlled	controlled	controlled	controlled	controlled
Fixed time	controlled	controlled	controlled	controlled	controlled
R-sq	0.289	0.6951	0.3550	0.0312	0.362
rho	0.9707	0.8994	0.8451	0.8595	0.9692
Observations	1672	1672	1672	1672	1672
Number of ids	152	152	152	152	152
Direct effect −0.0030		Indirect effect (aggregate) 0.0383			Total effect 0.0353

[1] Abbreviations are used for the following variables: logarithmic comprehensive index of agricultural air pollution (lnAAPI), logarithmic non-farm income (lnNI), logarithmic farmland scale(lnFS), cash crop ratio (CCR), labor migration ratio (LM), logarithmic rural population (lnRpop), logarithmic agricultural output (lnAoutput), logarithmic mechanical power (lnMpower), the same below. [2] The asterisks indicate different significance levels, *** for $p < 0.01$, ** for $p < 0.05$, and * for $p < 0.1$, the same below. [3] Content in brackets is the standard error, the same below.

4.1.3. Calculation of the Direct Effect and Indirect Effect

Model (1-5) includes rural labor migration and indirect variables, controlled variables into the regression model, testing the direct factor substitution effect, indirect effect, and total effect of labor force migration on the AAPI.

First, the study found that the direct effect, indirect income, and scale effects of labor migration have passed the statistical significance test, so the impact effect of labor migration on agricultural air pollution is manifested as a partial mediation effect.

Next, the report results indicate that the influence coefficient of the non-farm income variable is 0.0564, that is, the increase of non-farm income has a significantly positive correlation on the AAPI. This can be interpreted in that the increase of household income greatly increases the farming materials and chemical fertilizer, which aggravates agricultural air pollution. The mediation effect of scale variable has a significantly positively correlation, which is in line with Hypothesis 3. A peasant who manages more land than his/her capacity will lead to extensive operation, using more fertilizer and bringing more air pollution. In addition, the coefficient of crop-planting structure variable is significantly negative. According to Hypothesis 4, it can be interpreted that this non-grain trend of crop-planting structure will alleviate agricultural air pollution, but provided the result of Model (1-4), the coefficient of mediator structure variable is not significant, so the indirect effect of crop-planting structure is not included in the aggregate indirect effect calculation.

Finally, the premise of the total effect calculation is that the coefficient must be statistically significant. According to this screening, the coefficient of the direct effect is −0.0030; the indirect income effect is 0.0070 (indirect income effect is calculated as:

0.1248 × 0.0564 = 0.0070, the coefficients of non-farm income from Model (1-2) and (1-5); accordingly, indirect scale and structure effect are calculated in the same way), the scale indirect effect is 0.0313, the structure effect is not significant, so the aggregate indirect effect is 0.0383. The total effect is the sum of the direct effect and the aggregate indirect effect, that is, 0.0353.

It is also evident to analyze the other controlled variables to find that the influence coefficients of rural population, agricultural output value, and mechanical power on the AAPI are significantly positive, and the coefficients are 0.0869, 0.227, and 0.0724. Among them all, the output of agricultural production is undoubtedly the main culprit of agricultural air pollution.

4.2. Heterogeneity Analysis: The Regional Analysis of Rural Labor Migration in Hubei and Hunan

The original data show that the overall number of rural labor migration in Hubei Province is significantly lower than that in Hunan province, which supports the further exploring of the discriminative results of them. In this section, we divided the samples into two sub-sample sets and carried out the same test as the previous section. Tables 4 and 5 are the results of it in Hubei Province and Hunan Province, respectively.

Table 4. Heterogeneity analysis: results of mediating effect model analysis in Hubei Province.

Model	2-1	2-2	2-3	2-4	2-5
	Benchmark Model	**Mediation Test of the Income, Scale, and Crop-Planting Structure Effect of Rural Labor Force Migration**			**Direct and Indirect Effect of Labor Migration on AAPI**
Variable	lnAAPI	lnNI	lnFS	CCR	lnAAPI
lnNI					0.0285 (0.0232)
lnFS					0.238 *** (0.0412)
CCR					−0.0071 *** (0.0011)
LM	0.0002 (0.0007)	0.2241 ** (0.1101)	0.1813 *** (0.0481)	−0.1194 *** (0.0343)	−0.0055 *** (0.00125)
lnRpop	0.589 *** (0.153)	−0.1964 ** (0.0907)	−0.0283 (0.0396)	−0.0159 (0.0282)	0.639 *** (0.142)
lnAoutput	0.212 *** (0.0439)	0.8540 *** (0.07462)	0.2690 *** (0.0326)	0.0207 (0.0232)	0.186 *** (0.0406)
lnMpower	0.00257 (0.0148)	0.3238 *** (0.1084)	0.0898 * (0.0474)	−0.0176 (0.0338)	0.0093 (0.0130)
lnIrrigation	−0.0358 (0.0230)	0.2019 *** (0.0443)	0.0782 *** (0.0194)	−0.0163 (0.0138)	−0.0198 (0.0192)
Constant	−1.965 *** (0.454)	−0.9428 * (0.5652)	0.4365 ** (0.2471)	4.2857 *** (0.1761)	−2.111 *** (0.449)
Fixed location	control	control	control	control	control
Fixed time	control	control	control	control	control
R-sq	0.317	0.6038	0.3667	0.0214	0.369
rho	0.9665	0.7251	0.6746	0.8581	0.9525
Observations	759	759	759	759	913
Number of ids	69	69	69	69	83
Direct effect −0.0055		Indirect effect (aggregated) 0.0440			Total effect 0.0385

[1] Abbreviations are used for the following variables: logarithmic comprehensive index of agricultural air pollution (lnAAPI), logarithmic non-farm income (lnNI), logarithmic farmland scale(lnFS), cash crop ratio (CCR), labor migration ratio (LM), logarithmic rural population (lnRpop), logarithmic agricultural output (lnAoutput), logarithmic mechanical power (lnMpower), the same below. [2] The asterisks indicate different significance levels, *** for $p < 0.01$, ** for $p < 0.05$, and * for $p < 0.1$. [3] Content in brackets is the standard error.

Table 5. Heterogeneity analysis: results of mediating effect model analysis in Hunan Province.

Model	3-1	3-2	3-3	3-4	3-5
	Benchmark Model	**Mediation Test of the Income, Scale, and Crop-Planting Structure Effect of Rural Labor Force Migration**			**Direct and Indirect Effect of Labor Migration on AAPI**
variable	lnAAPI	lnNI	lnFS	CCR	lnAAPI
lnNI					0.393 **
					(0.173)
lnFS					0.0836
					(0.0710)
CCR					−0.0113 ***
					(0.0021)
LM	0.0028 **	0.0906 **	0.1482 ***	−0.0109	0.0001
	(0.0011)	(0.0371)	(0.0241)	(0.0175)	(0.0021)
lnRpop	0.0247	−0.0021	−0.0375 *	−0.2061	0.00644
	(0.0182)	(0.0409)	(0.0202)	(0.0193)	(0.0168)
lnAoutput	0.136	1.4673 ***	0.3131 ***	0.1509 ***	0.197 **
	(0.0878)	(0.0373)	(0.0242)	(0.0176)	(0.0828)
lnMpower	0.247 **	0.1833 ***	−0.0176	0.0035	0.253 ***
	(0.111)	(0.0446)	(0.0290)	(0.0211)	(0.0945)
lnIrrigation	0.0703 **	−0.0181	−0.0627 **	−0.0134	0.0428
	(0.0350)	(0.0395)	(0.0256)	(0.0187)	(0.0345)
Constant	−1.205*	−1.7712 ***	0.3872 ***	3.3162 ***	−2.151 ***
	(0.619)	(0.2092)	(0.1358)	(0.0988)	(0.788)
Fixed location	control	control	control	control	control
Fixed time	control	control	control	control	control
R-sq	0.488	0.8456	0.3429	0.1714	0.553
rho	0.9583	0.9723	0.8378	0.9007	0.9511
Observations	913	913	913	913	759
Number of ids	83	83	83	83	69
Direct effect		Indirect effect (aggregated)			Total effect
Not significant		0.0356			0.0356

[1] Abbreviations are used for the following variables: logarithmic comprehensive index of agricultural air pollution (lnAAPI), logarithmic non-farm income (lnNI), logarithmic farmland scale(lnFS), cash crop ratio (CCR), labor migration ratio (LM), logarithmic rural population (lnRpop), logarithmic agricultural output (lnAoutput), logarithmic mechanical power (lnMpower), the same below. [2] The asterisks indicate different significance levels, *** for $p < 0.01$, ** for $p < 0.05$, and * for $p < 0.1$. [3] Content in brackets is the standard error.

4.2.1. Benchmark Model Test of Hubei and Hunan Provinces, Separately

In the Model (2-1) and (3-1), the coefficients of the rural labor force migration variables on the AAPI are both positive, but in Hubei, it is not significant. Moreover, rural labor migration in counties in Hunan is higher than that in Hubei, the coefficient of rural labor migration variables in counties on the pollution index in Hubei Province is higher by 0.0026% than that in Hunan Province as a result.

4.2.2. Analysis of the Mediation Variables

In Model (2-2) and (2-3) of Hubei Province, we showed that the income intermediary variables and scale mediation variable coefficient of rural labor migration are significant, 0.2241 and 0.1813, and Model (3-2) and (3-3) of Hunan Province present that the coefficients of the income mediation variable and scale mediation variable are 0.0906 and 0.1482. The income effect and scale effect of rural labor migration in the two provinces are likewise significant, which indicates that the migration of rural labor force has also increased the household income and the rural per-capita cultivated land area. However, the difference lies in that the crop-planting structure effect of the counties in Hunan Province is negative but not significant; meanwhile, the structure effect coefficient of the counties in Hubei Province is significant negative. According to the study of Zhong (2016), it is mainly because the crop-planting structure of Hubei Province is relatively flexible, considering that the labor migration reduces the agricultural young and middle-aged labor force, and the convenience of mechanical operation for replacing manpower, labor migration substantially increases the proportion of food crops planted [20].

4.2.3. Calculation of the Mediation and Total Effect in Hubei and Hunan Provinces

The results of counties in Hubei show that although the direct effect is negative (−0.0055), the indirect effect is large enough to make the total effect positive (0.0385). In other words, the migration of labor force has generally improved the level of agricultural air pollution in Hubei Province. In addition, the income effect is not significant, and the farmland scale effect is dominant in the indirect effect (0.0431 out of 0.044).

In Hunan Province, the results are not all alike. Only the income effect on the pollution index is significantly positive (0.0356), while all other direct and indirect coefficients are not. Therefore, the total effect of labor migration on AAPI is entirely explained by income indirect effect. In other words, the migration of labor force has generally increased the agricultural air pollution in Hunan Province through the non-farm income indirect effect.

Comparing the results of these two provinces, we can draw such a conclusion: there are many practical factors, such as agricultural traditions, terrain, economic level, etc., which together lead to the difference in direct and indirect effects on pollution between the two. However, the effect of labor migration cannot be ignored, a higher migration ratio will lead to a more significant indirect and total effect.

Finally, in each province, the indirect and total effect of rural labor migration on agricultural air pollution is consistent with the results from the whole sample. This indicates that our previous analysis is generally applicable to each province, with minor differences in indirect effects.

4.3. *Preliminary Summary*

In this section, we introduced and calculated an index, AAPI, to evaluated the air pollutant emission from agricultural production activities and conducted research on the connection between agricultural air pollution and rural labor migration. The above studies can be summarized in Table 6.

Table 6. Summary of the pollution effect and decomposition conclusions of rural labor transfer.

Sample	Direct Effect	Indirect Effect				Total Effect
		Income Effect	Land Scale Effect	Crop-Planting Structure Effect	Indirect Effect	
Total sample	−0.0030	0.0070	0.0313	Not significant	0.0383	0.0353
Hubei province	−0.0055	Not significant	0.0431	0.0008	0.0440	0.0385
Hunan province	Not significant	0.0356	Not significant	Not significant	0.0356	0.0356

First, the increase of rural labor migration has a significant impact on both the mediator variables of non-farm income and farmland scale, but there is a difference in the indirect effect between the two provinces mentioned above according to our regional heterogeneity analysis. In general, the crop-planting structure effect of rural labor migration is not significant, but the income effect and scale effect are significant, indicating that a higher proportion of labor migration will not increase the proportion of cash crops obviously, but it does through the increase of non-farm income, which increases the opportunity cost of farming and discourages the labor force to farm, leading to an acceleration of the transfer of use rights of farmland, enlarging the per-capita farming area. In addition, certain differences occur in the rural labor migration on the mediator variables between two provinces. The income effect and scale effect of labor migration on the two provinces are both significant, but in Hubei, the impact of labor migration on the crop-planting structure is also significant, but the coefficient is negative; namely, in Hubei, labor migration actually increased the proportion of food crops.

Second, in the overall sample, a higher proportion of labor migration has exacerbated the level of agricultural air pollution, and it mainly comes from the indirect effect, specifically, from income effect and scale effect. The direct effect of migration on pollution is less significant compared with its indirect effect.

Third, there are heterogeneity differences between the two provinces. For the counties in Hubei province, the migration of rural labor has less of a direct effect on agricultural air pollution than that in Hunan; Figure 1 clearly shows this difference.

5. Conclusions and Discussion

5.1. Conclusions

Nowadays, air pollutant emissions from agriculture have increased tremendously, and it is imperative to contain this trend in China to ensure the sustainability of agricultural production, the health of the rural population, and the development of rural economy.

This paper mainly analyzed the effect of rural labor migration on agricultural air pollution. First, we conducted mechanism research on the internal connection between agricultural air pollution and rural labor migration and put forward some hypotheses. Second, we evaluated air pollutant emission from agricultural production activities by introducing a new index air pollution comprehensive index (AAPI), and we calculated it with the data from counties in Hubei and Hunan. Lastly, we used the empirical method to examine the hypotheses above and obtained some conclusions.

We found that the increase of labor migration could intensify the air pollution discharged from agricultural activity by changing the following factors: the supply of labor force in the agricultural sector, the budget line of rural residents, and the scale of crop farming and planting structure; among them, the income and scale factor plays the major role. However, there is a difference in the indirect total effect between two provinces according to our regional heterogeneity analysis.

We attempt to propose some policy implications from the studies above. The intrinsic driving force of rural labor force migration lies in the surplus of labor force and the gap between urban and rural development. Its flow trend will continue, but the rural environmental protection should also not be ignored. From the study, we already know that with the transfer of rural population, the income, and living conditions of rural residents have been improved (mainly from family members who emigrated to the city), and the per-capita farming scale has also increased, which will not only increase the application of chemical fertilizer but also promote the extensive management in the agricultural production process, resulting in more air pollution. Therefore, increasing the investment and promotion of green agricultural production technology and providing more technical training for the remaining farmers should be put on the agenda to alleviate the air pollution caused by agricultural production.

5.2. Discussion

The potential theoretical contributions of this paper are as follows. First, we applied the PCA method and used a wider catalogue to calculate a more comprehensive index to evaluate agricultural air pollution rather than focus on a certain agricultural sector or a specific pollutant (such as nitrogen oxides and ammonia). Furthermore, the existing studies have focused on exploring the factors influencing agricultural air pollution from a specific and ecological perspective, but there has been an absence of mechanism studies about sociological and economic factors behind the countryside and agricultural activities. This paper could be a necessary supplement to this field.

This paper also has some study limitations. The data collected from Hubei and Hunan provinces are still insufficient to comprehensively measure the air pollution of agricultural production, so are the relevant pollution calculating coefficients and index. Further research is needed in these aspects. In addition, the mediating effect model we used in this paper is still an imperfect method for this research. The causal steps approach model is the most commonly used and convenient method compared with other mediating effect models, but there must be a situation in which the direct and indirect effects are similar or have opposite symbols and even affect the overall relationship between dependent variables and independent variables. Therefore, the stepwise test method may miss some actual intermediary effects.

Author Contributions: S.Y., X.L. and J.R. conceptualized and designed the research. Y.L. and N.L. drafted the paper, acquired, and analyzed the data. J.F. and J.J. substantively revised the paper. All authors have read and agreed to the published version of the manuscript.

Funding: This study was supported by the Humanity and Social Science Youth Foundation of Ministry of Education in China (grant number: 20YJC790109) and School-level Scientific Research Project of Shenzhen Institute of Information Technology (grant number: SZIIT2020SK011).

Institutional Review Board Statement: Not applicable.

Informed Consent Statement: Not applicable.

Data Availability Statement: The data used in this paper come from the China Statistical Yearbook, China rural statistical yearbook, and China County Statistical Yearbook.

Conflicts of Interest: The authors declare that they have no known competing financial interest or personal relationship that could have appeared to influence the work reported in this paper.

References

1. Lovarelli, D.; Conti, C.; Finzi, A.; Bacenetti, J.; Guarino, M. Describing the trend of ammonia, participate matter and nitrogen oxides: The role of livestock activities in northern Italy during COVID-19 quarantine. *Environ. Res.* **2020**, *191*, 110048. [CrossRef]
2. Hellin, J.; McDonald, A.; Jat, M.L.; Shyamsundar, P.; Singh, A.K. Mitigating agriculture's contribution to air pollution in India. *Lancet Planet. Health* **2021**, *5*, e186. [CrossRef]
3. Zhao, H.; Zhang, X.; Zhang, S.; Chen, W.; Xiu, A.; Tong, D.Q. Effects of Agricultural Biomass Burning on Regional Haze in China: A Review. *Atmosphere* **2017**, *8*, 88. [CrossRef]
4. Zha, S.; Zhang, S.; Cheng, T.; Chen, J.; Huang, G.; Li, X.; Wang, Q. Agricultural Fires and their Potential Impacts on Regional Air Quality over China. *Aerosol Air Qual. Res.* **2013**, *13*, 992–1001. [CrossRef]
5. Long, X.; Luo, Y.; Sun, H.; Tian, G. Fertilizer using intensity and environmental efficiency for China's agriculture sector from 1997 to 2014. *Nat. Hazards* **2018**, *92*, 1573–1591. [CrossRef]
6. Wang, Y.; Lu, Y.; Yuan, J.; He, G. Evaluating the risks of nitrogen fertilizer-related grain production processes to ecosystem health in China. *Resour. Conserv. Recycl.* **2022**, *177*, 105982. [CrossRef]
7. Stacy, E.S. Effects of Environmental Regulation on Economic Activity and Pollution in Commercial Agriculture. *B. E. J. Econ. Anal. Policy* **2009**, *9*. [CrossRef]
8. Huang, J.; Xiong, R.; Fang, L.; Li, T.; Shen, W. Estimation of interannual trends of ammonia emissions from agriculture in Jiangsu Province from 2000 to 2017. *J. Atmos. Sci.* **2020**, *13*, 268–273. [CrossRef]
9. Guo, Y.Z.; Qiao, W.F. Rural Migration and Urbanization in China: Historical Evolution and Coupling Pattern. *Sustainability* **2020**, *12*, 7307. [CrossRef]
10. Xu, D.D.; Guo, S.; Xie, F.; Liu, S.; Cao, S. The impact of rural laborer migration and household structure on household land use arrangements in mountainous areas of Sichuan Province, China. *Habitat Int.* **2017**, *70*, 72–80. [CrossRef]
11. Bhandari, P.; Ghimire, D. Rural Agricultural Change and Individual Out-migration. *Rural Sociol.* **2016**, *4*, 572–600. [CrossRef] [PubMed]
12. Tian, X.; Yi, F.J.; Yu, X.H. Rising cost of labor and transformations in grain production in China. *China Agric. Econ. Rev.* **2020**, *1*, 158–172. [CrossRef]
13. Luan, J. Does Agricultural Labor Migration Have Substitution Relation with Fertilizer Use? Experience and Evidence from China' s Main Provinces and Municipalities Planting Crops. *J. Chongqing Technol. Bus. Univ.* **2017**, *4*, 12–21. [CrossRef]
14. Barry, K.G.; Ashok, K.M. Farming Efficiency and the Determinants of Multiple Job Holding by Farm Operators. *Am. J. Agric. Econ.* **2004**, *86*, 722–729. [CrossRef]
15. Luan, J.; Li, T.T.; Ma, K. Study on the impact of labor transfer on the pollution of agricultural fertilizer sources in China. *World Agric.* **2016**, *2*, 63–69. [CrossRef]
16. Cheng, M.; Shi, Q.; Jin, Y.; Gai, Q. The income gap and its root causes: Model and empirical. *Manage. World* **2015**, *7*, 17–28. [CrossRef]
17. Carletto, C.; Savastano, S.; Zezza, A. Fact or artifact: The impact of measurement errors on the farm size–productivity relationship. *J. Dev. Econ.* **2013**, *103*, 254–261. [CrossRef]
18. Qiu, T.W. Will the non-farm transfer of rural labor reduce the rate of agricultural land output? *J. Zhongnan Univ. Econ. Law* **2018**, *5*, 151–160. [CrossRef]

19. Gao, J.; Song, G.; Sun, X.Q. Does labor migration affect rural land transfer? Evidence from China. *Land Use Pol.* **2020**, *99*, 105096. [CrossRef]
20. Zhong, F.N.; Lu, W.Y.; Xu, Z.G. Is it not conducive to food production for rural labor force to go out to work?—The analysis of the factor substitution and planting structural adjustment behavior and constraints of farmers. *Chin. Rural Econ.* **2016**, *7*, 36–47. Available online: http://www.nssd.cn/articles/Article_Read.aspx?id=669630074 (accessed on 20 December 2021).
21. Wan, B.R. The current trend and suggestions of China's agricultural development. *Iss. Agric. Econ.* **2014**, *4*, 4–7. [CrossRef]
22. Chen, M.P.; Chen, J.N.; Lai, S.Y. Inventory analysis and spatial feature identification of agricultural and rural pollution in China. *Zhongguo Huanjing Kexue* **2006**, *6*, 751–755. [CrossRef]
23. Fan, B. Overall management and comprehensive management break through the difficult problem of rural water pollution control. *Huanjing Baohu* **2014**, *15*, 15–19. [CrossRef]
24. Cao, G.L.; Zhang, X.Y.; Wang, D.; Zheng, F.C. Inventory of Emissions of Pollutants from Open Burning Crop Residue in China. *J. Agric. Environ. Sci.* **2005**, *24*, 800–804. [CrossRef]
25. Wang, H.D.; Zhao, Z.Q.; Winiwarter, W.; Bai, Z.H.; Wang, X.; Fan, X.W.; Zhu, Z.P.; Hu, C.S.; Ma, L. Strategies to reduce ammonia emissions from livestock and their cost-benefit analysis: A case study of Sheyang county. *Environ. Pollut.* **2021**, *290*, 118045. [CrossRef]
26. Ministry of Ecological and Environment of P. R. China. Notice on Reduction and Exemption of Sewage Charges for Poultry Industry (Huanfa [2004] No.43). Available online: http://www.mee.gov.cn/gkml/zj/wj/200910/t20091022_172271.htm (accessed on 25 December 2021).
27. Wen, Z.L.; Ye, B.J. Analyses of Mediating Effects: The Development of Methods and Models. *Adv. Psychol. Sci.* **2014**, *22*, 731–745. [CrossRef]

atmosphere

MDPI

Article

Investigating Spatial Heterogeneity of the Environmental Kuznets Curve for Haze Pollution in China

Abdul Samad Abdul-Rahim [1], Yoomi Kim [2,*] and Long Yue [3]

[1] Institute of Tropical Agriculture & Food Security (ITAFoS), School of Business and Economics (SPE), Universiti Putra Malaysia, Serdang 43400, Malaysia; abdrahim_as@upm.edu.my

[2] Department of Public Administration, Ewha Womans University, Seoul 03760, Korea

[3] Faculty of Economics and Management, Universiti Putra Malaysia, Serdang 43400, Malaysia; longyue26@163.com

* Correspondence: kimym@ewha.ac.kr; Tel.: +82-2-3277-4030

Abstract: This study investigates the environmental Kuznets curve (EKC) for haze in 31 cities and provinces across China using the spatial data for a period of 15 years, from 2000 to 2014. We utilized the geographically weighted regression (GWR) model to consider the spatial non-stationary characteristics of the air quality in a vast territory. This approach allowed us to verify the region-specific characteristics, while the global model estimated the average relationship across the entire nation. Although the EKC for haze was statistically significant in the global models, the results only confirmed the existence of an EKC between the overall air quality and economic performance. Thus, it was difficult to determine the regional differences in an EKC. The results of the GWR model found the spatial variability of each variable and showed significant spatial heterogeneity in the EKC across regions. Although six regions—Beijing, Gansu, Heilongjiang, Jiangxi, Jilin, Liaoning, Shanghai, Tianjin, Xinjiang, and Zhejiang—showed inverted U-shaped EKCs, these were only statistically significant in three big cities—Beijing, Tianjin, and Shanghai. The results demonstrated no EKCs in the other 25 provinces and cities. These results provide strong empirical evidence that there is significant spatial heterogeneity in the EKC of China. Thus, a more regionally specialized air pollution control policy is required to create an effective policy for balanced economic growth in China.

Keywords: China; environmental Kuznets curve; geographically weighted regression; haze; spatial heterogeneity

Citation: Abdul-Rahim, A.S.; Kim, Y.; Yue, L. Investigating Spatial Heterogeneity of the Environmental Kuznets Curve for Haze Pollution in China. *Atmosphere* **2022**, *13*, 806. https://doi.org/10.3390/atmos13050806

Academic Editors: Zengyun Hu, Xuguang Tang and Qinchuan Xin

Received: 18 April 2022
Accepted: 10 May 2022
Published: 14 May 2022

Publisher's Note: MDPI stays neutral with regard to jurisdictional claims in published maps and institutional affiliations.

1. Introduction

China has emerged as one of the largest and most robust economies in the world. The ruling Chinese Communist Party made the decision to open up the country for trade and investment, as well as adopting a limited number of capitalist strategies. This is a natural progression toward a better model for regional trade and local growth [1]; however, regional discrepancies in sustainable development in China are among the barriers to future success [2].

The rapid industrialization in China has caused side effects, with the leading issue being environmental challenges. Currently, the most prominent environmental issue is air pollution. The haze phenomenon has significantly and negatively affected economic performance and contributes to environmental pollution in China. Haze is defined as the introduction of particulate matter (PM), biological molecules, and other contaminants into the atmosphere. The presence of excess suspended matter in the atmosphere can disrupt the daily life of humans. These contaminants usually negatively affect the health of humans and other living organisms. Furthermore, they damage crops, natural environments, and built environments [3,4]. It can also affect the economic performance of a region by causing declining employee performance, haze-related illnesses, slower crop growth, low visibility, and other issues [4].

In recent years, large cities in China have frequently experienced severe fog and haze episodes that disrupt visibility. There have been rapid changes in the patterns of haze and fog occurrence in these areas over the past several decades. For example, in research about the occurrence of haze and fog in the North China Plain, the occurrence of haze and fog was quite low from the 1950s to the 1980s but reached a peak from 1981 to 1998 [4,5]. Since 2000, haze pollution has shown a rapid increase; one of the reasons for this is that China's policy of opening up stimulates the evolution of industries. The haze problem became more noticeable in the mid-2000s, when environmentalists observed that the haze was not associated with the changing of the seasons. Instead, these particles remained in the air perpetually. In 2013, the Chinese Academy of Sciences documented the occurrence of a severe episode of haze, which affected northern and eastern China.

One of the most affected places was Beijing, which is the administrative, economic, and cultural center of China, meaning that the negative effects there far outweighed those in other regions. Beijing's high population density means that the people affected outnumber those affected anywhere else in the country. In Beijing, air pollution index (API) values of 900 have been recorded, making the environment toxic to its local residents (an API of 0–50 is considered good, 50–100 is moderate, 101–200 is unhealthy, 201–300 is very unhealthy, and anything above 300 is classified as hazardous). In terms of economic losses, Beijing lost approximately 254 million USD, which was about 0.008% of the total gross domestic product (GDP) of Beijing in 2013 [6]. Previous studies have examined the effects of haze on China's economy and most found that haze reduces economic performance as well as human resources [7–10].

Previous studies have tried to investigate the existence of an environmental Kuznets curve (EKC) in China to show the relationship between economic development and environmental performance and understand the current pollution status [11–19]. Scholars have analyzed the EKC for haze using data regarding the concentrations of PM, carbon dioxide (CO_2), sulfur dioxide (SO_2), and nitrous oxide in the atmosphere [20–23]. (PM is classified into different sizes, the most commonly measured of which are PM_{10} and $PM_{2.5}$. There has been a significant debate between researchers as to which is the better measurement system, but measuring $PM_{2.5}$ appears to be the preferred method in China as it has been found to be the dominant polluting agent). To investigate the heterogeneity of the EKC across regions in China, some studies have considered the spatial differences in the EKC and found significant spatial heterogeneity [16,20–25]. However, most studies focused on a certain pollutant, such as SO_2 or PM; thus, it was difficult to discover the EKC between overall air quality and economic performance.

Scholars have also attempted to identify the determinants of haze [7–10,26,27]. The energy structure is one of the crucial factors affecting haze. Scholars have suggested that the heavy usage of coal, renewable energy, and energy consumption have an influence on haze; however, the empirical evidence is still controversial [27–29]. Other socio-economic factors, such as income, human capital, trade openness, transportation, and population density, have also been considered as determinants of haze [16–19]. Moreover, if we widen the perspective, we should not overlook the impact of meteorology and stable atmospheric conditions on atmospheric pollution. For example, the amount of rainfall and temperature have been regarded as factors that influence the haze problem [25].

This study seeks to determine which economic factors influence haze and which regions have an EKC and turning point. Considering the spatial heterogeneity of economic growth and the haze phenomenon, this study applies the geographically weighted regression (GWR) model to investigate the EKC between haze and economic determinants for 31 Chinese cities. The overall air-quality data API and factors including the gross regional product (GRP), trade openness, and population density are used for analysis. After determining that the EKC has an inverted U, we verify and calculate the turning point, i.e., the decline in pollutants in line with increased economic performance [30]. The results from the GWR approach, which is suitable for considering the spatial non-stationary characteristics in vast territories, can provide a better understanding of the spatial heterogeneity of the haze problem in provinces in China with visual evidence. Moreover, the estimation of the turning points of the EKCs using the GWR estimations can be used to identify trends in sustainable development after the analysis periods. Through these processes, this study determines whether any stronger measures should be undertaken by the Chinese government to curb the haze issue, as it may affect the livelihood of the nation.

2. Materials and Methods

In this study, we utilized an EKC model to study the haze in China using a database of 31 cities and provinces across China from the period 2000 to 2014. To verify whether there was a significant difference in the regional economic capacity or innovation ability during the same period, we first applied global regression models to grasp the overall EKC of the entire country. We adopted the EKC model, as modified by Kang et al., to study the relationships among CO_2 emissions in China [31]. Interestingly, this study adopted the API, which is an overall measure of pollution that encompasses particles of all sizes, to determine the level of haze for each region, unlike previous studies, which used data regarding each pollutant. The API has been recommended as a sustainable indicator in air-quality assessment since it can overcome difficulties in interpretation of research results and problems of subjective classification of air quality [32].

In addition to the EKC analysis of haze in the form of API readings, we included the possible determinants of the environmental elements of haze—per capita GRP, trade openness, and population density—-in the analysis. To verify the inverted U shape, per capita GRP and its square were included in the model. Trade openness represents the degree of economic openness, which reflects economic competitiveness and the flow of goods and services. Previous research insisted that the opening up of the economy by the Chinese government affected the degree of pollution [33]. This study also considers population density as the indirect measure of active economic activities [25]. Data regarding the GRP, trade openness, and population density were taken from the China Compendium of Statistics for the period 2000–2014 and the China Statistical Yearbook [34]. The API levels for all 31 cities in China were obtained from China's Environmental Agency.

The EKC model used in this study is given as follows:

$$\ln(Y_{it}) = \alpha_{it} + \beta_1 \ln(\text{PGRP}_{it}) + \beta_2 \ln(\text{PGRP}_{it})^2 + \beta_3 \ln(\text{TR}_{it}) + \beta_4 \ln(\text{PD}_{it}) + \varepsilon_{it} \quad (1)$$

where Y_{it} is the haze level measured in API per capita, PGRP stands for per capita GRP, TR is the trade openness expressed as ratio of the gross export and import value to the GRP, and PD represents the population density in each of the 31 cities in China. i and t represent the region- and time-specific data.

The following models can be utilized for analysis to consider autocorrelation within panels and cross-sectional correlation and heteroskedasticity across panels: the pooled ordinary least squares (OLS) model, the fixed- and random-effects models, and the generalized least squares (GLS) model. This is because there can be subtle or no changes in a geographical area over time, which can cause homogeneity in the parameters of each spatial unit. An F-test, a Breush–Pagan Lagrange multiplier (LM) test, and a Hausman diagnostic test were conducted to determine which model provided the best fit for the data. Pesaran's test was also conducted to determine cross-sectional dependence.

Next, considering the spatial heterogeneity of each city and province, we performed a local linear regression using the GWR model. The GWR model has been used in earlier studies, such as in the Hedonic Pricing Model. The GWR model is also known as the local model and shows regression coefficients that vary across space, showing the spatial variations and relationships between the environmental determinants and their environmental factors via a local estimation done. The results allow us to understand how the regional determinants at a provincial level affect EKCs [2,21].

According to Wheeler [35], the GWR model is based on the linear regression model and assumes that the regression coefficient is a function of the observation point location. The GWR model focuses on the local geography, making it locally weighted. Thus, its performance from the perspective of spatial econometrics can be illustrated by using spatial heterogeneity. For the GWR model, we followed Fotheringham et al.'s approach [36] to consider regions expressed differently from the global models. This model expands the global model as follows:

$$Y_i = \beta_0(u_i, v_i) + \sum \beta_k(u_i, v_i)X_{ik} + \varepsilon_i \tag{2}$$

where Y_i and X_{ik} are the dependent and independent variables, respectively. (u_i, v_i) represents the location i of the observation and ε_i is the error term. The estimation for the GWR model is given as follows:

$$\beta' = \left[X^T W(u_i, v_i) X\right]^{-1} X^T W(u_i, v_i) Y. \tag{3}$$

where $W(u_i, v_i)$ is the square matrix of the weight assigned to a location (u_i, v_i). X and Y are the geographically weighted matrices of the values of the independent and dependent variables. The matrix of the geographical weights, $W(u_i, v_i)$, is as follows:

$$W(u_i, v_i) = \begin{bmatrix} W_1(u_1, v_1) & 0 & 0 \\ 0 & \cdots & 0 \\ 0 & 0 & W_n(u_1, v_1) \end{bmatrix} \tag{4}$$

The GWR model estimates the coefficients for each region, assuming that a closer observation has a greater effect on the parameters than a more distant observation. Based on the equations above, the specific model for the EKC can be expressed as follows:

$$Y_{it} = \alpha_{it} + \beta_1(u_i, v_i)\ln(\mathrm{PGRP}_{it}) + \beta_2(u_i, v_i)\ln(\mathrm{PGRP}_{it})^2 + \beta_3(u_i, v_i)\ln(\mathrm{TR}_{it}) + \beta_4(u_i, v_i)\ln(\mathrm{PD}_{it}) + \varepsilon_{it} \tag{5}$$

where Y_{it} is the per capita volume of the API in province or city i for the year t. (u_i, v_i) denotes the location i's longitude and latitude coordinates. PGRP and PGRPSQ denote the provincial gross regional product per capita and its squared term. TR is the trade openness expressed as the ratio of gross export and import value to the GRP. PD is population density (in people/m^2). This study utilized STATA for the global models, GWR 4 software for the spatial models, and QGIS for the mapping process. Table 1 presents the descriptive statistics.

Table 1. Descriptive Statistics of Variables.

Variable	Unit	Obs.	Mean	Median	Maximum	Minimum	Std. Dev.
API	Index	465	294.518	307.000	365.000	168.000	48.783
PGRP	100 million CNY	465	25,076.280	19,580.230	103,671.300	2742.070	19,679.650
TR	%	465	13.929	3.600	100.220	0.001	22.425
PD	Person/m^2	465	414.479	263.000	4349.000	2.000	621.088

Note: The data observations for each region for 15 years are consistent with the calculation of 15 years multiplied by 31 regions, which equals 465.

The determinant with the largest standard deviation is the PGRP, which indicates how varied the per capita income is for different regions in China. This is also seen in the trade openness variable, where some regions have almost no trade whatsoever and other regions trade almost all of their products. There is also a surprisingly significant difference in the population density across the regions, which can be attributed to the fact that China has a large land area and some places have a more inhospitable topography, which makes it difficult to farm and earn a living. Most of China's population is concentrated along coastal cities, with lower population density further inland.

3. Results

3.1. Subsection Results of Global Models

In this study, first, we tested the global models using the EKC formulas to understand the determinants of haze. Table 2 shows the results of the global models and the verifications of their fitness. The results demonstrate that the F-test value rejects the null hypothesis with a value of 210.19 with high statistical significance, and the Breush–Pagan LM test value also rejects the null hypothesis with a value of 2727.87 with a *p*-value of 0.000. Thus, both the fixed- and the random-effects model showed a better fit than the simple OLS model. However, the results of the Pesaran's test for the cross-sectional dependence show cross-sectional dependence in the models. In this regard, this study applied the GLS model to consider the autocorrelation within panels, cross-sectional correlation, and heteroskedasticity across panels.

Table 2. Estimation Results of Global Models.

Variable	Pool OLS	FE	RE	GLS
Intercept	5.518 *** (1.167)	6.356 *** (0.539)	5.938 *** (0.376)	5.675 *** (0.165)
$\ln(PGRP_{it})$	0.156 (0.236)	−0.008 (0.078)	0.032 (0.069)	0.122 *** (0.034)
$\ln(PGRP_{it})^2$	−0.013 (0.007094)	−0.003 (0.004)	−0.005 (0.004)	−0.011 *** (0.002)
$\ln(TR_{it})$	0.014 * (0.007)	0.001 (0.004)	0.001 (0.004)	0.013 *** (0.001)
$\ln(PD_{it})$	−0.012 (0.009)	−0.059 (0.046)	−0.013 (0.019)	−0.011 *** (0.001)
R^2	0.168	0.090	0.150	
F-test/Breush–Pagan LM test		210.19 ***	2727.87 ***	
Pesaran's test		4.041 ***	4.948 ***	

Note: ***, and * indicate significance at the 1% and 10% levels, respectively.

The results from the GLS model confirm the existence of an EKC in overall cities and provinces across China for the period 2000–2014. The PGRP variable shows a positive coefficient of 0.122, while the PGRP square's coefficient is negative, and both variables are statistically significant. These results indicate that significant EKCs have inverted U-shapes. Moreover, the positive value of the PGDP coefficient indicates that production activities contribute to haze concentrations; this finding supports those of previous studies, which suggested that production activities are the main contributors to haze concentrations [25,37]. This result shows the necessity of balanced environmental regulation, even if China's internal policy still prioritizes economic growth.

The trade openness has a positive coefficient at the 1% confidence level. The empirical evidence reveals that increasing trade openness results in an increase in API across China. This result is in line with previous research, which found a negative impact of the opening-up policy in China on haze pollution [33]. The population density variable has a negative coefficient at the 1% confidence level. The population density result shows that increases in population density cause a drop in API. This was an unexpected result, since population density is one of the indirect measures of active economic activity, which contributes to the increase in industrial emissions, including the haze phenomenon [25].

Even though the GLM model found the existence of the EKC for haze pollution, the global model only confirmed the existence of an EKC between overall air quality and economic performance in China. Thus, it is difficult to grasp the regional differences in the EKC [2,21,25]. Considering the possibility of significant spatial heterogeneity in the economic status of regions in China, it is necessary to investigate the heterogeneity of the EKC for each region.

3.2. Results of Geographical Weighted Regression Models

To verify the necessity of the spatial approach, we tested the spatial variability. Table 3 presents the results of the test for the spatial variability of the GWR coefficients. The differential criterion values are negative and highly significant, meaning that the variance of the regression values across the different provinces in China are also extremely high. These results demonstrate that there is significant spatial variability in terms of the model's selection criteria.

Table 3. Spatial Variability Test.

Variable	F	Diff of Criterion
Intercept	68,192.880 ***	−3960.470
$\ln(PGRP_{it})$	250.132 ***	−1049.020
$\ln(PGRP_{it})^2$	1473.370 ***	−1915.418
$\ln(TR_{it})$	54.240 ***	−379.452
$\ln(PD_{it})$	1516.960 ***	−1548.354

Note: Positive value of diff-criterion indicates that there is no spatial variability in terms of model selection criteria. *** indicates significance at the 1% level.

The spatial variability test results provide strong evidence that the EKCs are not always constant but, instead, vary among provinces or cities in China. The main contributors to the $PM_{2.5}$ in haze are coal and fossil-fuel burning, especially from vehicles and energy-intensive industries, and they are thus likely to be linked to the different levels of economic activity in a certain city [38]. However, the previous global OLS models could not capture these spatial differences, since the global model estimates the average relationships among all the provinces or cities in China to be in line with the findings of previous studies [21,25].

Table 4 presents the estimation results of the GWR model. The GWR parameter estimates for the independent variables show the distribution of the coefficients across regions that vary widely over space. The Akaike information criterion (AIC) value of the GWR model is also lower, at −1461.025, than the OLS value, of −324.234. Thus, the GWR model is superior to the global model.

Table 4. Estimation Results of GWR Model.

Variable	Min	LQ	Med	UQ	Max
Intercept	0.196	4.240	7.348	9.667	16.167
$\ln(PGRP_{it})$	−1.585	−0.726	−0.206	0.443	1.120
$\ln(PGRP_{it})^2$	−0.053	−0.026	0.008	0.032	0.079
$\ln(TR_{it})$	0.031	−1.099	−0.074	0.013	0.622
$\ln(PD_{it})$	−1.099	−0.325	0.016	0.047	0.147
N			465		
Adjusted R^2			0.939		
AIC			−1461.025		

The GWR model results confirm the spatial variability and indicate that there is significant spatial heterogeneity in the EKCs. In this regard, we investigated the existence of the EKC between the economic performance and the haze using the local coefficients from the GWR model for each of the 31 regions. One of the advantages of the GWR model is that the estimated results are spatially displayed, based on the resolution of the data used in the study. Since, in this this study, we conducted province- or city-level local regression, it is possible to display all the results using a provincial or municipal map of China generated by QGIS. Figure 1 presents the illustrated spatial variations in the EKC for haze in China.

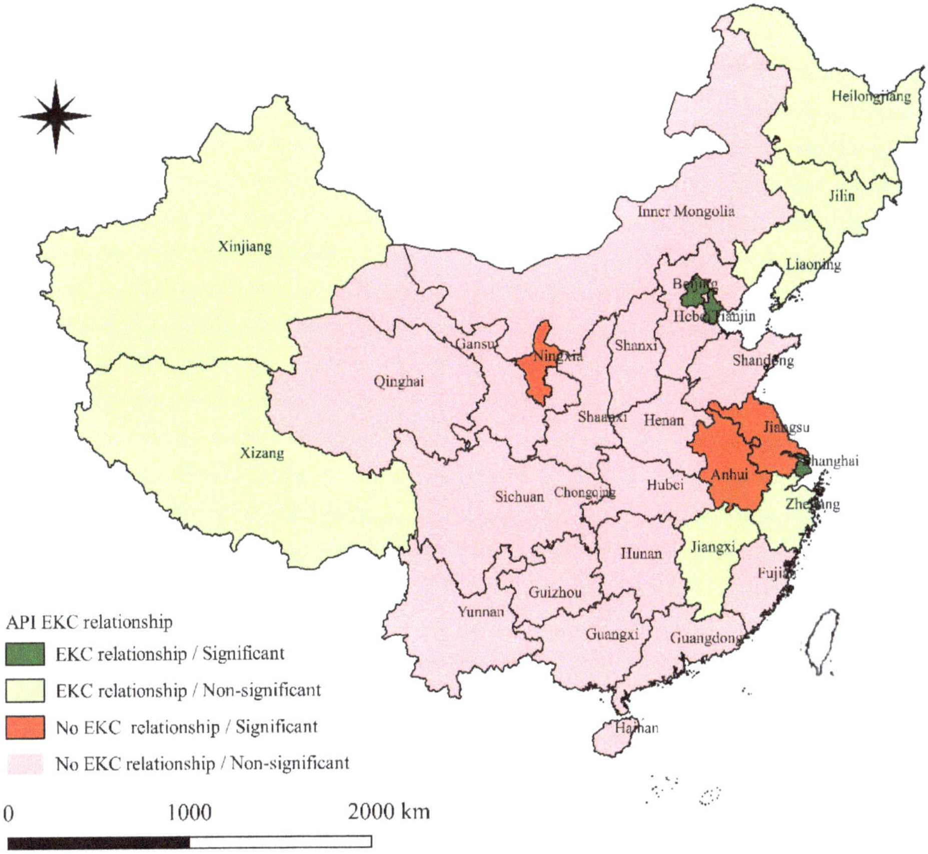

Figure 1. EKC for the haze in China.

Figure 1 shows the significant spatial heterogeneity of the EKC for haze in China. We colored the regions based on the significance and shape of their EKC (i.e., the symbols β_1 and β_2 in Equation (5)), which was calculated by the GWR estimation results. We verified the EKCs based on the sign and significance of the coefficients of the PGRP and the squared term of the PGRP variables. There were four cases based on the statistical significance and sign of the coefficients. Specifically, if the coefficient of β_1 had a plus sign and that of β_2 had a minus sign, the EKC had an inverted U-shape. Moreover, if β_1 and β_2 were statistically significant, it was possible to confirm the existence of EKCs in certain regions.

Six regions—Beijing, Gansu, Heilongjiang, Jiangxi, Jilin, Liaoning, Shanghai, Tianjin, Xinjiang, and Zhejiang—show inverted U-shaped EKCs. However, only three big cities (dark green colored regions in Figure 1)—Beijing, Tianjin, and Shanghai—show a statistically significant inverted U-shape between air quality and economic performance. According to the results, there are no EKCs in the other 25 provinces and cities, and only three regions—Anhui, Jiangsu, and Ningxia—are statistically significant. These findings support the necessity of balanced environmental regulation, even if China's internal policy still prioritizes economic growth. When considering these spatial analysis results, it is important to note that the initial control policies are currently working in certain big cities, such as, Beijing, Shanghai, and Tianjin, albeit to a lesser degree of effectiveness. However, it is necessary to perform a long-term analysis of the EKCs for haze, especially in statistically non-significant regions.

3.3. Estimation of the Turning Points of the EKCs for Haze

Based on the empirical results from the GWR analyses, it was possible to calculate the turning points of the EKCs for the haze in China. We used local coefficients of the three regions that showed a significant EKC for haze. Table 5 presents the ratio of the actual GRP per population in 2017 to the calculated turning point based on the analytical results regarding the EKC for each province or city.

Table 5. Calculated Turning Points of EKC; unit: 100 million CNY.

Region	2017 PGRP	Ln (2017 PGRP)	Turning Point	Ratio
Beijing	128,994	11.768	2.944	1.334
Tianjin	118,944	11.686	2.735	1.306
Shanghai	126,634	11.749	1.205	1.114

Note: The turning point is the calculated inflection point of EKCs from the GWR results. The ratio is calculated by equation: 1 + (PCGRP2017 − Turning point)/Turning point. The ratio 1.000 means that the PGRP in 2017 reached the turning point. The turning point is calculated only for the regions with significant EKCs—Beijing, Tianjin, and Shanghai—that show a statistically significant inverted U-shape.

From these estimates, it is possible to simulate the trend in sustainable development after the analysis periods. Specifically, a ratio over 1.00 indicates that the GRP per population reached a calculated turning point. The pollutants start to decrease after this point, along with economic growth; therefore, it is possible to understand the condition of pollutants on the EKC of each province or city. Since the GWR results indicate significant spatial heterogeneity, only three big cities—Beijing, Tianjin, and Shanghai—passed their turning points. Their ratios were 1.334, 1.306, and 1.114, respectively. In these developed cities, significant political efforts and investment were made to improve the environmental quality. Consequently, these cities are more likely to achieve sustainable development than other regions.

In light of these findings, it is still difficult to define the relationship between economic growth and haze for other provinces or cities. It is possible that there are different shapes, not only for EKCs, but also for other relationships, such as linear relationships. Thus, a long-term analysis is needed for these provinces or cities, in order to capture the graph shapes of the relationship between economic growth and environmental pollutants.

4. Conclusions and Policy Recommendation

This study examined the EKC in China, focusing on both average relationships and spatial heterogeneity. The study applied the global models and the GWR model using the data regarding the haze in about 31 provinces or cities in China from 2000 to 2014. Using the GWR model, we explored the non-stationary spatial characteristics, since this model can be fitted to each data point and weighs all the observations as functions of distance from the regression point. As a result, it was possible to obtain local coefficients that varied spatially [20–25]. Based on the results of the GWR model, this study verified the existence of the EKC among regions and calculated the turning point by applying the real 2017 GRP per population data.

This study provided convincing evidence that the GWR model is valid for explaining the spatial heterogeneity of the EKC in China more effectively than the global models. The spatial variability test also confirmed that all the variables used in this study have spatial variability. The GWR model could determine the existence of an EKC for the haze and its spatial variations in China. Based on the empirical results from the GWR analysis, this study visualized the local coefficients and estimated the turning points of the EKCs. The results provided empirical evidence that only the three most developed regions in China—Beijing, Tianjin, and Shanghai—have statistically significant EKCs. Furthermore, the estimated turning points of the EKCs show that Beijing, Tianjin, and Shanghai have already passed their turning points. However, it was difficult to find an EKC for the remaining 28 provinces and cities. Moreover, the results indicated that there was no EKC for the haze in the three regions—Anhui, Jiangsu, and Ningxia.

The empirical results from the research period show that although three developed cities—Beijing, Tianjin, and Shanghai—showed a positive relationship between economic and environmental performance, while other regions were still suffering from the vicious cycle of air pollution. To achieve harmony between t environmental and economic development, it is necessary to find an effective way to achieve balanced economic growth by considering the spatial heterogeneity of the levels of development and pollution. The Chinese authorities are trying to find a way to balance productivity with pressing environmental issues through their policies and subsequent economic plans, in order to ensure the sustainability of China's economy. For example, there are environmental regulations regarding air pollution in China, such as the Two-Control Zone policy, which was established in 1998 to define acid rain control zones in the southern regions and SO_2 control zones in the north-eastern regions in China. Moreover, the Chinese government announced regulations on environmental protection, such as the Taxation Law on Environmental Protection and the Law on the Prevention and Treatment of Water Pollution to consider environmental degradation [39].

The empirical results of the current study show that the policies and laws regarding the prevention and control of atmospheric pollution need to be improved. Stronger macroeconomic regulations and control measures to promote the transformation of enterprise development can be effective. This study found significant spatial heterogeneity, confirming the EKC relationship between air quality and economic growth only in three big cities. In other words, most regions in China are still far from achieving sustainable development. From these findings, region-specific policies should be established to consider regional differences. For example, it would be effective for policymakers to tighten environmental regulations on air pollution, especially in statistically significant no EKC regions, such as Anhui, Jiangsu, and Ningxia. Additionally, monitoring and evaluation systems, including the management of haze indexes, are needed to ensure long-term effectiveness. Enhancing financial support and updating haze indexes in a timely manner can be also effective methods to improve air quality.

In addition, the Chinese government is trying to shift the development focus to balanced economic growth by converting the country's energy structure [40]. As part of these efforts, renewable energy laws and policies were established by the National Energy Administration to improve China's energy consumption structure. The Renewable Energy Law is an example of these efforts [41]. Since one of the main sources of haze is coal consumption, reforming the old pattern of long-term economic transformation from a "high-carbon" to a "low-carbon" economy will be effective [28]. To change the original energy consumption structure, active step-by-step actions must be taken to explore new low-pollution energy sources to reduce the consumption of fossil fuels [42].

Although this study provides empirical evidence for the spatial heterogeneity of the EKC for the haze in China, there is still room for improvement. This study applied the haze levels measured in API per capita in one representative station of each region. Estimation using API microdata in specific monitoring stations can be more effective for grasping the spatial heterogeneity of the EKC in China. Moreover, the factors that affect haze were not comprehensively evaluated in this study. For example, energy structure, industrial structure, capital intensity, and environmental management capacities can be considered. These factors were not evaluated mostly due to the existence of multicollinearity and data availability. Thus, it may be necessary to investigate more socio-economic factors and specify them both theoretically and statistically to design effective policies to achieve balanced economic growth in China. Furthermore, further studies could investigate the relationship between environmental quality and economic performance with more advanced models and data, such as through the use of the Distance between Indices of Simulation and Observation, which is a new comprehensive index presenting the overall performances of different models [43,44]. A comprehensive approach to determining regional trends and the current state of pollution will help to inform future policy on sustainable development in China.

Author Contributions: Conceptualization, A.S.A.-R., Y.K. and L.Y.; methodology, A.S.A.-R., Y.K. and L.Y.; software, Y.K. and L.Y.; validation, Y.K.; formal analysis, A.S.A.-R., Y.K. and L.Y.; investigation, A.S.A.-R., Y.K. and L.Y.; resources, A.S.A.-R. and L.Y.; data curation, A.S.A.-R., Y.K. and L.Y.; writing—original draft preparation, A.S.A.-R., Y.K. and L.Y.; writing—review and editing, Y.K.; visualization, Y.K.; supervision, A.S.A.-R.; project administration, A.S.A.-R. and Y.K. All authors have read and agreed to the published version of the manuscript.

Funding: No funding was received for conducting this study.

Institutional Review Board Statement: Not applicable.

Informed Consent Statement: Not applicable.

Data Availability Statement: The data in this study are available upon reasonable request.

Conflicts of Interest: The authors declare that they have no conflict of interest.

References

1. Ma, L.J. Urban transformation in China, 1949–2000: A review and research agenda. *Environ. Plan A* **2002**, *34*, 1545–1569. [CrossRef]
2. Kim, Y. Spatial analysis of factors affecting location decisions of Korean companies in China. *Reg. Sci. Policy Pract.* **2019**, *11*, 383–401. [CrossRef]
3. World Health Organization (WHO). *Ambient Air Pollution: A Global Assessment of Exposure and Burden of Disease*; World Health Organization: Geneva, Switzerland, 2016; Available online: https://apps.who.int/iris/handle/10665/250141 (accessed on 15 April 2022).
4. Zhuang, X.; Wang, Y.; He, H.; Liu, J.; Wang, X.; Zhu, T.; Ge, M.; Zhou, J.; Tang, G.; Ma, J. Haze insights and mitigation in China: An overview. *J. Environ. Sci.* **2014**, *26*, 2–12. [CrossRef]
5. Quan, J.; Zhang, Q.; He, H.; Liu, J.; Huang, M.; Jin, H. Analysis of the formation of fog and haze in North China Plain (NCP). *Atmos. Chem. Phys.* **2011**, *11*, 8205–8214. [CrossRef]
6. Gao, M.; Guttikunda, S.K.; Carmichael, G.R.; Wang, Y.; Liu, Z.; Stanier, C.O.; Saide, P.E.; Yu, M. Health impacts and economic losses assessment of the 2013 severe haze event in Beijing area. *Sci. Total Environ.* **2015**, *511*, 553–561. [CrossRef]
7. Jalil, A.; Mahmud, S.F. Environment Kuznets curve for CO_2 emissions: A cointegration analysis for China. *Energy Policy* **2009**, *37*, 5167–5172. [CrossRef]
8. Hao, Y.; Niu, X.; Wang, J. Impacts of haze pollution on China's tourism industry: A system of economic loss analysis. *J. Environ. Manag.* **2020**, *295*, 113051. [CrossRef]
9. Long, X.; Naminse, E.Y.; Du, J.; Zhuang, J. Nonrenewable energy, renewable energy, carbon dioxide emissions and economic growth in China from 1952 to 2012. *Renew. Sustain. Energy Rev.* **2015**, *52*, 680–688. [CrossRef]
10. Wang, Y.; Han, R.; Kubota, J. Is there an environmental Kuznets curve for SO_2 emissions? A semi-parametric panel data analysis for China. *Renew. Sustain. Energy Rev.* **2016**, *54*, 1182–1188. [CrossRef]
11. Shen, J. A simultaneous estimation of environmental Kuznets curve: Evidence from China. *China Econ. Rev.* **2006**, *17*, 383–394. [CrossRef]
12. Tao, S.; Zheng, T.; Lianjun, T. An empirical test of the environmental Kuznets curve in China: A panel cointegration approach. *China Econ. Rev.* **2008**, *19*, 381–392.
13. Shu, L.; Fantang, Z.; Huaiyang, F.; Zhencheng, X. An empirical test of the environmental Kuznets curve in Guangdong Province, China. *APCBEE Procedia* **2012**, *1*, 204–209. [CrossRef]
14. Wang, Q.; Yuan, X.; Lai, Y.; Ma, C.; Ren, W. Research on interactive coupling mechanism and regularity between urbanization and atmospheric environment: A case study in Shandong Province, China. *Stoch. Environ. Res. Risk Assess.* **2012**, *26*, 887–898. [CrossRef]
15. Song, M.-L.; Zhang, W.; Wang, S.-H. Inflection point of environmental Kuznets curve in Mainland China. *Energy Policy* **2013**, *57*, 14–20. [CrossRef]
16. Du, G.; Liu, S.; Lei, N.; Huang, Y. A test of environmental Kuznets curve for haze pollution in China: Evidence from the penal data of 27 capital cities. *J. Clean. Prod.* **2018**, *205*, 821–827. [CrossRef]
17. Xu, T. Investigating environmental Kuznets curve in China—Aggregation bias and policy implications. *Energy Policy* **2018**, *114*, 315–322. [CrossRef]
18. Xie, Q.; Xu, X.; Liu, X. Is there an EKC between economic growth and smog pollution in China? New evidence from semiparametric spatial autoregressive models. *J. Clean. Prod.* **2019**, *220*, 873–883. [CrossRef]
19. Pata, U.K.; Caglar, A.E. Investigating the EKC hypothesis with renewable energy consumption, human capital, globalization and trade openness for China: Evidence from augmented ARDL approach with a structural break. *Energy* **2021**, *216*, 119220. [CrossRef]
20. Hao, Y.; Liu, Y.-M. The influential factors of urban $PM_{2.5}$ concentrations in China: A spatial econometric analysis. *J. Clean. Prod.* **2016**, *112*, 1443–1453. [CrossRef]

21. Kim, Y.; Tanaka, K.; Ge, C. Estimating the provincial environmental Kuznets curve in China: A geographically weighted regression approach. *Stoch. Environ. Res. Risk Assess.* **2018**, *32*, 2147–2163. [CrossRef]
22. Ma, Y.-R.; Ji, Q.; Fan, Y. Spatial linkage analysis of the impact of regional economic activities on $PM_{2.5}$ pollution in China. *J. Clean. Prod.* **2016**, *139*, 1157–1167. [CrossRef]
23. Zeraibi, A.; Ahmed, Z.; Shehzad, K.; Murshed, M.; Nathaniel, S.P.; Mahmood, H. Revisiting the EKC hypothesis by assessing the complementarities between fiscal, monetary, and environmental development policies in China. *Environ. Sci. Pollut. Res.* **2022**, *29*, 23545–23560. [CrossRef] [PubMed]
24. Chuai, X.; Huang, X.; Wang, W.; Wen, J.; Chen, Q.; Peng, J. Spatial econometric analysis of carbon emissions from energy consumption in China. *J. Geogr. Sci.* **2012**, *22*, 630–642. [CrossRef]
25. Kim, Y.; Tanaka, K.; Zhang, X. A spatial analysis of the causal factors influencing China's air pollution. *Asian J. Atmos. Environ.* **2017**, *11*, 194–201. [CrossRef]
26. Brajer, V.; Mead, R.W.; Xiao, F. Searching for an environmental Kuznets curve in China's air pollution. *China Econ. Rev.* **2011**, *22*, 383–397. [CrossRef]
27. Fang, M.; Chan, C.K.; Yao, X. Managing air quality in a rapidly developing nation: China. *Atmos. Environ.* **2009**, *43*, 79–86. [CrossRef]
28. Beşe, E.; Friday, H.S.; Özden, C. Coal consumption environmental Kuznets curve (EKC) in China and Australia: Evidence from ARDL model. *J. Appl. Bus. Econ.* **2020**, *22*, 25–36.
29. Koshta, N.; Bashir, H.A.; Samad, T.A. Foreign trade, financial development, agriculture, energy consumption and CO_2 emission: Testing EKC among emerging economies. *Indian Growth Dev. Rev.* **2020**, *14*, 50–80. [CrossRef]
30. Stern, D.I. The rise and fall of the environmental Kuznets curve. *World Dev.* **2004**, *32*, 1419–1439. [CrossRef]
31. Kang, Y.Q.; Zhao, T.; Yang, Y.Y. Environmental Kuznets curve for CO_2 emissions in China: A spatial panel data approach. *Ecol. Indic.* **2016**, *63*, 231–239. [CrossRef]
32. Payus, C.M.; Syazni, M.N.; Sentian, J. Extended air pollution index (API) as tool of sustainable indicator in the air quality assessment: El-Nino events with climate change driven. *Heliyon* **2022**, *8*, e09157. [CrossRef]
33. Dong, F.; Zhang, S.; Long, R.; Zhang, X.; Sun, Z. Determinants of haze pollution: An analysis from the perspective of spatiotemporal heterogeneity. *J. Clean. Prod.* **2019**, *222*, 768–783. [CrossRef]
34. National Bureau of Statistics of China. Available online: http://www.stats.gov.cn (accessed on 15 April 2022).
35. Wheeler, D.C. Geographically weighted regression. In *Handbook of Regional Science*; Fischer, M.M., Nijkamp, P., Eds.; Springer: Berlin/Heidelberg, Switzerland, 2014; pp. 1435–1459.
36. Fotheringham, A.S.; Brunsdon, C.; Charlton, M. *Geographically Weighted Regression: The Analysis of Spatially Varying Relationships*; John Wiley & Sons: Chichester, UK, 2002.
37. Shi, T.; Zhang, W.; Zhou, Q.; Wang, K. Industrial structure, urban governance and haze pollution: Spatiotemporal evidence from China. *Sci. Total Environ.* **2020**, *742*, 139228. [CrossRef] [PubMed]
38. Huang, R.-J.; Zhang, Y.; Bozzetti, C.; Ho, K.-F.; Cao, J.-J.; Han, Y.; Daellenbach, K.R.; Slowik, J.G.; Platt, S.M.; Canonaco, F.; et al. High secondary aerosol contribution to particulate pollution during haze events in China. *Nature* **2014**, *514*, 218–222. [CrossRef] [PubMed]
39. Sun, W.; Yang, Q.; Ni, Q.; Kim, Y. The impact of environmental regulation on employment: An empirical study of China's Two Control Zone policy. *Environ. Sci. Pollut. Res.* **2019**, *26*, 26472–26487. [CrossRef]
40. Zhang, K.; Dearing, J.A.; Tong, S.L.; Hughes, T.P. China's degraded environment enters a new normal. *Trends Ecol. Evol.* **2016**, *31*, 175–177. [CrossRef]
41. Liu, J. China's renewable energy law and policy: A critical review. *Renew. Sustain. Energy Rev.* **2019**, *99*, 212–219. [CrossRef]
42. Bekun, F.V. Mitigating emissions in India: Accounting for the role of real income, renewable energy consumption and investment in energy. *Int. J. Energy Econ. Policy.* **2022**, *12*, 188–192. [CrossRef]
43. Zhou, Q.; Chen, D.; Hu, Z.; Chen, X. Decompositions of Taylor diagram and DISO performance criteria. *Int. J. Climatol.* **2021**, *41*, 5726–5732. [CrossRef]
44. Hu, Z.; Chen, X.; Zhou, Q.; Chen, D.; Li, J. DISO: A rethink of Taylor diagram. *Int. J. Climatol.* **2019**, *39*, 2825–2832. [CrossRef]

atmosphere

MDPI

Article

Increased Exposure of China's Cropland to Droughts under 1.5 °C and 2 °C Global Warming

Lijuan Miao [1,*], Jing Zhang [1], Giri Raj Kattel [1,2,3] and Ran Liu [1]

1 School of Geographical Sciences, Nanjing University of Information Science and Technology, Ningliu Road 219, Nanjing 210044, China; 18360931642@163.com (J.Z.); gkattel@unimelb.edu.au (G.R.K.); liuran9909@163.com (R.L.)
2 Department of Infrastructure Engineering, The University of Melbourne, Melbourne 3010, Australia
3 Department of Hydraulic Engineering, Tsinghua University, Beijing 100084, China
* Correspondence: miaolijuan1111@gmail.com

Citation: Miao, L.; Zhang, J.; Kattel, G.R.; Liu, R. Increased Exposure of China's Cropland to Droughts under 1.5 °C and 2 °C Global Warming. *Atmosphere* **2022**, *13*, 1035. https://doi.org/10.3390/atmos13071035

Academic Editors: Zengyun Hu, Xuguang Tang and Qinchuan Xin

Received: 11 May 2022
Accepted: 22 June 2022
Published: 29 June 2022

Abstract: Global warming and human activities have intensified the duration, frequency, and extent of climatic extremes. The projected rise in global mean annual temperature of 1.5 °C/2 °C is thought to have severe impacts on the population exposed to droughts. Although these impacts on humans have been widely explored, the impacts associated with the cropland exposed to droughts have not been widely investigated. Here, we have examined the spatiotemporal pattern of China's drought conditions and cropland exposure to droughts under global warming of 1.5 °C and 2 °C, along with the avoided impacts (as evaluated by the cropland exposure to droughts) when limiting the global warming to 1.5 °C instead of 2 °C. Results suggest that compared to the reference period (1995–2014), drought conditions will be alleviated when the projected rise in mean global temperature is limited to 1.5 °C rather than 2.0 °C. Although severe droughts tend to be mainly distributed in northwestern China, drought severities are increasing in southern China, especially in the southeastern region. In addition, the total cropland exposure to droughts across China exhibits an increasing trend in response to the 0.5 °C of additional global warming, especially in northwestern China and Huang−Huai−Hai region. If global warming could be limited to 1.5 °C, the avoided impact will exceed 30%, especially in northwestern China, southwestern China, and the Huang−Huai−Hai Plain. Furthermore, the rising cropland exposure to droughts under the 2 °C global warming is likely to be triggered by the rising frequencies of moderate and extreme droughts. Therefore, climate mitigation strategies are urgently needed to keep the global temperature rise below 1.5 °C, for the future sustainability of China's cropland.

Keywords: drought; cropland; CMIP6; exposure; scPDSI; China

1. Introduction

A 1.09 °C increase in global surface temperature was observed in 2011–2020, as compared to 1850–1900 [1]. The rate of global warming is believed to exceed the bounds of natural variability [2]. Substantial changes brought by this unprecedented rate of global warming are happening in the climatic extremes (e.g., droughts, floods, and typhoon) [1,3]. For example, the widespread occurrences of droughts around the world in the 21st century due to occasional anomalies in climatic variables, as well as non−climatic factors, are becoming increasingly disastrous to mankind (e.g., causing damage to crops and raising serious concerns about agricultural production) [4,5]. These high drought risks accelerated by rapid global warming around the world are consequently challenging water and food security, especially in those regions with dense populations [6]. This concern is not only limited to historical observations but also to future projections. The latest CMIP6 models projections reaffirm the increase of up to 200% in the widespread drying and severe drought over most of the world under moderate–high emissions scenarios [7]. Meanwhile, according to several model projections, greenhouse gas−induced global warming may lead to more severe and widespread drought conditions [8,9].

As a complex phenomenon, drought can be influenced by a variety of factors and may occur almost anywhere in the world. Due to the complex terrain and climate characteristics, China has been frequently threatened by drought events on multiple timescales [10–12]. Yu, et al. [13] reported that the drought severity in China had been aggravating since the late 1990s, and the dry areas have been expanding. Zhou, et al. [14] found that more frequent recurrence of extreme droughts in large geographical regions of China has gravely affected the livelihoods of farming communities by severing agricultural yields. In the context of global warming, the frequency and duration of severe droughts are increasing not only in dry regions, but also in humid and sub−humid regions. For example, in the spring of 2011, the middle and lower reaches of the Yangtze River experienced the worst drought since 1954, causing serious damage to local agriculture [15]. In 2011–2012, Yunnan suffered from the most severe drought, which lasted for almost three years and resulted in huge economic losses [16]. The frequent occurrence and intensification of droughts are increasingly threatening food security and the ecological environment, as well as the development of the economy in our country. Therefore, a comprehensive assessment of climate change-induced droughts is urgently needed for providing reliable information for policymakers in climate mitigation and adaptation [17].

However, until recently, not much potentially reliable information on drought mitigation and adaptation has been suggested to combat the damage of drought to China's agriculture. Some scholars developed drought indices to explore the mechanism and factors of drought occurrence and evolution in China in order to provide more accurate and comparable information on spatiotemporal variations in droughts for the farming communities [10,18–20]. A drought index is a necessary tool for drought assessment, for its importance in defining drought parameters and quantifying drought on different time scales. Until now, hundreds of drought indices have been proposed in the world [21]. Among the numerous drought indices, the Palmer Drought Severity Index (PDSI), the Standardized Precipitation Index (SPI), and the Standardized Precipitation Evapotranspiration Index (SPEI) can precisely reflect the meteorological drought and evaluate the agricultural drought [22–24]. The SPI, as a single variable index, is easy to compute and has a variety of time scales, which allows it to monitor both short−term drought and long-term drought [22,25]. However, it is also hard for the SPI to identify diverse categories of drought because it only considers the P and runoff (RO) [26]. The SPEI has been broadly used for characterizing multi-category drought for drought monitoring and projection, however, there are still some limitations for SPEI applications in different climate regions [27]. An improvement on SPI, SPEI measures drought severity mainly in terms of the P and PET, but research indicates that variation exists in the relations between P and PET in different climate regions [28]. Although PET plays an essential role in detecting water deficit, merely relying on it for detailing droughts can lead to biases in water-limited regions [29]. The PDSI has always been one of the most prominent indices for its ability of considering multiple surface water variants and precisely quantifying long-term changes in drought and aridity in the world [21,22,30]. Unlike the simple water balance represented in SPEI, PDSI adopts a two-layer bucket model to quantify the cumulative moisture departure in estimating the surface water−energy balance. Further, Dai, et al. [31] showed that the PDSI values are significantly correlated with measured soil moisture. Most importantly, actual evaporation is often determined, to a large degree, by the availability of soil moisture, not by PET. However, the original PDSI index still has some shortcomings. For example, it has strong dependence on data calibration and limitations in spatial comparability [32–34]. To overcome these deficiencies, the self-calibrating PDSI (scPDSI) was created in 2004 [34]. In comparison to the original PDSI, scPDSI can calibrate the PDSI by using local conditions. The scPDSI's superiority of reducing value range has already been proved on a global scale since 2011 [30], and it was found to perform better than the original PDSI in Europe and North America [35,36]. Moreover, by comparing seven drought indices in

China, Yang, et al. [37] found the scPDSI was the best at capturing the droughts in China. Therefore, in this study, we chose the scPDSI to detect the drought conditions.

Drought has multiple eco-hydrological and socioeconomic impacts on human society, such as increasing wildfire risks, water scarcity, loss of crops and livestock, and raising food prices [38]. Meanwhile, with the acceleration of population growth and urbanization, human settlements and livelihoods are expected to be more frequently exposed to droughts than any time in the past [39]. Exposure to droughts is not only restricted to people but also affects the entire ecosystem, for example, environmental resources, economic, social, and cultural assets that could be adversely affected [40]. However, more attention from previous studies has been paid to drought and its impacts on population [17,41,42]. Although droughts are causing severe impacts on agriculture in the 21st century, cropland exposure to droughts in China is far less studied. The cropland affected by drought was larger than 200 thousand km^2 per year and the annual direct economic loss was larger than 34 billion yuan during 1984–2018 [43]. Food security will continue to be a global concern in the future, given crop yield failure as well as increasing water scarcity [44,45]. Some of the drought-prone regions of China where livelihoods are highly dependent on rain-fed agriculture would directly face challenges of water and food security [46]. Thus, investigating the cropland exposure to droughts in such drought-prone areas in China is potentially significant to ensure the future food security.

Global warming has become an increasing concern in recent years. Previous studies showed that climate change would have a significant impact on global food production and water resources, extreme weather and climate events, as well as on human health, when global warming reached 1–2 °C [17]. Moreover, under the global warming of 2.0 °C, extreme heat will frequently reach the tolerance thresholds of human health and agriculture production, posing widespread and serious threats to human livelihoods as well as to the ecological environment [6]. To reduce the climate risks, the Paris Agreement proposed to hold the global temperature rise well below 2.0 °C above preindustrial levels and to pursue efforts to limit the warming to 1.5 °C [47]. To meet this commitment, there have been multiple efforts devoted to investigate the variations in climatic extremes under the 1.5 °C and 2.0 °C global warming scenarios in recent years [48–50]. The Coupled Model Intercomparison Project (CMIP), organized by the World Climate Research Programme's (WCRP) Working Group on Coupled Modelling (WGCM) 20 years ago [51], is aimed at better understanding the past, present, and future climate change rising from natural, unforced variability or in response to changes in radiative forcings in a multi-model context [52]. Compared with Coupled Model Intercomparison Project Phase 5 (CMIP5), models involved in Coupled Model Intercomparison Project Phase 6 (CMIP6) additionally considered socioeconomic factors, i.e., the shared socioeconomic pathways (SSPs). Instead of the single representative concentration pathways (RCPs) in CMIP5, the SSPs work in harmony with RCPs in CMIP6 under shared policy assumptions, making future scenarios more reasonable [19,53]. Moreover, the great improvements and suitability of CMIP6 models in simulating temperature and precipitation have already been proven in China [54]. Therefore, the recently released and designed CMIP6 models are reliable sources to reveal future climate changes in China.

In this study, we focused on three objectives: (1) to figure out the spatiotemporal variations in droughts in China under global warming of 1.5 °C and 2.0 °C; (2) to explore the impacts of droughts on cropland in a warmer world, namely, the variation in cropland exposure to droughts in a 1.5 °C/2.0 °C warmer climate; and (3) to what extent the cropland exposure to droughts could be avoided if the global warming target is limited to 1.5 °C instead of 2.0 °C. Addressing these critical issues is useful for understanding droughts and their long-term impacts on China's cropland. It would further help policymakers to develop adaptation and mitigation strategies and to strengthen societal resilience to future drought-induced emergencies.

2. Data and Methods

2.1. Datasets

2.1.1. Climate Observations

We obtained the monthly gridded climate variables from the National Climate Centre of China Meteorological Administration (available from: http://data.cma.cn/ (accessed on 1 September 2020)). These variables include temperature, precipitation, wind speed, relative humidity, and shortwave radiation. This dataset is interpolated from more than 2400 ground-based observations, featuring a spatial resolution of 0.5° and spanning from the period 1961 to 2014.

2.1.2. CMIP6 Model Simulations

Climate simulations are downloaded from the World Research Programmer's (WCRP) Coupled Model Intercomparison Project Phase 6 (CMIP6) (available from: https://esgf-node.llnl.gov/projects/cmip6/ (accessed on 1 September 2020)). This involves historical simulations for 1961–2014 and future simulations for 2015–2100 under different SSPs−RCPs. In this study, we selected three scenarios, i.e., SSP1−2.6 (denoting a green/low-gas-emission pathway in a sustainable world), SSP2−4.5 (denoting an intermediate−gas−emission pathway in a moderate world), and SSP5−8.5 (denoting a high−gas−emission pathway in a rapid−fossil fuel−evolution world) [55,56]. Here, model outputs include monthly precipitation, temperature, wind speed, downward shortwave radiation, and relative humidity (as listed in Table 1). Considering the availability of the required variables and climate change scenarios, four models (CanESM5, IPSL−CM6A−LR, MIROC6, and MRI−ESM2−0) under SSP1−2.6, SSP2−4.5, and SSP5−8.5 were applied in this research (as listed in Table 2). CMIP6 outputs will be bias-corrected against observational data by applying the equidistant cumulative distribution function (EDCDF, refer to Section 2.2.1 for more detail) method and resampled to a regular 0.5° spatial resolution through a spatial disaggregation method [4,57,58].

Table 1. Detailed information of parameters in climatic datasets.

Period	Parameter	Unit
History (1961–2014)	Monthly temperature	°C
	Monthly precipitation	mm
	Monthly wind speed	m/s
Future (2015–2100)	Monthly downward shortwave radiation	W/m^2
	Monthly relative humidity	%

Table 2. Detailed information on CMIP6 models.

Model Name	Modeling Group	Original Resolution
CanESM5	Centre for Climate Modeling and Analysis, Canada	2.8125° × 2.7906°
IPSL−CM6A−LR	Institute Pierre Simon Laplace, France	2.5° × 1.2676°
MIROC6	Atmosphere and Ocean Research Institute, Japan	1.4063° × 1.4008°
MRI-ESM2-0	Planck Meteorological Institute, Germany	1.125° × 1.1215°

2.1.3. Historical and Future Land Use

The land use map of 2010 from the Data Centre of Resources and Environment, Chinese Academy of Sciences (CAS) was used to represent the historical land use conditions during the reference period (1995–2014) (available at: https://www.resdc.cn/ (accessed on 1 September 2020)), and its spatial resolution is 1 km. The harmonized set of scenarios developed by the Land−Use Harmonization project 2 (LUH2) was also applied to indicate future land use (available at: https://luh.umd.edu/index.shtml (accessed on 1 September

2020)). This dataset aims to estimate the fractional land-use patterns and land-use transitions and the key agricultural management information from the year 850 to 2100 at 0.25° resolution [59]. The land−use type of the LUH2 dataset could be divided into five main classes: primary vegetation (never impacted by human activities) and secondary vegetation (recovering from human disturbance), urban land, croplands, and pastures; the cropland includes all five crop types (e.g., C_3 annual and perennial, C_4 annual and perennial, and C_3 nitrogen−fixing) [59]. As mentioned earlier, to spatially match the land use data with the climate datasets, we resampled the land use data to 0.5° spatial resolution.

2.2. Methods

2.2.1. Bias Correction

In this study, the systematic bias between climate simulations and climate observations is corrected by the Equidistant Cumulative Distribution Functions (EDCDF) method [57]. The EDCDF method can be written as:

$$x_{corrected} = x + F_{oc}^{-1}(F_{ms}(x)) - F_{mc}^{-1}(F_{ms}(x)) \quad (1)$$

Here, x is the climate variable, F is the cumulative distribution (CDF), *oc* denotes observations in the training period, *mc* denotes model outputs in the training period, and *ms* denotes model outputs in a correction period.

GCM outputs are downscaled to a common resolution of 0.5° by the Spatial Disaggregation (SD) method [4,58]. In this study, the bilinear interpolation method was applied to interpolate the observational variables over China to GCM coarse resolution. Anomaly fields of temperature between observational data and bias-corrected model outputs are defined as the difference between them. For precipitation, wind speed, relative humidity, and shortwave radiation, the anomaly field is the ratio of GCM output to observational data. After the above process, the downscaled GCM simulations are obtained eventually.

2.2.2. The Self-Calibrating Palmer Drought Severity Index

Similar to PDSI, the computation of scPDSI involves four surface water fluxes (i.e., evapotranspiration, soil recharge, runoff, and water loss to the soil) [24]. Discrepancies between the PDSI and scPDSI are the empirical constants and the duration factors. In contrast to PDSI, these values of the scPDSI are generated automatically based on the historical climate information of a location, and thus have better spatial comparability. In addition, according to previous studies, biases in the estimation of PET can lead to an overestimation of drying trends [60,61]. Due to the strong recommendation for drought analysis in China, we used the Penman–Monteith method to evaluate the potential evapotranspiration (PET) in this study. More details about the computational procedures of scPDSI and PET based on the Penman–Monteith method can refer to Wells et al. [34] and Burke et al. [8]. The drought severity calculated by scPDSI can be categorized into four groups: near-normal dry (−1.99 to 0), moderately dry (−2.99 to −2), severely dry (−3.99 to −3.0), and extremely dry conditions (≤−4.0) [24]. It is universally accepted that the scPDSI ≤ -2 denotes a drought event. Meanwhile, the drought area in this study is extracted as the ratio of the sum of pixels where the scPDSI ≤ -2 to total pixels. The drought frequency is defined as the ratio of the dry months (monthly scPDSI ≤ -2) to the total months [41,62].

2.2.3. The Cropland Exposure to Droughts

Cropland exposure to droughts is defined as the cropland area exposed to moderate, severe, and extreme droughts, respectively (i.e., the frequencies of these droughts multiplied by drought-affected cropland area, as described by various authors) [41,63]. In this study, we compared the changes in cropland exposure to drought under 1.5 °C and 2 °C global warming levels with the reference period (1995–2014).

2.2.4. Avoided Impacts of Cropland Exposure to Droughts

The impact of cropland exposure to droughts that are avoided under a 1.5 °C global warming period compared with a 2 °C global warming period is defined as AI [41], which is estimated as below:

$$AI = \frac{C_{2.0} - C_{1.5}}{C_{2.0}} \times 100\% \quad (2)$$

where AI is the avoided impact and $C_{1.5}$ and $C_{2.0}$ are the changes under 1.5 °C and 2 °C global warming levels, respectively, compared with the reference period (1995–2014).

3. Results

3.1. Bias Correction of CMIP6 Models

As a result of the proven advantage of comparing different model data in previous studies [64,65], the Taylor diagram was applied in this study to evaluate the performance of the bias-corrected CMIP6 data against the climate observation data.

After the bias correction, annual mean temperature and annual total precipitation derived from CMIP6 models are relatively consistent with climate observation data (Figure 1), with correlation coefficients above 0.9 and a RMS (Root Mean Square) error of less than 0.4. A better simulation accuracy in both annual mean temperature and the annual total precipitation was achieved by Multi−Model Ensemble (MME) based on four CMIP6 models (Figure 1a,b). We also compared monthly mean temperature and monthly total precipitation over 1961–2014, derived from climate observations, with climate simulations (Figure S1). Results show that the method of MME could capture monthly variations in temperature and precipitation quite well.

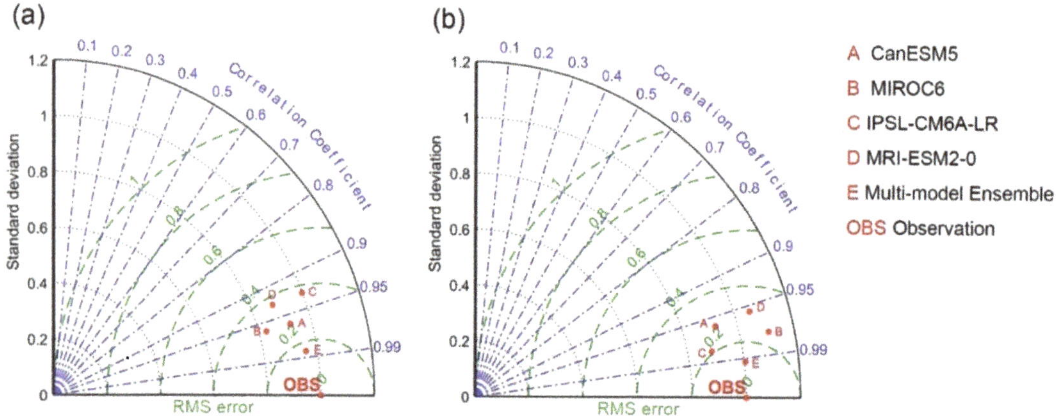

Figure 1. Taylor diagram of (**a**) annual mean temperature and (**b**) annual total precipitation for CMIP6 outputs, muti-model ensembles, and climate observations (1961–2014) in China.

3.2. Variations and Projections of Temperature and Precipitation from 1995 to 2100 in China

Results show that the annual mean temperature of China during 1995–2014 increased at a rate of 0.39 °C/10a (Figure 2a). Meanwhile, the annual total precipitation slightly increased at a rate of 5.52 mm/10a, from 554 mm to 563 mm (Figure 2b). Projections in future temperature and precipitation show different results under the three scenarios. To be specific, under SSP1−2.6, the annual mean temperature is projected to rise at a relatively higher rate (0.23 °C/10a) in 2015–2070 than in 2071–2100 (0.15 °C/10a) (Figure 2a). Under SSP2−4.5, the annual mean temperature is projected to rise faster than that under SSP1−2.6 (0.30 °C/10a for 2015–2100 (Figure 2a). A continuously amplified warming trend is monitored with an ongoing warming rate of 0.75 °C/10a in 2015–2100 under SSP5−8.5 (Figure 2a). During the same period, the annual total precipitation is also projected to increase at a rate of 6.34 mm/10a, 9.73 mm/10a, and 19.02 mm/10a, respectively under SSP1−2.6, SSP2−4.5, and SSP5−8.5 (Figure 2b). Overall, both annual mean temperature and annual total precipitation are detected to increase in China during the reference period (1995–2014) and thereafter, especially under SSP5−8.5.

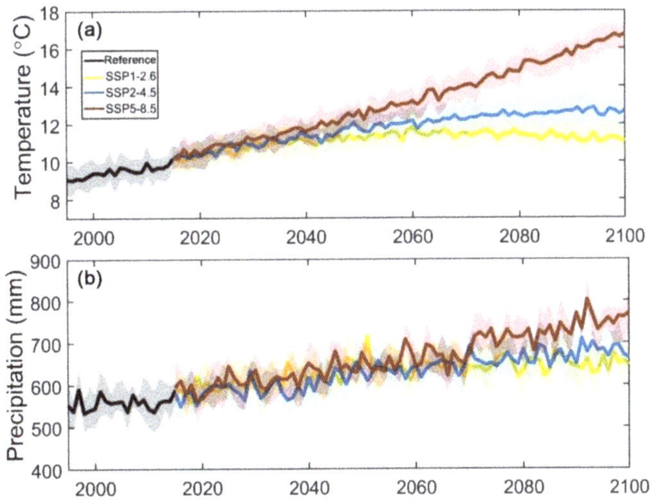

Figure 2. Variations and projections in (**a**) annual mean temperature and (**b**) annual total precipitation from 1995 to 2100 in China. Both annual mean temperature and annual total precipitation are detected to increase in China during 1995–2100, especially under SSP5−8.5.

3.3. Variations and Projections of Drought Conditions from 1995 to 2100 in China

According to CMIP6 model simulations, a stable increase of 1.5 °C (2.0 °C) global average temperature (above the preindustrial level) will occur in the years of 2025 (2056), 2026 (2043), and 2024 (2038) under SSP1−2.6, SSP2−4.5, and SSP5−8.5, respectively [66]. Both drought severity (scPDSI ≤ -2) in dry regions and drought areas in China are identified by scPDSI for the reference period (1995–2014) and the 1.5 °C/2.0 °C global warming periods under SSP1−2.6, SSP2−4.5, and SSP5−8.5, respectively (Figure 3).

Relative to the reference period (when the drought severity was estimated to be −2.9), the general drought severity in the 1.5 °C global warming period will be alleviated, albeit there were some differences in the result under SSP1−2.6, SSP2−4.5, and SSP5−8.5 (Figure 3a). The average drought severity of all three scenarios in the 1.5 °C global warming period was −2.8, which is slightly lower than the drought severity in the reference period. However, the drought severity during the 2 °C global warming period is higher than the general drought severity during the 1.5 °C global warming period (as the averaged drought severity of the three scenarios during the 2 °C global warming period was estimated to be −2.86). Therefore, the overall drought situation in China will become worse in the global warming period of 2.0 °C compared with the global warming period of 1.5 °C.

Changes in drought area in China (Figure 3b) are consistent with the variations in drought severity. The drought area is projected to shrink in the global warming period of 1.5 °C (as the averaged drought area of the three scenarios was estimated to be 19.6%) and then slightly increase in the global warming period of 2.0 °C (as the averaged drought area of the three scenarios was estimated to be 20.2%). The percentages of the drought area in the 1.5 °C and 2.0 °C global warming periods are slightly lower than the reference period (as the drought area was estimated to be 21.2%). In addition, compared with the global warming period of 1.5 °C, the drought area is detected to expand in the global warming period of 2.0 °C under both SSP1−2.6 and SSP2−4.5, yet the change in drought area between the 1.5 °C and 2.0 °C global warming periods under SSP5−8.5 is not obvious.

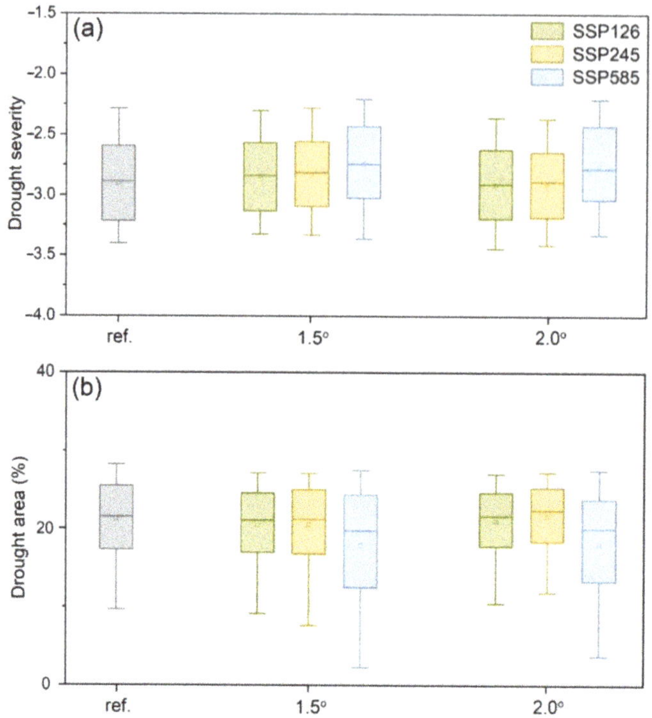

Figure 3. Drought severity (**a**) and drought area (**b**) in China for the reference period (1995–2014) and the 1.5 °C/2.0 °C global warming periods under SSP1−2.6, SSP2−4.5, and SSP5−8.5.

Severe droughts (scPDSI ≤ −3) were found to be mainly distributed in northwestern China during the reference period (1995–2014) (Figure 4a). The spatial distributions of dry regions (defined as regions with scPDSI ≤ −2) was found to differ from each other under the three SSPs−RCPs and two global warming periods. In the 1.5 °C global warming period (Figure 4b–d), the drought severity tends to become slightly increased in northeastern China under SSP1-2.6, but under SSP2−4.5 and SSP5−8.5, the drought severities tend to increase more in most regions of southern China, relative to the reference period. In the 2 °C global warming period (Figure 4e–g), a slight increase in drought severity is detected in northeastern China under SSP1−2.6, relative to the reference period. Note that compared to the reference period and the 1.5 °C global warming period, drought severities in the 2.0 °C global warming period under SSP2−4.5 and SSP5−8.5 will be maintained in southern China, especially in southeastern China.

Figure 5 shows the spatial distribution of drought frequency during different global warming periods. In general, the drought frequencies will both be enhanced in northwestern China in the 1.5 °C (Figure 5a–c) and 2 °C (Figure 5d–f) global warming periods, relative to the reference period. Compared with the 1.5 °C global warming period, the drought frequency will increase more in the 2 °C global warming period, especially in northwestern China (e.g., Xinjiang, Qinghai, and Inner Mongolia). In terms of different scenarios, under SSP1−2.6 and SSP2−4.5, the drought frequency in the northwest, southwest, and Huang−Huai−Hai Plain will increase significantly as a whole. Under SSP5−8.5, compared with the 1.5 °C global warming period, the drought frequencies in parts of northwestern China (e.g., Inner Mongolia and Qinghai) will decrease, while the drought frequencies in southwestern China and the Yellow River Basin will increase more.

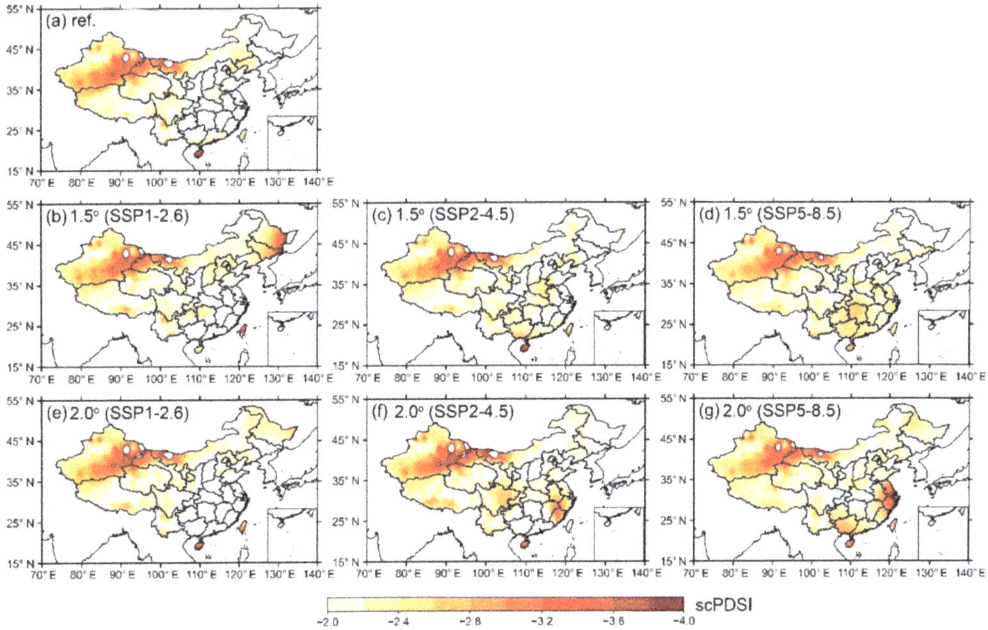

Figure 4. Spatial patterns of drought severity during (**a**) the reference period (1995–2014), (**b**–**d**) the 1.5 °C global warming period, and (**e**–**g**) the 2.0 °C global warming period.

3.4. Changes in Cropland Exposure to Droughts

Figure 6a shows the cropland exposure to droughts in the reference period (1995–2014), and Figure 6b–g show the changes in cropland exposure to droughts during global warming periods of 1.5 °C and 2.0 °C, relative to the reference period. Higher cropland exposure to droughts is generally observed in northwestern and southwestern China during the reference period. In addition, in comparison to the reference period, enhanced cropland exposure to droughts during the two global warming periods is detected, especially during the 2.0 °C global warming period. To be specific, compared to SSP1−2.6, the cropland exposure to droughts under SSP2−4.5 will greatly increase in northwestern China under 1.5 °C global warming. Such an increasing trend will be amplified in northwestern China and the Yellow River Basin under SSP5−8.5. The spatial distribution of the cropland exposure to droughts under global warming conditions of 2.0 °C is similar to the situation under global warming conditions of 1.5 °C, but with a more significant increasing trend, especially under SSP2−4.5. It should be noted that the future cropland exposure will also increase more in the Huang−Huai−Hai Plain, especially during the 2.0 °C global warming period.

The total cropland exposure to droughts in the reference period is 25,347 km^2 (Figure 7a). Under global warming of 1.5 °C, the cropland exposure to droughts under SSP1−2.6, SSP2−4.5, and SSP5-8.5 is 23,789 km^2, 26,617 km^2, and 23,256 km^2, respectively (Figure 7a). However, under the global warming of 2 °C, the cropland exposure to droughts decreased to 23,141 km^2 under SSP1−2.6, but increased to 32,854 km^2 under SSP2−4.5 and 24,888 km^2 under SSP5−8.5 (Figure 7a). In comparison to the reference period, the cropland exposure to droughts will decrease under SSP1−2.6 and SSP5−8.5, but increase under SSP2−4.5 in both 1.5 °C and 2.0 °C global warming periods. Furthermore, if the rise in global mean temperature is limited to 1.5 °C instead of 2 °C, the avoided impact will exceed 30%, especially in northwestern China, southwestern China, and the Huang−Huai−Hai region (Figure 7b). In addition, Figure 7c shows the cropland exposure to different kinds of droughts. We also found that the projected increase in cropland exposure mostly resulted from the moderate drought, which accounts for nearly 83.2% in the 1.5 °C global warming period and 81.8% in the 2 °C global warming period. The projected cropland exposure to extreme droughts only accounts for about 1.6% in the 1.5 °C global warming period and 2.3% in the 2 °C global warming period. That means, compared with the reference period (80.6% cropland exposure to moderate drought and 1.4% cropland exposure to extreme drought), the cropland exposure to moderate droughts and extreme droughts will increase in both the 1.5 °C and 2 °C global warming periods.

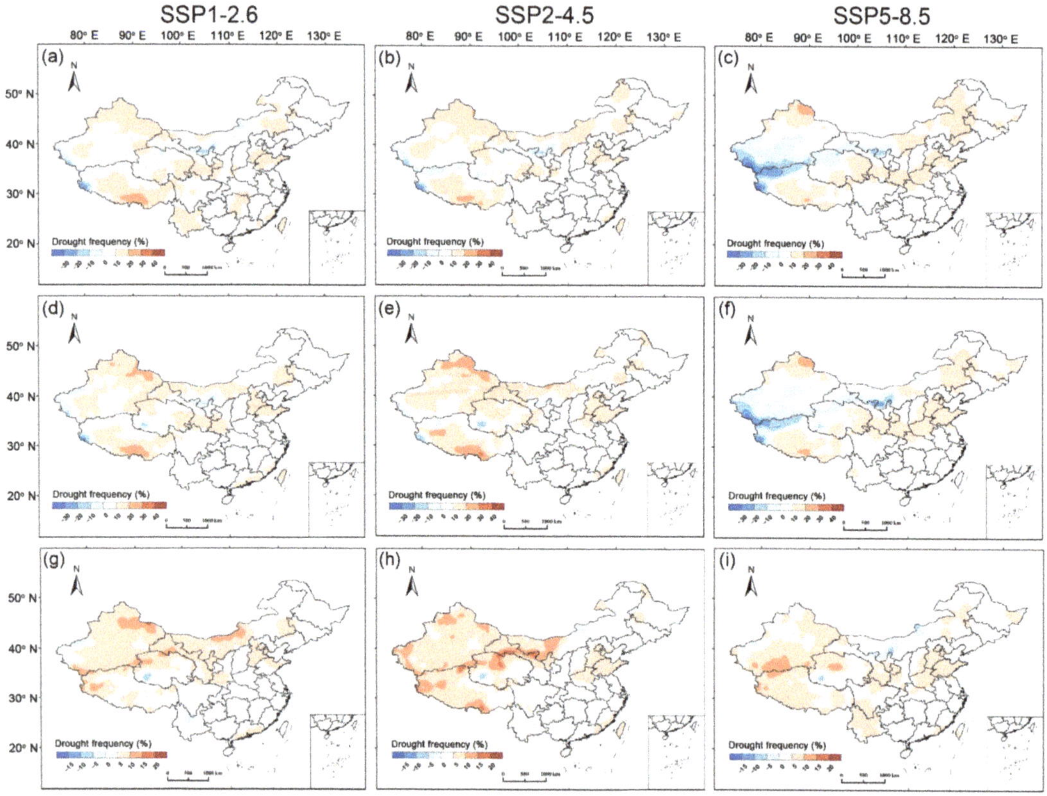

Figure 5. Spatial patterns of drought frequency change. (**a–c**) The 1.5 °C global warming period relative to the reference period (1995–2014), (**d–f**) the 2.0 °C global warming period relative to the reference period (1995–2014), and (**g–i**) the 2.0 °C global warming period relative to the 1.5 °C global warming period. Unit: %.

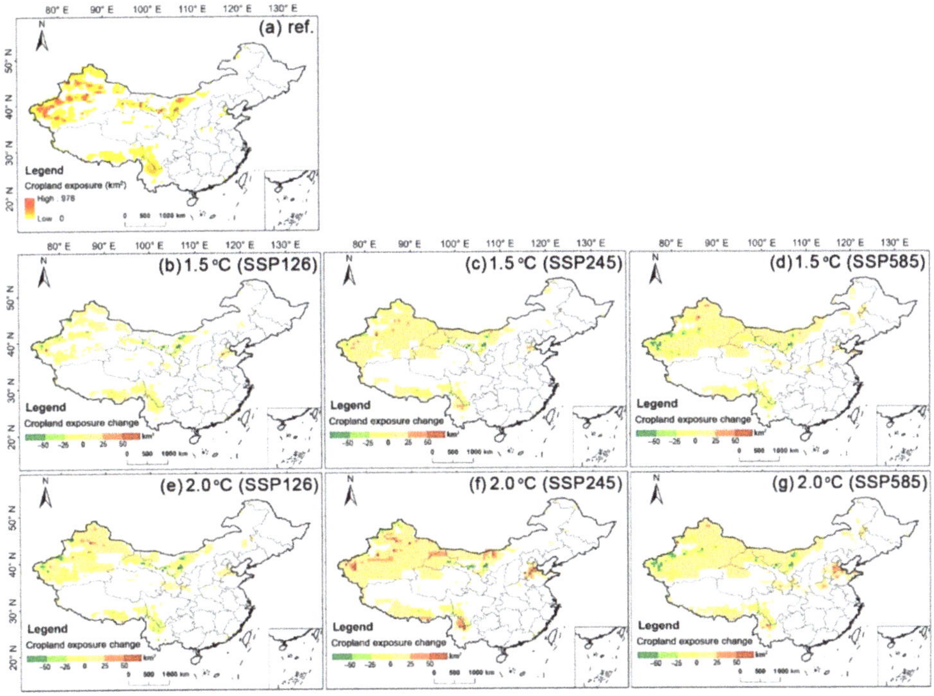

Figure 6. Spatial distributions of the cropland exposure to droughts in the reference period (**a**), the 1.5 °C global warming period (**b–d**), and the 2.0 °C global warming period (**e–g**). Units: km^2.

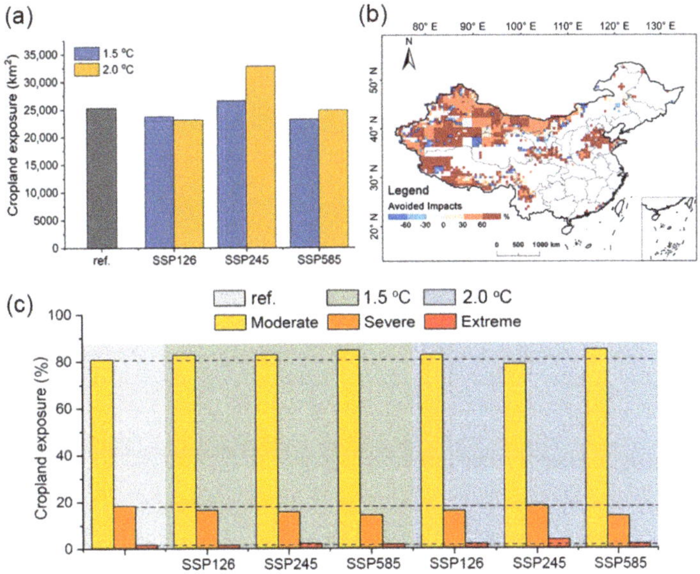

Figure 7. (**a**) Cropland exposure to droughts in reference period and the 1.5/2.0 °C global warming periods, unit: km^2; (**b**) Spatial distributions of the avoided impacts (the potential reduction in the cropland exposure to droughts) in China due to 0.5 °C less warming (limiting the global warming to 1.5 °C instead of 2.0 °C), unit: %; (**c**) Cropland exposure to moderate, severe, and extreme droughts in China in reference period and the 1.5/2.0 °C global warming periods, unit: %.

4. Discussion

Most CMIP6 models suggest that there will be a rise of 1.5 °C and 2 °C in global mean annual temperature by 2030 and 2050, respectively [66,67]. As the time is approaching, it is urgent to determine future drought conditions and their potential impacts on agricultural land. Our results indicate that both temperature and precipitation will increase rapidly in the mid−to−late 21st century, especially under the moderate–high emissions scenarios. Compared with 1.5 °C global warming, the overall drought severity and drought area in China will increase more in 2 °C global warming, especially under SSP2−4.5 and SSP5−8.5. This conclusion is in line with Su et al. [4] and Chen and Sun [17]. Our results also demonstrated that droughts will still be severe in northwestern China during 1.5 °C and 2 °C global warming periods. Qin et al. [68] reached a similar conclusion, that moderate and severe droughts will dominate in most of northwestern China during 2015–2100. This is mainly due to the rise in temperature in northern China [67]. We, therefore, should pay more attention to the occurrence of drought and its impacts on agricultural production particularly in northwestern China [69]. Moreover, southern China (especially southeastern China) is projected to witness even worse droughts during 1.5 °C and 2 °C global warming periods, especially under the SSP2−4.5 and SSP5−8.5. This is also consistent with the finding of Su et al. [19], which mentioned that increased changes in drought intensity were also found in humid regions (e.g., southeastern China). As a primary component of the water cycle, the evapotranspiration may influence the severe drought in a warmer context, to a large extent [70]. In addition, Su et al. [4] found that the evapotranspiration increased significantly in southern China during the 2 °C global warming period. Therefore, a possible explanation is that the growth of evapotranspiration in southern China exceed the increase in precipitation under the 2 °C global warming, although precipitation is projected to increase rapidly in the global warming world [19]. It should also be noted that the drought severity in southwestern China will also escalate in global warming periods of 1.5 °C and 2 °C. Considering the negative impact of the 2009–2012 extreme drought events in southwestern China on the local ecosystem [71] as well as our projections, China needs to preplan to combat the impacts posed by droughts in this region.

Agriculture is one of the most valuable fields among economic sectors, but climate change is altering the weather and, thus, it has a direct, biophysical effect on agricultural productivity [72]. To determine the potential impacts of drought on cropland, we calculated the cropland exposure to drought during global warming periods of 1.5 °C and 2 °C. Similar to the study of Spinoni et al. [73], our study also suggested that the cropland exposure to droughts will overall increase in the global warming period of 2.0 °C compared with the global warming period of 1.5 °C. A possible explanation is the increase in drought frequencies in the 2 °C global warming period, especially in northwestern China. In addition, we found that the spatial patterns of drought severities are distinct from those of cropland exposure to droughts in global warming periods of 1.5 °C and 2 °C. The overall cropland exposure to droughts exhibits an increasing pattern, especially in northwestern China, southwestern China, and the Huang−Huai−Hai region, while the drought severities are projected to increase noticeably in southwestern China and southeastern China. This may be related to the fact that drought severities are projected to increase more in southeastern China while drought frequencies are projected to be relatively lower. Considering there is no apparent difference in cropland area between the future and reference period (Figure S2), the increased cropland exposure to droughts in northwestern China, southwestern China, and the Huang-Huai-Hai region is probably due to the increasing drought frequencies.

We also found that the projected increase in cropland exposure in the 1.5 °C and 2 °C global warming periods mostly resulted from the moderate and extreme droughts, which is probably rooted in the increase in drought frequencies of moderate and extreme droughts. Considering the high incidence of moderate droughts in China and the destructive effects brought about by extreme droughts, China urgently needs to limit the adverse impacts on agriculture and develop strong measures to minimize the occurrences of droughts. In addition, if the rise in global mean temperature is limited to 1.5 °C, instead of 2 °C, the avoided impact will exceed 30%, especially in northwestern China, southwestern China, and Huang−Huai−Hai plain. From this perspective, the mitigation of warming by 0.5 °C is crucial to reduce cropland exposure, especially exposure to extreme droughts.

5. Conclusions

Taking advantage of the bias-corrected CMIP6 model simulations and the land-use datasets, we examined the spatiotemporal variations in drought conditions and cropland exposure to droughts in China under 1.5 °C/2 °C global warming scenarios. Results show that both the overall drought severity and frequency are projected to increase during the global warming period of 2 °C, compared with the global warming period of 1.5 °C. In terms of the distribution, the projected droughts will still dominate in northwestern China during global warming periods of 1.5 °C and 2 °C. Interestingly, a sudden increase in the drought severity is projected in humid, southeastern China, especially under SSP2−4.5 and SSP5−8.5, which may be related to the growth of evapotranspiration and the increase in precipitation in southern China under the 2 °C global warming condition. Meanwhile, cropland exposure to droughts exhibits an increasing trend in northwestern China, southwestern China, and the Huang−Huai−Hai region, in response to the 0.5 °C additional warming. We also found that the growing cropland exposure to droughts in the 2 °C global warming period is probably induced by the increased frequencies of moderate and extreme droughts. In addition, if the rise in global mean temperature is limited to 1.5 °C, instead of 2 °C, the avoided impact (the potential reduction in the cropland exposure to droughts) will exceed 30% in most areas in China. This study also proves the importance of mitigating global warming by 0.5 °C, which is crucial for climate adaptation policies and strategies in the 21st century.

Supplementary Materials: The following supporting information can be downloaded at: https://www.mdpi.com/article/10.3390/atmos13071035/s1. Figure S1. Comparison of muti-year monthly average temperature (a) and total precipitation (b) over China based on climate observations and CMIP6 outputs for 1961–2014; Figure S2. Spatial distribution of cropland area (km^2) for the reference period (1995–2014) (a), the 1.5 °C global warming period (b–d), and the 2 °C global warming period (e–g).

Author Contributions: Conceptualization, L.M. and J.Z.; methodology, L.M. and J.Z.; supervision, L.M.; writing—original draft, L.M., J.Z., G.R.K. and R.L.; writing—review and editing, L.M. and G.R.K. All authors have read and agreed to the published version of the manuscript.

Funding: This study was financially supported by the National Key Research and Development Program of China (2019YFC1510200), the Key Project of National Social and Scientific Fund Program (16ZDA047), the Natural Science Foundation of Jiangsu Province (SBK2020040616), the Natural Science Foundation of the Jiangsu Higher Education Institutions of China (21KJB170003), and the National Natural Science Foundation of China (4210010673).

Data Availability Statement: All data resources are provided in the manuscript with links.

Acknowledgments: Giri Raj Kattel and Lijuan Miao would like to acknowledge Longshan at Nanjing University of Information Science and Technology.

Conflicts of Interest: The authors declare no conflict of interest. All the authors declare that they have no known competing financial interests that could have appeared to influence the scientific work reported in this manuscript.

References

1. IPCC. In Proceedings of the Summary for Policymakers. In *Climate Change 2021: The Physical Science Basis. Contribution of Working Group I to the Sixth Assessment Report of the Intergovernmental Panel on Climate Change*; Cambridge University Press: Cambridge, UK; New York, NY, USA, 2021.
2. Karl, T.R.; Trenberth, K.E. Modern Global Climate Change. *Science* **2003**, *302*, 1719–1723. [CrossRef] [PubMed]
3. IPCC. Managing the risks of extreme events and disasters to advance climate change adaptation. In *A Special Report of Working Groups I and II of the Intergovernmental Panel on Climate Change*; Cambridge University Press: Cambridge, UK; New York, NY, USA, 2012.

4. Su, B.; Huang, J.; Fischer, T.; Wang, Y.; Kundzewicz, Z.W.; Zhai, J.; Sun, H.; Wang, A.; Zeng, X.; Wang, G. Drought losses in China might double between the 1.5 °C and 2.0 °C warming. *Proc. Natl. Acad. Sci. USA* **2018**, *115*, 10600–10605. [CrossRef] [PubMed]
5. IPCC. *Global Warming of 1.5 °C. An IPCC Special Report on the Impacts of Global Warming of 1.5 °C above Pre-Industrial Levels and Related Global Greenhouse Gas Emission Pathways, in the Context of Strengthening the Global Response to the Threat of Climate Change, Sustainable Development, and Efforts to Eradicate Poverty*; Cambridge University Press: Cambridge, UK; New York, NY, USA, 2018.
6. Fahad, S.; Saud, S.; Chen, Y.; Wu, C.; Wang, D.E. *Abiotic Stress in Plants*; IntechOpen: London, UK, 2021.
7. Zhao, T.; Dai, A. CMIP6 Model-Projected Hydroclimatic and Drought Changes and Their Causes in the Twenty-First Century. *J. Clim.* **2022**, *35*, 897–921.
8. Burke, E.J.; Brown, S.J.; Christidis, N. Modeling the recent evolution of global drought and projections for the twenty-first century with the Hadley Centre climate model. *J. Hydrometeorol.* **2006**, *7*, 1113–1125. [CrossRef]
9. Dai, A. Increasing drought under global warming in observations and models. *Nat. Clim. Chang.* **2013**, *3*, 52–58. [CrossRef]
10. Chen, H.; Sun, J. Changes in Drought Characteristics over China Using the Standardized Precipitation Evapotranspiration Index. *J. Clim.* **2015**, *28*, 5430–5447. [CrossRef]
11. Han, X.; Wu, J.; Zhou, H.; Liu, L.; Yang, J.; Shen, Q.; Wu, J. Intensification of historical drought over China based on a multi-model drought index. *Int. J. Climatol.* **2020**, *40*, 5407–5419. [CrossRef]
12. Zhai, J.; Huang, J.; Su, B.; Cao, L.; Wang, Y.; Jiang, T.; Fischer, T. Intensity–area–duration analysis of droughts in China 1960–2013. *Clim. Dyn.* **2016**, *48*, 151–168. [CrossRef]
13. Yu, M.; Li, Q.; Hayes, M.J.; Svoboda, M.D.; Heim, R.R. Are droughts becoming more frequent or severe in China based on the Standardized Precipitation Evapotranspiration Index: 1951–2010? *Int. J. Climatol.* **2014**, *34*, 545–558. [CrossRef]
14. Zhou, L.; Wu, J.; Mo, X.; Zhou, H.; Diao, C.; Wang, Q.; Chen, Y.; Zhang, F. Quantitative and detailed spatiotemporal patterns of drought in China during 2001–2013. *Sci. Total Environ.* **2017**, *589*, 136–145. [CrossRef]
15. Zhang, H.B.; Zhang, H.Y.; Chen, L.; Guo, J.P.; Shan, H.T.; Shang, J.L.; Liu, D.J. Anomalous Circulation of Droughts over the Middle and Lower Reaches of the Yangtze River in Spring of 2011. *Mod. Agric. Sci. Technol.* **2018**, *7*, 227–230.
16. Qiu, J. China drought highlights future climate threats: Yunnan's worst drought for many years has been exacerbated by destruction of forest cover and a history of poor water management. *Nature* **2010**, *465*, 142–144. [CrossRef] [PubMed]
17. Chen, H.; Sun, J. Increased population exposure to extreme droughts in China due to 0.5 °C of additional warming. *Environ. Res. Lett.* **2019**, *14*, 064011. [CrossRef]
18. Guo, H.; Bao, A.; Liu, T.; Jiapaer, G.; Ndayisaba, F.; Jiang, L.; Kurban, A.; De Maeyer, P. Spatial and temporal characteristics of droughts in Central Asia during 1966–2015. *Sci. Total Environ.* **2018**, *624*, 1523–1538. [CrossRef] [PubMed]
19. Su, B.; Huang, J.; Mondal, S.K.; Zhai, J.; Wang, Y.; Wen, S.; Gao, M.; Lv, Y.; Jiang, S.; Jiang, T.; et al. Insight from CMIP6 SSP-RCP scenarios for future drought characteristics in China. *Atmos. Res.* **2021**, *250*, 105375. [CrossRef]
20. Li, S.; Miao, L.; Jiang, Z.; Wang, G.; Gnyawali, K.R.; Zhang, J.; Zhang, H.; Fang, K.; He, Y.; Li, C. Projected drought conditions in Northwest China with CMIP6 models under combined SSPs and RCPs for 2015–2099. *Adv. Clim. Change Res.* **2020**, *11*, 210–217. [CrossRef]
21. Dai, A. Drought under global warming: A review. *Wiley Interdiscip. Rev. Clim. Chang.* **2011**, *2*, 45–65. [CrossRef]
22. McKee, T.B.; Doesken, N.J.; Kleist, J. The relationship of drought frequency and duration to time scales. In Proceedings of the 8th Conference on Applied Climatology, Anaheim, CA, USA, 17–22 January 1993; pp. 179–183.
23. Vicente-Serrano, S.M.; Beguería, S.; López-Moreno, J.I. A multiscalar drought index sensitive to global warming: The standardized precipitation evapotranspiration index. *J. Clim.* **2010**, *23*, 1696–1718. [CrossRef]
24. Palmer, W.C. *Meteorological Drought*; US Department of Commerce, Weather Bureau: Washington, DC, USA, 1965.
25. Mishra, A.K.; Singh, V.P. A review of drought concepts. *J. Hydrol.* **2010**, *391*, 202–216. [CrossRef]
26. Farahmand, A.; AghaKouchak, A. A generalized framework for deriving nonparametric standardized drought indicators. *Adv. Water Resour.* **2015**, *76*, 140–145. [CrossRef]
27. Ayantobo, O.O.; Wei, J. Appraising regional multi-category and multi-scalar drought monitoring using standardized moisture anomaly index (SZI): A water-energy balance approach. *J. Hydrol.* **2019**, *579*, 124139. [CrossRef]
28. Yang, Y.; Donohue, R.J.; McVicar, T.R.; Roderick, M.L. An analytical model for relating global terrestrial carbon assimilation with climate and surface conditions using a rate limitation framework. *Geophys. Res. Lett.* **2015**, *42*, 9825–9835. [CrossRef]
29. Paulo, A.; Rosa, R.; Pereira, L. Climate trends and behaviour of drought indices based on precipitation and evapotranspiration in Portugal. *Nat. Hazards Earth Syst. Sci.* **2012**, *12*, 1481–1491. [CrossRef]
30. Dai, A. Characteristics and trends in various forms of the Palmer Drought Severity Index during 1900–2008. *J. Geophys. Res.* **2011**, *116*, D12. [CrossRef]
31. Dai, A.; Trenberth, K.E.; Qian, T. A global dataset of Palmer Drought Severity Index for 1870–2002: Relationship with soil moisture and effects of surface warming. *J. Hydrometeorol.* **2004**, *5*, 1117–1130. [CrossRef]
32. Andreadis, K.M.; Clark, E.A.; Wood, A.W.; Hamlet, A.F.; Lettenmaier, D.P. Twentieth-Century Drought in the Conterminous United States. *J. Hydrometeorol.* **2005**, *6*, 985–1001. [CrossRef]
33. Sheffield, J.; Andreadis, K.M.; Wood, E.F.; Lettenmaier, D.P. Global and Continental Drought in the Second Half of the Twentieth Century: Severity-Area-Duration Analysis and Temporal Variability of Large-Scale Events J. *Clim.* **2009**, *22*, 1962–1981. [CrossRef]
34. Wells, N.; Goddard, S.; Hayes, M.J. A self-calibrating Palmer drought severity index. *J. Clim.* **2004**, *17*, 2335–2351. [CrossRef]

35. Van der Schrier, G.; Briffa, K.; Osborn, T.; Cook, E. Summer moisture availability across North America. *J. Geophys. Res. Atmos.* **2006**, *111*, D11. [CrossRef]
36. Van der Schrier, G.; Efthymiadis, D.; Briffa, K.; Jones, P. European Alpine moisture variability for 1800–2003. *Int. J. Climatol.* **2007**, *27*, 415–427. [CrossRef]
37. Yang, Q.; Li, M.; Zheng, Z.; Ma, Z. Regional applicability of seven meteorological drought indices in China. *Sci. China Earth Sci.* **2017**, *47*, 337–353. [CrossRef]
38. Mukherjee, S.; Mishra, A.; Trenberth, K.E. Climate Change and Drought: A Perspective on Drought Indices. *Curr. Clim. Chang. Rep.* **2018**, *4*, 145–163. [CrossRef]
39. UNDESA. *The World's Cities in 2016*; United Nations: New York, NY, USA, 2016.
40. Field, C.B.; Barros, V.R. *Climate Change 2014–Impacts, Adaptation and Vulnerability: Regional Aspects*; Cambridge University Press: Cambridge, UK; New York, NY, USA, 2014.
41. Chen, J.; Liu, Y.; Pan, T.; Liu, Y.; Sun, F.; Ge, Q. Population exposure to droughts in China under the 1.5 °C global warming target. *Earth Syst. Dyn.* **2018**, *9*, 1097–1106. [CrossRef]
42. Sun, H.; Wang, Y.; Chen, J.; Zhai, J.; Jing, C.; Zeng, X.; Ju, H.; Zhao, N.; Zhan, M.; Luo, L.; et al. Exposure of population to droughts in the Haihe River Basin under global warming of 1.5 and 2.0 °C scenarios. *Quat. Int.* **2017**, *453*, 74–84. [CrossRef]
43. China Meteorological Administration. *Yearbook of Meteorological Disasters in China*; Meteorological Press of China: Beijing, China, 2019.
44. Rosegrant Mark, W.; Cline Sarah, A. Global Food Security: Challenges and Policies. *Science* **2003**, *302*, 1917–1919. [CrossRef]
45. Kang, Y.; Khan, S.; Ma, X. Climate change impacts on crop yield, crop water productivity and food security—A review. *Prog. Nat. Sci.* **2009**, *19*, 1665–1674. [CrossRef]
46. He, X.; Estes, L.; Konar, M.; Tian, D.; Anghileri, D.; Baylis, K.; Evans, T.P.; Sheffield, J. Integrated approaches to understanding and reducing drought impact on food security across scales. *Curr. Opin. Environ. Sustain.* **2019**, *40*, 43–54. [CrossRef]
47. UNFCCC. Report of the ad hoc working group on the Durban platform for enhanced action on the eighth part of its second session, held in Geneva from 8 to 13 February 2015. In Proceedings of the United Nations Framework Convention on Climate Change, Geneva, Switzerland, 8–13 February 2015; United Nations: Geneva, Switzerland, 2015.
48. Jiang, D.; Sui, Y.; Lang, X. Timing and associated climate change of a 2 °C global warming. *Int. J. Climatol.* **2016**, *36*, 4512–4522. [CrossRef]
49. King, A.D.; Karoly, D.J.; Henley, B.J. Australian climate extremes at 1.5 °C and 2 °C of global warming. *Nat. Clim. Chang.* **2017**, *7*, 412–416. [CrossRef]
50. Paltan, H.; Allen, M.; Haustein, K.; Fuldauer, L.; Dadson, S. Global implications of 1.5 °C and 2 °C warmer worlds on extreme river flows. *Environ. Res. Lett.* **2018**, *13*, 094003. [CrossRef]
51. Meehl, G.A.; Boer, G.J.; Covey, C.; Latif, M.; Stouffer, R.J. Intercomparison makes for a better climate model. *Eos Trans. Am. Geophys. Union* **1997**, *78*, 445–451. [CrossRef]
52. Eyring, V.; Bony, S.; Meehl, G.A.; Senior, C.A.; Stevens, B.; Stouffer, R.J.; Taylor, K.E. Overview of the Coupled Model Intercomparison Project Phase 6 (CMIP6) experimental design and organization. *Geosci. Model Dev.* **2016**, *9*, 1937–1958. [CrossRef]
53. O'Neill, B.C.; Tebaldi, C.; Van Vuuren, D.P.; Eyring, V.; Friedlingstein, P.; Hurtt, G.; Knutti, R.; Kriegler, E.; Lamarque, J.-F.; Lowe, J. The scenario model intercomparison project (ScenarioMIP) for CMIP6. *Geosci. Model Dev.* **2016**, *9*, 3461–3482. [CrossRef]
54. Zhu, H.; Jiang, Z.; Li, J.; Li, W.; Sun, C.; Li, L. Does CMIP6 Inspire More Confidence in Simulating Climate Extremes over China? *Adv. Atmos. Sci.* **2020**, *37*, 1119–1132. [CrossRef]
55. O'Neill, B.C.; Kriegler, E.; Riahi, K.; Ebi, K.L.; Hallegatte, S.; Carter, T.R.; Mathur, R.; van Vuuren, D.P. A new scenario framework for climate change research: The concept of shared socioeconomic pathways. *Clim. Chang.* **2014**, *122*, 387–400. [CrossRef]
56. O'Neill, B.C.; Kriegler, E.; Ebi, K.L.; Kemp-Benedict, E.; Riahi, K.; Rothman, D.S.; van Ruijven, B.J.; van Vuuren, D.P.; Birkmann, J.; Kok, K. The roads ahead: Narratives for shared socioeconomic pathways describing world futures in the 21st century. *Glob. Environ. Chang.* **2017**, *42*, 169–180. [CrossRef]
57. Li, H.; Sheffield, J.; Wood, E.F. Bias correction of monthly precipitation and temperature fields from Intergovernmental Panel on Climate Change AR4 models using equidistant quantile matching. *J. Geophys. Res. Atmos.* **2010**, *115*, D10. [CrossRef]
58. Wood, A.W.; Leung, L.R.; Sridhar, V.; Lettenmaier, D.P. Hydrologic Implications of Dynamical and Statistical Approaches to Downscaling Climate Model Outputs. *Clim. Chang.* **2004**, *62*, 189–216. [CrossRef]
59. Hurtt, G.C.; Chini, L.; Sahajpal, R.; Frolking, S.; Bodirsky, B.L.; Calvin, K.; Doelman, J.C.; Fisk, J.; Fujimori, S.; Klein Goldewijk, K.; et al. Harmonization of global land use change and management for the period 850–2100 (LUH2) for CMIP6. *Geosci. Model Dev.* **2020**, *13*, 5425–5464. [CrossRef]
60. Sheffield, J.; Wood, E.F.; Roderick, M.L. Little change in global drought over the past 60 years. *Nature* **2012**, *491*, 435–438. [CrossRef]
61. Milly, P.; Dunne, K.A. A hydrologic drying bias in water-resource impact analyses of anthropogenic climate change. *J. Am. Water Resour. Assoc.* **2017**, *53*, 822–838. [CrossRef]
62. Wang, G.; Gong, T.; Lu, J.; Lou, D.; Hagan, D.F.T.; Chen, T. On the long-term changes of drought over China (1948–2012) from different methods of potential evapotranspiration estimations. *Int. J. Climatol.* **2018**, *38*, 2954–2966. [CrossRef]
63. Jones, B.; O'Neill, B.C.; McDaniel, L.; McGinnis, S.; Mearns, L.O.; Tebaldi, C. Future population exposure to US heat extremes. *Nat. Clim. Chang.* **2015**, *5*, 652–655. [CrossRef]

64. Li, X.; Ting, M. Understanding the Asian summer monsoon response to greenhouse warming: The relative roles of direct radiative forcing and sea surface temperature change. *Clim. Dyn.* **2017**, *49*, 2863–2880. [CrossRef]
65. Li, Q.; Zhang, L.; Xu, W.; Zhou, T.; Wang, J.; Zhai, P.; Jones, P. Comparisons of time series of annual mean surface air temperature for China since the 1900s: Observations, model simulations, and extended reanalysis. *Bull. Am. Meteorol. Soc.* **2017**, *98*, 699–711. [CrossRef]
66. Tebaldi, C.; Debeire, K.; Eyring, V.; Fischer, E.; Fyfe, J.; Friedlingstein, P.; Knutti, R.; Lowe, J.; O'Neill, B.; Sanderson, B.; et al. Climate model projections from the Scenario Model Intercomparison Project (ScenarioMIP) of CMIP6. *Earth Syst. Dynam.* **2021**, *12*, 253–293. [CrossRef]
67. Zhang, G.; Zeng, G.; Yang, X.; Jiang, Z. Future Changes in Extreme High Temperature over China at 1.5 °C–5 °C Global Warming Based on CMIP6 Simulations. *Adv. Atmos. Sci.* **2021**, *38*, 253–267. [CrossRef]
68. Qin, J.; Su, B.; Tao, H.; Wang, Y.; Huang, J.; Li, Z.; Jiang, T. Spatio-temporal variations of dryness/wetness over Northwest China under different SSPs-RCPs. *Atmos. Res.* **2021**, *259*, 105672. [CrossRef]
69. Wang, H.L.; Gan, Y.; Wang, R.Y.; Niu, J.-y.; Zhao, H.; Yang, Q.; Li, G.C. Phenological trends in winter wheat and spring cotton in response to climate changes in northwest China. *Agric. For. Meteorol.* **2008**, *148*, 1242–1251. [CrossRef]
70. Miao, L.; Li, S.; Zhang, F.; Chen, T.; Shan, Y.; Zhang, Y. Future drought in the drylands of Asia under the 1.5 °C and 2.0 °C warming scenarios. *Earth's Future* **2020**, *8*, e2019EF001337. [CrossRef]
71. Lin, W.; Wen, C.; Wen, Z.; Gang, H. Drought in Southwest China: A review. *Atmos. Ocean. Sci. Lett.* **2015**, *8*, 339–344.
72. Fahad, S.; Hasanuzzaman, M.; Alam, M.; Ullah, H.; Saeed, M.; Khan, I.A.; Adnan, M.E. *Environment, Climate, Plant and Vegetation Growth*; Springer: Cham, Switzerland, 2020.
73. Spinoni, J.; Barbosa, P.; Bucchignani, E.; Cassano, J.; Cavazos, T.; Cescatti, A.; Christensen, J.H.; Christensen, O.B.; Coppola, E.; Evans, J.P. Global exposure of population and land-use to meteorological droughts under different warming levels and SSPs: A CORDEX-based study. *Int. J. Climatol.* **2021**, *41*, 6825–6853. [CrossRef]

atmosphere

MDPI

Article

Impact of Environmental Regulation on Efficiency of Green Innovation in China

Tongtong Shen [1], Dongju Li [2,*], Yuanyuan Jin [2] and Jie Li [2]

1 School of Statistics, Dongbei University of Finance and Economics, Dalian 116021, China; shenttdufe@163.com
2 School of Statistics and Big Data, Henan University of Economics and Law, Zhengzhou 450046, China; 20192300055@stu.huel.edu.cn (Y.J.); 20212000653@stu.huel.edu.cn (J.L.)
* Correspondence: 20100452@huel.edu.cn

Abstract: The implementation of a reasonable and effective environmental regulation policy can compensate for the dual externalities of green technology innovation and improve green innovation efficiency. Therefore, environmental regulation policy has gradually become an effective means of solving ecological environment problems and achieving green industrial transformation. This paper measures the green innovation efficiency of 30 provinces in China from 2009 to 2019 using the SBM (slacks-based measure) of super-efficiency based on the undesirable output. The dynamic panel regression model is established to explore the impact of different environmental regulations on green innovation efficiency and regional differences. The results reveal that the green innovation efficiency of the 30 provinces shows a fluctuating upward trend, but that differences among provinces are relatively significant. There is a nonlinear relationship between environmental regulation and green innovation efficiency. The impact of command-control and market incentive environmental regulations on green innovation efficiency shows inverted N-shaped and U-shaped patterns, respectively. In different regions, the impact of environmental regulation on green innovation efficiency is also different. In order to ensure that environmental regulation promotes green innovation efficiency, some recommendations are proposed for the government, enterprises, and three regions, respectively.

Keywords: environmental regulation; green innovation efficiency; SBM of super-efficiency; system GMM estimation

Citation: Shen, T.; Li, D.; Jin, Y.; Li, J. Impact of Environmental Regulation on Efficiency of Green Innovation in China. *Atmosphere* **2022**, *13*, 767. https://doi.org/10.3390/atmos13050767

Academic Editors: Zengyun Hu, Xuguang Tang and Qinchuan Xin

Received: 7 April 2022
Accepted: 6 May 2022
Published: 9 May 2022

1. Introduction

The public nature of environmental resources and the externalities of ecological damage have long made it difficult to solve the problem of sustainable economic development solely by market mechanisms. However, green innovation technology can effectively alleviate increasingly severe ecological and environmental problems, reduce pollution emissions, and save energy consumption. It has become an essential means to promote sustainable and green economic development in China. Green innovation efficiency is used as a measure of green innovation technology. It is characterized by positive knowledge spillover externalities and negative environmental externalities. However, the effect is minimal, relying only on the autonomous allocation of regional innovation resources. The implementation of a reasonable and effective environmental regulation policy can compensate for the dual externalities of green technology innovation and improve green innovation efficiency [1]. Therefore, environmental regulation policy has gradually become an effective means of solving ecological environment problems and achieving green industrial transformation.

The impact mechanism of environmental regulation on green innovation efficiency is shown in Figure 1. In order to maintain a good state of the local environment, the government needs to adopt some environmental regulation policies, such as limiting emission standards and technical standards for enterprises. Suppose the intensity of environmental regulation is relatively low; in that case, the requirement of pollutant restrictions is not high, and the increased cost of pollutant discharge may be less than the cost of technological innovation. Therefore, enterprises will choose to increase pollution control expenditure to cope with environmental regulations, which will occupy the funds initially planned for innovation, to the detriment of the efficiency of green innovation. In the meantime, some enterprises from areas with high environmental regulation intensity or foreign enterprises may be attracted to move in. These enterprises will compete with existing enterprises, reducing the share of green innovation investment of local enterprises. Moreover, the excessive concentration of enterprises will generate additional undesirable outputs and cause a so-called "pollution paradise", which is not conducive to green development.

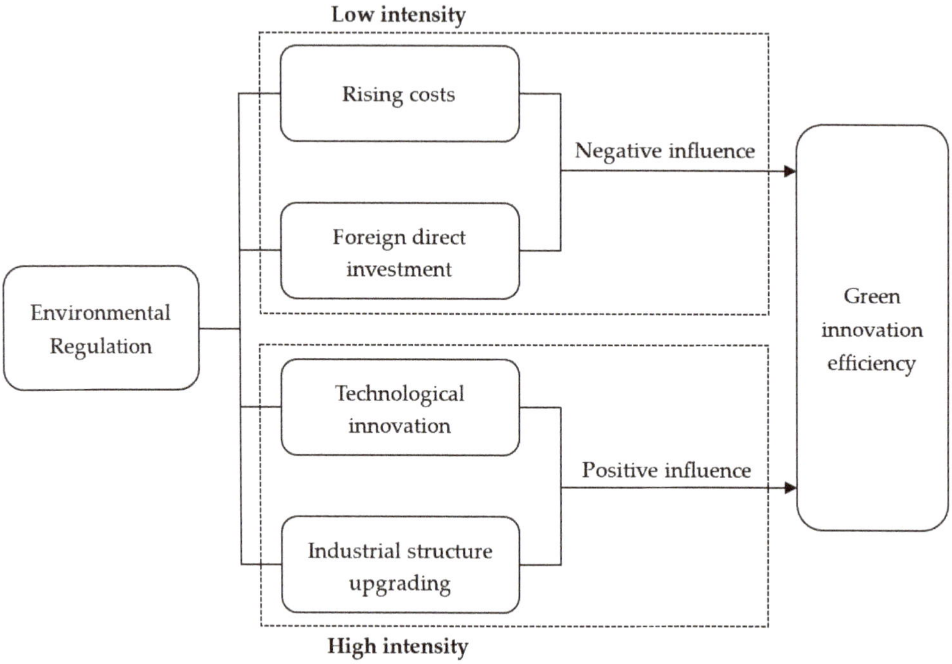

Figure 1. The impact mechanism of environmental regulation on green innovation efficiency.

Technological innovation is a better choice if the emission standards are set high in the long run. By improving their production processes, enterprises can, on the one hand, reduce the consumption of resources and the emission of pollutants and meet the emission standards required by the government, and, on the other hand, improve production efficiency, promote upgrading industrial structure, and enhance competitiveness, which is conducive to the sustainable development of enterprises and the development of local green innovation in the long run. Nevertheless, higher intensity of environmental regulation may not continue to promote green innovation efficiency, but rather, lead to some unintended consequences. The government must grasp the intensity of environmental regulations based on local conditions, which can promote green innovation efficiency without damaging the interests of local enterprises.

2. Literature Review

2.1. Review of Green Innovation Efficiency

Research on green innovation efficiency mainly takes the form of measurements of green innovation efficiency and analyses of influencing factors. For the measurement methods of green innovation efficiency, there are mainly data envelopment analyses (DEA) [2–4], stochastic frontier analyses [5,6], and multivariate statistical analyses [7]. Most of the literature measuring China's green innovation efficiency shows that the overall efficiency value is gradually increasing, but that there are significant differences among regions [8,9]. There are influencing factors to positively promote green innovation efficiency, such as government scientific research sponsorship [10], public participation [11], and industrial agglomeration [12]. Since the process of green innovation will produce both desirable and undesirable outputs, this paper chooses the SBM of super-efficiency based on undesirable outputs to measure green innovation efficiency.

2.2. Review of Environmental Regulation

Environmental regulation means that the government standardizes the behavior of enterprises with a view to protect the environment. Environmental regulation can encourage enterprises to reform and innovate pollution control technology at a micro level. Environmental regulation can also guide enterprises to restructure and promote the upgrading of industrial structure at a macro level. At present, there are three types of environmental regulation policies in China, including command-control, market incentive, and public voluntary. The first two types of environmental regulations are called formal environmental regulations, and the last one is called informal environmental regulations. There are no unified indicators for measuring the intensity of environmental regulation. Scholars generally develop the indicators of environmental regulation intensity according to research issues and data availability. The main types of measurement are as follows: (a) the proportion of the cost of pollution control in industry added value [13] or GDP [14]; (b) indexes of the removal rate of pollutants, such as the standard discharge rate of industrial wastewater and the removal rate of sulfur dioxide [15], or the weighted composite index of various removal rate index of pollutants [16]; (c) pollutant emissions per unit output value [17,18]; (d) the ratio of pollution treatment investment to pollutant emissions [19,20].

Studies have shown that various environmental regulations can significantly affect energy saving and emission reduction [21]. Moreover, environmental regulation can promote technological and green innovation, thus boosting regional economic development [22–24]. The joint development of command-control environmental regulations and market incentive environmental regulations can play a better role [25]. Most of the literature analyses environmental regulation using only a single means without considering the differences in the effects of different environmental regulation means. This paper selects two significant environmental regulation policies, namely command-control and market incentive environmental regulation, to analyze the difference in effect.

2.3. Research on Impact of Environmental Regulation on Green Innovation Efficiency

Research on the impact of environmental regulation on green innovation efficiency can be divided into the following three perspectives:

(1) Scholars who support the "Porter hypothesis" believe that environmental regulation will encourage enterprises to embrace technological innovation, which, in turn, improves the green innovation efficiency of society as a whole. Brunnermeier and Cohen found that for every $1 increase in environmental governance costs, green innovation efficiency would increase by 0.4% on average [26]. Castellacci and Lie found that mandatory environmental regulations had a potent positive effect on green innovation efficiency [27]. Singh et al. found that environmental regulation policies in Japan drove green innovation efficiency for society as a whole [28]. Zhang and Wang found that both environmental regulation policies and government financial support have a positive effect on green innovation efficiency, but that the former has a greater impact [29]. Wang and Zhang (2018) argued that different environmental regulation policies would positively promote green innovation efficiency, and command-control environmental regulation has a more significant promoting effect [30]. Wang and He (2022) believed that environmental regulation could promote green innovation, and green innovation can promote the upgrading of industrial structures [31].

(2) Other scholars argue that environmental regulations are not conducive to green innovation efficiency because strict restrictions on environmental emissions may increase

the costs of enterprises. Domazlicky and Weber argued that the benefits of technological change brought about by environmental regulations could not compensate for the increased costs to the enterprises [32]. Sinn proposed the "green paradox", arguing that environmental regulations increased the expenditure of enterprises on emission reduction and reduced the efficiency of green innovation [33]. Li and Bi believed that environmental regulation is not conducive to the technological progress of enterprises and green innovation [34].

(3) Another view is that the impact of environmental regulation on green innovation efficiency is uncertain. Kneller and Manderson argued that mandatory environmental regulation policies would increase the costs of pollutant reduction and R&D (Research and Development) in the UK while having little impact on the total capital accumulation [35]. Peuckert pointed out that environmental regulation would squeeze expenditure, inhibit technological innovation in the short run, and promote development in the long run [36]. Peng et al. found that formal and informal environmental regulation policies showed a U-shaped and inverted U-shaped relationship with green innovation efficiency, respectively [37]. Luo and Chen found that environmental regulations have a non-linear relationship with green efficiency through the threshold regression model [38]. Gao and Xiao believed that autonomous environmental regulations have a U-shaped impact on improving the green innovation efficiency of industrial enterprises [39].

Based on the above analysis, existing studies have stated that environmental regulations have positive, negative, and uncertain effects on green innovation efficiency. This effect is considered from a national perspective, with little consideration for inter-regional heterogeneity. This paper constructs a dynamic panel system GMM model to analyze the impact of different types of environmental regulations on green innovation efficiency. A regression model was established for the eastern, central, and western regions to analyze the regional differences.

3. Research Method

3.1. SBM of Super-Efficiency

The traditional models of Data Envelopment Analysis (DEA) are radial measures of efficiency, requiring that the input and output change in the same proportion. However, it is not easy to meet this condition in actual production. Based on the traditional DEA model, Tone proposed a slacks-based measure of efficiency (SBM), which is non-radial and deals with input/output slacks directly [40]. Tone proposed the SBM of super-efficiency, an evolutionary form of SBM in the following year, which can further evaluate Decision Making Units (DMUs) with an efficiency value greater than 1 to obtain more accurate efficiency results [41]. This paper uses the SBM of super-efficiency to measure the green innovation efficiency of provinces in China. The model is as follows.

$$\rho^* = \min \frac{\frac{1}{m}\sum_{i=1}^{m}\frac{\bar{x}_i}{x_{ik}}}{\frac{1}{s_1+s_2}\left(\sum_{r=1}^{s_1}\frac{\bar{y}_r^d}{y_{rk}^d}+\sum_{q=1}^{s_2}\frac{\bar{y}_q^u}{y_{qk}^u}\right)}$$

$$s.t.\begin{cases} \bar{x} \geq \sum_{j=1,\neq k}^{n} \lambda_j x_{ij} \\ \bar{y}^d \leq \sum_{j=1,\neq k}^{n} \lambda_j y_{rj}^g \\ \bar{y}^u \geq \sum_{j=1,\neq k}^{n} \lambda_j y_{qj}^b \\ \lambda_j \geq 0, i=1,2,\ldots,m; j=1,2,\ldots,n; r=1,2,\ldots,s_1; q=1,2,\ldots,s_2 \end{cases} \quad (1)$$

where ρ^* is the efficiency of green innovation, whereby the higher the efficiency value, the higher the level of green innovation; x, y^d, and y^u represent the necessary elements in the input matrix, the desirable output matrix, and the undesirable output matrix, respectively;

and n represents the number of DMU, namely the number of provinces in this paper ($n = 30$). Each DMU has m inputs, s_1 desirable outputs, and s_2 undesirable outputs. λ is the weight vector.

3.2. Kernel Density Estimation

The Kernel Density Estimation (KDE) method, first proposed by Parzen, is a non-parametric test method for solving the probability density function of random variables [42]. It can analyze the dynamic evolution characteristics of the sample distribution according to the sample data. Compared with the parameter estimation method, the functional form of the kernel density can be set flexibly with few restrictions on the data, which is one of the common methods to study the unbalanced distribution. The expression of kernel density estimation is as follows.

$$f_h(x) = \frac{1}{nh}\sum_{i-1}^{n} K\left(\frac{x - x_i}{h}\right) \tag{2}$$

In Formula (2), $f(x)$ represents the density function of green innovation efficiency. n is the number of observed provinces. h represents the bandwidth, and its value affects the shape and smoothness of the KDE curve. The smaller the bandwidth, the higher the estimation accuracy. x_i represents the green innovation efficiency of i province, and x represents the mean value of the green innovation efficiency. $K(\cdot)$ is the kernel function. The Gaussian kernel function is uesde in this paper to estimate the dynamic evolution of the distribution of green innovation efficiency, as shown in Formula (3).

$$K(x) = \frac{1}{\sqrt{2\pi}}\exp(-\frac{x^2}{2}) \tag{3}$$

3.3. System-GMM

This paper adopts the dynamic GMM (Generalized Method of Moments) model to measure the impact of environmental regulation on green innovation efficiency. The dynamic GMM model can effectively overcome the biased regression results caused by autocorrelation in the regression process. GMM estimation mainly includes the difference GMM and the system GMM. In contrast, the system GMM reduces some omission errors caused by the difference GMM, effectively improving the estimation efficiency [43]. In empirical research, the system GMM method is preferred for estimates. From the above analysis, it can be seen that there is not only a linear relationship between environmental regulation and green innovation efficiency; in order to verify whether there is a more complex linear relationship between environmental regulation and green innovation efficiency, a cubic term of environmental regulation intensity is introduced. The production process of green innovation is a process of continuous accumulation and dynamic adjustment, which may be affected by the previous period of green technology innovation. It is necessary to introduce the lag term of green innovation efficiency. Therefore, the dynamic panel GMM model is established as follows.

$$GI_{i,t} = \alpha_0 + \alpha_1 GI_{i,t-1} + \beta_1 ER_{i,t} + \beta_2 ER_{i,t}^2 + \beta_3 ER_{i,t}^3 + \gamma_m K_{i,t} + u_i + \varepsilon_{i,t} \tag{4}$$

where $GI_{i,t}$ represents the green innovation efficiency of i province in t year; $GI_{i,t-1}$ is the first-order lag term of green innovation efficiency; α_0 is a constant term; ER represents the intensity of environmental regulation, which is the core explanatory variable; β_1, β_2, and β_3 are the coefficient terms of the core explanatory variables; $K_{i,t}$ is the control variable and γ_m is the coefficient term of the control variable; finally, u_i represents the individual effect and $\varepsilon_{i,t}$ represents the random error term. Different β coefficient values lead to different trend characteristics of regression curve, as shown in Table 1.

Table 1. Characteristics of some common regression curve.

Coefficient Value	Shape of Regression Curve	Indication
$\beta_2 = \beta_3 = 0, \beta_1 \neq 0$	Monotonically increasing or decreasing	The intensity of environmental regulation promotes or inhibits green innovation efficiency.
$\beta_3 = 0, \beta_2 > 0, \beta_1 < 0$	U-shaped	The green innovation efficiency first decreases and then increases with the increase of the intensity of environmental regulation.
$\beta_3 = 0, \beta_2 < 0, \beta_1 > 0$	Inverted U-shaped	The green innovation efficiency first increases and then decreases with the increase of the intensity of environmental regulation.
$\Delta = 4\beta_2^2 - 12\beta_1\beta_3 > 0,$ $\beta_1 > 0, \beta_2 < 0, \beta_3 > 0$	N-shaped	The green innovation efficiency first increases, decreases to a certain level, and finally increases again with the increase of the intensity of environmental regulation.
$\Delta = 4\beta_2^2 - 12\beta_1\beta_3 > 0,$ $\beta_1 < 0, \beta_2 > 0, \beta_3 < 0$	Inverted N-shaped	The green innovation efficiency first decreases, increases to a certain level, and finally decreases again with the increase of the intensity of environmental regulation.

3.4. *Indicator Selection and Variable Description*

3.4.1. Construction of Green Innovation Efficiency System

Based on relevant literature [7,8], this paper comprehensively considers the entire input and output process of green innovation and constructs the system of green innovation efficiency from the perspectives of input, desirable output, and undesirable output, as shown in Table 2.

Table 2. Input-output system of green innovation efficiency.

Type	Indicator	Definition	Source
Input	Green input	Energy consumption (ten thousand tons of standard coal)	China Energy Statistical Yearbook
	Innovation input	Full-time equivalent of R&D personnel (ten thousand man-years)	China Statistical Yearbook on Science and Technology
		Internal expenditure of R&D funds (ten thousand yuan)	China Statistical Yearbook on Science and Technology
Output	Innovation desirable output	New product sales revenue (ten thousand yuan)	China Statistical Yearbook on Science and Technology
		Regional GDP (billion yuan)	China Statistical Yearbook
		Number of domestic patent applications accepted (piece)	China Statistical Yearbook on Science and Technology
	Green undesirable output	Total industrial sulfur dioxide emissions (ton)	China Statistics Yearbook on Environment
		Organic matter content in industrial wastewater (ton)	China Statistics Yearbook on Environment

3.4.2. Measurement of Environmental Regulation Intensity

There are currently three main types of environmental regulation policies in China. The two most important types concern command-control and market incentives. The third type is the public-voluntary environmental regulation policy which lacks data. Therefore, this paper selects the first two types of environmental regulation policies as the research objects and measures the intensity of the two environmental regulation policies, respectively.

Command-control environmental regulation refers to the use of mandatory measures by the government on enterprises to achieve a specific environmental goal and the formulation of a series of standards to regulate the behavior of enterprises, including emission standards, technical standards, etc. Based on emission standards, this paper looks for indicators to measure the intensity of command-control environmental regulation [17,18]. According to the current situation of pollutant discharge in China and the availability of data, the comprehensive index of three types of pollutant is measured as the intensity of command-controlled environmental regulation, including industrial soot emissions, industrial sulfur dioxide emissions, and industrial wastewater emissions. The data source for the three types of pollutant is China Statistics Yearbook on Environment. The calculation formulas are as follows.

$$R_{ij}^{s} = \frac{R_{ij} - \min R_j}{\max R_j - \min R_j} \tag{5}$$

$$W_j = R_{ij}/\overline{R}_j \tag{6}$$

$$ER1 = \frac{1}{3}\sum_{j=1}^{3} W_j \times R_{ij}^{s} \tag{7}$$

In Formula (5), R_{ij} represents the emission of the pollutant j in province i, and R_{ij}^{s} represents the standardized result; $\min R_j$ and $\max R_j$ represent the maximum and maximum value of the emissions of the pollutant j in all provinces, respectively. In Formula (6), W_j represents the weight of the pollutants j, and $\overline{R}_j$ represents the average emission of the pollutants j in all provinces. In Formula (7), $ER1$ represents the intensity of command-control environmental regulation, which is the weighted average of the three pollutants.

Market incentive environmental regulation encourages enterprises to find technologies and methods to reduce pollutant emissions through economic means in order to minimize the degree of environmental pollution. Currently, the measures include levying an emissions tax on enterprises and establishing a system of emission rights trading and emission fee. The systems of emissions tax and emission rights have not played a good role in China, but the pollutant discharge fee system was implemented earlier. Therefore, the intensity of market-incentive environmental regulation ($ER2$) is measured by the proportion of pollutant discharge fees (PF) to regional GDP (GDP), as shown in Formula (8). The data on pollutant discharge fees and regional GDP respectively come from the China Environmental Statistical Yearbook and the China Statistical Yearbook.

$$ER2 = \frac{PF}{GDP} \tag{8}$$

3.4.3. Variable Description

This paper selects the panel data of 30 provinces from 2009 to 2019 to construct the dynamic GMM model. The variable description is shown in Table 3. The data of control variables came from the China Statistical Yearbook, China Statistical Yearbook on Science and Technology, and China Industrial Economy Statistical Yearbook.

Table 3. Variable Description.

Type	Variable Name	Definition
Explained variable	Green innovation efficiency (GI)	Efficiency measured by the SBM of super-efficiency
Explanatory variable	Environmental Regulation Intensity (ER)	Intensity of command-control environmental regulation ($ER1$) and market incentive environmental regulation ($ER2$)
Control variable	Government support (GS)	Proportion of local fiscal expenditure in regional GDP
	Urbanization (UR)	Urbanization rate
	Technical progress (TP)	Turnover of technology market
	Openness (OP)	Ratio of total import and export trade to GDP
	Human capital (HC)	Full-time equivalent of R&D personnel
	Foreign direct investment (FDI)	Total amount of foreign investment actually used
	Optimization of industrial structure (IS)	Proportion of tertiary industry value in regional GDP

4. Analysis of Empirical Results

4.1. Measurement and Analysis of Green Innovation Efficiency in China

4.1.1. Evolution of Green Innovation Efficiency

This paper uses MATLAB to measure the green innovation efficiency of 30 provinces. The results are shown in Appendix A, Table A1. In order to more intuitively see the evolution of green innovation efficiency in 30 provinces, a spatial distribution map of green innovation efficiency was drawn for 30 provinces in 2009 and 2019 by ArcGIS, as shown in Figure 2.

The evolution of green innovation efficiency from 2009 to 2019 shows that the value of efficiency is gradually improving in China. The number of provinces with low green innovation efficiency is gradually decreasing. In 2009, the number of provinces with low green innovation efficiency was 12, but only six remained in 2019. There are 12 provinces whose average value of green innovation efficiency is greater than one over the ten years. Most provinces are developed coastal provinces and key provinces supported by the government with a better natural environment. The green innovation efficiency of developed provinces is higher because they have better development opportunities, attract more talents, and produce more creative output. Beijing has the highest average green innovation efficiency at 1.9873. Research institutes in Beijing, with the largest number, attract many talents every year. Beijing enjoys excellent development conditions, and its innovation level is at the forefront in China. Beijing is also the first province to implement an environmental policy in China. It was better to control pollution discharge as a pilot of pollution discharge rights. Some remote provinces with a beautiful environment may not have much input in innovation but enjoy more government support policies. Their values of green innovation efficiency are relatively high due to the low green undesirable output. The average value of green innovation efficiency in Xinjiang ranks second at 1.2849. The green innovation efficiency increased the most in Qinghai, from 0.2332 in 2009 to 1.0104 in 2019, increasing more than four times.

From the perspective of various regions, the green innovation efficiency of eastern provinces is higher than that of other regions, especially some coastal provinces, such as Jiangsu, Zhejiang, and Guangdong, whose green innovation efficiency stays at a high level. The green innovation efficiency of provinces in central China is at a middle level, but there are still some provinces with low green innovation efficiency. For example, the average innovation efficiency in Heilongjiang is the lowest, mainly because the traditional industry is relatively developed in the early years, causing sizeable environmental pollution, and the transformation process to the new technology industry is still slow. Although the economy in western China is less developed, the green innovation efficiency is not at a low level, especially the green innovation efficiency in Xinjiang is at the forefront of the country.

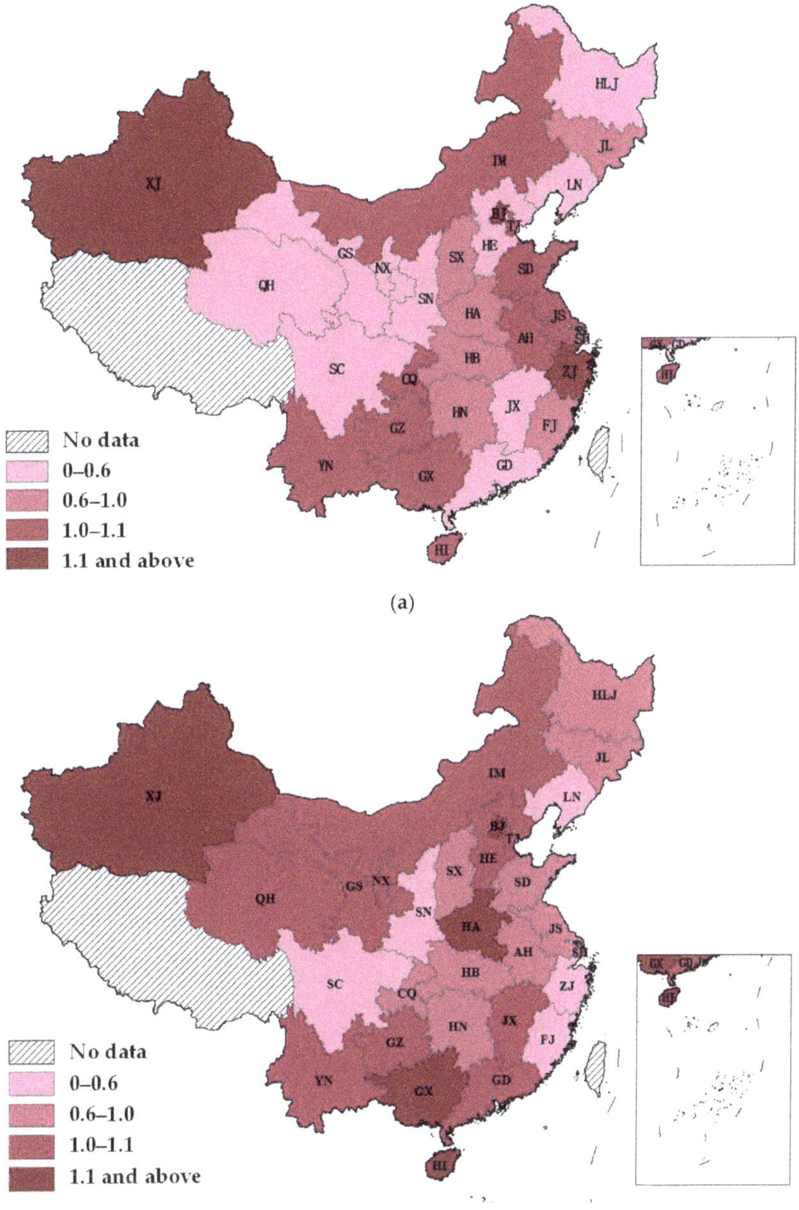

Figure 2. Spatial evolution of green innovation efficiency in China from 2009 to 2019. (**a**) 2009; (**b**) 2019.

4.1.2. Dynamic Evolution of Green Innovation Efficiency

This paper uses the KDE method to analyze the dynamic distribution situation of green innovation efficiency for 30 provinces in China. The years 2010, 2013, 2016, and 2019 were selected as observation time points to draw Kernel density curves for the four years, as shown in Figure 3.

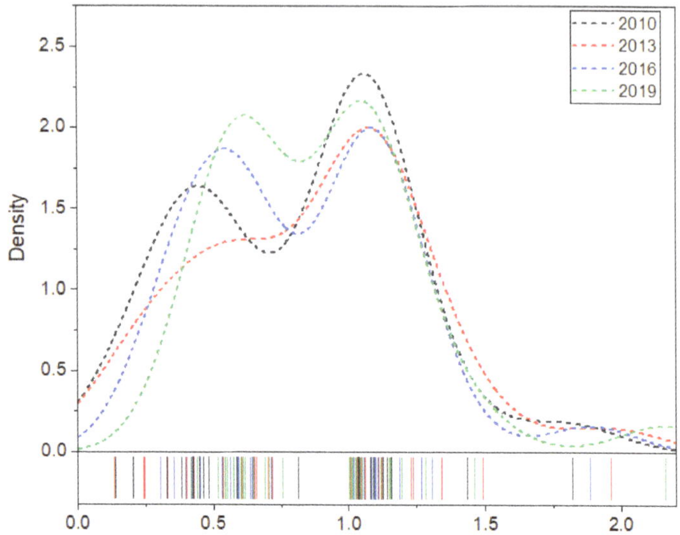

Figure 3. Kernel density distribution of green innovation efficiency in China.

As shown in Figure 3, the center of the green innovation efficiency distribution curve in China has shifted significantly to the right, indicating that the green innovation efficiency of each province has gradually improved over time. In terms of distribution shape, the four curves show a broad distribution, and the right tails tend to be elongated, indicating that there are significant differences in the green innovation efficiency among provinces. The kernel density curve in 2013 shows a unimodal distribution, while the curves in 2016 and 2019 are characterized by a clear bimodal distribution, indicating a trend toward polarization of the green innovation efficiency in China. As there are many provinces with green innovation efficiency of around 1, the main peak of the four curves increases significantly, while the height of the small peak in the right tail tends to decline slowly. The difference between provinces with low green innovation efficiency is expanding, while the difference between provinces with high green innovation efficiency is narrowing. The evolution of green innovation efficiency in China is not coordinated among different regions. The four curves are right-skewed distribution. The vertical height of the peak increases, and the horizontal width decreases over time. Although there are regional differences in green innovation efficiency in China, such differences are gradually narrowing and have the characteristics of dynamic convergence.

4.2. Impact of Environmental Regulation on Green Innovation Efficiency

4.2.1. Empirical Analysis of the Impact Effect of Different Environmental Regulations

This paper adopts the dynamic panel system GMM model for regression analysis to analyze the impact of different environmental regulations on green innovation efficiency. In order to prevent heteroscedasticity, non-proportional indicators were logarithmically treated. Stata was used to obtain regression results, as shown in Table 4. First of all, the rationality of variables and regression model selection is analyzed. ADF tests are performed on the variables, all of which are stationary. The *p*-value of AR(2) of the two models is greater than 0.05, indicating that there was no serial correlation, which proved the rationality of the model selection. The *p*-value of Sargan's test is greater than 0.05, which makes it difficult to reject the null hypothesis that all instrumental variables are valid, indicating that the instrumental variables selected by the system GMM model are reasonable.

Table 4. Regression results of impact of environmental regulation on green innovation efficiency.

Variable	Dynamic Panel System GMM Model	
	(1) Command-Control Environmental Regulation	(2) Market Incentive Environmental Regulation
$L.GI$	0.3888 *** (7.22)	0.3239 *** (5.93)
ER	−1.0215 ** (−2.33)	−0.2415 *** (−3.56)
ER^2	1.3781 ** (2.3)	0.0969 ** (2.48)
ER^3	−0.4908 ** (−2.2)	- -
GS	1.6306 * (1.89)	2.5172 ** (2.42)
UR	1.2206 *** (4.92)	0.7418 *** (2.92)
$LNTP$	0.1426 *** (3.3)	0.1067 *** (−2.72)
$LNOP$	0.1292 *** (3.59)	0.1143 *** (2.76)
$LNHC$	−0.1953 *** (−14.06)	−0.1955 *** (−10.41)
$LNFDI$	0.025 (1.39)	0.0538 ** (2.41)
IS	0.5402 * (1.68)	0.4316 * (1.67)
$Cons$	−0.8971 (−1.39)	−0.2862 (−0.6)
Curve type	Inverted N-shaped	U-shaped
Inflection point	0.5071 1.3677	1.2461
AR(1)	−3.8475	−4.0413
p-value	0.0001	0.0001
AR(2)	−1.7165	−1.7925
p-value	0.0861	0.0731
Sargan	24.4225	24.8043
p-value	0.4951	0.4734

Note: * a significance level of 10%, ** a significance level of 5%, *** a significance level of 1%. The values of Z-statistic are in parentheses.

To verify the effect of time variation in green innovation efficiency on the accuracy of the regression model, a static panel regression model is also constructed and compared with the dynamic panel regression model in this paper. A new index (DISO, distance between indices of simulation and observation) can comprehensively describe the overall performance of different models, with smaller values of DISO indicating higher model prediction accuracy. See Hu et al. and Zhou et al. for the calculation method of DISO [44,45]. The DISO values for the static panel models of command-control and market incentive environ-mental regulation are 0.93 and 0.96, respectively. The DISO values for the dynamic panel regression models of command-control and market incentive environ-mental regulation are 0.57 and 0.54, respectively. The comparison results show that the dynamic panel model outperforms the static panel model for both the command-control and market incentive environmental regulation. Adding time-varying factors of dependent variables can indeed improve model prediction accuracy. As can be seen from Table 4, the lag term $L.GI$ of green innovation efficiency passes the significance test, indicating that the green innovation efficiency of the previous period has a very significant promoting effect on the current period. The green innovation activity is a continuous and dynamic accumulation process. The technological progress and innovation activity in the previous period will influence green innovation in the subsequent period.

The two types of environmental regulation intensity have different effects on green innovation efficiency. Specifically, the regression curve between the intensity of command-control environmental regulation and the green innovation efficiency shows an inverted N-shaped pattern, with a downward-upward-downward trend. In the first stage, before reaching the inflection point of 0.5071, the increase in pollution control costs drains the funds for technological innovation. The green innovation efficiency decreases with the increase of environmental regulation intensity. When the intensity of command-control environmental regulation exceeds 0.5071, the second stage is reached. At this stage, most enterprises begin to choose to carry out technological innovation, and command-control environmental regulation plays a role in promoting green innovation efficiency. When the intensity of command-control environmental regulation reaches the second inflection point 1.3677, the green innovation efficiency decreases again. The possible reason is that when the intensity of environmental regulations is too high, some enterprises fail to meet the standards or turn to operate in areas with less stringent environmental regulations. In this case, the green innovation efficiency will decline. In other words, if command-control environmental regulation is to promote green innovation efficiency, the intensity of environmental regulation should be controlled between the two inflection points.

The regression curve between market incentive environmental regulation intensity and green innovation efficiency shows a U-shaped pattern. When the intensity of market incentive environmental regulation does not reach the inflection point 1.2461, the cost of enterprises to pollutant discharge is low. Therefore, most enterprises do not choose to carry out technological innovation. As the intensity of environmental regulation increases, the cost of pollutant discharge increases, while the fund for technological innovation and the green innovation efficiency decreases. When the environmental regulation intensity index rises again, reaching 1.2461, the cost of pollutant discharge is already very high. Most enterprises tend to innovate to reduce pollution at the source. On the right side of the inflection point, green innovation efficiency increases with the increase of environmental regulation intensity.

Government support, urbanization, technological progress, openness, and industrial structure optimization positively affect green innovation efficiency. Government support has the greatest impact. Innovation is inseparable from high-tech development industries, for which government support is vital. A place with a high level of urbanization will have better conditions for innovation and development, attracting more high-tech industries to promote the development of green innovation efficiency. There is no doubt that technological progress can contribute to green innovation efficiency and enable enterprises to produce more output with less input. The impact of openness on green innovation efficiency is significantly positive. Places with a high degree of openness to trade are likely to attract capital and talent inflows, which lead to technological innovation and promote green efficiency development. Industrial structure optimization refers to the transformation of knowledge-intensive industries, which is conducive to improving energy efficiency, reducing the impact on the ecological environment, and promoting green innovation efficiency. Human capital has a negative impact on green innovation efficiency, indicating that the negative effect of pollution brought by human input is higher than the positive effect of innovation.

4.2.2. Regional Model Estimation

The economic development and resource endowments of 30 provinces in China are not uniform. It would be inappropriate to formulate the same policies on environmental regulation according to the national situation. In order to explore whether there are differences in the impact of environmental regulation on green innovation efficiency in different regions, 30 provinces are divided into three regions (eastern, central, and western regions). The regressions were conducted separately for the three regions, and the estimated results for the core explanatory variables are shown in Table 5.

Table 5. Regional regression results.

Variable	Dynamic Panel System GMM Model		
	Eastern	Central	Western
$ER1$	0.2102 ***	-	−0.1654 ***
	(7.64)	-	(14.9)
$ER1^2$	−0.0964 ***	-	0.0922 ***
	(3.87)	-	(8.07)
Curve type	Inverted U-shaped	-	U-shaped
Inflection point	1.619	-	0.8970
$ER2$	1.8050 ***	−0.1112 **	−0.1043 ***
	(7.22)	(−2.16)	(−3.76)
$ER2^2$	-	0.0735 ***	0.063 ***
	-	(7.56)	(8.32)
Curve type	Straight line	U-shaped	U-shaped
Inflection point	-	0.7565	0.6020

Note: ** a significance level of 5%, *** a significance level of 1%. The values of Z-statistic are in parentheses.

As shown in Table 5, the impacts of environmental regulations on green innovation efficiency among the eastern, central, and western regions are different. The regression curve between the intensity of command-control environmental regulation and green innovation efficiency in eastern China shows an inverted U-shaped pattern. When the intensity of environmental regulation does not exceed inflection point 1.619, improving the intensity of command-control environmental regulation can positively promote green innovation efficiency. If the intensity of environmental regulation is set too high, it will cause great pressure on some enterprises and reduce their profits. Some enterprises will choose to move to places with less stringent environmental regulations, which is not conducive to the development of local green innovation efficiency. Market incentive environmental regulation plays a direct role in promoting green innovation efficiency, which is related to the sufficient capital and talent reserve in eastern China. When the intensity of market incentive environmental regulation increases, there are more opportunities to enhance technological innovation in eastern China to promote the improvement of green innovation efficiency.

The intensity of command-control environmental regulation has no significant impact on green innovation efficiency in central China. The relationship between market incentive environmental regulation intensity and green innovation efficiency is U-shaped. When the intensity of market incentive environmental regulation reaches 0.7565, the improvement of green innovation efficiency can be promoted.

There is a U-shaped relationship between green innovation efficiency and environmental regulation in western China, whether command-control or market incentive. When the intensity of environmental regulation is relatively low, most enterprises choose to increase pollution discharge fees for waste treatment, which takes up the cost of technological innovation and is not conducive to the development of green innovation efficiency. When the intensity of environmental regulation reaches a certain value, enterprises will increase pollution discharge fees. By contrast, it is more cost-effective to use technological innovation to tackle pollution at the source. As technological innovation improves, so will green innovation efficiency.

5. Conclusions and Recommendations

5.1. Conclusions

This paper uses the SBM of a super-efficiency model based on the undesirable output to measure green innovation efficiency, and establishes the dynamic panel system GMM model to analyze the impact of environmental regulation intensity thereon. The main conclusions of the study are as follows:

(1) The green innovation efficiency in China is showing a rising trend over time and is at a high level overall. However, it varies greatly among different regions in China. The green innovation efficiency in eastern China is higher than the national average, while that in central and western China is lower than the national average.
(2) The impact of command-control environmental regulation on green innovation efficiency follows an inverted N-shaped pattern, with the trend of downward-upward-downward. The market incentive environmental regulation has a U-shaped influence on green innovation efficiency, with a downward-upward trend. The intensity of command-control environmental regulation in most provinces of China is in a range that can effectively promote the improvement of green innovation efficiency. However, the intensity of market incentive environmental regulation in most provinces has not reached the threshold that can effectively promote the improvement of green innovation efficiency.
(3) The impact of environmental regulations on green innovation efficiency also varies across regions. Command-control environmental regulation has an inverted U-shaped impact on green innovation efficiency in eastern China. Additionally, market incentive environmental regulations have a direct positive impact on green innovation efficiency. The impact of market incentive environmental regulations on green innovation efficiency follows a U-shaped pattern in central China. Both types of environmental regulation have a U-shaped effect on green innovation efficiency in western China.

5.2. Recommendations

Recommendations are made for different subjects, aiming to protect the ecological environment and promote the efficiency of green innovation through environmental regulation.

(1) For environmental regulation to contribute to green innovation efficiency, the government must ensure that the intensity of environmental regulation reaches the threshold for technological innovation. However, command-control environmental regulations should not be so severe that enterprises are pressured to close or move out. Therefore, the government should control the pollution discharge standard so that the pollution discharge fee is close to or even greater than the cost of enterprises to prevent and control pollution. Encourage enterprises to carry out technological innovation, improve the industrial structure and prevent pollution from the source. The market incentive environmental regulation policies in most provinces of China have not worked well. The government should provide better guidance regarding market-incentive environmental regulations and make them work hand in hand with command-control environmental regulations to jointly achieve good policy effects. Additionally, the government can adopt a combination of incentives and mandatory measures to manage enterprises. Enterprises that do a good job in terms of discharging pollutants should be given some incentive subsidies or appropriate tax reductions. For some heavily polluting enterprises, compulsory policies can be adopted. The government should urge them to rectify the situation and force them to optimize their industrial structure. Additionally, the government should better guide enterprises which are seeking to engage in technological innovation and focus on environmental protection.
(2) The contribution of technological innovation to green innovation is significant. As the primary creators in innovation activities, enterprises have the responsibility to promote the innovation of the whole industry. The role of enterprises is crucial. First of all, enterprises should fully understand the government's environmental regulation policies and implement pollution prevention and control policies. Secondly, enterprises choose the most appropriate way to control pollution according to the needs of their development and based on maximizing their benefits. Enterprises should adjust their industrial structure and use more environmentally friendly raw materials for production. The concept of green production runs through the whole production

process, and enterprises try to minimize the pollution from the source. Finally, enterprises should reduce investment in industries that produce more pollutants, develop green industries, and play the role of sustainable incentive for green industries.

(3) Environmental regulation policies in different regions have different impacts on the efficiency of green innovation. The government should improve the environmental regulation policy system and formulate policies according to the development needs and the resource endowment of different regions and the conditions for policy implementation. The previous development strategy can be continued in eastern China to attract talents for technological innovation and promote regional innovation while developing the economy. More incentive policies and measures should be implemented to accelerate green innovation efficiency. The implementation of command-control environmental regulation policies should not be too strict to prevent the emergence of a "pollution paradise". The intensity of environmental regulation should be increased, and policies should be actively implemented so that the intensity of environmental regulation reaches a threshold in central and western China. The government should force enterprises to meet emission standards through innovation, thereby promoting green innovation and efficiency.

Author Contributions: T.S. drafted the paper and conducted the data interpretation. D.L. conceptualized and designed the research. Y.J. constructed empirical models and obtained the results. J.L. collected data. All authors have read and agreed to the published version of the manuscript.

Funding: This paper is supported by the Key Program of the National Philosophy and Social Science Foundation of China (Grant No. 21ATJ003), and Innovation Team of Philosophy and Social Sciences in Henan Colleges and Universities (2017–CXTD-07).

Institutional Review Board Statement: Not applicable.

Informed Consent Statement: Not applicable.

Data Availability Statement: Although the data used in this paper are from various publicly available yearbooks in China, some yearbooks are not freely available. The data for each variable can be obtained from the corresponding yearbook provided in the paper. Alternatively, the datasets used in this paper are available from the corresponding authors on reasonable request.

Conflicts of Interest: The authors declare no conflict of interest.

Appendix A

Table A1. Green innovation efficiency in China from 2009 to 2019.

Region	Province	2009	2010	2011	2012	2013	2014	2015	2016	2017	2018	2019
Eastern	Beijing/BJ	1.9572	1.8197	1.8218	1.8959	1.9637	1.9703	2.0041	1.8847	2.151	2.2307	2.1611
	Tianjin/TJ	1.0432	1.008	1.0156	1.115	1.1118	1.0916	1.043	1.0827	1.0102	0.574	1.0248
	Liaoning/LN	0.5077	1.0265	0.5523	0.4831	0.5506	0.4076	0.4659	0.4505	0.4091	0.4399	0.4404
	Shanghai/SH	0.5847	0.4837	1.0068	0.7096	0.6054	0.5692	0.5518	0.595	0.603	1.0022	0.6007
	Jiangsu/JS	1.0617	1.0995	1.1009	1.22	1.2307	1.0745	1.0469	1.0983	1.1237	0.6883	0.608
	Zhejiang/ZJ	1.1632	1.0811	1.0977	1.0869	1.0618	1.0534	1.0184	1.0993	1.0229	1.0031	0.5523
	Fujian/FJ	0.6163	0.4626	0.6567	1.0642	0.701	0.5996	0.651	0.636	0.5627	0.5023	0.5155
	Shandong/SD	1.0479	0.5871	0.7005	0.6813	0.6519	0.59	0.5697	0.5338	0.5505	0.6339	0.7032
	Guangdong/GD	0.4596	1.4324	1.1643	1.1879	1.3403	1.3958	1.0932	1.1078	1.2896	1.1453	1.148
	Hainan/HI	1.1198	1.1249	1.1264	1.1027	1.0944	1.0981	1.1044	1.1002	1.0971	1.0823	1.1544
	Hebei/HE	0.4378	0.4232	0.3995	0.6433	0.4287	0.4405	0.4087	0.4184	0.4435	1.0415	1.1192
Central	Shanxi/SX	1.0017	1.0974	1.0077	0.6922	1.0091	1.0398	1.0821	0.3565	0.4801	1.0197	1.0025
	Neimenggu/NM	1.0847	1.1426	1.1865	1.1151	1.1107	1.0925	1.0759	0.5613	1.15	1.0586	1.0264
	Jilin/JL	0.6162	1.0332	0.4285	0.569	1.0155	0.6002	0.7803	0.6009	0.4933	0.5547	0.756
	Heilongjiang/HL	0.1885	0.1397	0.2617	0.4167	0.2475	0.2729	0.2434	0.3059	0.2439	0.4226	0.6098
	Anhui/AH	1.0859	1.0897	1.0444	1.1056	1.1103	1.0775	1.0504	1.055	1.0403	1.0271	0.6167
	Jiangxi/JX	0.4168	0.4401	0.5707	1.0217	1.0151	0.7156	0.5694	1.0113	0.6394	0.5451	1.0252
	Henan/HA	0.7061	0.8123	1.0163	1.0445	1.239	1.2112	1.226	1.1861	1.1764	1.2795	1.1949
	Hubei/HB	0.5823	0.5631	0.5178	0.5573	0.5369	0.6268	0.6499	0.6433	0.681	0.6611	0.576
	Hunan/HN	0.6415	0.8124	0.7734	1.0348	1.0281	0.7775	0.6548	0.7115	0.5935	0.5653	0.6473
	Guangxi/GX	1.0241	1.0362	1.0101	1.0437	1.1224	1.1462	1.1995	1.2687	1.2829	1.2648	1.4608
Western	Chongqin/CQ	1.06	1.0481	1.3738	1.0802	1.0465	1.135	1.1728	1.0117	1.052	0.6824	0.6914
	Sichuan/SC	0.5509	0.3274	1.075	0.7411	0.7185	0.7345	0.7533	0.6367	0.6462	0.6376	0.5428
	Guizhou/GZ	1.0632	0.2046	0.2731	0.3017	0.2446	0.2774	0.3156	1.0167	1.028	1.0386	1.0411
	Yunnan/YN	1.108	1.0577	1.0106	0.7998	1.006	1.0201	0.6446	1.0406	1.0604	1.0505	1.049
	Shaanxi/SN	0.296	0.4182	0.4339	0.4051	0.3303	0.4254	0.3213	0.3975	0.403	0.3707	0.4159
	Gansu/GS	0.4665	0.4478	0.6368	0.7601	0.6596	0.6091	0.718	0.589	0.557	0.543	1.0143
	Qinghai/QH	0.2332	1.0174	1.3932	1.0073	0.1332	1.0246	1.0395	1.0925	1.005	1.0188	1.0104
	Ningxia/NX	0.465	0.3801	0.4091	0.4048	0.4001	0.379	0.3727	0.3537	0.2703	0.4079	1.0145
	Xinjiang/XJ	1.4049	1.1558	1.0418	1.5436	1.4918	1.4577	1.1934	1.3076	1.064	1.1919	1.2816

References

1. Fan, F.; Zhang, X.; Lian, H. Impact of Environmental Regulations on Green Innovation Efficiency of the Yangtze River Economic Belt: Intermediary Effect Test Based on Foreign Direct Investment. *Sci. Technol. Manag. Res.* **2021**, *41*, 191–196.
2. Adhikari, C.B.; Bjorndal, T. Analyses of technical efficiency using SDF and DEA models: Evidence from Nepalese agriculture. *Appl. Econ.* **2012**, *44*, 3297–3308. [CrossRef]
3. Han, J. Research on Efficiency of Regional Green Innovation in China. *Res. Financ. Econ. Issues* **2012**, *11*, 130–137.
4. Feng, Z. Research on Green Innovation Efficiency of Chinese Industrial Enterprises. *Forum Sci. Technol. China* **2013**, *2*, 82–88.
5. Faria, A.; Fenn, P.; Bruce, A. Production technologies and technical efficiency: Evidence from Portuguese manufacturing industry. *Appl. Econ.* **2005**, *37*, 1037–1046. [CrossRef]
6. Xiao, W.; Lin, B. Government support, R&D Management and Technological Innovation Efficiency: An Empirical Analysis Based on China's Industry. *Manag. World* **2014**, *4*, 71–80.
7. Li, X.; Zeng, Y. The Measurement of China's Regional innovative Green Development Efficiency and Its Influencing Factors. *Sci. Technol. Prog. Policy* **2020**, *37*, 33–42.
8. Xu, W.; Li, L.; Zheng, J.; Liu, C. Comparison of regional differences between green innovation efficiency and Ecological Welfare performance in China. *Stat. Decis.* **2021**, *37*, 56–59.
9. Xu, Y.; Liu, S. Spatial pattern evolution and influencing factors of green innovation efficiency in the Yellow River Basin. *J. Nat. Resour.* **2022**, *37*, 627–644. [CrossRef]
10. Bai, J.; Jiang, F. Collaborative Innovation, Spatial Relevance and Regional Innovation Performance. *Econ. Res. J.* **2015**, *50*, 174–187.
11. Qin, B.; Guo, Y.; Ge, L. How Public Participation Affects Corporate Green Technology Innovation? Analysis Based on Intermediary Effect and Spatial Effect. *J. Technol. Econ.* **2022**, *41*, 50–61.
12. Zhu, D.; Luo, Y.; Lu, Z. Environmental regulation, industrial agglomeration and green innovation efficiency. *Stat. Decis.* **2021**, *37*, 53–57.
13. Ederington, J.; Minier, J. Is environmental policy a secondary trade barrier? An empirical analysis. *Can. J. Econ.* **2003**, *36*, 137–154. [CrossRef]
14. Song, S. Research on Regional Difference in the Effect of Different Environmental Regulation Tools on Polluting Industry Investment-empirical Analysis of the Four Regions of China from the Perspective of Investment Location. *West Forum* **2017**, *27*, 90–99.

15. Zhang, Z.; Zhao, G. FDI, Environmental Regulation and Technological Progress. *J. Quant. Tech. Econ.* **2012**, *29*, 19–32.
16. Fu, J.; Li, L. A Case Study on the Environmental Regulation, the Factor Endowment and the International Competitiveness in Industries. *Manag. World* **2010**, *10*, 87–98.
17. Hao, S.; Zhang, Y. The Impact of Environmental Regulation on Economic Agglomeration: From the Perspective of New Economic Geography. *Soft Sci.* **2016**, *30*, 27–30.
18. Ren, X.; Liu, Y.; Zhao, G. The impact and transmission mechanism of economic agglomeration on carbon intensity. *China Popul. Resour. Environ.* **2020**, *30*, 95–106.
19. Yan, W.; Guo, S.; Shi, Y. Environmental Regulation, Industrial Structure Upgrading, and Employment Effect: Linear or nonlinear? *Econ. Sci.* **2012**, *6*, 23–32.
20. Xu, K.; Yuan, Y. Empirical Study of Environmental Regulation and Industrial Structure Adjustment: A Sys-GMM Estimation Based on Different Pollutants Control. *J. Ind. Technol. Econ.* **2014**, *33*, 101–109.
21. Huang, Q.; Gao, M. Quantity and quality effects of environmental regulation on economic growth: Test based on simultaneous equations. *Economist* **2016**, *4*, 53–62.
22. Wang, Z.; Wen, Y.; Zou, C. The Analysis of the Impact of Environmental Regulation on Technological Innovation Mechanism. *J. Nanjing Univ. Financ. Econ.* **2012**, *3*, 8–13.
23. Deng, Y.; Wang, L.; Zhou, W. Does Environmental Regulation Promote Green Innovation Capability?—Evidence from China. *Stat. Res.* **2021**, *38*, 76–86.
24. Zhang, W.; Chen, Y. Research on the Impact of Environmental Regulation Tools and R&D Subsidies on Green Technology Innovation. *Ecol. Econ.* **2022**, *38*, 36–46.
25. Sun, Y.; Liu, Y. Comparison between environmental tax and pollutant discharge permit and policy choice of pollution. *Public Financ. Res.* **2015**, *4*, 96–99.
26. Brunnermeier, S.B.; Cohen, M.A. Determinants of environmental innovation in US manufacturing industries. *J. Environ. Econ. Manag.* **2003**, *45*, 278–293. [CrossRef]
27. Castellacci, F.; Lie, C.M. A taxonomy of green innovators: Empirical evidence from South Korea. *J. Clean. Prod.* **2017**, *143*, 1036–1047. [CrossRef]
28. Singh, R.K.; Yabar, H.; Nozaki, N.; Niraula, B.; Mizunoya, T. Comparative study of linkage between environmental policy instruments and technological innovation: Case study on end-of-life vehicles technologies in Japan and EU. *Waste Manag.* **2017**, *4*, 114–122. [CrossRef]
29. Zhang, X.; Wang, Y. Effects of Environmental Regulation and R&D Investment on Green Technology Innovation. *Sci. Technol. Prog. Policy* **2017**, *17*, 111–119.
30. Wang, J.; Zhang, Y. Environmental regulation, green technological innovative intention and green technological innovative behavior. *Stud. Sci. Sci.* **2018**, *2*, 352–360.
31. Wang, F.; He, Z. Environmental regulation, green innovation and upgrading of industrial structure. *Stat. Decis.* **2022**, *38*, 73–76.
32. Domazlicky, B.R.; Weber, W.L. Does environmental protection lead to slower productivity growth in the chemical industry? *Environ. Resour. Econ.* **2004**, *28*, 301–324. [CrossRef]
33. Sinn, H.W. Public policies against global warming: A supply side approach. *Int. Tax Public Financ.* **2008**, *15*, 360–394. [CrossRef]
34. Li, W.; Bi, K. Study on impact of environmental regulation intensity on green technology innovation in pollution intensive industries: An empirical test based on panel data from 2003 to 2010. *RD Manag.* **2013**, *6*, 72–81.
35. Kneller, R.; Manderson, E. Environmental regulations and innovation activity in UK manufacturing industries. *Resour. Energy Econ.* **2012**, *34*, 211–235. [CrossRef]
36. Peuckert, J. What shapes the impact of environmental regulation on competitiveness? Evidence from Executive Opinion Surveys. *Environ. Innov. Soc. Transit.* **2014**, *10*, 77–94. [CrossRef]
37. Peng, W.; Cheng, F.; Lu, J. Research on threshold effect of environmental regulation on provincial green innovation efficiency. *South China J. Econ.* **2017**, *9*, 73–84.
38. Luo, Y.; Chen, P. Research on threshold effect of environmental regulation on improvement of China's industrial green innovation efficiency. *J. Northeast. Univ.* **2018**, *2*, 147–154.
39. Gao, H.; Xiao, T. Can heterogeneous environmental regulation force the optimization of industrial structure? *Jianghan Trib.* **2022**, *3*, 13–21.
40. Tone, K. A slacks-based measure of efficiency in data envelopment analysis. *Eur. J. Oper. Res.* **2001**, *130*, 498–509. [CrossRef]
41. Tone, K. A slacks-based measure of super-efficiency in data envelopment analysis. *Eur. J. Oper. Res.* **2002**, *143*, 32–41. [CrossRef]
42. Parzen, E. On Estimation of a Probability Density Function and Mode. *Ann. Math. Stat.* **1962**, *33*, 1065–1076. [CrossRef]
43. Zhang, H.; Li, L.; Zhao, M. Digital Inclusive Finance and Green Total Factor Productivity: Empirical Test Based on System GMM. *New Financ.* **2022**, *3*, 31–40.

44. Hu, Z.; Chen, X.; Zhou, Q.; Chen, D.; Li, J. DISO: A rethink of Taylor diagram, International Journal of Climatology. *Int. J. Climatol.* **2019**, *39*, 2825–2832. [CrossRef]
45. Zhou, Q.; Chen, D.; Hu, Z.; Chen, X. Decompositions of Taylor diagram and DISO performance criteria. *Int. J. Climatol.* **2021**, *41*, 5726–5732. [CrossRef]

atmosphere

MDPI

Article

Projected Elevated [CO_2] and Warming Result in Overestimation of SPAD-Based Rice Leaf Nitrogen Status for Nitrogen Management

Ye Tao [1,2,†], Jishuang Zhang [1,2,†], Lian Song [1,*], Chuang Cai [1], Dongming Wang [1,2], Wei Wei [1,2], Xinyue Gu [1,2], Xiong Yang [1,2] and Chunwu Zhu [1]

1 State Key Laboratory of Soil and Sustainable Agriculture, Institute of Soil Science, Chinese Academy of Sciences, Nanjing 210008, China; taoyetoy@gmail.com (Y.T.); jszhang@issas.ac.cn (J.Z.); chuangcai@issas.ac.cn (C.C.); dmwang@issas.ac.cn (D.W.); weiwei@issas.ac.cn (W.W.); guxinyue20@mails.ucas.ac.cn (X.G.); yangx@issas.ac.cn (X.Y.); cwzhu@issas.ac.cn (C.Z.)

2 University of Chinese Academy of Sciences, Beijing 100010, China

* Correspondence: songlian@issas.ac.cn; Tel.: +86-25-8688-1161

† These authors contributed equally to the article.

Abstract: Nitrogen (N) has a unique place in agricultural systems with large requirements. To achieve optimal nitrogen management that meets the needs of agricultural systems without causing potential environmental risks, it is of great significance to increase N use efficiency (NUE) in agricultural systems. A chlorophyll meter, for example, the SPAD-502, can provide a simple, nondestructive, and quick method for monitoring leaf N status and NUE. However, the SPAD-based crop leaf's N status varies greatly due to environmental factors such as CO_2 concentration ([CO_2]) or temperature variations. In this study, we conducted [CO_2] (ambient and enriched up to 500 μmol moL^1) and temperature (ambient and increased by 1.5~2.0 °C) controlled experiments from 2015 to 2017 and in 2020 in two Free-Air CO_2 Enrichment (FACE) sites. Leaf characters (SPAD readings, chlorophyll a + b, N content, etc.) of seven rice cultivars were measured in this four year experiment. Here, we provide evidence that SPAD readings are significantly linearly correlated with rice leaf chlorophyll a + b content (chl a + b) and N content, while the relationships are profoundly affected by elevated [CO_2] and warming. Under elevated [CO_2] treatment (E), the relationship between chl a + b content and N content remains unchanged, but SPAD readings and chl a + b content show a significant difference to those under ambient (A) treatment, which distorts the SPAD-based N monitoring. Under warming (T), and combined elevated [CO_2] and warming (ET) treatments, both of the relationships between SPAD and leaf a + b content and between leaf a + b content and N content show a significant difference to those under A treatment. To deal with this issue under the background of global climate change dominated by warming and elevated [CO_2] in the future, we need to increase the SPAD reading's threshold value by at least 5% to adjust for applying N fertilizer within the rice cropping system by mid-century.

Keywords: elevated [CO_2]; warming; SPAD; leaf nitrogen monitoring; nitrogen management

Citation: Tao, Y.; Zhang, J.; Song, L.; Cai, C.; Wang, D.; Wei, W.; Gu, X.; Yang, X.; Zhu, C. Projected Elevated [CO_2] and Warming Result in Overestimation of SPAD-Based Rice Leaf Nitrogen Status for Nitrogen Management. *Atmosphere* **2021**, *12*, 1571. https://doi.org/10.3390/atmos12121571

Academic Editors: Zengyun Hu, Xuguang Tang and Qinchuan Xin

Received: 16 August 2021
Accepted: 25 November 2021
Published: 27 November 2021

1. Introduction

Due to its critical role in plants' metabolic activities and heavy losses associated with soil–plant systems, nitrogen (N) is an essential element in plant metabolic functions [1]. Insufficient nitrogen application will result in rice production remaining below its potential [2,3]. Consequently, there is an urgent need to balance N input and rice production. Nitrogen (N) has a unique place in agricultural systems with significant requirements for cropping systems' efficient N management and NUE. Great efforts have been made to enhance the NUE of crops using soil-based or plant-based strategies for identifying appropriate time-splitting applications and optimizing fertilizer deposition methods [4–7]. Based on monitoring the N status of rice by measuring chlorophyll content per leaf area, plant-based strategies can improve NUE significantly [8,9]. Due to the incongruence between N supply and crop demand, soil-based strategies have rarely been employed.

Currently, several methods can measure chlorophyll content in crops' leaves, which can be divided into destructive and nondestructive methods. The destructive methods estimate chlorophyll using solvents and spectrophotometer. These methods can be laborious and time-consuming [10]. To determine chlorophyll content near-instantly, hand-held chlorophyll meters (SPAD-502, Konica Minolta Sensing, Inc., Sakai, Osaka, Japan) have been used to estimate relative chlorophyll content on the same leaf over time. Based on the difference in chlorophyll absorption of two wavelengths of light (650 nm, which is absorbed by chlorophyll; 940 nm, in which no chlorophyll absorption occurs), this method can provide a quick, and nondestructive method for estimating leaf chlorophyll content [11]. SPAD is widely used in scientific experiments and agricultural production to rapidly obtain chlorophyll content and the health status of different kinds of crops and to guide N fertilizer management [9,12–15]. For rice, numerous studies have shown that SPAD-based nitrogen management can improve agronomic N use efficiency (the increase in grain yield per unit of N applied) by 200–276% [8,16,17]. However, other studies have found that SPAD, chlorophyll content, and leaf N status were substantially affected by environmental factors such as crops' growing temperature, CO_2 concentration, etc. [18,19]. No obvious effect of elevated [CO_2] or temperature elevation on leaf SPAD was detected in the early and middle growth stages [16,20–22]. However, a decrease in leaf N content was often found in elevated [CO_2] studies [23,24], and the combination of E and T treatment further enhances the decreased magnitude [25]. In view of the fact that SPAD, chlorophyll, and leaf nitrogen have changed differently in elevated [CO_2] or temperature, their relationship is bound to change and needs to be reassessed.

According to the latest IPCC AR6 report, [CO_2] will continue to increase during this century, reaching 600–1000 ppm by 2100; the global surface temperature will also continue to increase, and global warming of 1.5 °C and 2 °C will be exceeded during the twenty-first century unless there are deep reductions in CO_2 and other greenhouse gas emissions [26,27]. This rapid increase in [CO_2] and temperature alters several basic biological functions in plants. Changes in leaf morphology, including thickness and density, can, in turn, impact the optical path of the SPAD device and then the SPAD value. Overall, these results suggest that the relationship between SPAD readings and chlorophyll leaf content will be altered as [CO_2] and temperature continue to increase. Elevated [CO_2] has also been shown in multiple studies to reduce leaf N concentration [19,28–33]. Since SPAD determines the relative value of leaf chlorophyll content, many studies have sought to establish critical SPAD threshold values below which rice corresponds to N fertilization deficiency. Generally, SPAD readings below 35 units were defined as a critical value of the need for N fertilization in rice [8,34–36]. However, the variations in relationship between SPAD readings and nitrogen content under elevated [CO_2] and warming remain unexplored, and it is unclear whether the current critical value can still work for N fertilizer management in the future. Thus, in this work, at two Free-Air CO_2 Enrichment (FACE) facilities, we explored the variations in rice leaf SPAD readings, chl a + b content and N content and the relationship between each of them under elevated [CO_2] and warming treatments. Specifically, we aim to address the following questions: (1) how does the relationship between rice leaf SPAD readings and N content vary under elevated [CO_2] and warming treatments? (2) What is the reason for the variations in their relationships? (3) Finally, how can we deal with this issue in the future?

2. Materials and Methods

2.1. Experimental Site and Growth Condition

We conducted CO_2 and temperature-controlled field experiments in 2020 in the FACE 1 system located in Kangbo village (31°30′ N, 120°33′ E), Changsu Municipality, Jiangsu Province, China. As a further verification and supplement, we also include CO_2-controlled experimental data from 2015 to 2017 in the FACE 2 system in our study. FACE2 is located in Zongcun village (32°35″ N, 119°42″ E), Jiangdu District, Jiangsu Province in China, and both are typical rice growing regions (Figure 1).

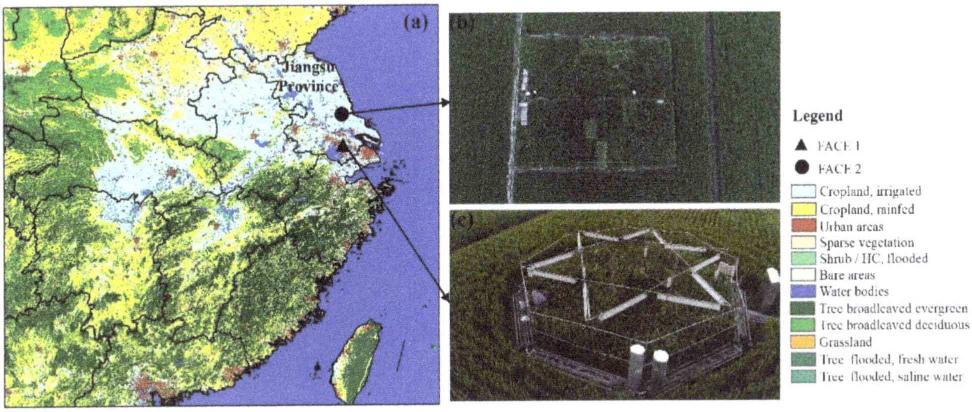

Figure 1. (**a**) Location of the experimental site in Jiangsu Province in the east of China and the ESA Climate Change Initiative (CCI) land cover map of the study site in 2010 at 300 m spatial resolution. Additionally, included are the on-site photos of the study site of FACE2 (**b**) and FACE1 (**c**).

The region of the FACE1 system has a subtropical monsoon climate with a rice growing season average temperature of 24 °C in 2020. The soil is classified as a gleyic stagnic anthrosol. Properties of the topsoil at a depth of 15 cm are as follows: soil pH 7.0, soil organic carbon 1.6%, and total nitrogen 1.9 g kg^{-1}. The region of the FACE 2 system is typical of a northern subtropical monsoon climate. The soil is classified as a Shajiang-Aquic Cambiosol. Soil properties at a depth of 15 cm are as follows: bulk density 1.16 g cm^{-3}, soil organic carbon 18.4 g kg^{-1}, total nitrogen 1.45 g kg^{-1}, total phosphorous 0.63 g kg^{-1}, total potassium14.0 g kg^{-1}, available phosphorous 10.1 mg kg^{-1}, available potassium 70.5 mg kg^{-1}, and pH 6.8.

The operation and control systems for both FACE facilities were the same as those used at the Japan FACE site [37,38]. Each FACE plot was encircled with a ring (8 m diameter for FACE1 and 14 m diameter for FACE2 system) with emission tubes that injected pure CO_2 at around 30 cm above the plant canopy controlled by Li-820 CO_2 sensors (LI-COR Inc., Lincoln, NE, USA). The rings were raised as the canopy grew to maintain the CO_2 set point. Ambient control plots did not receive any supplemental CO_2. The CO_2 set value in FACE plots was ~200 μmol moL^{-1} above that of ambient control plots. CO_2 release was controlled by a program with an algorithm based on the speed and direction of the wind to maintain the target CO_2 concentration ([CO_2]).

In the FACE1 system, 12 octagonal plots were located in uniform paddy fields with similar soils and agronomic histories [39], and the facility also included 12 infrared heaters (detailed information can be found in Cai et al., 2016). The canopy warming magnitude under the controlled plot in 2020 was +1.6~1.7 °C as compared with ambient plot. In the FACE2 system, three rectangular and uniform paddy fields were established. A FACE treatment was paired with an ambient plot within each area, and plot centers were 90 m apart to avoid additional CO_2 diffusion.

2.2. Sampling and Measurements

In the FACE1 system, four rice (*Oryza sativa* L.) varieties including cvs. Xiangliangyou 143, Changyou 5, Wuyungeng 23 and Yangdao 6 were planted. Seeds of each line were sown on 27 May 2020 and transplanted on 22 June 2020. The basal nitrogen (N) and two top-dressed N were applied at 69, 60, and 52 kg·ha^{-1}, respectively. For FACE2 experiments, five rice (*Oryza sativa* L.) varieties including cvs. Yangdao 6 (YD6), Y Liangyou 900 (Y900), Wuyungeng 23 (WYG23), Wuyungeng 27 (WYG27), Nangeng 9108 (NG9108) were chosen (YD6, WYG23, NG9108 for 2015; YD6, Y900, WYG23, NG9108 for 2016; YD6, Y900, WYG23, WYG27 for 2017). Seeds of each line were sown on 20 May 2015 and 2016, 22 May 2017, and transplanted on 20 June, 17 June and 21 June in 2015, 2016 and 2017, respectively. The spacing of the hills was 16.7 cm × 25 cm (equivalent to 24 hills·m^{-2}). A fertilizer dose of 225:90:90 kg ha^{-1} nitrogen (N)–phosphorus (P)–potassium (K) was applied to all plots over the season; 40% of the N and whole of the P and K was applied as the basal starter dose, while residual N was equally split at middle tillering and the panicle initiation stage.

SPAD readings and leaf samples were collected on 15 August, 6 September, and 25 September in 2020, respectively, in FACE1. In FACE2 system, SPAD readings and leaf samples were collected on September 1, 8, and 23 in 2015. In 2016 and 2017, SPAD readings and leaf samples were collected every ten days from 10 August to 23 September and from 3 August to 5 September, respectively. SPAD measurements were conducted on the newest fully expanded leaf (flag leaf during grain fill following full expansion). Measurements occurred between 07:00 a.m. and 09:00 a.m. to minimize the potential effects of light intensity on SPAD readings [40]. Four to eight SPAD readings were taken from around the midpoint of each leaf blade, and both sides of the midrib for each leaf. In FACE1 system, four leaves were selected in each plot and the SPAD values of each leaf were recorded. Then, these four leaves were kept separately in liquid nitrogen tank. In the lab, a hole punch was used to remove a fixed area of the middle third of leaves to determine chlorophyll content, and the remaining middle third of leaves was measured for leaf area and N content (mg cm^{-2}). In the FACE2 system, ten leaves were selected in each plot, and then the SPAD readings were averaged. After SPAD readings were recorded, these leaves were collected. Half of the leaves were randomly selected and put into a portable tank with liquid nitrogen for back up, while the remaining leaves were photographed on a graduated scale on a white background and processed with Image J software (The National Institutes of Health) to obtain the leaf area. Then, photographed leaves were used for determination of N. Chlorophyll concentration, measurements were conducted using a spectrophotometer (UV2102, Unico, NJ, America) and 95% (*v/v*) alcohol extracts of leaf tissue [41]. The samples for leaf N measurement were oven-dried at 80 °C to constant weight and digested using the micro-Kjeldahl method, after which the N concentration was measured with a discrete wet chemistry analyzer. The concentration of leaf N was calculated as area-based (N weight per unit leaf area).

2.3. Data Analysis

To better compare the variations of SPAD, chl a + b content, and nitrogen content, we calculated their normalized values using minimum–maximum value-based normalization method as follows [42]

$$Y' = \frac{Y - Y_{min}}{Y_{max} - Y_{min}} \quad (1)$$

where Y' denotes the normalized SPAD/chl a + b content/N content; Y is the original SPAD/chl a + b content/N content values; Y_{max} and Y_{min} are, respectively, the maximum or minimum values of the original SPAD/chl a + b/N content values. ANOVA was used to determine differences between treatment means. All results reported as significant had a $p < 0.05$ unless stated otherwise. To show the distributions of all measured leaf SPAD, chl a + b content, and N content under different controls, violin plot is employed here. The violin plot synergistically combines the box plot and density trace into a single display that can reveal structure found within the data [43].

3. Results

3.1. Variations of Rice Leaf SPAD Readings and N Content under Different Treatments

We analyzed the measured rice leaf SPAD readings and N content to assess its variations under different controls (A, E, T, and ET) in the FACE1 system. Figure 2 shows that leaf SPAD and N content decrease under elevated [CO_2] (E), warming (T), and elevated [CO_2] and warming (ET) controls compared to ambient (A) conditions with different magnitudes. For rice leaf N content (Figure 2b), the average N content under ambient conditions is 15.63 mg cm^{-2}. Under E and T controls, the average N content shows a small decrease of 15.24 and 15.09 mg cm^{-2}, respectively. When rice is grown under ET control, the average N content is 13.49 mg cm^{-2}, which is −13.69% smaller than that under A. The decreasing magnitudes for leaf SPAD (Figure 2a) under controlled conditions (E, T, and ET) show desynchronized variations with N content. The average SPAD readings under E, T, and ET conditions are 43.17, 42.77, and 42.63 units, which is −0.16%, −1.09%, and −1.41%, respectively, smaller than under the A condition (43.24 unit). The linear fit results of the normalized parameters also indicate a similar conclusion (insert plot in Figure 2a,b). The slope of the linear fit of normalized SPAD readings for A, E, T, and ET is −0.01, and the slope of normalized N content is −0.04. The linear fit results of normalized parameters indicate that under E and T controls, leaf N content shows a larger decrease magnitude than that of leaf SPAD readings. Under ET control, leaf SPAD readings and N content all indicate the largest decrease as compared with A, E, and T treatments. To assess the influence of rice subspecies on the SPAD and nitrogen content response to different controls, we also analyzed the indica varieties (including cvs Xiangliangyou 143 and Yangdao 6, Figure S1) and japonica varieties (including cvs Changyou 5 and Wuyungeng 23, Figure S2) of rice leaf SPAD readings and N content under A, E, T, and ET in the FACE1 system. Indica and Japonica rice indicate similar results. Under controlled conditions, the decreasing magnitude of SPAD readings is smaller than that of N content. The slopes of normalized SPAD readings of indica and japonica rice are −0.01 (r = −0.37) and −0.02 (r = −0.67), respectively; the values for N content are −0.02 (r = −0.49) and −0.03 (r = −0.74), respectively.

3.2. Relationship between Rice Leaf SPAD Readings and N Content Per Leaf Area under Different Treatments

To further analyze the influence of the discrepancy among the variations of SPAD readings and N content under different treatments, Figure 3 shows the scatter plots and linear fit results between SPAD readings and N content under A, E, T, and ET treatments in the FACE1 system. Under both A and controlled treatments, SPAD readings significantly correlate with N content (R^2 under A: 0.56, R^2 under E: 0.58, R^2 under T: 0.60, R^2 under ET: 0.58, all $p < 0.01$). Nonetheless, the slope and intercept of the regression lines between SAPD readings and N content vary significantly under different treatments (insert plot in Figure 3). Under the A condition, the slope is 1.07 (±0.16) mg cm^{-2}, and slope values increase substantially under controlled treatments (E, T, and ET), with the largest values under ET. The initial intercept value under A is −31.49 (±1.07), and, under E, T, and ET controls, the intercept value is 21%, 25%, and 30% smaller than that under the A treatment, respectively.

The scatter plot results between SPAD and N content under A and E treatments in the FACE2 experiment also indicate similar trends as the FACE1 experiment (Figure 4). Under both A and E treatments, SPAD shows a middle strong correlation with N content (R^2 under A: 0.42, R^2 under E: 0.51, both $p < 0.01$). The slope value increases significantly from 0.59 (±0.07) to 0.668 (±0.07) from A to E treatments. For intercept values, the intercept under E treatment is 51% smaller than that under the A treatment.

3.3. Variations of Rice Leaf N Content Corresponding to Rice N Demand SPAD Value

In the paddy field SPAD-based nitrogen treatment, nitrogen was top-dressed when rice leaf SPAD readings fell below 35 [8,34–36]. We then estimated and compared N content corresponding to SPAD = 35 under A, E, T, and ET treatments in the FACE1 experiment (Figure 5). According to linear fit regression results in Figure 3, when rice SPAD = 35, N content under A treatment is 5.96 (±0.46) mg cm^{-2}. The estimated N content under E and T treatment, respectively, is 14.8% and 15.8% smaller than that under A, with a significant difference. Under ET treatments, the estimated N content is 3.73 (±0.63) mg cm^{-2}, the decreased magnitude is near twice as many as E and T (37.4% smaller than that under

A). Overall, the results indicate that the estimated N content corresponding to SPAD = 35 under controlled treatments is smaller than under A treatment.

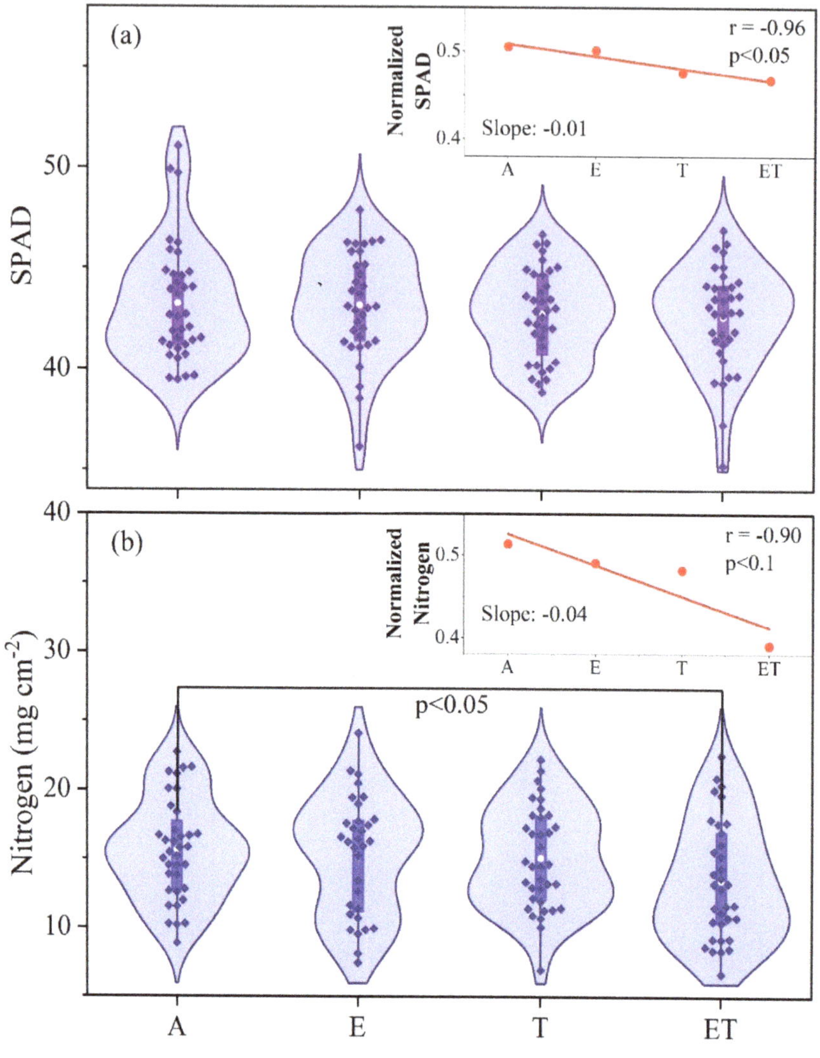

Figure 2. Violin plots of the rice leaf SPAD values (**a**) and nitrogen content (**b**) in different controls (A: ambient, E: elevated [CO_2], T: warming, and ET: elevated [CO_2] and warming). The violin plot is a box plot with the width of the box proportional to the estimated density of the observed SPAD readings and N content. The maximum density of the group-specific data distribution is indicated by the largest width of the violins. The white dot in the violin plot is the median, the thick vertical bar indicates the interquartile range, and the thin vertical bar indicates 95% confidence intervals. A linear fit between normalized SPAD readings, N content and different controls is presented in insert.

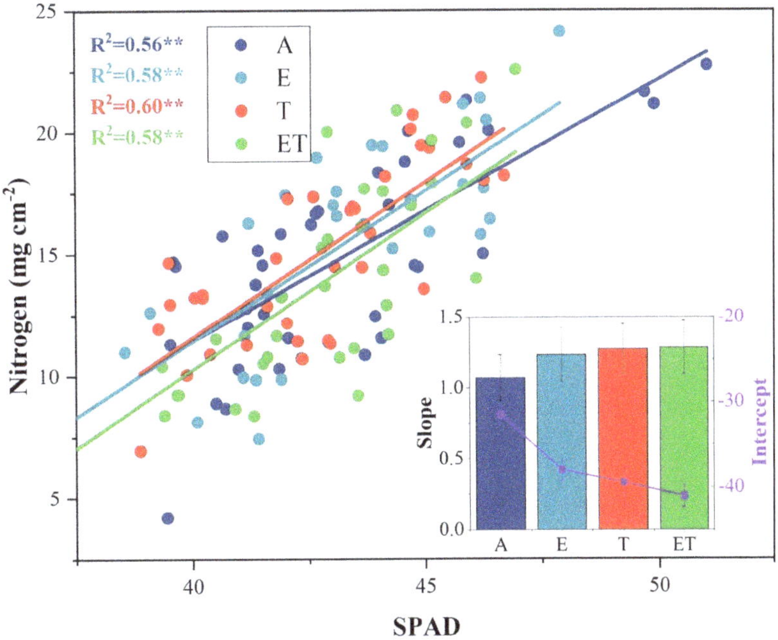

Figure 3. Scatter plots and linear fit results between rice leaf SPAD values and nitrogen content within different controls in FACE1 system. The estimated slope and intercept of the linear fit are presented in the inserted bar plot. ** in the plot indicates $p < 0.01$.

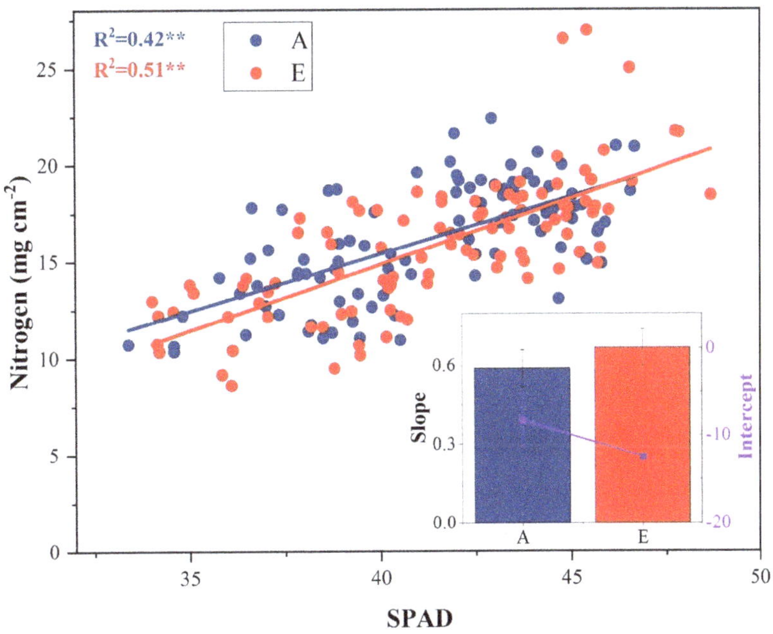

Figure 4. Scatter plots and linear fit results between rice leaf SPAD values and nitrogen content under ambient and elevated [CO_2] conditions in FACE2 system. The estimated slope and intercept of the linear fit are presented in the inserted bar plot. ** in the plot indicates $p < 0.01$.

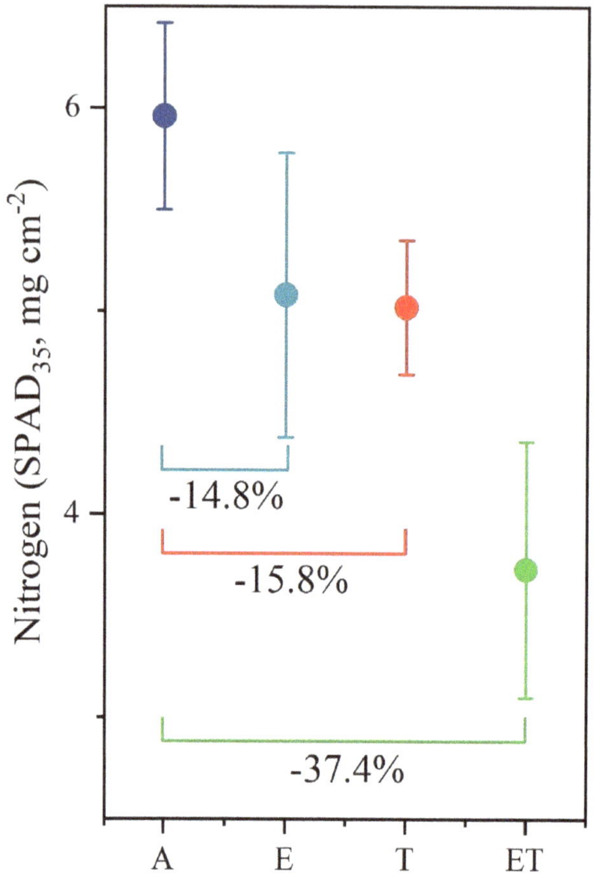

Figure 5. Estimated rice leaf N content per leaf area corresponding to SPAD = 35 using the linear fit results of Figure 3 in different controls. The horizontal lines indicate the change percent of N content of E, T, and ET controls as compared with the values under A.

3.4. Attributions of the Discrepancy between Rice Leaf SPAD Readings and N Content under Different Treatments

To determine the reason for the discrepancy in SPAD reading responses to N content by changes in [CO_2] and temperature, we analyzed variations in the chl a + b content and relationships with SPAD under different treatments in the FACE1 experiment site. The average chl a + b content under E, T, and ET controls are −0.06%, −3.5%, and −4.64% smaller than under A treatment (64.25 μg cm^{-2}), respectively. However, the decreasing trends of chl a + b content from A to controlled treatments are smaller than that of N content as indicated by Figure 6a. The slope of normalized chl a + b content is −0.02, which is also smaller than the value of normalized N content (−0.04). Results from the indica and japonica indicate similar trends (Figures S3 and S4). The decreasing trends in both indica and japonica rice leaf chl a + b content are smaller than that of N content.

As indicated by Figure 6b, under all of the treatments (A, E, T, and ET), SPAD shows a strong correlation with chl a + b content. The slope of the linear fit results indicates increasing trends from A to controlled treatments, and the intercept indicates decreasing trends, which are similar to the trends of N content. For the indica and japonica subspecies, both of the leaf SPAD readings of these two subspecies show a strong correlation with chl a + b and N content, and the slope and intercept values indicate increasing and decreasing trends as a result for all samples (Figures S5 and S6). Based on the linear fit results, the chl a + b content corresponding to SPAD = 35 under A, E, T, and ET treatments was calculated (Figure 6c). Under A treatment, chl a + b content is 47.55 (±2.24) μg cm^{-2}, and the values under E, T, and ET controls are 9.2%, 10%, and 17.6% smaller than that under A treatment, respectively. Thus, although chl a + b content shows similar trends with N content, the decreasing magnitude of chl a + b is smaller than that of N content from A to controlled treatments.

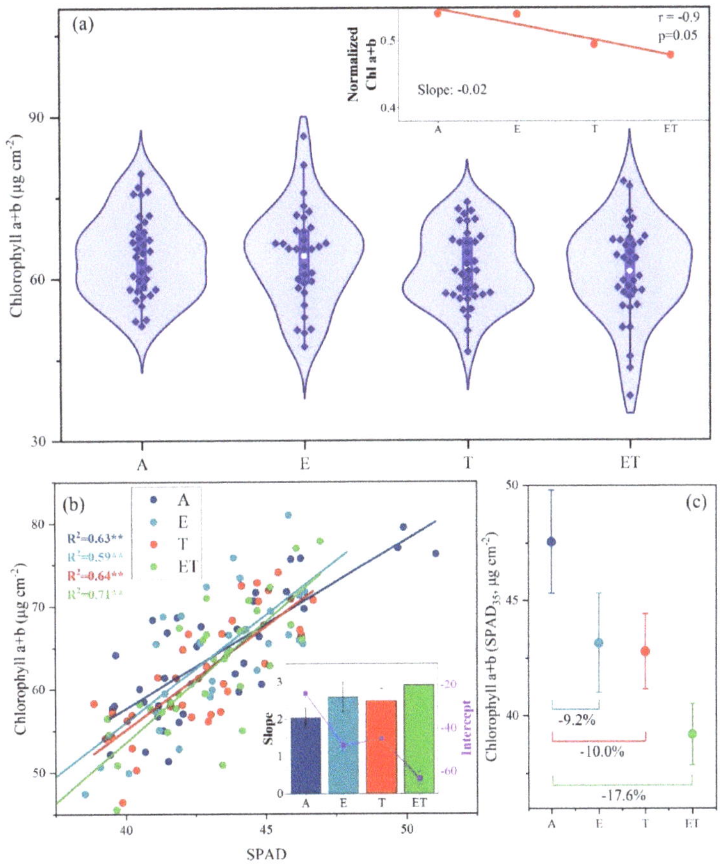

Figure 6. (**a**) Variations in rice leaf chl a + b content under different controls at FACE1 experiment site. (**b**) Scatter plots and linear fit results between rice leaf SPAD values and chl a + b content under different conditions at FACE1 experiment site. (**c**) Estimated rice leaf chl a + b content per leaf area corresponding to SPAD = 35 using the linear fit results of Figure 3 under different controls at FACE1 experiment site. ** in the plot indicates $p < 0.01$.

Figure 7 indicates that under E treatment, the normalized slope and intercept between SPAD readings and N content and SPAD readings and chl a + b content vary significantly, while the relationship between chl a + b and N content remains unchanged compared with A. Under T treatment, the slope between normalized SPAD and N content increases from 0.81 (A) to 1.16 (T), and the intercept decreases from 0.1 (A) to −0.07 (T). The response of normalized SPAD to chl a + b content and normalized chl a + b content to nitrogen content show similar trends; both of the normalized slope values show increasing trends and normalized intercept values show decreasing trends. Under ET control (Figure 8), the slope between normalized SPAD and nitrogen content increases by 43%, and the normalized intercept decreases by −250% compared with values under A treatment. For the variations in normalized SPAD and chl a + b content and chl a + b content and nitrogen content, the slope values under ET increase by 43% and 11%, respectively, and the intercept values decrease by −85% and −550%, respectively.

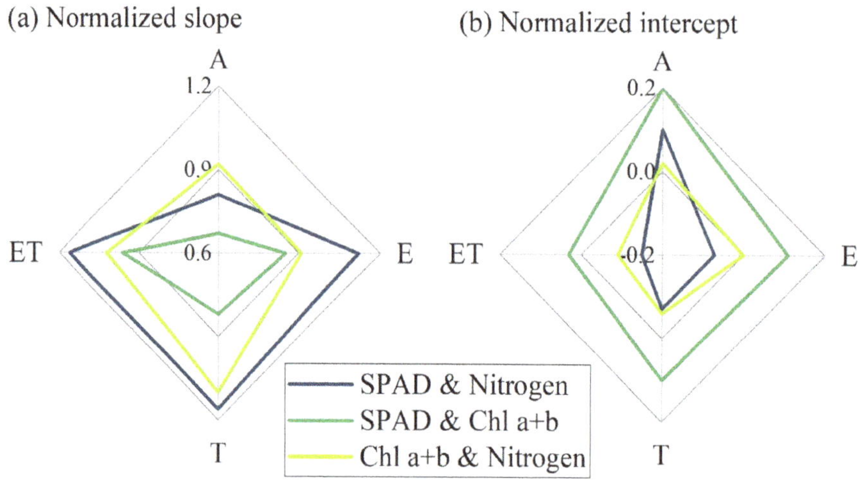

Figure 7. Wind rose diagram of (**a**) normalized slope and (**b**) normalized intercept under different controls. The normalized slope and intercept is the linear fit result of normalized SPAD, chl a + b content, and nitrogen content.

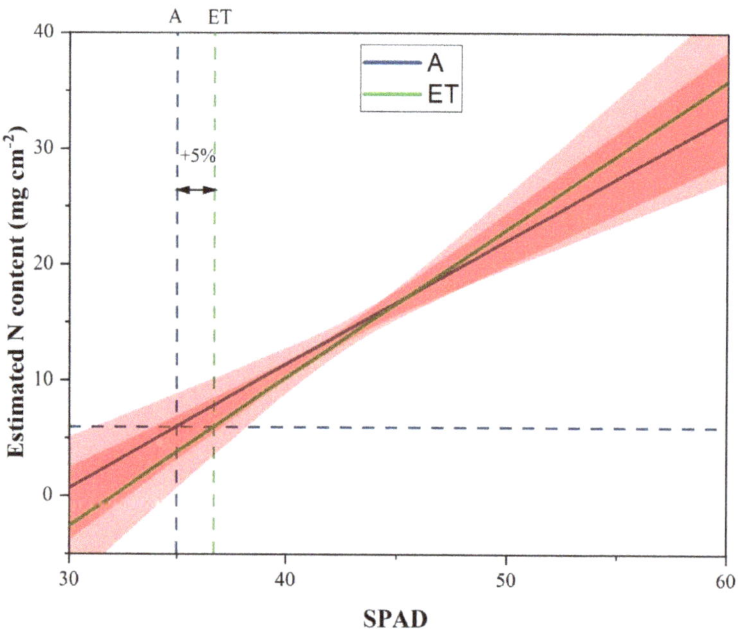

Figure 8. Regression lines based on linear fit results of Figure 3 at FACE1 site under A and ET conditions. The transparent red area indicates 95% confidence bands. The vertical dark blue dotted line (vb) is at the position of SPAD = 35; the horizontal dark blue dotted line (hb) is the estimated rice leaf N content using regression result under A condition. The vertical green dotted line (vg) is the SPAD value corresponding to the estimated N content of hb using regression result under ET condition.

4. Discussion

Under E or T treatments, rice leaf N content after heading decreases compared with that under A treatment, and the combination of E and T treatment further enhances the decreased magnitude (Figure 2). The decrease in rice leaf N content under E and T treatments has been found in many previous studies. For example, Cai et.al. reported that rice leaf N content decreased 16.7% under E control and 33.3% under ET control, respectively, after the heading stage in a field experiment [25]. This decrease is generally considered to be the dilution effect of increased biomass due to elevated [CO_2] [23,24]. Additionally, some studies suggested that a decrease in transpiration rate due to restricted stomatal opening under E control results in poorer N absorption capacity [20,44]. After anthesis, warming, elevated [CO_2], or both might accelerate leaf senescence [25]. Faster leaf senescence is often accompanied by faster N remobilization from green leaves to grains in cereal crops [45,46].

The average SPAD readings only decrease 0.6 units under ET control (decrease of 1.41% compared with the value under A treatment), which is far smaller than the decreased magnitude of leaf N content (−14% compared with the value under A treatment, Figure 2). The variations in this decline result in the distortion of the relationship between SPAD readings and leaf N content under controlled treatments (E, T, and ET). This is confirmed by another result from the meta-analysis of FACE experiments; nitrogen reported on an area basis was reduced by 12% in plants under E treatment, while there was no significant change in chlorophyll content per leaf area [47].

Compared with the relationship between SPAD readings and leaf N content under A treatment, the relationships under all three controls indicate a higher slope and lower intercept (Figures 3 and 4). This means that if the relationships between SPAD and leaf N content under A treatment continue to be employed to estimate leaf N status under E and T treatments in the future, rice leaf N status may be under/overestimated under relatively high/low leaf N statuses, respectively. Previous studies have already reported that SPAD-based estimations of rice leaf N status are influenced by multiple environmental factors; leaf's N status may be under/overestimated by SPAD readings due to changes in environmental factors [18].

To analyze the attributions of distortion of the SPAD-based leaf N content estimation under different treatments, we investigated the relationship between SPAD and chl a + b content per leaf area and chl a + b content and N content under three controls (E, T, and ET, Figure 7). Under E control, there is no significant difference compared with A treatment for the relationship between chl a + b content and N content, while there is a significant difference between SPAD readings and chl a + b content. Under T and ET controls, both relationships between SPAD readings and chl a + b content and chl a + b and N content show significant differences to those under A treatment. This result indicates that the discrepancies between SPAD and leaf N status under E control mainly result from the difference between chl a + b and leaf N status, while for the T and ET controls, both significant differences in the relationship between SPAD and chl a + b content and chl a + b content and N content with that under A, lead to the significant discrepancy in SPAD and N status.

The SPAD value in SPAD-based N management could be absolute or relative SPAD [48], and the absolute value was used in the present study. The relative SPAD value calculated by multiple indicators is more advanced in evaluating leaf nitrogen content and the N fertilizer management [49], but it increases the workload in field determination, which is contrary to the rapid measurement and restricts its application. On the other side, by mid-century, global [CO_2] and warming magnitude are predicted to increase by 200 μmol mol^{-1} and 2 °C under the high emissions RCP8.5 scenario [50]. Here, we monitored similarly elevated [CO_2] and warming scenes in the FACE site, and our results indicate that not only do these indicators change differently but also the relationship between SPAD readings and leaf N content varies greatly under E and T controls, which means that this may cause great potential risk to SPAD-based N management, especially when leaves are of low leaf

N status in future. The key factor in SPAD-based N management is the SPAD threshold readings for N application. For example, when SPAD = 35 is considered as the threshold value for applying N fertilization in rice, the leaf N content estimated by relationship under ET control is 37.4% lower than the value based on the relationship under the current A treatment. The lower leaf N status of the same SPAD readings means that if the threshold of the current condition continues to be employed in future elevated [CO_2] and warming environments, it will delay the application of N fertilization, which may result in a great potential risk of yield reduction. The present study is the first reminder of the risk of insufficient nitrogen supply that may be hidden under unchanged performance (SPAD-based). Certainly, such results are not enough to complete the correction of SPAD-based N management in climate change. The SPAD threshold and the amount of fertilization need to be determined according to specific rice cultivars, environmental factors, etc. [51], and this requires more experimental studies.

Although the application of N fertilizer plays an important role in global food security, excessive nitrogen fertilizer use in agriculture production can also have negative environmental effects such as increased greenhouse gas (GHG) emissions [52,53], soil acidification [54], and eutrophication of groundwater [55]. Compared with the possible yield loss, the additional cost is tolerable with the relatively low nitrogen fertilizer price [56], which is the main reason for the excessive nitrogen input. It is undeniable that a rising nitrogen fertilizer price will encourage farmers to use nitrogen more cautiously [56]. Multiple studies have documented that expediting the development of optimal nitrogen management is an effective technique to confront the challenges of food security and environmental sustainability [15,16,35]. The aim of SPAD-based nitrogen management is to achieve higher nitrogen use efficiency by greatly reducing nitrogen fertilizer with no yield loss rather than achieving higher yield. Maintaining the balance between nitrogen input and crops' needs will drastically reduce nitrogen loss during agricultural production.

5. Conclusions

Chlorophyll meters, for example, the SPAD-502, can provide a simple, nondestructive and quick method for monitoring leaf N status in agricultural systems.

In this study, we provide evidence that under E and T controls, current SPAD-based monitoring of rice leaf N status can be overestimated under relatively low leaf N status, which may cause great potential risk to fertilization management and rice yield decline. The distortion of the SPAD-based N monitoring under E control is caused by the discrepancy of the relationship between leaf chl a + b figure and N content; under T and ET controls, the distortions are caused by the discrepancy of the relationships between SPAD and leaf chl a + b content and leaf chl a + b and N content. To deal with this issue under the background of global climate change dominated by warming and elevated [CO_2] by mid-century, we need to increase the threshold SPAD readings by at least 5% to adjust N management.

Supplementary Materials: The following are available online at https://www.mdpi.com/article/10.3390/atmos12121571/s1, Figure S1: Violin plots of the indica rice leaf SPAD values (a) and nitrogen content (b) in different controls (A: ambient, E: elevated [CO_2], T: warming and ET: elevated [CO_2] & warming). Figure S2: Violin plots of the japonica rice leaf SPAD values (a) and nitrogen content (b) in different controls (A: ambient, E: elevated [CO_2], T: warming and ET: elevated [CO_2] & warming). Figure S3: Violin plots of the indica rice leaf chlorophyll a+b content in different controls (A: ambient, E: elevated [CO_2], T: warming and ET: elevated [CO_2] & warming). Figure S4: Violin plots of the japonica rice leaf chlorophyll a+b content in different controls (A: ambient, E: elevated [CO_2], T: warming and ET: elevated [CO_2] & warming). Figure S5: Scatter plots and linear fit results between indica rice leaf (a) SPAD values & chlorophyll a+b content and (b) SPAD values & nitrogen content within different controls. Figure S6: Scatter plots and linear fit results between japonica rice leaf (a) SPAD values & chlorophyll a+b content and (b) SPAD values & nitrogen content within different controls. The estimated slope and intercept of the linear fit is presented in the inserted bar plot. ** in the plot indicates $p < 0.01$. Figure S7: The relationship between the changes of SPAD and leaf Ndw (blue dots, N concentration by unit dry weight), Na (orange dots, N concentration by unit leaf area) in E relative to A of 2016 and 2017. The orange and blue regression lines represent the regression of Na and Ndw with SPAD, respectively.

Author Contributions: Conceptualization, C.Z.; Data curation, Y.T. and J.Z.; Funding acquisition, L.S., C.C. and C.Z.; Investigation, Y.T., J.Z., L.S., D.W., W.W., X.G. and X.Y.; Methodology, L.S.; Software, L.S.; Supervision, C.Z.; Writing—original draft, L.S.; Writing—review and editing, Y.T., J.Z., L.S., C.C., D.W. and C.Z. All authors have read and agreed to the published version of the manuscript.

Funding: National Natural Science Foundation Grant of China 32001191; National Natural Science Foundation Grant of China 31800359; the Doctor of Mass Entrepreneurship and Innovation in Jiangsu Province; the "0 to 1" original innovation project of the Basic Frontier Science Research Program of CAS ZDBS-LY-DQC020; Natural Science Foundation of Jiangsu province in China BK2018402.

Institutional Review Board Statement: Not applicable.

Informed Consent Statement: Not applicable.

Data Availability Statement: Not applicable.

Acknowledgments: The work was supported by the National Natural Science Foundation Grant of China (Grant NO. 32001191, 31800359), the Doctor of Mass Entrepreneurship and Innovation in Jiangsu Province, the "0 to 1" original innovation project of the Basic Frontier Science Research Program of CAS (ZDBS-LY-DQC020), the Natural Science Foundation of Jiangsu province in China (BK2018402).

Conflicts of Interest: The authors declare no conflict of interest.

References

1. Ladha, J.K.; Reddy, P.M. Nitrogen fixation in rice systems: State of knowledge and future prospects. *Plant Soil* **2003**, *252*, 151–167. [CrossRef]
2. Carpenter, S.R.; Caraco, N.F.; Correll, D.L.; Howarth, R.W.; Sharpley, A.N.; Smith, V.H. Nonpoint pollution of surface waters with phosphorus and nitrogen. *Ecol. Appl.* **1998**, *8*, 559–568. [CrossRef]
3. Bodirsky, B.L.; Popp, A.; Lotze-Campen, H.; Dietrich, J.P.; Rolinski, S.; Weindl, I.; Schmitz, C.; Mueller, C.; Bonsch, M.; Humpenoeder, F.; et al. Reactive nitrogen requirements to feed the world in 2050 and potential to mitigate nitrogen pollution. *Nat. Commun.* **2014**, *5*, 3858. [CrossRef]
4. Bausch, W.C.; Diker, K. Innovative remote sensing techniques to increase nitrogen use efficiency of corn. *Commun. Soil Sci. Plant Anal.* **2001**, *32*, 1371–1390. [CrossRef]
5. Fontes, P.C.R.; de Araujo, C. Use of a chlorophyll meter and plant visual aspect for nitrogen management in tomato fertigation. *J. Appl. Hort* **2006**, *8*, 8–11.
6. De Datta, S.; Samson, M.; Kai-Rong, W.; Buresh, R. Nitrogen use efficiency and nitrogen-15 balances in broadcast-seeded flooded and transplanted rice. *Soil Sci. Soc. Am. J.* **1988**, *52*, 849–855. [CrossRef]
7. Strong, W.M. Effects of nitrogen application before sowing, compared with effects of split applications before and after sowing, for irrigated wheat on the Darling Downs. *Aust. J. Exp. Agric.* **1986**, *26*, 201–207. [CrossRef]
8. Peng, S.; Garcia, F.V.; Laza, R.C.; Sanico, A.L.; Visperas, R.M.; Cassman, K.G. Increased N-use efficiency using a chlorophyll meter on high-yielding irrigated rice. *Field Crop. Res.* **1996**, *47*, 243–252. [CrossRef]
9. Piekielek, W.P.; Fox, R.H.; Toth, J.D.; Macneal, K.E. Use of a chlorophyll meter at the early dent stage of corn to evaluate nitrogen sufficiency. *Agron. J.* **1995**, *87*, 403–408. [CrossRef]
10. Pinkard, E.A.; Patel, V.; Mohammed, C. Chlorophyll and nitrogen determination for plantation-grown Eucalyptus nitens and E-globulus using a non-destructive meter. *For. Ecol. Manag.* **2006**, *223*, 211–217. [CrossRef]
11. Yadava, U.L. A rapid and nondestructive method to determine chlorophyll in intact leaves. *Hortscience* **1986**, *21*, 1449–1450.
12. Wu, J.; Wang, D.; Rosen, C.J.; Bauer, M.E. Comparison of petiole nitrate concentrations, SPAD chlorophyll readings, and QuickBird satellite imagery in detecting nitrogen status of potato canopies. *Field Crop. Res.* **2007**, *101*, 96–103. [CrossRef]
13. Pagola, M.; Ortiz, R.; Irigoyen, I.; Bustince, H.; Barrenechea, E.; Aparicio-Tejo, P.; Lamsfus, C.; Lasa, B. New method to assess barley nitrogen nutrition status based on image colour analysis Comparison with SPAD-502. *Comput. Electron. Agric.* **2009**, *65*, 213–218. [CrossRef]
14. Chang, S.X.; Robison, D.J. Nondestructive and rapid estimation of hardwood foliar nitrogen status using the SPAD-502 chlorophyll meter. *For. Ecol. Manag.* **2003**, *181*, 331–338. [CrossRef]
15. Arregui, L.M.; Lasa, B.; Lafarga, A.; Iraneta, I.; Baroja, E.; Quemada, M. Evaluation of chlorophyll meters as tools for N fertilization in winter wheat under humid Mediterranean conditions. *Eur. J. Agron.* **2006**, *24*, 140–148. [CrossRef]
16. Liu, L.; Sang, D.; Liu, C.; Wang, Z.; Yang, J.; Zhu, Q. Effects of Real-time and Site-specific Nitrogen Managements on Rice Yield and Nitrogen Use Efficiency. *Sci. Agric. Sin.* **2003**, *36*, 1456–1461.

17. Hassan, M.S.; Khair, A.; Haque, M.M.; Azad, A.K.; Hamid, A. Genotypic variation in traditional rice varieties for chlorophyll content, SPAD value and nitrogen use efficiency. *Bangladesh J. Agric. Res.* **2009**, *34*, 505–515. [CrossRef]
18. Xiong, D.; Chen, J.; Yu, T.; Gao, W.; Ling, X.; Li, Y.; Peng, S.; Huang, J. SPAD-based leaf nitrogen estimation is impacted by environmental factors and crop leaf characteristics. *Sci. Rep.* **2015**, *5*, 13389. [CrossRef] [PubMed]
19. Zhu, C.; Zhu, J.; Cao, J.; Jiang, Q.; Liu, G.; Ziska, L.H. Biochemical and molecular characteristics of leaf photosynthesis and relative seed yield of two contrasting rice cultivars in response to elevated CO_2. *J. Exp. Bot.* **2014**, *65*, 6049–6056. [CrossRef] [PubMed]
20. Pang, J.; Zhu, J.-G.; Xie, Z.-B.; Liu, G.; Zhang, Y.-L.; Chen, G.-P.; Zeng, Q.; Cheng, L. A new explanation of the N concentration decrease in tissues of rice (*Oryza sativa* L.) exposed to elevated atmospheric pCO_2. *Environ. Exp. Bot.* **2006**, *57*, 98–105. [CrossRef]
21. Huang, J.; Yang, H.; Yang, L.; Wang, Y.; Zhu, J.; Liu, H.; Dong, G.; Shan, Y. Effects of Free-air CO_2 Enrichment(FACE) on Nitrogen Nutrition at Different Growth Stages in Rice(*Oryza sativa* L.) Cultivar Wuxiangjing 14. *Acta Agron. Sin.* **2004**, *30*, 1237–1243.
22. Zhou, N.; Jing, L.; Wang, Y.; Zhu, J.; Yang, L.; Wang, Y. Effects of Elevated Atmospheric CO_2 and Temperature on Dynamics of Leaf Chlorophyll Contents and SPAD Value of Rice in Open-Air Field Conditions. *Chin. J. Rice Sci.* **2017**, *31*, 524–532.
23. Pang, J.; Zhu, J.; Xie, Z.; Chen, G.; Liu, G.; Zhang, Y. Effects of Elevated pCO_2 on Nutrient Uptake by Rice and Nutrient Contents in Rice Grain. *Chin. J. Rice Sci.* **2005**, *19*, 350–354.
24. Coleman, J.S.; McConnaughay, K.D.M.; Bazzaz, F.A. Elevated CO_2 and plant nitrogen-use: Is reduced tissue nitrogen concentration size-dependent? *Oecologia* **1993**, *93*, 195–200. [CrossRef]
25. Cai, C.; Yin, X.; He, S.; Jiang, W.; Si, C.; Struik, P.C.; Luo, W.; Li, G.; Xie, Y.; Xiong, Y.; et al. Responses of wheat and rice to factorial combinations of ambient and elevated CO_2 and temperature in FACE experiments. *Glob. Chang. Biol.* **2016**, *22*, 856–874. [CrossRef] [PubMed]
26. Meinshausen, M.; Smith, S.J.; Calvin, K.; Daniel, J.S.; Kainuma, M.L.T.; Lamarque, J.F.; Matsumoto, K.; Montzka, S.A.; Raper, S.C.B.; Riahi, K.; et al. The RCP greenhouse gas concentrations and their extensions from 1765 to 2300. *Clim. Chang.* **2011**, *109*, 213–241. [CrossRef]
27. Masson-Delmotte, V.; Zhai, P.; Pirani, A.; Connors, S.L.; Péan, S.B.C.; Caud, N.; Chen, Y.; Goldfarb, L.; Gomis, M.I. Climate Change 2021: The Physical Science Basis. Contribution of Working Group I to the Sixth Assessment Report of the Intergovernmental Panel on Climate Change. *Press* **2021**. Available online: https://www.ipcc.ch/site/assets/uploads/2021/11/IPCCWG1ar6 slides_SBSTA_MASTER.pdf (accessed on 5 May 2021).
28. Cotrufo, M.F.; Ineson, P.; Scott, A. Elevated CO_2 reduces the nitrogen concentration of plant tissues. *Glob. Chang. Biol.* **1998**, *4*, 43–54. [CrossRef]
29. Cao, B.; Dang, Q.-L.; Zhang, S. Relationship between photosynthesis and leaf nitrogen concentration in ambient and elevated CO_2 in white birch seedlings. *Tree Physiol.* **2007**, *27*, 891–899. [CrossRef]
30. Sakai, H.; Hasegawa, T.; Kobayashi, K. Enhancement of rice canopy carbon gain by elevated CO_2 is sensitive to growth stage and leaf nitrogen concentration. *New Phytol.* **2006**, *170*, 321–332. [CrossRef] [PubMed]
31. Ma, H.; Zhu, J.; Xie, Z.; Liu, G.; Zeng, Q.; Han, Y. Responses of rice and winter wheat to free-air CO_2 enrichment (China FACE) at rice/wheat rotation system. *Plant Soil* **2007**, *294*, 137–146. [CrossRef]
32. Liang, W.; Jianguo, Z.H.U.; Chunwu, Z.H.U.; Jiling, C.A.O.; Mingna, W.; Qing, Z.; Zubin, X.I.E.; Gang, L.I.U. Relationship between Decrease in Nitrogen Content and Activities of Key Enzymes Related to Nitrogen Metabolism in Rice Leaves under Elevated CO_2 Concentration. *Chin. J. Rice Sci.* **2008**, *22*, 499–506.
33. Wang, J.; Wang, C.; Chen, N.; Xiong, Z.; Wolfe, D.; Zou, J. Response of rice production to elevated CO_2 and its interaction with rising temperature or nitrogen supply: A meta-analysis. *Clim. Chang.* **2015**, *130*, 529–543. [CrossRef]
34. Islam, M.S.; Bhuiya, M.; Rahman, S.; Hussain, M. Evaluation of SPAD and LCC based nitrogen management in rice (*Oryza sativa* L.). *Bangladesh J. Agric. Res.* **2009**, *34*, 661–672. [CrossRef]
35. Maiti, D.; Das, D.K.; Karak, T.; Banerjee, M. Management of nitrogen through the use of leaf color chart (LCC) and soil plant analysis development (SPAD) or chlorophyll meter in rice under irrigated ecosystem. *Sci. World J.* **2004**, *4*, 838–846. [CrossRef] [PubMed]
36. Fan, H.E.; Jianliang, H.; Kehui, C.U.I.; Jianmin, Z.; Bo, X.U.; Shaobing, P.; Buresh, R.J. Effect of Real-Time and Site-Specific Nitrigen Management on Rice Yield and Quality. *Sci. Agric. Sin.* **2007**, *40*, 123–132.
37. Zhu, C.; Zeng, Q.; Ziska, L.H.; Zhu, J.; Xie, Z.; Liu, G. Effect of nitrogen supply on carbon dioxide-induced changes in competition between rice and barnyardgrass (*Echinochloa crus-galli*). *Weed Sci.* **2008**, *56*, 66–71. [CrossRef]
38. Okada, M.; Lieffering, M.; Nakamura, H.; Yoshimoto, M.; Kim, H.Y.; Kobayashi, K. Free-air CO_2 enrichment (FACE) using pure CO_2 injection: System description. *New Phytol.* **2001**, *150*, 251–260. [CrossRef]
39. Liu, Y.; Li, M.; Zheng, J.; Li, L.; Zhang, X.; Zheng, J.; Pan, G.; Yu, X.; Wang, J. Short-term responses of microbial community and functioning to experimental CO_2 enrichment and warming in a Chinese paddy field. *Soil Biol. Biochem.* **2014**, *77*, 58–68. [CrossRef]
40. Hoel, B.O.; Solhaug, K.A. Effect of irradiance on chlorophyll estimation with the Minolta SPAD-502 leaf chlorophyll meter. *Ann. Bot.* **1998**, *82*, 389–392. [CrossRef]
41. Porra, R.J.; Thompson, W.A.; Kriedemann, P.E. Determination of accurate extinction coefficients and simultaneous equations for assaying chlorophylls a and *b* extracted with four different solvents: Verification of the concentration of chlorophyll standards by atomic absorption spectroscopy. *Biochim. Et Biophys. Acta* **1989**, *975*, 384–394. [CrossRef]

42. Singh, D.; Singh, B. Investigating the impact of data normalization on classification performance. *Appl. Soft Comput.* **2020**, *97*. [CrossRef]
43. Hintze, J.L.; Nelson, R.D. Violin plots: A box plot-density trace synergism. *Am. Stat.* **1998**, *52*, 181–184. [CrossRef]
44. Conroy, J.P. Influence of elevated atmospheric CO_2 concentrations on plant nutrition. *Aust. J. Bot.* **1992**, *40*, 445–456. [CrossRef]
45. Yoshida, S. *Fundamentals of Rice Crop Science*; The International Rice Research Institute: Los Baños, Philippines, 1981.
46. Evans, L.T. *Crop Evolution, Adaptation and Yield*; Cambridge University Press: Cambridge, UK, 1996.
47. Ainsworth, E.A.; Long, S.P. What have we learned from 15 years of free-air CO_2 enrichment (FACE)? A meta-analytic review of the responses of photosynthesis, canopy. *New Phytol.* **2005**, *165*, 351–371. [CrossRef]
48. Jiang, J.; Wang, C.; Wang, H.; Fu, Z.; Cao, Q.; Tian, Y.; Zhu, Y.; Cao, W.; Liu, X. Evaluation of Three Portable Optical Sensors for Non-Destructive Diagnosis of Nitrogen Status in Winter Wheat. *Sensors* **2021**, *21*, 5579. [CrossRef] [PubMed]
49. Rhezali, A.; Aissaoui, A.E. Feasibility Study of Using Absolute SPAD Values for Standardized Evaluation of Corn Nitrogen Status. *Nitrogen* **2021**, *2*, 298–307. [CrossRef]
50. Sutton, R.T.; Hawkins, E. ESD Ideas: Global climate response scenarios for IPCC AR6. *Earth Syst. Dyn. Discuss.* **2020**, *2020*, 1–4.
51. Hasegawa, T.; Li, T.; Yin, X.; Zhu, Y.; Boote, K.; Baker, J.; Bregaglio, S.; Buis, S.; Confalonieri, R.; Fugice, J.; et al. Causes of variation among rice models in yield response to CO_2 examined with Free-Ai_r CO_2 Enrichment and growth chamber experiments. *Sci. Rep.* **2017**, *7*, 14858. [CrossRef] [PubMed]
52. Zhao, H.; Li, X.; Jiang, Y. Response of Nitrogen Losses to Excessive Nitrogen Fertilizer Application in Intensive Greenhouse Vegetable Production. *Sustainability* **2019**, *11*, 1513. [CrossRef]
53. Yang, L.; Deng, Y.; Wang, X.; Zhang, W.; Shi, X.; Chen, X.; Lakshmanan, P.; Zhang, F. Global direct nitrous oxide emissions from the bioenergy crop sugarcane (*Saccharum* spp. inter-specific hybrids). *Sci. Total Environ.* **2021**, *752*, 141795. [CrossRef] [PubMed]
54. Guo, J.H.; Liu, X.J.; Zhang, Y.; Shen, J.L.; Han, W.X.; Zhang, W.F.; Christie, P.; Goulding, K.W.T.; Vitousek, P.M.; Zhang, F.S. Significant Acidification in Major Chinese Croplands. *Science* **2010**, *327*, 1008–1010. [CrossRef] [PubMed]
55. Le, C.; Zha, Y.; Li, Y.; Sun, D.; Lu, H.; Yin, B. Eutrophication of Lake Waters in China: Cost, Causes, and Control. *Environ. Manag.* **2010**, *45*, 662–668. [CrossRef] [PubMed]
56. Williamson, J.M. The Role of Information and Prices in the Nitrogen Fertilizer Management Decision: New Evidence from the Agricultural Resource Management Survey. *J. Agric. Resour. Econ.* **2011**, *36*, 552–572.

atmosphere

MDPI

Article

The Impact of Atmospheric Pollutants on Human Health and Economic Loss Assessment

Houli Zhang [1], Shibing You [1,*], Miao Zhang [1], Difei Liu [1], Xuyan Wang [2], Jingru Ren [3] and Chuanhua Yu [4,*]

[1] Economics and Management School, Wuhan University, Wuhan 430072, China; hlzhang@whu.edu.cn (H.Z.); zhangmiao123@whu.edu.cn (M.Z.); difeiliu@whu.edu.cn (D.L.)
[2] Union Hospital, Tongji Medical College, Huazhong University of Science and Technology, Wuhan 430022, China; wangxy6021@163.com
[3] Shenzhen Institute of Information Technology, Shenzhen 518172, China; jingrur124@gmail.com
[4] Department of Epidemiology and Health Statistics, School of Health Sciences, Wuhan University, Wuhan 430071, China
* Correspondence: sbyou@whu.edu.cn (S.Y.); YuCHua@whu.edu.cn (C.Y.)

Abstract: The impact of air pollution on human health is becoming increasingly severe, and economic losses are a significant impediment to economic and social development. This paper investigates the impact of air pollutants on the respiratory system and its action mechanism by using information on inpatients with respiratory diseases from two IIIA (highest) hospitals in Wuhan from 2015 to 2019, information on air pollutants, and meteorological data, as well as relevant demographic and economic data in China. This paper describes the specific conditions of air pollutant concentrations and respiratory diseases, quantifies the degree of correlation between the two, and then provides a more comprehensive assessment of the economic losses using descriptive statistical methods, the generalized additive model (GAM), cost of illness approach (COI), and scenario analysis. According to the findings, the economic losses caused by $PM_{2.5}$, PM_{10}, SO_2, NO_2, and CO exposure are USD 103.17 million, USD 70.54 million, USD 98.02 million, USD 40.35 million, and USD 142.38 million, for a total of USD 454.46 billion, or approximately 0.20% of Wuhan's GDP in 2019. If the government tightens control of major air pollutants and meets the WHO-recommended criterion values, the annual evitable economic losses would be approximately USD 69.4 million or approximately 0.03% of Wuhan's GDP in 2019. As a result, the relevant government departments must strengthen air pollution control to mitigate the impact of air pollution on population health and the associated economic losses.

Keywords: air pollution; respiratory disease; generalized additive model; scenario analysis; assessment of economic losses

Citation: Zhang, H.; You, S.; Zhang, M.; Liu, D.; Wang, X.; Ren, J.; Yu, C. The Impact of Atmospheric Pollutants on Human Health and Economic Loss Assessment. *Atmosphere* **2021**, *12*, 1628. https://doi.org/10.3390/atmos12121628

Academic Editors: Zengyun Hu, Xuguang Tang and Qinchuan Xin

Received: 28 September 2021
Accepted: 3 December 2021
Published: 6 December 2021

1. Introduction

Air pollution, also known as atmospheric pollution, is defined by the International Standardization Organization (ISO) as the entry of certain substances into the atmosphere as a result of human activities or natural processes that present a sufficient concentration for a sufficient period and, thus, endanger human comfort, health, and welfare and the environment. The primary sources of air pollutants are industrial production, home furnaces, heating boilers, transportation, and smoke from forest fires. The six common categories of pollutants are fine particulate matter ($PM_{2.5}$), inhalable particulate matter (PM_{10}), sulfur dioxide (SO_2), nitrogen dioxide (NO_2), ozone (O_3) and carbon monoxide (CO). These six significant pollutants are known as "criteria pollutants". National ambient air quality standards have been set. Air quality is often evaluated using the Air Quality Index (AQI), a dimensionless index describing the overall condition of urban ambient air quality. The AQI takes into account the pollution levels of the six air pollutants, namely, $PM_{2.5}$, PM_{10}, SO_2, NO_2, O_3, and CO, as specified in the "Technical Regulation on Ambient Air Quality Index (on trial) (HJ633-2012)" [1], with larger AQI values indicating heavier pollution levels. In recent years, health problems caused by air pollution have become increasingly severe and seriously endanger public health, with the burden of disease caused by it rising every year, especially in low-income and middle-income countries. It is estimated that outdoor air pollution exposure caused approximately 4.9 million deaths worldwide in 2017, an increase of 5.8% compared to 2007 [2]. Among the effects on human health, the correlation between air pollution and respiratory diseases is the strongest, with numerous epidemiological and toxicological findings showing that short-term exposure to lowly concentrated air pollutants, continuous exposure to lowly concentrated air pollutants, and acute exposure to highly concentrated air pollutants can cause damage to the respiratory system [3–6].

Respiratory disease is a common and frequently occurring disease mainly in the trachea, bronchi, lungs, and thorax, with coughing, chest pain, and affected breathing in mild cases and respiratory distress, hypoxia, and even respiratory failure in severe cases. According to a study conducted by the University of Washington's Institute for Health Metrics and Evaluation (IHME), an estimated 545 million people worldwide suffered from chronic respiratory disease in 2017, a 39.8% increase from 1900, and approximately 3,914,200 people died as a result of chronic respiratory diseases. Chronic respiratory diseases accounted for 7.0% of all deaths worldwide, ranking third after cardiovascular diseases (31.8%) and tumors (17.1%) [5]. According to the analysis of disease burden, the global growth rate of disability-adjusted life years (DALYs) due to chronic respiratory diseases was 14.5% from 2007 to 2017, with 112 million DALYs in 2017, with chronic obstructive pulmonary disease (COPD) as one of the top five leading causes of DALYs worldwide, rising from eighth place in 2007 to fifth place in 2017 [6]. According to data from the China Statistical Yearbook (2021), the mortality rate from respiratory diseases ranked fourth among the major disease mortality rates for both urban and rural residents in China in 2020, at 55.36 per 100,000 residents and 63.64 per 100,000 residents, respectively [7].

Previous studies have demonstrated a strong association between air pollutant exposure and respiratory disease, with air pollutant exposure leading to an increased incidence of respiratory disease and mortality. Studies have demonstrated an increased risk of respiratory disease hospital visits when air concentrations of $PM_{2.5}$, PM_{10}, SO_2, NO_2, and CO are elevated [8], and a significant increase in hospital admissions for respiratory disease in children (0–14 years) [9]. Air pollutants have also been associated with respiratory disease deaths, with 25,600, 252,000, 22,300, 57,300, 31,600, and 45,700 deaths attributable to $PM_{2.5}$, $PM_{2.5\text{-}10}$, SO_2, NO_2, CO, and O_3, respectively, in China [10], and other countries, such as Mexico [11], India [12], and South Korea [13], have also found a link between air pollution and deaths from respiratory diseases. In terms of specific diseases, the more affected respiratory diseases include COPD, asthma, pulmonary infections, and pulmonary tuberculosis. Long-term inhalation of air pollutants reduces pulmonary function in COPD patients, leading to disease progression, more hospital admissions, and higher mortality [14]. For every 10 $\mu g/m^3$ rise in $PM_{2.5}$ concentration, years of life lost (YLLs) in COPD increase by 0.91 person-years [15]. Furthermore, COPD visits increased when $PM_{2.5}$ and $PM_{2.5\text{-}10}$ concentrations increased [16], and there was a substantial positive association between COPD hospital admissions and SO_2, NO_2, O_3, and CO concentrations in the air [17,18]. Air pollution is one of the most important factors contributing to the exacerbation of childhood asthma, especially in Europe and the Caribbean, where the prevalence of asthma is high [19,20]. A prospective cohort study carried out in ten European cities found a link between the onset of asthma symptoms and short-term $PM_{2.5}$ and PM_{10} exposure [21]. A strong positive relationship has also been shown between airborne CO concentrations and the number of asthma outpatient visits [22], and NO_2 and O_3 exposure can influence asthma development and increase the likelihood of hospitalization for asthma [23,24]. When $PM_{2.5}$ and $PM_{2.5\text{-}10}$ concentrations increased by 10 $\mu g/m^3$, the excess relative risk (ERR) for upper respiratory infections (URI) and pneumonia increased by 5.40% and 6.37%, respectively, according to a study conducted in China [16]. Most bronchiectasis exacerbations are assumed to be caused by pathogenic infections. However, pathogens can be challenging to identify, and studies have shown that air pollution is a substantial risk factor for bronchiectasis. A study of 432 patients with a clinical diagnosis of bronchiectasis after high-resolution computed tomography (HRCT) found that when PM_{10} and NO_2 concentrations increased by 10 μg/m^3, the chance of exacerbation increased by 4.5% and 3.2%, respectively [25]; when CO concentrations increased, the number of outpatient visits for bronchiectasis increased [22]; and when SO_2 concentrations increased, it caused an increase in hospital admissions for bronchiectasis [26]. There is also a correlation between air quality and the pathogenesis of pulmonary tuberculosis, with studies linking $PM_{2.5}$, PM_{10}, NO_2, and SO_2 exposure to the likelihood of acquiring active pulmonary tuberculosis [27,28], and Korean research found a 1.20-fold rise in tuberculosis detection rates with a substantial delayed effect when PM_{10} concentrations increased by one standard deviation (5.63 μg/m^3) [29].

The available literature has primarily concentrated on studies of air pollution's health effects on the respiratory system (e.g., mortality, morbidity, and hospital admissions), with little research on air pollution's economic losses. Therefore, we take the strong association between air pollutant exposure and respiratory disease recognized by the literature mentioned above as the hypothesis of this research. Considering geographical, air pollutant, and climatic characteristics of Wuhan city, based on data from monitoring sites of six major air pollutants and inpatients with respiratory diseases from two IIIA (highest) hospitals in Wuhan, this study assesses the health effects and economic losses attributable to $PM_{2.5}$, PM_{10}, SO_2, NO_2, O_3, and CO in the population. The study provides a reference for the future assessment of the health effects of air pollution and cost–benefit evaluation for the development of environmental management policies.

2. Materials and Methods

2.1. Sampling Sites and Sample Collection

The city of Wuhan is chosen as the study area in this paper. Wuhan is the largest and only sub-provincial city in central China and the capital of Hubei Province. It has an area of 8494 square kilometers and is located in the eastern half of the Jianghan Plain at longitude 113°41′115°05′ E and latitude 29°58′31°22′ N. With the rapid growth of the economy in recent years, the Wuhan industry scale has expanded, and industrial production not only consumes energy but also emits a large number of pollutants, including particulate matter (PM), sulfur oxides (SO), nitrogen oxides (NO), carbon monoxide (CO), and hydrocarbons; combined with a large number of vehicle emissions, air pollution in Wuhan is more severe and has caused adverse effects on the population's health.

Three types of data are used in this study to examine the impact of air pollution on the number of hospital admissions for respiratory diseases. The first type is data from the Hospital Information System (HIS) on hospital admissions for respiratory diseases; the second type is data from ambient air pollutant monitoring; and the third type is some meteorological data.

The hospitalization data for respiratory diseases are obtained from the HIS of two IIIA (highest) hospitals in Wuhan, China. This study gathers inpatient cases with respiratory disease between 1 January 2015 and 31 December 2019. The inpatient's gender, age, date of admission, date of discharge, disease diagnosis, length of stay, and inpatient expenditure are all included in the case information. According to the 10th edition of the International Classification of Diseases (ICD-10), ICD-10 codes for respiratory diseases are J00~J99, J12~J18 for pneumonia, and J40~J99 for chronic obstructive pulmonary disease (COPD). Furthermore, the research object is divided into three age groups: 0–14 years, 15–64 years, and 65+ years, and the cold and warm seasons were divided based on the month of inpatient admission, with the warm season lasting from April to October and the cold season lasting from November to March [30].

The Department of Ecology and Environment of Hubei Province provides data on air pollutant concentrations [31]. This paper collects data on air pollution monitoring in Wuhan City from 1 January 2015 to 31 December 2019, including the concentrations of six major air pollutants: $PM_{2.5}$, PM_{10}, SO_2, NO_2, O_3, and CO. Except O_3, which has a daily maximum 8 h average concentration, $PM_{2.5}$, PM_{10}, SO_2, NO_2, and CO are all 24 h average concentrations.

Meteorological information is obtained from the China Meteorological Data website [32]. This article collects data on Wuhan's average temperature (°C) and relative humidity (%) from 1 January 2015 to 31 December 2019.

2.2. *Sample Analysis*

2.2.1. Statistical Descriptive Analysis

From 1 January 2015 to 31 December 2019, the daily number of hospital admissions for respiratory diseases, inpatient expenditure, length of stay, and air pollutant concentrations are presented as $\overline{X} \pm S$; extremum and the percentile are statistically described for the pneumology department of these two IIIA (highest) hospitals. Furthermore, hospital admissions for respiratory diseases are statistically described using frequencies and percentages based on disease subgroups, gender groups, age groups, and season groups.

2.2.2. Time Series Analysis

The data for each period in the time series are the combined result of multiple elements. In this study, the additive model in the time series decomposition method is used to analyze the daily hospital admissions for respiratory diseases and each air pollutant concentration from 1 January 2015 to 31 December 2019, including the long-term trend and seasonal trend and random fluctuation elements. The long-term trend element represents the long-term trend characteristics of the time series, which can be characterized as a continuous upward, continuous downward, or smooth trend during the study period. The seasonal trend element is a cyclical fluctuation influenced by seasonal changes, characterized as a recurring cyclical change every year during the study period; random events usually cause the random fluctuation, and its changes are generally irregular. The expression is shown in Equation (1):

$$Y_t = T_t + S_t + R_t, t = 1, 2, \cdots, n \quad (1)$$

In Equation (1), Y_t is the time series, T_t represents the long-term trend, S_t is the seasonal trend, and R_t is the random fluctuation.

2.2.3. Generalized Additive Model

The generalized additive model (GAM) extends the generalized linear model (GLM) and explains the complex non-linear correlation between the independent and dependent variables. GAM is widely used in environmental epidemiology to explore the correlation between air pollutant exposure and disease mortality or morbidity. The occurrence of hospital admission for respiratory disease is a small probability event for the total number of people in an area, and its distribution approximately follows the Poisson distribution. Since the daily hospital admissions for respiratory diseases are tested to have overdispersion, a GAM based on a quasi-Poisson distribution is developed to analyze the effect of six major air pollutants—$PM_{2.5}$, PM_{10}, SO_2, NO_2, O_3, and CO—on changes in the number of hospital admissions for respiratory diseases. Assuming a log-linear distribution of pollutants and diseases, we use the number of daily hospitalization events as the outcome and the daily average pollutant concentration as the predictor, and we smooth the time trend and meteorological elements, correcting for the weekend effect and the holiday effect. The model is shown in Equation (2):

$$\log(E_i) = \beta_i(C_i) + ns(Time, df) + ns(MT, df) + ns(RH, df) + DOW + Holiday + \alpha \quad (2)$$

In Equation (2), E_i represents the expected value of the number of hospital admissions for respiratory diseases on the *i-th* day, and C_i represents the average concentration of air pollutants on the *i-th* day. The regression coefficient β_i is calculated using the model. *Time* is the date variable; *MT* is the daily mean temperature; *RH* is the daily mean relative humidity; *DOW* is the weekend effect (0 = working day, 1 = weekend); and *Holiday* is the holiday effect (0 = non-statutory holiday and 1 = statutory holiday, both of which are incorporated as factor variables into the model). *ns* is the natural smooth spline function, and *df* is its degree of freedom, which was selected to be 4 for the *Time* variable [17] and 3 for the *MT* and *RH* variables [33]. α is the intercept.

Previous research has shown that there is a considerable hysteresis effect of ambient air pollutant exposure on population health, which means that daily air pollutant exposure may still impact population health after several days. As a result, the effect on the admission day (Lag0) and the hysteresis effects on the first day (Lag1), second day (Lag2), third day (Lag3), fourth day (Lag4), fifth day (Lag5), sixth day (Lag6), and seventh day (Lag7) are examined. In addition, separate models are developed based on gender groups, age groups (0–14 year group, 15–64 year group, and 65+ year group), season groups (cold season, warm season), and disease subgroups (pneumonia, COPD) to examine the associations between each air pollutant and the number of hospital admissions for respiratory diseases on the admission day and lag days.

The following Equation (3) is used to calculate the percentage change (PC) in the number of hospital admissions for respiratory diseases and its 95% confidence interval (CI) for each 10 g/m^3 increment in air pollutant concentration:

$$\mathrm{PC} = [\exp(\beta_\mathrm{i} \times 10) - 1] \times 100 \tag{3}$$

In Equation (3), β_i refers to the regression coefficient of each air pollutant derived from Equation (2).

2.3. Methodology of Economic Loss Assessment—Attributable Risk and Cost of Illness Approach

Attributable fraction (AF) and attributable number (AN) are the fundamental indicators of attributable risk (AR). In this study, AN represents the number of excessive hospital admissions due to air pollutant exposure; AF represents the proportion of excessive hospital admissions due to air pollutant exposure to the total number of hospital admissions. It may alternatively be understood as the proportion of the equivalent reduction in hospital admissions to the overall number of hospital admissions if the population is exposed to air pollution concentrations below a health-affecting threshold level [7]. Both are computed using past research, as indicated in Equations (4) and (5).

$$AF = \sum_{i=0}^{n}\left\{1 - \frac{1}{\exp[\beta_i \times (C_i - C_0)]}\right\} \tag{4}$$

In Equation (4), AF refers to attributable fraction; β_i is the regression coefficient of each air pollutant derived from Equation (2); C_i is the average concentration of each air pollutant on the *i-th* day; C_0 refers to a threshold concentration of each air pollutant. Existing studies do not provide evidence for a defined threshold concentration in the exposure–response relationship between air pollutant concentrations and health effects [9]. Threshold concentrations for the acute health effects of O_3, PM_{10}, SO_2, and NO_2 exposure have also not been determined. As a result, a threshold concentration of 0 is used for each air pollutant in this investigation.

$$AN = AF \times \sum_{j=1}^{n}(Pop_j \times Pro_j) \tag{5}$$

In Equation (5), AN refers to the attributable number; Pop_j is the annual resident population in Wuhan from 2015 to 2019, which is 10,607,700, 10,767,200, 10,892,900, 11,081,100, and 12,210,000, respectively. Pro_j refers to the hospital admission rate for respiratory diseases. Because the data particular to Wuhan are unavailable, the 2017 hospital admission rate for respiratory diseases in China is utilized universally, which is 810.22 per 100,000 people [10].

According to Kennelly and Zhang's relevant research [34,35], the cost of illness (COI) approach is used to assess the economic losses of hospital admission for respiratory diseases caused by air pollutant exposure. COI considers the direct inpatient expenditure for hospital admissions (the direct cost of the illness) and the losses of productivity caused by hospital admissions (indirect cost of the illness). Daily per capita gross domestic product (PGDP) is utilized in Wuhan instead of the daily per capita productivity losses [35]. Economic losses are calculated using Equations (6) and (7):

$$ECO_{loss} = Cost_{mean} + Day_{mean} \times PGDP_{day} \quad (6)$$

$$TECO_{loss} = AN \times ECO_{loss} \quad (7)$$

In Equation (6), ECO_{loss} is the economic losses of an individual inpatient with respiratory disease. The term $Cost_{mean}$ refers to the average inpatient expenditure for respiratory disease. The term $PGDP_{day}$ refers to the daily GDP per capita of Wuhan city. The daily GDP per capita in Wuhan in 2015, 2016, 2017, 2018, and 2019 is USD 56.61, USD 53.51, USD 55.24, USD 59.18, USD 57.46, respectively, using the 2019 Gross Domestic Product (GDP) index as the base period and deflating the daily GDP per capita from 2015 to 2018 (see Table 1). In Equation (7), $TECO_{loss}$ is the overall economic losses, and AN is the attributable number of inpatients.

Table 1. GDP indicator and GDP deflator in China for 2015–2019.

Year	GDP Indicator (1978 = 100)	GDP Deflator *
2015	3035.9	1.294
2016	3243.5	1.211
2017	3468.8	1.133
2018	3703.0	1.061
2019	3929.2	1.000

Note: * The 2019 GDP indicator is used as the basic period.

The GDP indicator is a relative number reflecting the trend and extent of changes in GDP over a certain period of time. The GDP indicator is calculated at constant prices, and this paper uses 1978 as the basic period to calculate the GDP indicator for 2015–2019. Based on the GDP indicator, we calculate the GDP deflator for 2015–2019 with 2019 as the price basic period and adjust nominal GDP to real GDP for the corresponding year, which eliminates the effect of price volatility on GDP per capita.

Furthermore, it is assumed that air pollutant concentrations could be kept reasonably low during the research period, and the evitable economic losses are calculated using Equations (4)–(7).

3. Results

3.1. Statistical Description of Admission Data for Respiratory Diseases and Air Pollution Concentrations

A total of 45,699 inpatients with respiratory diseases were included in the study, that is, 27,725 male inpatients (60.67%) and 17,974 female inpatients (39.33%). According to the analysis of the number of respiratory inpatients in different age groups, nearly half of the inpatients were aged 65 and above, accounting for 44.42% (20,285 cases), while the inpatients in the 0–14 years old and 15–64 years old groups accounted for 18.26% (8340 cases) and 37.39% (17,074 cases), respectively. In the patients with respiratory diseases, pneumonia and chronic obstructive pulmonary disease (COPD) were the majority, among which 10,724 patients with pneumonia (23.47%) and 11,517 patients with COPD (25.20%) were hospitalized. The total proportion of the two was 48.67%. Table 2 shows the results.

Table 2. Basic information for inpatients with respiratory diseases.

Variable	n (%)
Gender	
Male	27,725 (60.67)
Female	17,974 (39.33)
Age (years)	
0~14	8340 (18.26)
15~64	17,074 (37.39)
65+	20,285 (44.42)
Disease subgroups	
Pneumonia	10,724 (23.47)
COPD	11,517 (25.20)
Other diseases	23,458 (51.33)

According to the gender and age distribution of inpatient hospital admissions with respiratory diseases, the findings of this paper show that males predominate among inpatients with pneumonia and COPD, accounting for 56.13% and 70.83%, respectively; when the age groups are examined, more inpatients with pneumonia are aged 0–14 years and 65+ years, accounting for 32.94% and 39.08%, respectively. See Table 3 for further information.

Table 3. Gender and age distribution of inpatients with respiratory diseases.

Variable	Pneumonia (n = 10,724)		COPD (n = 11,517)		Others (n = 23,458)	
	Cases	Percentage (%)	Cases	Percentage (%)	Cases	Percentage (%)
Gender						
Male	6019	56.13	8154	70.80	13,552	57.77
Female	4705	43.87	3363	29.20	9906	42.23
Age (years)						
0~14	3532	32.94	0	0.00	4808	20.50
15~64	3001	27.98	3138	27.25	10,935	46.62
65+	4191	39.08	8379	72.75	7715	32.89

3.1.1. Characteristics of Numbers of Hospital Admissions for Respiratory Diseases

An analysis of the daily number of hospital admissions for respiratory diseases in this article indicates that the average daily number of hospital admissions is 25.05 people; when particular types of diseases are examined, the average daily number of hospital admissions is 5.87 people for pneumonia and 6.31 people for COPD. Males have a greater average daily number of hospital admissions than females, with 15.18 people for the former and 9.84 people for the latter. The average daily number of hospital admissions rises with age. The 65+ age group has the greatest average daily number of hospital admissions (11.11 people), followed by the 15–64 age group (9.41 people) and the 0–14 age group (4.62 people). According to a season analysis, the average daily hospital admissions in the warm and cold seasons are 25.58 and 24.52, respectively. Table 4 shows the results.

A time series decomposition analysis of hospital admissions from 2015 to 2019 indicates an increasing tendency in the number of day-to-day hospital admissions for respiratory diseases, with significant seasonal fluctuations, with more admissions in winter and spring and fewer in summer and fall. Long-term trends and seasonal fluctuations in the number of daily hospital admissions for pneumonia and COPD are broadly consistent with the patterns observed in the disease-specific analysis for all respiratory diseases.

3.1.2. Characteristics of Inpatient Expenditure for Respiratory Diseases

In order to eliminate the impact of price fluctuations on the inpatient expenditures of respiratory diseases, the price deflator is applied to the inpatient expenditures from

2015 to 2018 based on the Consumer Price Index (CPI) published on the official website of the National Bureau of Statistics [36], using the CPI of 2019 as the base period. The CPI deflators for 2015–2019 are calculated to be 1.089, 1.067, 1.051, 1.029, and 1.000, respectively.

Table 4. Basic information for the period of 2015–2019 on the number of daily hospital admissions for respiratory diseases.

Variable	$\overline{X} \pm S$	Min [1]	P_{25} [2]	P_{50} [3]	P_{75} [4]	Max [5]
All inpatients	25.05 ± 12.07	0	16	24	33	77
Disease subgroups						
Pneumonia	5.87 ± 3.87	0	3	5	8	23
COPD	6.31 ± 3.57	0	4	6	8	30
Gender						
Male	15.18 ± 7.84	0	10	15	20	49
Female	9.84 ± 5.29	0	6	9	13	36
Age (years)						
0~14	4.62 ± 3.13	0	2	4	6	20
15~64	9.41 ± 5.34	0	5	8	13	32
65+	11.11 ± 6.08	0	7	10	15	53
Season						
Cold season	24.52 ± 13.28	0	15	24	33	77
Warm season	25.58 ± 10.74	0	18	25	32	67

[1] Minimum; [2] 25th percentile; [3] 50th percentile (median); [4] 75th percentile; [5] maximum.

This study looked at the inpatient expenditures for respiratory diseases and discovered that the median of inpatient expenditures is USD 1334.18. When particular disease categories were examined, inpatient expenditures for pneumonia were lower than those for COPD, with the former having a median of USD 1131.18 and the latter having a median of USD 1486.65. Males had greater inpatient expenditures than that of females, with a median of USD 1488 and USD 1162.77, respectively. The higher the age, the higher the inpatient expenditures. The 65+ age group had the greatest inpatient expenditures, with a median of USD 1843.32, followed by the 15–64 age group, with a median of USD 1296.84, and the 0–14 age group, with a median of USD 622.33. The examination of inpatient expenditures by season revealed that the median was greater for inpatients in the cold season than for those in the warm season, with the former costing USD 2200.55 and the latter costing USD 1613.34. See Table 5 for further information.

Table 5. Basic information for the period of 2015–2019 on the inpatient expenditure for respiratory diseases (unit: USD).

Variable	$\overline{X} \pm S$	Min [1]	P_{25} [2]	P_{50} [3]	P_{75} [4]	Max [5]
All inpatients	2370.76 ± 967.37	1.29	774.49	1334.18	2489.42	142,717.11
Disease subgroups						
Pneumonia	2286 ± 2230.45	12.04	705.73	1131.18	2258.07	98,337.68
COPD	2390.32 ± 1492.43	1.86	1045.09	1486.65	2538.34	79,745.31
Gender						
Male	2655.08 ± 1374.98	1.29	825.82	1488.00	2818.49	142,717.11
Female	1930.95 ± 1261.26	1.29	723.32	1162.77	2019.55	93,152.57
Age (years)						
0~14	680.03 ± 417.34	11.19	514.12	622.33	789.19	10,817.72
15~64	2121.07 ± 1560.15	1.86	830.85	1296.84	2173.50	92,695.38
65+	3234.6 ± 1912.49	1.29	1163.24	1843.32	3715.42	142,717.11
Season						
Cold season	2364.84 ± 1024.72	2.48	1719.98	2200.55	2786.48	98,337.68
Warm season	1711.03 ± 929.33	1.29	1043.22	1613.34	2248.95	142,717.11

[1] Minimum; [2] 25th percentile; [3] 50th percentile (median); [4] 75th percentile; [5] maximum.

An examination of inpatient expenditures by year from 2015 to 2019 reveals an upward trend for inpatients with respiratory diseases, with the greatest inpatient expenditure in 2019 at a median of USD 1451.05 and the lowest inpatient expenditure in 2015 at a median of USD 1277.43. Analysis of specific diseases revealed an upward trend in inpatient expenditures for inpatients suffering from pneumonia and COPD, with the median for pneumonia inpatients rising from USD 989.86 in 2015 to USD 1281.29 in 2019, and the median for COPD inpatients rising from USD 1439.08 in 2015 to USD 1523.13 in 2019. Figure 1 shows one example of this.

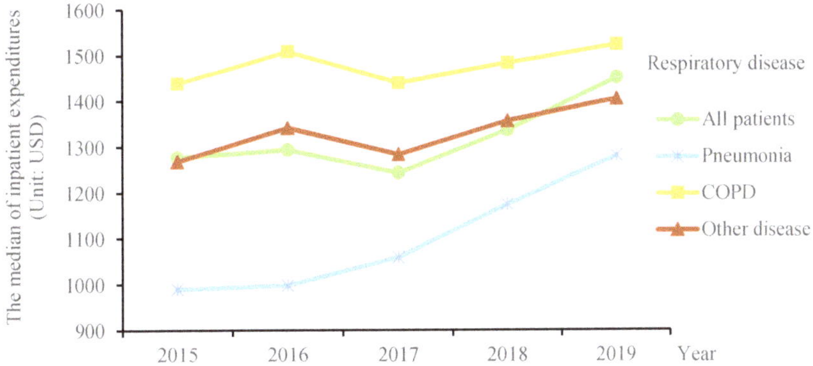

Figure 1. Trends in inpatient expenditure for respiratory diseases for 2015–2019.

We analyze inpatient expenditure trends by gender from 2015 to 2019 and reveal an upward tendency for both male and female inpatients. Male inpatients' median of inpatient expenditure trend grows from USD 1413.52 in 2015 to USD 1648.57 in 2019; female inpatients' median of inpatient expenditure increases from USD 1119.02 in 2015 to USD 1263.08 in 2019. Figure 2 shows an example of this.

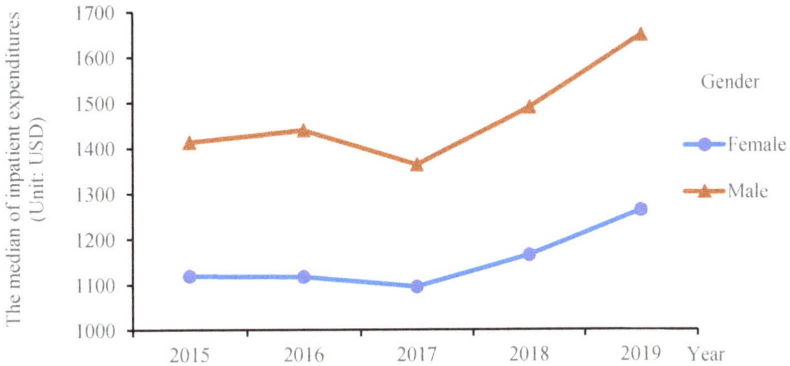

Figure 2. Trends in inpatient expenditure for different genders for respiratory diseases for 2015–2019.

The analysis of the trend of inpatient expenditures by age from 2015 to 2019 revealed that the trend of inpatient expenditure for the 0–14 year group is insignificant during the study period; inpatient expenditure for the 15–64 year group shows an increasing trend, with the median expenditure increasing from USD 1242.28 in 2015 to USD 1392.6 in 2019. In contrast, expenditure for the 65+ age group varies, the median remains continuously over USD 1500, reaching a low of USD 1765.87 in 2015 and climbing to USD 1829.59 and USD 1949.25 in 2018 and 2019, respectively. Figure 3 shows an example of this.

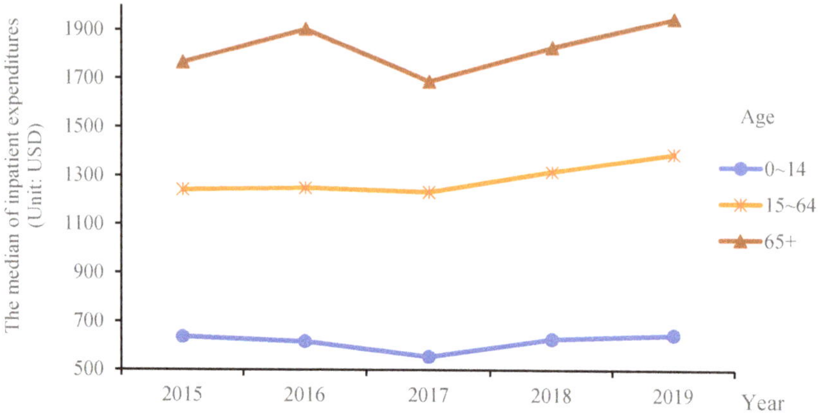

Figure 3. Trends in inpatient expenditure for different ages for respiratory diseases for 2015–2019.

3.1.3. Characteristics of Length of Stay for Respiratory Diseases

A study of the length of stay in hospital of inpatients with respiratory illnesses reveals a median of 9 days. In the disease-specific study, COPD patients have a greater median number of hospital days (10 days) than that of pneumonia inpatients (8 days). Males spend more time in the hospital than females, with a median stay of 9 days for the former and 8 days for the latter. The median number of hospital days increases with age, with the 65+ age group having the most, with a median of 11 days, and the 15–64 and 0–14 age groups having a median of 8 days and 5 days, respectively. According to an analysis of the length of stay by season, the median number of length of stay for inpatients hospitalized in both the cold and hot seasons is 9 days. Table 6 shows the detailed information.

Table 6. Basic information for the period of 2015–2019 for the length of stay for respiratory diseases.

Variable	$\overline{X} \pm S$	Min [1]	P_{25} [2]	P_{50} [3]	P_{75} [4]	Max [5]
All inpatients	9.31 ± 2.27	1	8	9	10	350
Disease subgroups						
Pneumonia	8.68 ± 4.99	1	7	8	10	180
COPD	10.17 ± 3.57	1	8	10	12	130
Gender						
Male	9.79 ± 3.17	1	8	9	11	76
Female	8.52 ± 2.73	1	7	8	10	39
Age (years)						
0~14	4.63 ± 2.08	1	4	5	6	39
15~64	8.50 ± 4.63	1	7	8	9	350
65+	11.53 ± 3.39	1	10	11	13	197
Season						
Cold season	9.38 ± 2.54	1	8	9	10	350
Warm season	9.26 ± 1.93	1	8	9	10	191

[1] Minimum; [2] 25th percentile; [3] 50th percentile (median); [4] 75th percentile; [5] maximum.

3.1.4. Characteristics of Changes in Air Pollutants

The average concentrations of pollutants in Wuhan from 2015 to 2019 are as follows: $PM_{2.5}$, 54.07 μg/m^3; PM_{10}, 87.32 μg/m^3; SO_2, 11.24 μg/m^3; NO_2, 44.97 μg/m^3; O_3, 92.83 μg/m^3; and CO, 1.020 mg/m^3, with SO_2, O_3, and CO meeting the national level one standard (the annual average concentration of SO_2: 15 μg/m^3; the maximum daily 8 h average concentration of O_3: 100 μg/m^3; and the 24 h average concentration of CO: 4 mg/m^3), while $PM_{2.5}$, PM_{10}, and NO_2 exceeded the national level two standard (annual average concentration of $PM_{2.5}$: 35 μg/m^3; the annual average concentration of PM_{10}: 70 μg/m^3; the annual average concentration of NO_2: 40 μg/m^3) (see Table 7). The country has set corresponding ambient air quality standards for these six categories of air pollutants. According to the revised Ambient Air Quality Standards (GB3095-2012) in 2012, the annual average values of the primary air quality standards for $PM_{2.5}$, PM_{10}, SO_2, NO_2, O_3, and CO, are 15 μg/m^3, 40 μg/m^3, 20 μg/m^3, 40 μg/m^3, 100 μg/m^3, and 4 mg/m^3, respectively, where higher levels indicate higher pollution levels as shown in Table 8. The number of days in Wuhan when the daily average concentration of pollutants did not meet the national level one standard accounted for 65.93% for $PM_{2.5}$, 77.60% for PM_{10}, 5.70% for NO_2, and 35.21% for O_3, and the number of days when the daily average concentration of pollutants did not meet the national level two standard accounted for 19.82% for $PM_{2.5}$, 9.20% for PM_{10}, 5.70% for NO_2, 13.53% for O_3 of the total.

Table 7. Basic Information for the period of 2015–2019 for air pollutants in Wuhan.

Air Pollutants	$\overline{X} \pm S$	Min [1]	P_{25} [2]	P_{50} [3]	P_{75} [4]	Max [5]
$PM_{2.5}$ (μg/m^3)	54.07 ± 34.99	4	30	46	67	281
PM_{10} (μg/m^3)	87.32 ± 47.75	3	53	81	112	618
SO_2 (μg/m^3)	11.24 ± 7.66	2	6	9	14	74
NO_2 (μg/m^3)	44.97 ± 18.94	11	30	42	56	119
O_3 (μg/m^3)	92.83 ± 53.51	4	53	85	122	281
CO (mg/m^3)	1.020 ± 0.301	0.416	0.810	0.990	1.170	2.672

[1] Minimum; [2] 25th percentile; [3] 50th percentile (median); [4] 75th percentile; [5] maximum.

Table 8. Ambient Air Quality Standards (GB 3095-2012) concentration value.

Pollutants	Average Time	Concentration Value		Concentration Unit
		Level One	Level Two	
$PM_{2.5}$	Annual average	15	35	μg/m^3
	Daily average	35	75	
PM_{10}	Annual average	40	70	
	Daily average	50	150	
SO_2	Annual average	20	60	
	Daily average	50	150	
NO_2	Annual average	40	40	
	Daily average	80	80	
O_3	Eight hour average	100	160	
CO	Daily average	4	4	mg/m^3

3.2. *Quantitative Analysis of the Impact of Air Pollutants on Hospital Admissions for Respiratory Diseases*

In this study, a generalized additive model (GAM) is developed for each air pollutant and the number of hospital admissions for respiratory diseases. Except for O_3, the results show that the daily average concentrations of $PM_{2.5}$, PM_{10}, SO_2, NO_2, and CO have an association with hospital admissions for respiratory disease, with $PM_{2.5}$, PM_{10}, SO_2, and CO all having the strongest hysteresis effect on the seventh day (Lag7), while NO_2 has the strongest hysteresis effect at Lag6. For $PM_{2.5}$, PM_{10}, SO_2, and CO concentration rises by 10 μg/m^3, the number of hospital admissions for respiratory disease increases by 0.90% (95% CI: 0.47, 1.32), 0.60% (0.18, 1.02), 0.49% (0.07, 0.90), 0.47% (0.05, 0.88), 0.98% (0.56, 1.39), and 1.71% (1.29, 2.12).

PM_{10} only has a hysteresis effect on hospital admissions for respiratory disease at Lag7. This shows that hospital admissions with an increase of 0.71% (0.45, 0.97) are associated with each 10 μg/m^3 increment of PM_{10} concentration. With each 10 μg/m^3 increment of SO_2 concentration at Lag1, Lag2, Lag3, Lag4, Lag6, and Lag7, hospital admissions for respiratory diseases increase by 5.80% (3.33, 8.33), 6.86% (4.39, 9.40), and 7.80% (5.30,

10.36). At Lag6 and Lag7, NO_2 and CO have a hysteresis effect on the number of hospital admissions for respiratory disease. This shows that hospital admissions with an increase of 0.76% (0.13, 1.40) and 0.74% (0.10, 1.39) are associated with each 10 $\mu g/m^3$ increment of NO_2 concentration, and hospital admissions with an increase of 0.07% (0.08, 0.19) and 0.13% (0.08, 0.19) are associated with each 10 $\mu g/m^3$ increment of CO concentration.

Inpatients are divided into two groups based on whether they have pneumonia or COPD, and there is a difference in the association of hospitalization due to air pollutants between the two groups. Inpatients with pneumonia are more susceptible to $PM_{2.5}$, PM_{10}, SO_2, NO_2, and O_3, in addition to CO. The average daily concentrations of $PM_{2.5}$, PM_{10}, SO_2, NO_2, and O_3 have a correlation with pneumonia hospital admissions, with $PM_{2.5}$ having an influence on pneumonia hospital admissions at Lag1, Lag2, Lag5, Lag6, and Lag7, with the biggest hysteresis effect at Lag7.

With each 10 $\mu g/m^3$ increment of $PM_{2.5}$ concentration, hospital admission increases by 2.11% (1.23, 2.98). SO_2 influences pneumonia hospital admissions at Lag1, Lag3, Lag4, Lag6, and Lag7, with the biggest hysteresis effect at Lag7. With each 10 $\mu g/m^3$ increment of SO_2 concentration, hospital admission increases by 13.33% (7.78, 19.16). NO_2 influences the pneumonia hospital admissions at Lag6 and Lag7, with the biggest hysteresis effect at Lag6. With each 10 $\mu g/m^3$ increment of NO_2 concentration, hospital admissions increase by 2.0% (0.69, 3.33). PM_{10} and O_3 only have a hysteresis effect on pneumonia hospital admissions at a single lag day. With each 10 $\mu g/m^3$ increment of PM_{10} concentration at Lag7, hospital admissions increase by 0.82% (0.30, 1.35). With each 10 $\mu g/m^3$ increment of O_3 concentration at Lag1, hospital admissions increase by 0.78% (0.04, 1.52). The analysis result of COPD shows that only the average daily concentrations of $PM_{2.5}$, SO_2, and CO have a relationship with COPD hospital admissions. $PM_{2.5}$ influences the COPD hospital admissions at Lag1 and Lag7, with the biggest hysteresis effect at Lag7. With each 10 $\mu g/m^3$ increment of $PM_{2.5}$ concentration, hospital admission increases by 1.9% (1.12, 2.69). SO_2 influences the COPD hospital admissions at Lag1, Lag2, and Lag7, with the biggest hysteresis effect at Lag1. With each 10 $\mu g/m^3$ increment of SO_2 concentration, hospital admission increases by 6.50% (1.93, 11.27). CO only influences the COPD hospital admissions at Lag7. With each 10 $\mu g/m^3$ increment of CO concentration, hospital admission increases by 0.19% (0.08, 0.30).

Similar to the categorization of disease type, this research develops the GAM of each air pollutant and the number of hospital admission for respiratory disease by gender, age, and season. The analysis results reveal gender disparities in the risk of hospital admissions for respiratory disease due to air pollutants, with males bearing a more significant effect; i.e., the percentage change in the number of hospital admissions for respiratory disease due to air pollution is more significant for males. There were age differences in the risk of hospital admissions for respiratory diseases caused by air pollutants, with $PM_{2.5}$, PM_{10}, SO_2, and NO_2 having a more significant effect on people aged 0–14 years, while O_3 and CO had a more significant effect on people aged 65+ years. The risk of hospitalization for respiratory diseases caused by air pollution was seasonally related, with $PM_{2.5}$, PM_{10}, O_3, and CO having a sizable effect during the summer.

3.3. *Assessment of Economic Losses*

3.3.1. Analysis Results of Attributable Risk

To make the results easier to understand, attributable analysis is applied for each air pollutant to determine the lag day with the greatest hysteresis effect (i.e., the biggest regression coefficient calculated by the GAM) on hospital admissions for respiratory disease. Table 9 shows the lag days with the greatest hysteresis effect of each air pollutant, as well as their regression coefficients (β). Lag1–Lag7 in the table represent the hysteresis effects of each air pollutant on the first day (Lag1), second day (Lag2), third day (Lag3), fourth day (Lag4), fifth day (Lag5), sixth day (Lag6), and seventh day (Lag7) of hospital admissions.

Table 9. The lag days with the greatest hysteresis effect of each air pollutant and regression coefficients (β).

Variable	$PM_{2.5}$	PM_{10}	SO_2	NO_2	O_3	CO
All inpatients	Lag7 (0.0017)	Lag7 (0.0007)	Lag7 (0.0080)	Lag6 (0.0008)	-	Lag7 (0.0001)
Disease Subgroups						
Pneumonia	Lag7 (0.0021)	Lag7 (0.0008)	Lag7 (0.0125)	Lag6 (0.0020)	Lag1 (0.0008)	-
COPD	Lag7 (0.0019)	-	Lag1 (0.0063)	-	-	Lag7 (0.0002)
Gender						
Male	Lag7 (0.0020)	Lag7 (0.0009)	Lag7 (0.0100)	Lag7 (0.0011)	Lag6 (0.0005)	Lag7 (0.0001)
Female	Lag7 (0.0013)	Lag7 (0.0005)	Lag2 (0.0069)	-	-	Lag7 (0.0001)
Age (years)						
0~14	Lag7 (0.0026)	Lag7 (0.0009)	Lag7 (0.0163)	Lag6 (0.0018)	-	-
15~64	Lag7 (0.0015)	Lag7 (0.0006)	Lag3 (0.0086)	-	-	Lag7 (0.0001)
65+	Lag7 (0.0014)	Lag7 (0.0008)	Lag7 (0.0074)	Lag7 (0.0011)	Lag4 (0.0006)	Lag7 (0.0002)

Air pollutant exposure has a negative impact on the population's respiratory health. Our research shows the attributable fractions, which represent the proportion of excessive hospital admissions due to air pollutant ($PM_{2.5}$, PM_{10}, SO_2, NO_2, and CO) exposure to the total number of hospital admissions, are 8.50%, 5.81%, 8.80%, 3.33%, and 11.73%, respectively. Based on the attributable fraction, further estimation shows that the number of hospital admissions for respiratory diseases attributable to $PM_{2.5}$, PM_{10}, SO_2, NO_2, and CO exposure is 37,600, 25,700, 35,700, 14,700, and 51,800, respectively. The attributable risk differed by disease subgroups, gender groups, and age groups; for example, more males than females are hospitalized for respiratory disease due to $PM_{2.5}$, 22,800 for the former and 14,800 for the latter with attributable fractions of 9.70% and 6.50%, respectively. The number of hospital admissions for respiratory diseases due to PM_{10} rose with age, with 5700 cases (ages 0–14), 7500 cases (15–64), and 12,600 cases (65+) with attributable fractions of 6.98%, 4.55%, and 6.40%, respectively. SO_2 caused more hospital admissions for pneumonia (12,300 cases) than COPD (7600 cases), with attributable fractions of 11.84% and 6.81%, respectively. See Table 10 for details.

Table 10. Attributable analysis results of hospital admissions for respiratory diseases due to air pollutants.

Variable	$PM_{2.5}$	PM_{10}	SO_2	NO_2	O_3	CO
			Attributable Fraction (%, 95% CI)			
All inpatients	8.50 (6.54, 10.40)	5.81 (3.73, 7.82)	8.80 (5.80, 10.27)	3.33 (0.57, 6.00)	-	11.73 (3.42, 19.27)
Disease Subgroups						
Pneumonia	10.19 (6.18, 13.97)	6.60 (2.47, 10.51)	11.84 (7.35, 16.04)	8.35 (3.01, 13.34)	6.29 (0.34, 11.76)	-
COPD	9.91 (6.05, 13.55)	-	6.81 (2.15, 11.15)	-	-	16.05 (8.01, 23.33)
Gender						
Male	9.70 (7.24, 12.07)	7.10 (4.51, 9.60)	9.95 (7.13, 12.66)	4.83 (1.32, 8.20)	4.29 (0.59, 7.80)	12.76 (7.40, 17.79)
Female	6.50 (3.22, 9.62)	4.05 (0.59, 7.36)	7.10 (3.36, 10.63)	-	-	8.78 (1.68, 15.32)
Age (years)						
0~14	12.11 (7.79, 16.16)	6.98 (2.34, 11.34)	15.18 (10.22, 19.77)	7.64 (1.47, 13.36)	-	-
15~64	7.53 (4.21, 10.70)	4.55 (0.99, 7.95)	8.74 (4.94, 12.34)	-	-	12.99 (6.05, 19.38)
65+	7.42 (4.40, 10.32)	6.40 (3.27, 9.41)	7.75 (4.32, 11.01)	4.82 (0.68, 8.76)	4.95 (0.71, 8.94)	9.79 (3.28, 15.83)
			Attributable Number (1000 cases, 95% CI)			
All inpatients	37.6 (28.9, 46.)	25.7 (16.5, 34.6)	35.7 (25.7, 45.4)	14.7 (2.5, 26.5)	-	51.8 (15.1, 85.2)
Disease Subgroups						
Pneumonia	8.8 (6.4, 14.5)	6.8 (2.6, 10.9)	12.3 (7.6, 16.6)	8.7 (3.1, 13.8)	6.5 (0.4, 12.2)	-
COPD	9.5 (6.7, 15.1)	-	7.6 (2.4, 12.4)	-	-	17.9 (8.9, 26.0)
Gender						
Male	22.8 (19.4, 32.4)	19.0 (12.1, 25.8)	26.7 (19.1, 33.9)	13.0 (3.5, 22.)	11.5 (1.6, 20.9)	34.2 (19.8, 47.7)
Female	14.8 (5.6, 16.7)	7.0 (1, 12.8)	12.3 (5.8, 18.5)	-	-	15.3 (2.9, 26.6)
Age (years)						
0~14	6.9 (6.4, 13.2)	5.7 (1.9, 9.2)	12.4 (8.3, 16.1)	6.2 (1.2, 10.9)	-	-
15~64	14.0 (6.9, 17.6)	7.5 (1.6, 13.1)	14.4 (8.1, 20.2)	-	-	21.3 (9.9, 31.8)
65+	16.7 (8.6, 20.2)	12.6 (6.4, 18.5)	15.2 (8.5, 21.6)	9.5 (1.3, 17.2)	9.7 (1.4, 17.5)	19.2 (6.4, 31.1)

3.3.2. Results of Economic Loss Assessment

Hospital admission for respiratory diseases due to air pollutant exposure causes economic losses to both society and individuals. According to the analysis in this paper, the per capita economic losses of inpatient expenditure is approximately USD 2746.35, and the economic losses attributable to $PM_{2.5}$, PM_{10}, SO_2, NO_2, and CO exposure are USD 103.17 million, USD 70.54 million, USD 98.02 million, USD 40.35 million, and USD 142.38 million, respectively, accounting for approximately 0.20% of Wuhan's GDP in 2019. As demonstrated in Table 11, the economic losses from hospital admissions due to air pollution differed by disease subgroups, gender groups, and age groups.

Table 11. Economic losses of inpatients for respiratory disease due to air pollutants (95% CI) (unit: million USD).

Variable	$PM_{2.5}$	PM_{10}	SO_2	NO_2	O_3	CO
All inpatients	103 (79, 126)	71 (45, 95)	98 (70, 125)	40 (7, 73)	-	142 (41, 234)
Disease subgroups						
Pneumonia	24 (18, 40)	19 (7, 30)	34 (21, 46)	24 (9, 38)	17 (1, 32)	-
COPD	26 (18, 41)	-	21 (7, 34)	-	-	50 (25, 73)
Gender						
Male	63 (53, 89)	52 (33, 71)	73 (53, 93)	36 (10, 60)	35 (5, 64)	104 (60, 145)
Female	41 (15, 46)	19 (3, 35)	34 (16, 51)	-	-	35 (7, 61)
Age (years)						
0~14	19 (17, 36)	16 (5, 25)	34 (23, 44)	17 (3, 30)	-	-
15~64	38 (19, 48)	21 (4, 36)	39 (22, 56)	-	-	59 (27, 87)
65+	46 (24, 56)	34 (18, 51)	42 (23, 59)	26 (4, 47)	36 (5, 65)	53 (18, 85)

4. Discussion and Conclusions

The main conclusions and discussion from the study and analysis are as follows:

The frequency of respiratory hospitalizations in Wuhan has increased in recent years, and there is a link between changes in respiratory hospitalizations and exposure to the air pollutants $PM_{2.5}$, PM_{10}, SO_2, NO_2, O_3, and CO, particularly in those with pneumonia, males, and those aged 0–14 years. Furthermore, respiratory hospitalization caused by air pollution results in certain economic losses and burdens society and individuals financially. As a result, relevant government departments should be urged to improve air pollution management to reduce the impact of air pollution on population health and associated economic losses, hence improving health benefits and economic advantages.

Based on the quantitative analysis of the impact on hospital admissions for respiratory diseases due to air pollution, this paper found that the total economic losses on hospital admission for respiratory disease due to air pollution in Wuhan during the study period were USD 454.46 million, accounting for approximately 0.20% of Wuhan's GDP in 2019. $PM_{2.5}$, SO_2, and CO generate far greater economic losses than PM_{10}, NO_2, and O_3. As a result, the government should step up its prevention and control measures for $PM_{2.5}$, SO_2, and CO.

We may examine the essential scenario analysis of the economic losses of the impact of air pollution on human health based on the aforementioned findings. Assuming that the daily concentrations of $PM_{2.5}$, PM_{10}, SO_2, NO_2, O_3, and CO are at a relatively low level in Wuhan from 2015 to 2019, four scenarios are presented for keeping these six types of air pollutants at relatively low levels, and the evitable economic losses under different scenarios are further assessed. The scenario analysis findings are presented in the table below. In the best-case scenario, assuming that $PM_{2.5}$ and PM_{10} concentrations during the study period are within the WHO-recommended criterion values (annual average $PM_{2.5}$ concentration: 10 $\mu g/m^{3,}$ and annual average PM_{10} concentration: 20 $\mu g/m^3$), the annual evitable economic losses would be USD 16.88 million and USD 10.88, respectively. If the average SO_2 concentration limits are 4 $\mu g/m^3$, 6 $\mu g/m^3$, 8 $\mu g/m^3$, and 10 $\mu g/m^3$ over the research period, the annual evitable economic losses would be USD 12.45 million, USD 8.73 million, USD 5.01 million, and USD 1.14 million, respectively. If the average NO_2

concentration limits are 10 μg/m^3, 20 μg/m^3, 30 μg/m^3, and 40 μg/m^3 over the research period, the annual evitable economic losses would be USD 6.3 million, USD 4.44 million, USD 2.72 million, and USD 0.86 million, respectively. The study revealed a significant association between O_3 exposure and hospital admissions for pneumonia inpatients, male inpatients, and inpatients aged 65+ years with respiratory disease, with O_3 having the highest impact on hospital admissions for male respiratory inpatients. If the average O_3 concentration limits are 20 μg/m^3, 40 μg/m^3, 60 μg/m^3, and 80 μg/m^3 over the research period, the annual evitable economic losses would be USD 4.87 million, USD 3.43 million, USD 2 million, and USD 0.57 million, respectively. If the average CO concentration limits are 0.25 μg/m^3, 0.50 μg/m^3, 0.75 μg/m^3, and 1.00 μg/m^3 over the research period, the annual evitable economic losses would be USD 21.75 million, USD 14.74 million, USD 7.58 million, and USD 0.14 million, respectively. Table 12 shows the detailed findings of the study for the different disease subgroups, gender groups, and age groups for the four scenarios for each pollutant.

Table 12. Scenario analysis of evitable economic losses at relatively low levels of air pollutant concentrations in Wuhan.

Air Pollutants	Annual Average Concentrations of Pollutants in the Air under Four Scenarios	Evitable Economic Losses under Different Scenarios of Decreasing Air Pollutant Concentrations (Million USD)							
		All Inpatients	Disease Subgroups		Gender Groups		Age Groups		
			Pneumonia	COPD	Male	Female	0~14	15~64	65+
$PM_{2.5}$	10 μg/m^3	16.88	4.72	5.01	11.73	5.01	4.44	5.58	6.58
	20 μg/m^3	13.02	3.58	4.01	9.01	3.86	3.29	4.29	5.15
	30 μg/m^3	9.01	2.58	2.86	6.30	2.72	2.15	3.00	3.72
	40 μg/m^3	5.01	1.29	1.72	3.43	1.57	1.14	1.72	2.15
PM_{10}	20 μg/m^3	10.88	2.86	-	8.01	3.00	2.43	3.15	5.29
	40 μg/m^3	7.58	2	-	5.58	2.00	1.57	2.15	3.72
	60 μg/m^3	4.15	1	-	3.15	1.14	0.86	1.29	2.15
	80 μg/m^3	0.86	0.14	-	0.57	0.29	0.14	0.29	0.57
SO_2	4 μg/m^3	12.45	4.15	2.72	9.30	4.29	4.29	5.01	5.29
	6 μg/m^3	8.73	2.86	2	6.44	3.00	2.86	3.58	3.86
	8 μg/m^3	5.01	1.43	1.29	3.72	1.72	1.43	2.00	2.29
	10 μg/m^3	1.14		0.43	0.72	0.43	0.00	0.43	0.72
NO_2	10 μg/m^3	6.3	3.72	-	5.58	-	2.72	-	4.01
	20 μg/m^3	4.44	2.58	-	4.01	-	1.86	-	2.86
	30 μg/m^3	2.72	1.57	-	2.29	-	1.14	-	1.72
	40 μg/m^3	0.86	0.43	-	0.72	-	0.29	-	0.57
O_3	20 μg/m^3	4.87	2.72	-	4.87	-	-	-	4.15
	40 μg/m^3	3.43	1.86	-	3.43	-	-	-	2.86
	60 μg/m^3	2.00	1	-	2.00	-	-	-	1.57
	80 μg/m^3	0.57	0.14	-	0.57	-	-	-	0.43
CO	0.25 mg/m^3	21.75	-	7.58	14.31	6.44	-	9.01	8.01
	0.50 mg/m^3	14.74	-	5.29	9.73	4.29	-	6.01	5.58
	0.75 mg/m^3	7.58	-	2.86	5.01	2.29	-	3.15	2.86
	1.00 mg/m^3	0.14	-	0.29	0.14	0.14	-	0	0.14

Based on the preceding scenario analysis and discussion, it is evident that the health and economic returns of improved air pollution control are considerable to some extent.

This paper's research methodology is relatively generalizable. When examining the influences of air pollution on human health and calculating economic losses, this research takes Wuhan as an example. As a result, by understanding the number of hospital admissions for respiratory diseases, inpatient expenditure, length of stay, and the general situation and temporal trends of air pollutant concentrations in a given region, it is possible to quantitatively assess the effects of air pollutants on the number of hospital admissions for respiratory diseases and the associated economic losses in that region, which can serve as a benchmark for assessing the health effects of air pollution.

The data for this study came from two IIIA (highest) hospitals in Wuhan, where the hospital information system is well established, ensuring data accuracy. Furthermore, in order to study the respiratory health effects of air pollutants, this study uses the COI to estimate the economic losses associated with air pollution, making the findings more relevant for policy guidance and providing a reference for cost–benefit analysis in formulating air pollution control policies. However, there are several limitations to this research. The study relies on air pollutant concentration data from fixed location monitoring stations rather than individual air pollutant exposure, and it excludes other personal data, such as lifestyle, socioeconomic status, and comorbidities, which could give bias to effect estimation. Future studies should focus on obtaining individual exposure data and incorporating questionnaires to obtain more personal information to assess the respiratory impacts of air pollution more accurately. Second, data from hospitals on respiratory inpatients include both unintentional and purposeful admissions. In the future, more data on patient hospitalization should be collected, with some planned inpatients eliminated and only the number of unintentional inpatients included for model fitting, resulting in more scientifically valid findings.

Author Contributions: Conceptualization, S.Y. and H.Z.; methodology, S.Y. and C.Y.; software, X.W. and J.R.; validation, H.Z.; formal analysis, M.Z.; investigation, C.Y.; resources, H.Z. and D.L.; data curation, D.L.; writing—original draft preparation, H.Z. and M.Z.; writing—review and editing, M.Z.; visualization, H.Z. and M.Z.; supervision, S.Y.; project administration, S.Y. All authors have read and agreed to the published version of the manuscript.

Funding: This research received no external funding.

Institutional Review Board Statement: Not applicable.

Informed Consent Statement: Not applicable.

Data Availability Statement: The data on respiratory hospital admissions in the study were obtained from the HIS of two tertiary hospitals in Wuhan. Air pollutant concentration data were obtained from the Hubei Provincial Department of Ecology and Environment.

Conflicts of Interest: The authors declare no conflict of interest.

References

1. Technical Regulation on Ambient Air Quality Index (on Trial) (HJ 633—2012). Available online: https://www.mee.gov.cn/ywgz/fgbz/bz/bzwb/jcffbz/201203/t20120302_224166.shtml (accessed on 20 November 2021).
2. Stanaway, J.D.; Afshin, A.; Gakidou, E.; Lim, S.S.; Abate, D.; Abate, K.H.; Abbafati, C.; Abbasi, H.; Abbastabar, H.; Abd-Allah, F.; et al. Global, regional, and national comparative risk assessment of 84 behavioural, environmental and occupational, and metabolic risks or clusters of risks for 195 countries and territories, 1990–2017: A systematic analysis for the Global Burden of Disease Study 2017. *Lancet* **2018**, *392*, 1923–1994. [CrossRef]
3. Lamichhane, D.K.; Kim, H.C.; Choi, C.M.; Shin, M.H.; Shim, Y.M.; Leem, J.H.; Ryu, J.S.; Nam, H.S.; Park, S.M. Lung Cancer Risk *Resid. Expo.* To air pollution: A Korean population-based case-control study. *Yonsei Med. J.* **2017**, *58*, 1111–1118. [CrossRef]
4. Abdo, N.; Khader, Y.S.; Abdelrahman, M.; Graboski-Bauetr, A.; Malkawi, M.; Al-Sharif, M.; Elbetieha, A.M. Respiratory health outcomes and air pollution in the Eastern Mediterranean Region: A systematic review. *Rev. Environ. Health* **2016**, *31*, 259–280. [CrossRef] [PubMed]

5. Soriano, J.B.; Kendrick, P.J.; Paulson, K.R.; Gupta, V.; Abrams, E.M.; Adedoyin, R.A.; Adhikari, T.B.; Advani, S.M.; Agrawal, A.; Ahmadian, E.; et al. Prevalence and attributable health burden of chronic respiratory diseases 1990–2017: A systematic analysis for the Global Burden of Disease Study 2017. *Lancet Respir. Med.* **2020**, *8*, 585–596. [CrossRef]
6. Kyu, H.H.; Abate, D.; Abate, K.H.; Abay, S.M.; Abbafati, C.; Abbasi, N.; Abbastabar, H.; Abd-Allah, F.; Abdela, J.; Abdelalim, A.; et al. Global, regional, and national disability-adjusted life-years (DALYs) for 359 diseases and injuries and healthy life expectancy (HALE) for 195 countries and territories, 1990-2017: A systematic analysis for the Global Burden of Disease Study 2017. *Lancet* **2018**, *392*, 1859–1922. [CrossRef]
7. China Statistic Yearbook 2021. Available online: http://www.stats.gov.cn/tjsj/ndsj/2021/indexch.htm (accessed on 15 September 2021).
8. Ma, Y.; Xiao, B.; Liu, C.; Zhao, Y.; Zheng, X. Association between ambient air pollution and emergency room visits for respiratory diseases in spring dust storm season in Lanzhou, China. *Int. J. Environ. Res. Public Health* **2016**, *13*, 613. [CrossRef]
9. Dong, J.; Liu, Y.; Bao, H. Revalue associations of short-term exposure to air pollution with respiratory hospital admissions in Lanzhou, China after the control and treatment of current pollution. *Int. J. Hyg. Environ. Health* **2021**, *231*, 113658. [CrossRef] [PubMed]
10. Yao, M.; Wu, G.; Zhao, X.; Zhang, J. Estimating health burden and economic loss attributable to short-term exposure to multiple air pollutants in China. *Environ. Res.* **2020**, *183*, 109184. [CrossRef]
11. Mamkhezri, J.; Bohara, A.K.; Camargo, A.I. Air pollution and daily mortality in Mexico city Metropolitan area. *Atmosfera* **2019**, *33*, 249–267. [CrossRef]
12. Afghan, F.R.; Patidar, S.K. Health impacts assessment due to $PM_{2.5}$, PM_{10} and NO_2 exposure in National Capital Territory (NCT). *Pollution* **2020**, *6*, 115–126. [CrossRef]
13. Hwang, J.; Kwon, J.; Yi, H.; Bae, H.-J.; Jang, M.; Kim, N. Association between long-term exposure to air pollutants and cardiopulmonary mortality rates in South Korea. *BMC Public Health* **2020**, *20*, 1402–1410. [CrossRef]
14. Hansel, N.N.; Paulin, L.M.; Gassett, A.J.; Peng, R.D.; Alexis, N.; Fan, V.S.; Bleecker, E.; Bowler, R.; Comellas, A.P.; Dransfield, M.; et al. Design of the Subpopulations and Intermediate Outcome Measures in COPD (SPIROMICS) AIR Study. *BMJ Open Respir. Res.* **2017**, *4*, e186. [CrossRef]
15. Li, G.; Huang, J.; Xu, G.; Pan, X.; Qian, X.; Xu, J.; Zhao, Y.; Zhang, T.; Liu, Q.; Guo, X.; et al. The short-term burden of ambient fine particulate matter on chronic obstructive pulmonary disease in Ningbo, China. *Environ. Health* **2017**, *16*, 54. [CrossRef] [PubMed]
16. Zhang, Z.; Chai, P.; Wang, J.; Ye, Z.; Shen, P.; Lu, H.; Jin, M.; Gu, M.; Li, D.; Lin, H.; et al. Association of particulate matter air pollution and hospital visits for respiratory diseases: A time-series study from China. *Environ. Sci. Pollut. Res. Int.* **2019**, *26*, 12280–12287. [CrossRef]
17. Luo, L.; Zhang, Y.; Jiang, J.; Luan, H.; Yu, C.; Nan, P.; Luo, B.; You, M. Short-Term effects of ambient air pollution on hospitalization for respiratory disease in Taiyuan, China: A time-series analysis. *Int. J. Environ. Res. Public Health* **2018**, *15*, 2160. [CrossRef] [PubMed]
18. Yang, Q.; Chen, Y.; Krewski, D.; Burnett, R.T.; Shi, Y.; McGrail, K.M. Effect of short-term exposure to low levels of gaseous pollutants on chronic obstructive pulmonary disease hospitalizations. *Environ. Res.* **2005**, *99*, 99–105. [CrossRef]
19. Cadelis, G.; Tourres, R.; Molinie, J. Short-Term effects of the particulate pollutants contained in saharan dust on the visits of children to the emergency department due to asthmatic conditions in guadeloupe (French archipelago of the Caribbean). *PLoS ONE* **2014**, *9*, e91136. [CrossRef]
20. Akpinar-Elci, M.; Martin, F.E.; Behr, J.G.; Diaz, R. Saharan dust, climate variability, and asthma in Grenada, the Caribbean. *Int. J. Biometeorol.* **2015**, *59*, 1667–1671. [CrossRef]
21. Guarnieri, M.; Balmes, J.R. Outdoor air pollution and asthma. *Lancet* **2014**, *383*, 1581–1592. [CrossRef]
22. Zhao, Y.; Hu, J.; Tan, Z.; Liu, T.; Zeng, W.; Li, X.; Huang, C.; Wang, S.; Huang, Z.; Ma, W. Ambient carbon monoxide and increased risk of daily hospital outpatient visits for respiratory diseases in Dongguan, China. *Sci. Total Environ.* **2019**, *668*, 254–260. [CrossRef]
23. Moore, K.; Neugebauer, R.; Lurmann, F.; Hall, J.; Brajer, V.; Alcorn, S.; Tager, I. Ambient ozone concentrations cause increased hospitalizations for asthma in children: An 18-year study in southern California. *Environ. Health Perspect.* **2008**, *116*, 1063–1070. [CrossRef]
24. Clark, N.A.; Demers, P.A.; Karr, C.J.; Koehoorn, M.; Lencar, C.; Tamburic, L.; Brauer, M. Effect of early life exposure to air pollution on development of childhood asthma. *Environ. Health Perspect.* **2010**, *118*, 284–290. [CrossRef] [PubMed]
25. Wang, H.; Gao, Z.; Ren, J.; Chang, L.T.-C.; Cheung, K.; Feng, Y.; Li, Y. An urban-rural and sex differences in cancer incidence and mortality and the relationship with $PM_{2.5}$ exposure: An ecological study in the southeastern side of Hu line. *Chemosphere* **2019**, *216*, 766–773. [CrossRef] [PubMed]
26. Katanoda, K.; Sobue, T.; Satoh, H.; Tajima, K.; Suzuki, T.; Nakatsuka, H.; Takezaki, T.; Nakayama, T.; Nitta, H.; Tanabe, K.; et al. An Association between long-term exposure to ambient air pollution and mortality from lung cancer and respiratory diseases in Japan. *J. Epidemiol.* **2011**, *21*, 132–143. [CrossRef]
27. Nyberg, F.; Gustavsson, P.; Järup, L.; Bellander, T.; Berglind, N.; Jakobsson, R.; Pershagen, G. Urban air pollution and lung cancer in Stockholm. *Epidemiology* **2000**, *11*, 487–495. [CrossRef]

28. Nsoh, M.; Mankollo, B.O.Y.; Ebongue, M.; Cyprien, K.N.; Likeng, J.L.N.; Islam, S.M.S.; Collier, A.; Tsoka-Gwegweni, J.M.; Cumber, S.N. Acute respiratory infection related to air pollution in Bamenda, North West Region of Cameroon. *Pan. Afr. Med. J.* **2019**, *32*, 1–8. [CrossRef]
29. Zheng, P.W.; Wang, J.B.; Zhang, Z.Y.; Shen, P.; Chai, P.-F.; Li, D.; Jin, M.-J.; Tang, M.-L.; Lu, H.-C.; Lin, H.-B.; et al. Air pollution and hospital visits for acute upper and lower respiratory infections among children in Ningbo, China: A time-series analysis. *Environ. Sci. Pollut. Res.* **2017**, *24* (Suppl. 4), 1–10. [CrossRef]
30. Nhung, N.T.T.; Schindler, C.; Dien, T.M.; Probst-Hensch, N.; Künzli, N. Association of ambient air pollution with lengths of hospital stay for hanoi children with acute lower-respiratory infection, 2007–2016. *Environ. Pollut.* **2019**, *247*, 752–762. [CrossRef]
31. Department of Ecology and Environment of Hubei Province. Available online: http://sthjt.hubei.gov.cn (accessed on 5 September 2021).
32. China Meteorological Data Sharing Service System. Available online: http://data.cma.cn (accessed on 8 September 2021).
33. Wu, Z.; Chen, X.; Li, G.; Tian, L.; Wang, Z.; Xiong, X.; Yang, C.; Zhou, Z.; Pan, X. Attributable risk and economic cost of hospital admissions for mental disorders due to $PM_{2.5}$ in Beijing. *Sci. Total Environ.* **2020**, *718*, 137274. [CrossRef]
34. Kennelly, B. How should cost-of-illness studies be interpreted? *Lancet Psychiatry* **2017**, *4*, 735–736. [CrossRef]
35. Zhang, P.; Zhou, X. Health and economic impacts of particulate matter pollution on hospital admissions for mental disorders in Chengdu, Southwestern China. *Sci. Total Environ.* **2020**, *733*, 139114. [CrossRef] [PubMed]
36. National Bureau of Statistics. Available online: http://www.stats.gov.cn/ (accessed on 10 September 2021).

atmosphere

MDPI

Article

Transmission Risk Prediction and Evaluation of Mountain-Type Zoonotic Visceral Leishmaniasis in China Based on Climatic and Environmental Variables

Yuwan Hao [1,†], Zhuowei Luo [1,†], Jian Zhao [2], Yanfeng Gong [1], Yuanyuan Li [1], Zelin Zhu [1], Tian Tian [1], Qiang Wang [1], Yi Zhang [1,3], Zhengbin Zhou [1,*], Zengyun Hu [4,*] and Shizhu Li [1,3,*]

1 National Institute of Parasitic Diseases, Chinese Center for Disease Control and Prevention (Chinese Center for Tropical Diseases Research), WHO Collaborating Centre for Tropical Diseases; National Center for International Research on Tropical Diseases, NHC Key Laboratory of Parasite and Vector Biology, Shanghai 200025, China; haoyw@nipd.chinacdc.cn (Y.H.); lzw1997lzw@163.com (Z.L.); peak_gong@163.com (Y.G.); liyy@nipd.chinacdc.cn (Y.L.); zhuzl@nipd.chinacdc.cn (Z.Z.); tiantian@nipd.chinacdc.cn (T.T.); wangqiang@nipd.chinacdc.cn (Q.W.); zhangyi@nipd.chinacdc.cn (Y.Z.)

2 College of Geography and Remote sensing Sciences, Xinjiang University, Urumqi 830046, China; qaz19147@126.com

3 Chinese Center for Tropical Diseases Research, School of Global Health, Shanghai Jiao Tong University School of Medicine, Shanghai 200025, China

4 State Key Laboratory of Desert and Oasis Ecology, Xinjiang Institute of Ecology and Geography, Chinese Academy of Sciences, Urumqi 830011, China

* Correspondence: zhouzb@nipd.chinacdc.cn (Z.Z.); huzengyun@ms.xjb.ac.cn (Z.H.); lisz@chinacdc.cn (S.L.)

† These authors contributed equally to this work.

Citation: Hao, Y.; Luo, Z.; Zhao, J.; Gong, Y.; Li, Y.; Zhu, Z.; Tian, T.; Wang, Q.; Zhang, Y.; Zhou, Z.; et al. Transmission Risk Prediction and Evaluation of Mountain-Type Zoonotic Visceral Leishmaniasis in China Based on Climatic and Environmental Variables. *Atmosphere* **2022**, *13*, 964. https://doi.org/10.3390/atmos13060964

Academic Editor: Ian Colbeck

Received: 27 May 2022
Accepted: 13 June 2022
Published: 14 June 2022

Publisher's Note: MDPI stays neutral with regard to jurisdictional claims in published maps and institutional affiliations.

Abstract: With global warming and socioeconomic developments, there is a tendency toward the emergence and spread of mountain-type zoonotic visceral leishmaniasis (MT-ZVL) in China. Timely identification of the transmission risk and spread of MT-ZVL is, therefore, of great significance for effectively interrupting the spread of MT-ZVL and eliminating the disease. In this study, 26 environmental variables—namely, climatic, geographical, and 2 socioeconomic indicators were collected from regions where MT-ZVL patients were detected during the period from 2019 to 2021, to create 10 ecological niche models. The performance of these ecological niche models was evaluated using the area under the receiver-operating characteristic curve (AUC) and true skill statistic (TSS), and ensemble models were created to predict the transmission risk of MT-ZVL in China. All ten ecological niche models were effective at predicting the transmission risk of MT-ZVL in China, and there were significant differences in the mean AUC ($H = 33.311$, $p < 0.05$) and TSS values among these ten models ($H = 26.344$, $p < 0.05$). The random forest, maximum entropy, generalized boosted, and multivariate adaptive regression splines showed high performance at predicting the transmission risk of MT-ZVL (AUC > 0.95, TSS > 0.85). Ensemble models predicted a transmission risk of MT-ZVL in the provinces of Shanxi, Shaanxi, Henan, Gansu, Sichuan, and Hebei, which was centered in Shanxi Province and presented high spatial clustering characteristics. Multiple ensemble ecological niche models created based on climatic and environmental variables are effective at predicting the transmission risk of MT-ZVL in China. This risk is centered in Shanxi Province and tends towards gradual radiation dispersion to surrounding regions. Our results provide insights into MT-ZVL surveillance in regions at high risk of MT-ZVL.

Keywords: mountain-type zoonotic visceral leishmaniasis; climate variables; environmental variables; ecological niche model; transmission risk prediction

1. Introduction

Visceral leishmaniasis (VL), also known as kala-azar, is a zoonotic infectious disease caused by protozoan parasites of the genus *Leishmania* spp. and transmitted by the bite of infected *Phlebotomus* [1]. Currently, this zoonosis is mainly prevalent in 88 countries across East Africa, South Asia, South America, and the Mediterranean, and it is estimated that 200,000 to 400,000 new cases are diagnosed with VL across the world each year, including approximately 60,000 deaths due to a failure of timely treatment [2]. Globally, VL ranks second only to malaria among all parasitic diseases in terms of mortality [3,4]. VL was once prevalent in 650 counties across 16 provinces north of the Yangtze River in China, and approximately 530,000 cases were diagnosed with VL in China in 1951, with the majority reported in young and middle-aged individuals [5–7]. In China, VL is classified into three types—anthroponotic VL (AVL), mountain-type zoonotic VL (MT-ZVL), and desert-type zoonotic VL (DT-ZVL)—according to geographical landscapes, epidemiological characteristics, and the species of sand fly vectors [8].

Following concerted efforts for decades, VL was almost eradicated in the eastern and central plain regions of China in the early 1960s [9]. However, VL cases continue to be detected in western China, including Kashgar in Xinjiang Uygur Autonomous Region, southern Gansu Province, and northern Sichuan Province, and local clusters of VL have been reported occasionally [10–12]. Since 2000, the exacerbation of global warming, the implementation of the Conversion of Cropland to Forest and Grassland Program, and the afforestation of hillsides in central and western China have resulted in gradual improvements in natural ecological environments and rapid increases in the population density of wild species of *Phlebotomus sinensis* [13,14]. This has resulted in the re-emergence and spread of VL in hilly regions. In particular, clusters of MT-ZVL have reemerged in the provinces of Shanxi, Shaanxi, Henan, and Hebei [15,16]. Timely identification of the transmission risk and spread of MT-ZVL is, therefore, of great significance to effectively interrupt the reemergence and spread of VL and ultimately eliminate the disease.

Ecological niche models combine species occurrence data with environmental variables to estimate the ecological conditions of species distribution, thereby predicting the true or potential distribution [17]. Recently, ecological niche models have widely been employed to predict the transmission risk and trends of the spread of infectious diseases by unraveling the interplay between environmental factors and disease transmission [18,19]. Vector-borne tropical diseases, such as schistosomiasis [20], malaria [21], and VL [22,23], are more likely to be affected by climatic and eco-environmental variables, so ecological niche models are particularly suited to predicting their spread. Based on the data of *Oncomelania hupensis* distributions, ecological niche models were used to investigate the trends for *O. hupensis* spread and predict the potential transmission risk of schistosomiasis in China [24]. Based on the data indicating pathogen distributions, three ecological niche models were created to predict the transmission risk of schistosomiasis in Yunnan Province [25]. In addition, three ecological niche models were generated to predict the trends for MT-ZVL spread in the provinces of Shanxi and Henan [26]. To our knowledge, however, no previous study has predicted the potential transmission risk of MT-ZVL nationwide in China. In this study, multiple ecological niche models were created by using environmental variables based on updated national epidemiological data of MT-ZVL in China. The models were used to predict the transmission risk of MT-ZVL in China and to provide insights into VL surveillance and control in the country.

2. Materials and Methods

2.1. Study Area

Based on previous distributions of sand flies and recently reported VL cases, Gansu, Sichuan, Shanxi, Shaanxi, Henan, Hebei, Qinghai, Ningxia, Liaoning, and Beijing were selected as potential transmission foci of MT-ZVL [27,28]. This study area covers 113 cities and 1002 counties in central China, approximately one-third of the total land area of China. The area is delimited by Jiuquan City in Gansu Province to the north (43°43′ N, 96°38′ E), Liangshan Yi Autonomous Prefecture in Sichuan Province to the south (26°24′ N, 102°8′ E), Yushu Tibetan Autonomous Prefecture in Qinghai Province to the west (35°16′ N, 89°29′ E), and Fushun City in Liaoning Province to the east (40°53′ N, 125°36′ E). The study area has diverse ecological and climatic conditions divided by the Qin Mountains and has both temperate monsoon climates and continental climates. The precipitation and relative humidity gradually increase from the north to the south of the study area, and the temperature gradually increases. The sand fly *Ph. chinensis*, a primary vector for transmission of MT-ZVL in China, is predominantly identified in plains, hilly, and loess plateau regions at 30° N to 43° N and 102° E to 121° E, with elevations of 10 to 2750 m [29]. Local cases of MT-ZVL reported during the past five years were all concentrated in this region. Therefore, it is possible to examine the correlation between the distribution of MT-ZVL patients and potential *P. sinensis* distribution in this study area.

2.2. Collection of Case Data

All epidemiological data pertaining to MT-ZVL cases were collected from the National Notifiable Diseases Reporting System of the Chinese Center for Disease Control and Prevention. The townships or streets where local MT-ZVL cases were reported during the 2019–2021 period were defined as sampling sites, and the longitude and latitude coordinates were measured for each site. Then, each sampling site was checked to ensure the accuracy of the geographical coordinates. A total of 187 valid sampling sites were finally selected. In addition, the databases of reported cases from 2015 to 2018 were collected and combined to describe the trends of the disease.

2.3. Collection of Climatic and Environmental Variables

The environmental variable data pertaining to the distributions of VL and its vector sand flies were collected and finalized based on a literature review [30,31]. Ultimately, we included 19 climatic variables, 7 geographical variables, and 2 socioeconomic variables (Table 1). The climatic variables were extracted from the WorldClim website. Available online: https://www.worldclim.org/data/worldclim21.html (accessed on 25 April 2022). Geographical variables included elevation, type of landform, type of land use, annual normalized difference vegetation index, and type of vegetation coverage. The socioeconomic variables included gross domestic product and population density, provided by the Resource and Environment Data Center, Chinese Academy of Sciences (Beijing, China). Available online: https://www.resdc.cn (accessed on 25 April 2022). All environmental variable data were clipped to the study area using ArcGIS 10.2 (ESRI, Redlands, CA, USA) and a bilinear resampling algorithm with 1×1 km spatial resolution.

Table 1. Environmental variables affecting the distribution of mountain-type zoonotic visceral leishmaniasis and *Ph. chinensis* vectors.

Variable Classification	Variable Name (Units)	Definition	Year
Bioclimatic Data	BIO1 (°C)	Annual mean temperature	1955–2000
	BIO2 (°C)	Mean diurnal range	
	BIO3 (%)	Isothermality	
	BIO4 (%)	Standard deviation of temperature seasonality	
	BIO5 (°C)	Max temperature of warmest month	
	BIO6 (°C)	Min temperature of coldest month	
	BIO7 (°C)	Temperature annual range	
	BIO8 (°C)	Mean temperature of wettest quarter	
	BIO9 (°C)	Mean temperature of driest quarter	
	BIO10 (°C)	Mean temperature of warmest quarter	
	BIO11 (mm)	Mean temperature of coldest quarter	
	BIO12 (mm)	Annual precipitation	
	BIO13 (mm)	Precipitation of wettest month	
	BIO14 (mm)	Precipitation of driest month	
	BIO15	Coefficient of variation for precipitation seasonality	
	BIO16 (mm)	Precipitation of wettest quarter	
	BIO17 (mm)	Precipitation of driest quarter	
	BIO18 (mm)	Precipitation of warmest quarter	
	BIO19 (mm)	Precipitation of coldest quarter	
Geographical Data	ElV	Elevation	2000
	LF	Landform	2010
	LU	Land use	2015
	NDVI	Normalized difference vegetation index	2019
	VEG	Vegetation	2015
Socioeconomic Data	GDP	Gross domestic product	2015
	DP	Density of population	2015

2.4. Ecological Niche Modeling

Multiple algorithms are available to create ecological niche models, and modeling with the same data may generate different results and predictive maps [32,33]. There is no consensus on an optimal single algorithm for ecological niche models [34,35]. Therefore, scientists are encouraged to overcome the uncertainty using different approaches. In our study, ten ecological niche models were generated based on four machine learning algorithms: the surface range envelope (SRE) model, based on an environmental envelope (threshold-based) algorithm; the generalized linear model (GLM), generalized additive model (GAM), and multivariate adaptive regression splines (MARS), based on statistical regression algorithms; the generalized boosted model (GBM), classification tree analysis (CTA), and flexible discriminant analysis (FDA), based on classification algorithms; and the maximum entropy (MaxEnt) model, artificial neural network (ANN), and random forest (RF) model, based on machine learning algorithms. All models were created using the BIOMOD2 package in R. As presence–absence data are required to create models, 500 absence points were randomly sampled at a ratio of 1:2 based on 186 presence points to create two sets of presence–absence data. Each set of data was classified into 75% training datasets and 25% validation datasets, which were input into the ten ecological niche models. Each run was repeated ten times for a single model with the same variables.

2.5. Assessment of the Performance of Ecological Niche Models

Currently, there are threshold-free and -associated measures to assess the performance of ecological niche models [19]. In this study, the area under the receiver-operating characteristic curve (AUC), a threshold-free index, and the true skill statistic (TSS), a threshold-related index, were employed to assess the performance of the ecological niche models. The AUC value, ranging from 0 to 1, indicates the predictive accuracy of the ecological niche models, and an AUC value approaching 1 indicates higher accuracy. The TSS value is not affected by the occurrence of the distribution, but it can accurately identify the accuracy of a model. TSS values range between -1 and 1, and a TSS value close to 1 indicates high accuracy. A TSS value of >0.7 is generally accepted as a satisfactory prediction. Both indices can be used to evaluate the predictive accuracy of ecological niche models. The AUC value can be used to assess the predictive accuracy of a model regardless of the data distribution [36], whereas the TSS measures the consistency between the prediction results by models and sample data [37]. Therefore, the combination of these two indices more comprehensively assesses the performance of ecological niche models.

2.6. Prediction of MT-ZVL Transmission Risk in China

Due to the uncertainty of predictions using a single ecological niche model, ensemble models were created based on a single model to improve the accuracy and reliability of predicting the transmission risk of MT-ZVL. Briefly, a single ecological niche model with a poor training efficiency was removed based on an AUC value of >0.90 and a TSS value of >0.85 to screen for ensemble models. Then, the prediction accuracy of a single ecological niche model was normalized such that the raster data ranged from 0 to 1. Following the definition of the weight according to TSS values, the predictive accuracy of ensemble models was estimated using a weighted mean of probabilities (WM). Four grades were classified based on the WM values: no-risk areas (WM, 0.5 and lower), low-risk areas (WM, 0.51 to 0.7), medium-risk areas (0.71 to 0.9), and high-risk areas (0.91 to 1). Finally, the model prediction results were loaded into ArcGIS software to map the transmission risk of MT-ZVL in China.

2.7. Statistical Analysis

In what follows, the AUC and TSS values are described as the mean $\pm$ standard deviation (SD). Comparisons of the AUC and TSS values were carried out using the Kruskal–Wallis H-test. A p-value of <0.05 was considered statistically significant.

3. Results

3.1. Epidemiological Characteristics of MT-ZVL in China from 2015 to 2021

A total of 990 MT-ZVL cases were reported in China during the period from 2015 to 2021: specifically, 82 in 2015, 95 in 2016, 113 in 2017, 121 in 2018, 151 in 2019, 204 in 2020, and 224 in 2021 [38–40] (Figure 1). The MT-ZVL cases were predominantly detected in four provinces—Gansu, Sichuan, Shaanxi, and Shanxi—and the geographical distribution of MT-ZVL cases gradually expanded over time. In 2015, MT-ZVL cases were predominantly detected in four provinces (Gansu, Sichuan, Shaanxi, and Shanxi), whereas MT-ZVL cases were reported in seven provinces by the end of 2021, with local cases identified in Henan Province in 2016 and Hebei Province and Beijing Municipality in 2019.

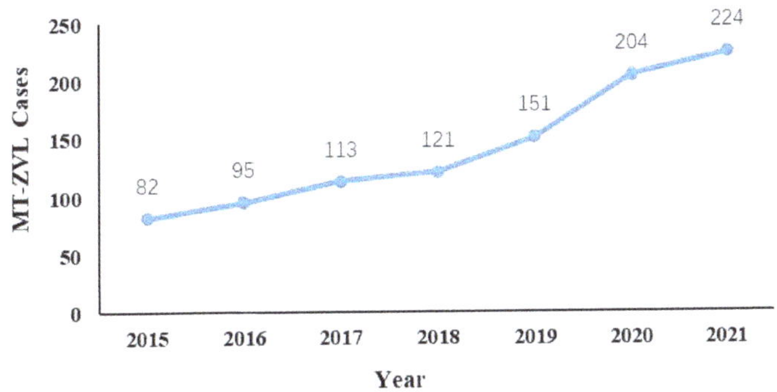

Figure 1. Changes in number of patients with mountain-type zoonotic visceral leishmaniasis in China from 2015 to 2021.

Since 2019, the highest number of MT-ZVL cases was reported in Shanxi Province. A total of 299 MT-ZVL cases were detected in Shanxi Province from 2019 to 2021, consisting of 51.64% of all cases reported in China. The number of MT-ZVL cases increased from 54 in 2019 to 141 in 2021 in Shanxi Province. The second-largest number of MT-ZVL cases was reported in Shaanxi Province. There, 118 cases were detected from 2019 to 2021, consisting of 20.38% of all cases reported in China. The third-largest number of MT-ZVL cases was reported in Gansu Province, where the number of cases gradually declined from 2019 to 2021. The number of MT-ZVL cases reported in the three provinces of Shanxi, Shaanxi, and Gansu consisted of 90.33% of all cases in China during the period from 2019 to 2021, whereas 5.83%, 1.89%, 1.38%, and 0.52% of cases were detected in Henan, Hebei, Sichuan, and Beijing, respectively (Figure 2).

3.2. Comparison of the MT-ZVL Transmission Risk Predicted by Using Ecological Niche Models

Based on the distributions of MT-ZVL cases and *Ph. chinensis* vectors, ten ecological niche models were created to predict the transmission risk of MT-ZVL. Our data showed that all ten ecological niche models were effective at predicting areas at risk of MT-ZVL transmission; however, the coverage of the predicted areas varied among the models. The ANN and MARS predicted the largest but most dispersed areas at risk of MT-ZVL transmission, and CTA predicted the second-largest areas. The other seven models predicted relatively concentrated areas at risk of MT-ZVL transmission, and the GAM and MaxEnt models predicted the smallest areas. The CTA and SRE models predicted large areas at risk of MT-ZVL transmission, and most areas were predicted to be at high or medium risk. The high-, medium-, and low-risk areas predicted by the other seven models were centered in Shanxi Province and showed a tendency to disperse to the surrounding regions, with southern Liaoning Province to the north and the southern edge of the hilly regions in northern Sichuan Province to the south (Figure 3).

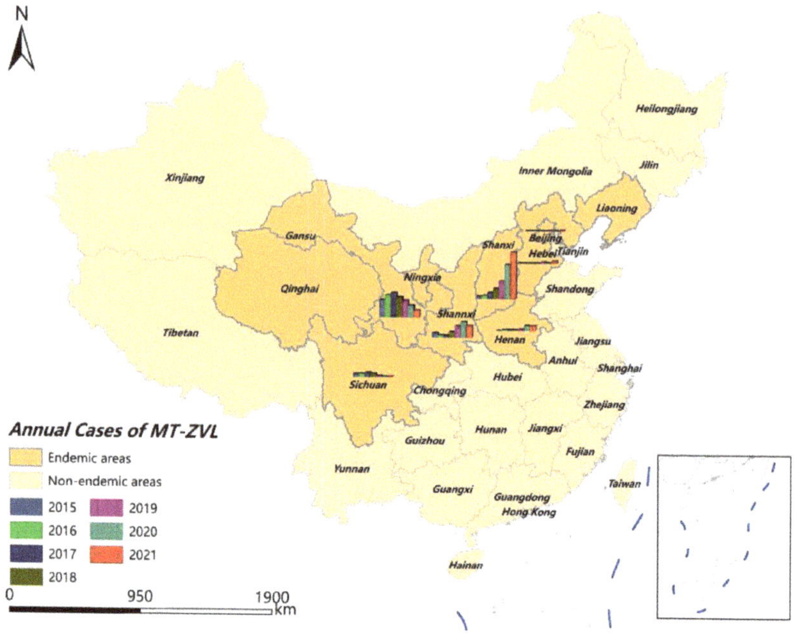

Figure 2. Geographical distribution of patients with mountain-type zoonotic visceral leishmaniasis in China from 2015 to 2021.

3.3. *Performance of Ecological Niche Models*

Ten ecological niche models were employed to predict the transmission risk areas of MT-ZVL in China, and the performance of the ecological niche models was evaluated. The RF model showed the highest mean AUC and TSS, and the SRE model exhibited the lowest mean AUC and TSS (Table 2). There were significant differences in the mean AUC (H = 33.311, $p < 0.05$) and TSS values among the ten ecological niche models (H = 26.344, $p < 0.05$). The results show that the RF, MaxEnt, GBM, and MARS models had the highest relative performance at predicting the risk of MT-ZVL transmission (mean AUC > 0.95, mean TSS > 0.85). The SRE model had the lowest performance (mean AUC < 0.8, mean TSS < 0.8).

Table 2. Performance of ten ecological niche models.

Model	AUC Value	TSS Value
ANN	0.912 ± 0.033	0.777 ± 0.074
CTA	0.902 ± 0.029	0.775 ± 0.054
FDA	0.963 ± 0.014	0.829 ± 0.044
GAM	0.940 ± 0.019	0.826 ± 0.065
GBM	0.965 ± 0.016	0.854 ± 0.045
GLM	0.943 ± 0.039	0.828 ± 0.068
MARS	0.961 ± 0.018	0.854 ± 0.073
MaxEnt	0.968 ± 0.019	0.856 ± 0.046
RF	0.971 ± 0.011	0.857 ± 0.043
SRE	0.790 ± 0.022	0.581 ± 0.044

ANN, artificial neural network; CTA, classification tree analysis; FDA, flexible discriminant analysis; GAM, generalized additive model; GBM, generalized boosted model; GLM, generalized linear model; MARS, multivariate adaptive regression spline; MaxEnt, maximum entropy model; RF, random forest; SRE, surface range envelope model.

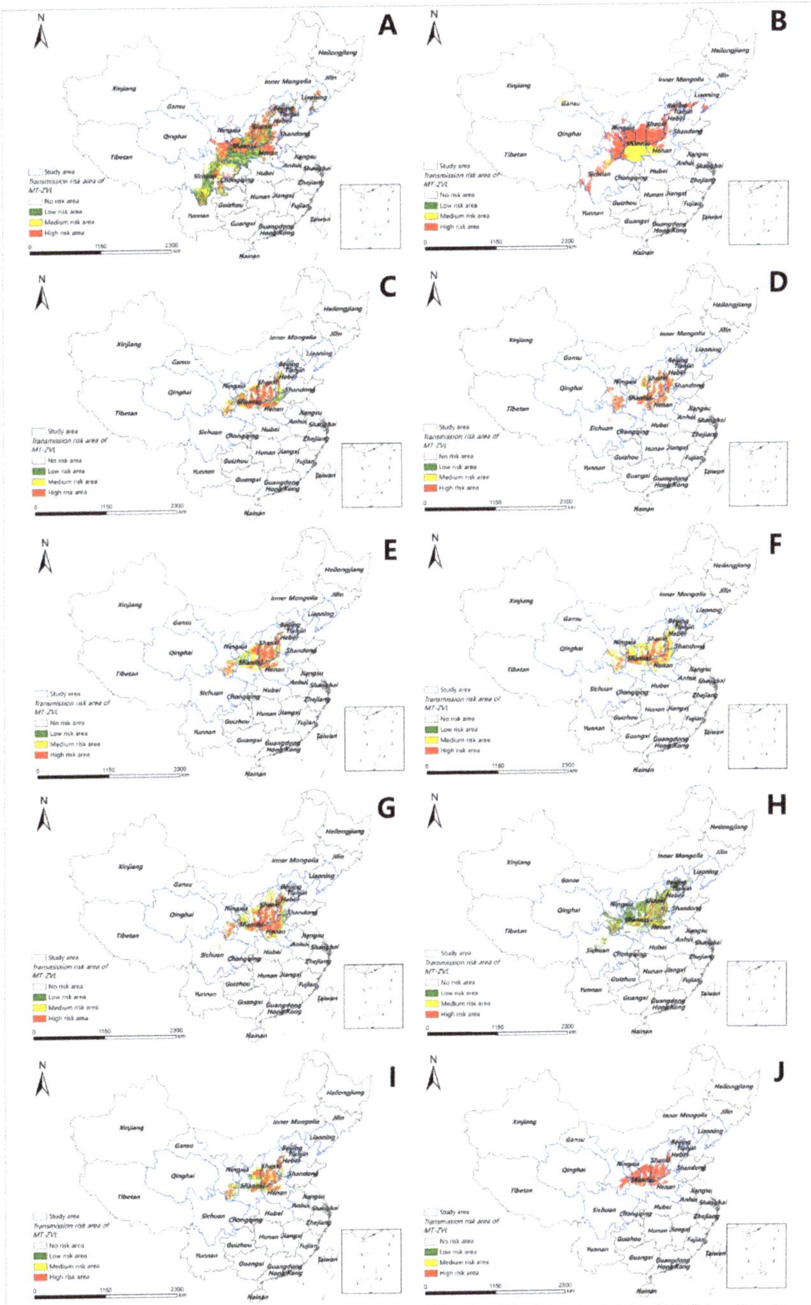

Figure 3. Prediction of areas at risk of mountain-type zoonotic visceral leishmaniasis transmission in China using ecological niche models based on climatic and environmental variables: (**A**) artificial neural network (ANN); (**B**) classification tree analysis (CTA); (**C**) flexible discriminant analysis (FDA); (**D**) generalized additive model (GAM); (**E**) generalized boosted model (GBM); (**F**) generalized linear model (GLM); (**G**) multivariate adaptive regression spline (MARS); (**H**) maximum entropy (MaxEnt) model; (**I**) random forest (RF); (**J**) surface range envelope (SRE) model.

3.4. Contributions of Environmental Variables to Ecological Niche Models

The contributions of the environmental variables to the ten ecological niche models were normalized (Table 3). The climatic variables that contributed most to the ecological niche models included the mean temperature of the coldest quarter, precipitation during the wettest quarter, the minimum temperature of the coldest month, and annual precipitation. The highest contributing geographical variables were elevation and normalized difference vegetation index, and the socioeconomic variables that contributed most were gross domestic product and population density.

Table 3. Contributions of environmental variables to ten ecological niche models (%).

Variable Name	ANN	CTA	FDA	GAM	GBM	GLM	MARS	MaxEnt	RF	SRE
bio_01	0.9	2.7	5.7	4.9	1.4	6.6	2.8	3.3	5.1	5.8
bio_02	0.2	0.0	4.4	0.8	0.1	4.1	0.0	11.6	1.6	4.2
bio_03	0.8	0.8	4.0	0.0	7.7	4.6	6.0	6.1	13.5	4.6
bio_04	2.8	0.0	7.0	0.3	0.6	5.1	4.3	0.2	2.7	2.8
bio_05	2.8	0.0	3.3	0.3	0.1	6.6	0.4	0.5	0.4	3.7
bio_06	2.0	0.0	11.5	0.7	4.9	6.3	18.3	24.1	6.8	6.7
bio_07	1.4	0.0	3.9	0.6	0.2	7.3	3.8	0.0	1.4	3.4
bio_08	1.5	0.0	6.5	6.8	2.0	4.6	3.7	5.1	7.4	4.4
bio_09	0.6	0.0	0.3	3.6	0.0	4.9	0.0	5.4	3.7	6.3
bio_10	1.2	0.0	7.2	3.7	0.3	5.5	8.6	1.5	1.2	4.4
bio_11	0.9	60.8	7.2	5.5	63.6	6.9	4.7	5.1	11.3	6.9
bio_12	17.8	6.7	10.5	7.5	3.5	3.1	14.9	7.2	2.9	5.6
bio_13	7.0	0.0	4.4	5.3	0.3	2.4	0.0	4.5	1.6	4.6
bio_14	0.2	0.7	0.2	7.4	0.0	1.0	1.8	6.5	0.0	1.2
bio_15	1.6	1.3	0.3	3.6	0.4	1.8	5.0	0.3	2.7	1.2
bio_16	11.1	20.9	10.3	5.1	3.6	5.0	11.5	6.3	3.7	5.1
bio_17	2.7	1.4	1.3	5.5	0.0	6.2	0.0	0.0	0.8	2.7
bio_18	8.7	0.0	9.5	7.5	2.1	4.5	3.7	1.6	2.5	4.8
bio_19	2.4	0.0	1.0	4.8	0.0	5.9	0.2	0.1	0.8	2.5
elv	15.1	2.9	0.4	1.4	1.5	4.6	4.5	4.8	3.1	4.7
gdp	11.0	0.0	0.0	0.9	0.2	0.5	0.5	0.1	1.9	4.0
lf	0.6	0.0	0.4	2.3	1.5	1.0	2.1	1.6	2.5	1.1
lu	0.5	0.0	0.0	5.3	0.0	0.2	0.0	0.3	0.2	1.6
ndvi	0.0	1.9	0.6	6.2	4.7	0.5	1.2	2.6	17.0	3.3
pop	5.0	0.0	0.0	4.1	0.6	0.2	0.1	0.0	2.5	4.0
veg	1.2	0.0	0.3	5.7	0.6	0.7	2.1	1.2	2.7	0.4

The full names of variable abbreviations are provided in Table 1.

In addition, the importance of each environmental variable varied in ecological niche models that performed best at predicting the transmission risk of MT-ZVL. The five most important contributing variables in the RF model were normalized difference vegetation index (17%), isothermality (13.5%), mean temperature of the coldest quarter (11.3%), mean temperature of the wettest quarter (7.4%), and the minimum temperature of the coldest month (6.8%), with cumulative contributions of 56%. The five most important contributing variables in the MaxEnt model were the minimum temperature of the coldest month (24.1%), mean diurnal range (11.6%), annual precipitation (7.2%), precipitation during the driest month (6.5%), and precipitation during the wettest quarter (6.3%), with cumulative contributions of 55.7%. The five most important contributing variables in the GBM model were the mean temperature of the coldest quarter (63.6%), isothermality (7.7%), the minimum temperature of the coldest month (4.9%), the normalized difference vegetation index (4.7%), and precipitation during the wettest quarter (3.6%), with cumulative contributions of 84.5% (Figure 4).

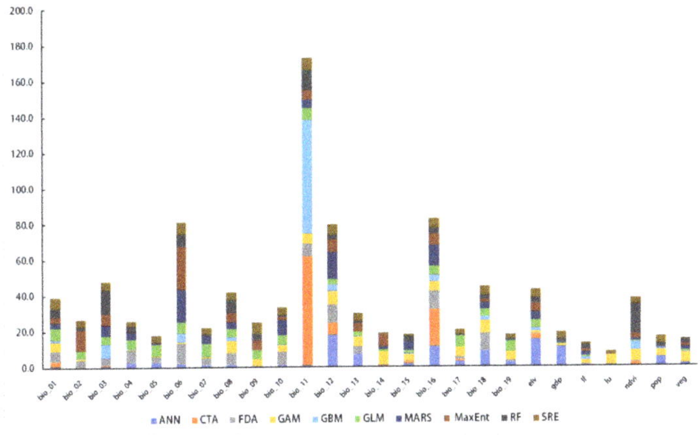

Figure 4. Contributions of environmental variables to ten ecological niche models.

3.5. Prediction of MT-ZVL Transmission Risk in China

Ensemble models predicted the risk of MT-ZVL transmission in a majority of the provinces of Shanxi and Shaanxi, southern Henan Province, the junction of Ningxia, southern Gansu, and northern Sichuan, the border areas of Hebei and Shanxi, and the hilly regions in northwestern Beijing. The areas at high risk of MT-ZVL transmission were concentrated in central and southern Shanxi, eastern Shaanxi, and northern Henan. Medium-risk areas surrounded the high-risk areas and were predominantly detected in Shanxi, Shaanxi, and Henan. Low-risk areas were mainly found on the border between southern Gansu and Sichuan, northeastern Henan, southern Hebei, and the hilly regions in northwestern Beijing. Overall, the areas at risk of MT-ZVL transmission were centered in Shanxi Province and distributed to the surrounding areas, which presented a high cluster (Figure 5).

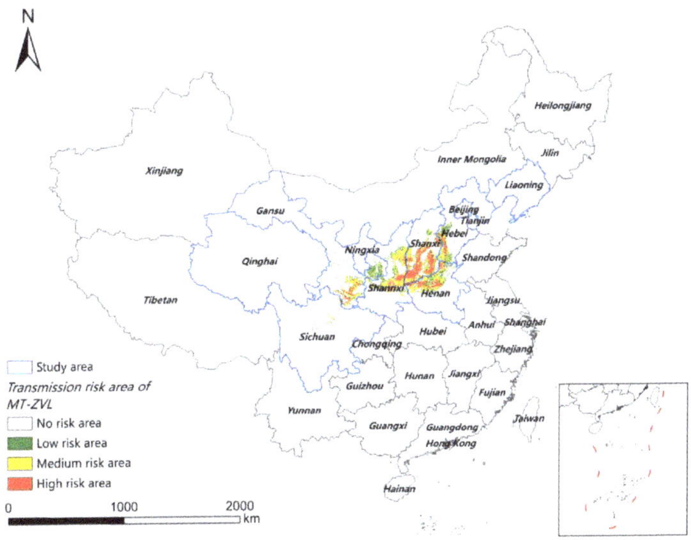

Figure 5. Areas at risk of mountain-type zoonotic visceral leishmaniasis transmission in China predicted with ensemble ecological niche models based on climatic and environmental variables.

4. Discussion

Historically, MT-ZVL was mainly prevalent in the hilly regions of Gansu, Qinghai, Ningxia, northern Sichuan, northern Shaanxi, Henan, Hebei, Beijing, and Liaoning—north of the Yangtze River in China [41]. The predominant source of infection is canines. Humans acquire infections from infected dogs, and vectors transmitting MT-ZVL are wild species of sand flies. During the early stage of the founding of the People's Republic of China, there were a large number of MT-ZVL cases, and most were detected in children and young individuals from 5 to 30 years old (74.4%), as well as in men (63%) [42]. The disease severely damaged socioeconomic development in China. In the 1950s and 1960s, a package of VL control measures was implemented including *Ph. chinensis* control, canine control, and case management, and VL was almost eradicated in eastern, central, and northern China [43,44]. Therefore, MT-ZVL was controlled effectively, but only sporadic cases were found in Gansu and Sichuan, in southern China [45,46]. With recent socioeconomic developments, the living conditions and the environment have greatly improved, and the number of dogs has increased rapidly. Due to suitable climatic conditions caused by global warming, there is a rapid expansion in the distribution and density of *P. chinensis* populations, resulting in reemerging VL or emerging natural foci of VL in Shaanxi, Shanxi, Henan, and Hebei [28]. Notably, the number of reported MT-ZVL cases has increased rapidly in Shanxi Province during the past five years [47]. In 2016, local cases were identified in Henan Province, which had not reported any cases since 1983 [48]. Since then, local cases of VL have been detected in Hebei and Beijing, posing an imminent threat to local human health [40]. Therefore, it is urgent to identify the areas at risk of MT-ZVL transmission to interrupt its transmission.

Since vectors and pathogens are greatly affected by climatic, ecological, and environmental factors [49] and show temporal and spatial clustering accordingly [50], ecological niche models show great promise at predicting the transmission risk of vector-borne diseases [51]. Ecological niche models have widely been employed to predict the transmission risk of schistosomiasis and malaria; however, there have been few reports pertaining to the application of ecological niche models to studies of VL. In this study, we collected the distributions of MT-ZVL cases in China and climatic, geographical, and socioeconomic variables that may affect VL transmission risk to create ten ecological niche models. Ensemble models were then screened to predict the areas at risk of MT-ZVL transmission in China.

All ten ecological niche models were found to be effective at predicting the areas at risk of MT-ZVL transmission, but with varying results among the models. The GAM, MaxEnt, and GBM models predicted relatively concentrated areas at risk of MT-ZVL transmission, while the ANN, MARS, and CTA models predicted widespread areas. The CTA and SRE models predicted large areas at risk of MT-ZVL transmission, and most areas were predicted to be at high or medium risk. The high-, medium-, and low-risk areas predicted by the ecological niche models were centered in Shanxi Province and tended toward dispersion to the surrounding regions, with southern Liaoning Province to the north and the southern edge of the hilly regions in northern Sichuan Province to the south. We also evaluated the performance of the ten ecological niche models: The RF, MaxEnt, GBM, and MARS models performed best (mean AUC > 0.95 and mean TSS > 0.85). As a single ecological niche model suffers from problems of spatial uncertainty, such as low accuracy and poor predictive value [35], ensemble models were screened and showed higher performance and predictive accuracy than any single ecological niche model, reducing the spatial uncertainty of predicting MT-ZVL transmission risk [52]. To investigate the model's performance, the AUC and TSS values were employed. However, AUC and TSS provide the accuracy of different aspects of models. A new system named CCHZ (C: Chen, C: Chen, H: Hu, and Z: Zhou)-DISO (distance between indices of simulation and observation) has the advantage of measuring the overall and comprehensive performance of different models [53–55]. It will be used to quantify the overall performance among different models in our future study.

Our findings showed that the areas at risk of MT-ZVL transmission predicted by our ensemble models covered the current actual distribution of MT-ZVL and extended to the surrounding regions. High- and medium-risk areas were predicted to be concentrated in central and southern Shanxi, eastern Shaanxi, southern Gansu, and northern Henan. Low-risk areas were mainly predicted in northern Sichuan, southern Hebei, and the hilly regions in northwestern Beijing. Overall, the primary high-risk areas for MT-ZVL transmission were centered in Shanxi Province and distributed to the surrounding regions, appearing as a high spatial cluster. The secondary high-risk areas were mainly distributed in southern Gansu and concentrated in several counties of Longnan City and the Gannan Tibetan Autonomous Region. The results mostly agreed with the current distribution of MT-ZVL. We predicted larger areas at risk of MT-ZVL transmission and larger high-risk areas in Shanxi and Henan provinces than the previous predictions of the nine ecological niche models. This was consistent with the current trends for MT-ZVL incidence in these two provinces [56]. In addition, our ensemble models predicted medium- and low-risk areas for MT-ZVL transmission in Ningxia, Henan, Hebei, and Beijing. To date, few cases of MT-ZVL have been reported in Henan and Hebei, and no local cases have been detected in Ningxia. Notably, local cases of MT-ZVL were identified in the suburbs of Beijing Municipality, where VL had been almost eradicated for decades. These reports demonstrate the likelihood of VL rebounding and reemerging in these provinces, implying the urgent need for the surveillance of *Ph. chinensis* vectors.

Next, we investigated the potential causes responsible for MT-ZVL rebounding and spreading. We identified climatic variables as the most important contributors to the ecological niche models, and four climatic variables (the mean temperature of the coldest quarter, precipitation during the wettest quarter, the minimum temperature of the coldest month, and annual precipitation) and two geographical variables (elevation and the normalized difference vegetation index) contributed the most to the models, emphasizing the rise of the minimum temperature and increased precipitation caused by global warming. In addition, we predicted the high-risk areas for MT-ZVL transmission to be in the Yanshan-Taihangshan mountain deciduous broad-leaved forest ecological zone and the Fenwei Basin Agro-ecological zone. With a temperate continental monsoon climate, these two ecological zones are hot and rainy in summer, with a hilly and frondent environment and a large number of wild animals. These are favorable climatic conditions for the breeding, reproduction, and habitation of wild sand flies. Furthermore, our findings showed the contribution of socioeconomic variables to ecological niche models. These variables are associated with the living environment of local residents. Local loess cave dwellings and architectural works made of cement, brick, and tile provide suitable habitats for *Ph. chinensis* breeding [57]. More importantly, the diagnosis, screening, and management of VL were neglected following the effective control of this disease in China, and this is partly responsible for the rebound of local VL epidemics [58]. In the context of a gradually increasing population density of wild species of *P. chinensis*, the introduction of animal sources of infections is very likely to cause VL transmission and reemergence. This may result in the spread of local VL to surrounding regions. With global warming, the duration and area of *Ph. chinensis* activities may further expand, which will further increase the VL transmission risk in these regions [59].

5. Conclusions

In summary, we generated ensemble ecological niche models based on environmental variables, and these ensemble models were shown to be effective at predicting the transmission risk of MT-ZVL in China. The findings by the ecological niche models agreed with the trends for MT-ZVL incidence in China and may provide technical support for MT-ZVL control and surveillance. Nevertheless, there are some limitations, such as the MT-ZVL case data were collected from the National Notifiable Diseases Reporting System, missing diagnoses or reports cannot be completely excluded, and the variable of the distribution of *Ph. chinensis* was not included, which may lead to risk underestimated. Further studies to acquire accurate distributions of MT-ZVL cases and vector species through field surveys are needed, and more quantitative variables should be included in the training datasets to increase the accuracy and effectiveness of ecological niche models for predicting the transmission risk of MT-ZVL.

Author Contributions: Conceptualization, Y.H., Z.L., Z.H., and S.L.; data curation, Y.H. and Z.L.; formal analysis, Y.H., Z.L., Y.G., Y.L., J.Z., and Z.Z. (Zhengbin Zhou); funding acquisition, Z.H. and S.L.; methodology, Y.H., Z.L., Y.G., and J.Z.; software, Z.L., Y.G., and J.Z.; writing—original draft preparation, Y.H., Z.L., and S.L.; writing—review and editing, Z.Z. (Zhengbin Zhou), Z.Z. (Zelin Zhu), T.T., Q.W., Y.Z., Z.H., and S.L. All authors have read and agreed to the published version of the manuscript.

Funding: This research was funded by the National Key Research and Development Program of China (Grant No. 2021YFC2300800, 2021YFC2300804), the National Natural Science Foundation of China (Grant No.32161143036), the National Youth Talent Program (Grant No. E1190301), the Alliance of International Science Organizations (Grant No. ANSO-CR-KP-2021-02), and the National Tutor System (Grant No. Qngg2021031).

Institutional Review Board Statement: Not applicable.

Informed Consent Statement: Not applicable.

Data Availability Statement: The data used and/or analyzed in the current study are publicly available in (the Information Center of National Institute of Parasitic Diseases, China CDC (Chinese Center for Tropical Diseases Research) at https://www.ipd.org.cn/ accessed on 8 May 2022; The China Meteorological Data Sharing Center at http://data.cma.cn accessed on 21 November 2021; The WorldClim website at https://www.worldclim.org/data/worldclim21.html/ accessed on 16 November 2021; and the Resource and Environment Data Center, Chinese Academy of Sciences at https://www.resdc.cn/ accessed on 12 November 2021.

Conflicts of Interest: The authors declare no conflict of interest.

References

1. Abdoli, A. Leishmaniasis. *Lancet* **2019**, *393*, 872. [CrossRef]
2. Alvar, J.; Yactayo, S.; Bern, C. Leishmaniasis and poverty. *Trends Parasitol.* **2006**, *22*, 552–557. [CrossRef] [PubMed]
3. Desjeux, P. Leishmaniasis: Current situation and new perspectives. *Comp. Immunol. Microbiol. Infect. Dis.* **2004**, *27*, 305–318. [CrossRef] [PubMed]
4. Al-Salem, W.; Herricks, J.R.; Hotez, P.J. A review of visceral leishmaniasis during the conflict in South Sudan and the consequences for East African countries. *Parasites Vectors* **2016**, *9*, 460. [CrossRef]
5. Lun, Z.-R.; Wu, M.-S.; Chen, Y.-F.; Wang, J.-Y.; Zhou, X.-N.; Liao, L.-F.; Chen, J.-P.; Chow, L.M.C.; Chang, K.P. Visceral Leishmaniasis in China: An endemic disease under control. *Clin. Microbiol. Rev.* **2015**, *28*, 987–1004. [CrossRef]
6. Guan, L.; Wu, Z. Historical experience in the elimination of visceral leishmaniasis in the plain region of Eastern and Central China. *Infect. Dis. Poverty* **2014**, *3*, 10. [CrossRef]
7. Qian, M.B.; Chen, J.; Bergquist, R.; Li, Z.J.; Li, S.Z.; Xiao, N.; Utzinger, J.; Zhou, X.N. Neglected tropical diseases in the People's Republic of China: Progress towards elimination. *Infect. Dis. Poverty* **2019**, *8*, 6–21. [CrossRef]
8. Guan, L.R.; Wang, J.; Hu, Y.D.; Xiong, G.H.; Zheng, S.C. Epidemiology of kala-azar in China and its significance in the prevention and treatment. *J. Control Res. Epidem. Dis.* **1976**, *3*, 225–231. (In Chinese)
9. Guan, L.R. Current status of kala-azar and vector control in China. *Bull. World Health Organ.* **1991**, *69*, 595–601.
10. Li, Y.F.; Zhong, W.X.; Zhao, G.H.; Wang, H.F. Prevalence and control status of kala-azar in China. *J. Pathog. Biol.* **2011**, *6*, 629–631. (In Chinese)
11. Jiang, D.; Ma, T.; Hao, M.; Qian, Y.; Chen, S.; Meng, Z.; Wang, L.; Zheng, C.; Qi, X.; Wang, Q.; et al. Spatiotemporal patterns and spatial risk factors for visceral leishmaniasis from 2007 to 2017 in Western and Central China: A modelling analysis. *Sci. Total Environ.* **2021**, *764*, 144275. [CrossRef] [PubMed]
12. Zhao, S.; Li, Z.; Zhou, S.; Zheng, C.; Ma, H. Epidemiological Feature of Visceral Leishmaniasis in China, 2004–2012. *Iran J. Public Health* **2015**, *44*, 51–59. [PubMed]
13. Han, S.; Wu, W.; Xue, C.; Ding, W.; Hao, Y.; Hou, Y.; Feng, Y.; Zhong, B.; Cao, L.; Zhang, Y.; et al. Endemic status of visceral leishmaniasis in China from 2004 to 2016. *Chin. J. Parasitol. Parasit. Dis.* **2019**, *37*, 189–195. (In Chinese)
14. Zhao, Y.; Jiang, D.; Ding, F.; Hao, M.; Wang, Q.; Chen, S.; Xie, X.; Zheng, C.; Ma, T. Recurrence and Driving Factors of Visceral Leishmaniasis in Central China. *Int. J. Environ. Res. Public Health* **2021**, *18*, 9535. [CrossRef] [PubMed]
15. Zhou, Z.; Lyu, S.; Zhang, Y.; Li, Y.; Li, S.; Zhou, X.N. Visceral Leishmaniasis-China, 2015–2019. *China CDC Wkly.* **2020**, *2*, 625–628. [CrossRef] [PubMed]

16. Wang, X.; Xia, S.; Xue, J.; Zhou, Z.; Li, Y.; Zhu, Z.; Zhang, Y.; Wang, Q.; Li, S. Transmission risks of mountain-type zoonotic visceral leishmaniasis-six endemic provincial-level administrative divisions, China, 2015–2020. *China CDC Wkly.* **2022**, *4*, 148–152. [CrossRef]
17. Chavy, A.; Nava, A.F.D.; Luz, S.L.B.; Ramírez, J.D.; Herrera, G.; dos Santos, T.V.; Ginouves, M.; Demar, M.; Prévot, G.; Guégan, J.-F.; et al. Ecological niche modelling for predicting the risk of cutaneous leishmaniasis in the Neotropical moist forest biome. *PLOS Negl. Trop. Dis.* **2019**, *13*, e0007629. [CrossRef]
18. Acharya, B.K.; Cao, C.; Xu, M.; Khanal, L.; Naeem, S.; Pandit, S. Present and Future of Dengue Fever in Nepal: Mapping Climatic Suitability by Ecological Niche Model. *Int. J. Environ. Res. Public Health* **2018**, *15*, 187. [CrossRef]
19. Leta, S.; Fetene, E.; Mulatu, T.; Amenu, K.; Jaleta, M.B.; Beyene, T.J.; Negussie, H.; Revie, C.W. Modeling the global distribution of Culicoides imicola: An ensemble approach. *Sci. Rep.* **2019**, *9*, 14187. [CrossRef]
20. Zhu, G.; Fan, J.; Peterson, A.T. Schistosoma japonicum transmission risk maps at present and under climate change in mainland China. *PLoS Negl. Trop. Dis.* **2017**, *11*, e0006021. [CrossRef]
21. Marques, R.; Krüger, R.F.; Cunha, S.K.; Silveira, A.S.; Alves, D.M.; Rodrigues, G.D.; Peterson, A.T.; Jiménez-García, D. Climate change impacts on Anopheles (K.) cruzii in urban areas of Atlantic Forest of Brazil: Challenges for malaria diseases. *Acta Trop.* **2021**, *224*, 106123. [CrossRef]
22. Abdullah, A.Y.M.; Dewan, A.; Shogib, M.R.I.; Rahman, M.M.; Hossain, M.F. Environmental factors associated with the distribution of visceral leishmaniasis in endemic areas of Bangladesh: Modeling the ecological niche. *Trop. Med. Health* **2017**, *45*, 13. [CrossRef] [PubMed]
23. Chalghaf, B.; Chemkhi, J.; Mayala, B.; Harrabi, M.; Benie, G.B.; Michael, E.; Ben Salah, A. Ecological niche modeling predicting the potential distribution of Leishmania vectors in the Mediterranean basin: Impact of climate change. *Parasit Vectors* **2018**, *11*, 461. [CrossRef] [PubMed]
24. Gong, Y.-F.; Zhu, L.-Q.; Li, Y.-L.; Zhang, L.-J.; Xue, J.-B.; Xia, S.; Lv, S.; Xu, J.; Li, S.-Z. Identification of the high-risk area for schistosomiasis transmission in China based on information value and machine learning: A newly data-driven modeling attempt. *Infect. Dis. Poverty* **2021**, *10*, 88. [CrossRef] [PubMed]
25. Hu, X.; Hao, Y.; Xia, S.; Guo, Y.; Xue, J.; Zhang, Y.; Wang, L.; Dong, Y.; Xu, J.; Li, S. Detection of schistosomiasis transmission risks in Yunnan Province based on ecological niche modeling. *Chin. J. Parasitol. Parasit. Dis.* **2020**, *38*, 80–87. (In Chinese)
26. Gong, Y.; Hu, X.; Zhou, Z.; Zhu, H.; Hao, Y.; Wang, Q.; Zhang, Y.; Li, S. Ecological niche modeling-based prediction on transmission risk of visceral leishmaniasis in the extension region of Loess Plateau, China. *Chin. J. Parasitol. Parasit. Dis.* **2021**, *39*, 218–225. (In Chinese)
27. Guan, L.; Zhou, Z.; Jin, C.; Fu, Q.; Chai, J. Phlebotomine sand flies (Diptera: Psychodidae) transmitting visceral leishmaniasis and their geographical distribution in China: A review. *Infect. Dis. Poverty* **2016**, *5*, 15. [CrossRef]
28. Wang, X.; Xue, J.; Xia, S.; Han, S.; Hu, X.; Zhou, Z.; Zhang, Y.; Li, S. Distribution of Suitable Environments for Phlebotomus chinensis as the Vector for Mountain-Type Zoonotic Visceral Leishmaniasis-Six Provinces, China. *China CDC Wkly.* **2020**, *2*, 815–819. [CrossRef]
29. Wu, Y.; Li, G.; Liao, Q.; Chen, Y.; Gao, B. Epidemiology and control of kala-azar in Sichuan Province. *End Dis. Bull* **1995**, *10*, 25–29. (In Chinese)
30. Sofizadeh, A.; Rassi, Y.; Vatandoost, H.; Hanafi-Bojd, A.A.; Mollalo, A.; Rafizadeh, S.; Akhavan, A.A. Predicting the Distribution of Phlebotomus papatasi (Diptera: Psychodidae), the Primary Vector of Zoonotic Cutaneous Leishmaniasis, in Golestan Province of Iran Using Ecological Niche Modeling: Comparison of MaxEnt and GARP Models. *J. Med. Entomol.* **2017**, *54*, 312–320. [CrossRef]
31. Andrade-Filho, J.D.; Scholte, R.G.C.; Amaral, A.L.G.; Shimabukuro, P.H.F.; Carvalho, O.S.; Caldeira, R.L. Occurrence and Probability Maps of Lutzomyia longipalpis and Lutzomyia cruzi (Diptera: Psychodidae: Phlebotominae) in Brazil. *J. Med. Entomol.* **2017**, *54*, 1430–1434. [CrossRef] [PubMed]
32. Diniz-Filho, J.A.F.; Bini, L.M.; Rangel, T.F.; Loyola, R.D.; Hof, C.; Nogue´s-Bravo, D.; Araújo, M.B. Partitioning and mapping uncertainties in ensembles of forecasts of species under climate change. *Ecography* **2009**, *32*, 897–906. [CrossRef]
33. Buisson, L.; Thuiller, W.; Casajus, N.; Lek, S.; Grenouillette, G. Uncertainty in ensemble forecasting of species distribution. *Glob. Chang. Biol.* **2009**, *16*, 1145–1157. [CrossRef]
34. Carvalho, B.M.; Rangel, E.F.; Vale, M.M. Evaluation of the impacts of climate change on disease vectors through ecological niche modelling. *Bull Entomol. Res.* **2017**, *107*, 419–430. [CrossRef] [PubMed]
35. Qiao, H.; Sobero´n, J.; Peterson, A.T. No silver bullets in correlative ecological niche modelling: Insights from testing among many potential algorithms for niche estimation. *Methods Ecol. Evol.* **2015**, *6*, 1126–1136. [CrossRef]
36. Heikkinen, R.K.; Marmion, M.; Luoto, M. Does the interpolation accuracy of species distribution models come at the expense of transferability? *Ecography* **2012**, *35*, 276–288. [CrossRef]
37. Allouche, O.; Tsoar, A.; Kadmon, R. Assessing the accuracy of species distribution models: Prevalence, kappa and the true skill statistic (TSS). *J. Appl. Ecol.* **2006**, *43*, 1223–1232. [CrossRef]
38. Zhou, Z.; Li, Y.; Zhang, Y.; Li, S. Prevalence of visceral leishmaniasis in China during 2015–2018. *Chin. J. Parasitol. Parasit. Dis.* **2020**, *38*, 339–345. (In Chinese)
39. Zhou, Z.; Li, Y.; Zhang, Y.; Li, S. Prevalence of visceral leishmaniasis in China in 2019. *Chin. J. Parasitol. Parasit. Dis.* **2020**, *38*, 602–607. (In Chinese)

40. Luo, Z.; Zhou, Z.; Gong, Y.; Feng, J.; Li, Y.; Zhang, Y.; Li, S. Current status and challenges of visceral leishmaniasis in China. *Chin. J. Parasitol. Parasit. Dis.* **2022**, *40*, 146–152. (In Chinese)
41. Zhou, Z.-B.; Wang, J.-Y.; Gao, C.-H.; Han, S.; Li, Y.-Y.; Zhang, Y.; Zhou, X.-N. Contributions of the National Institute of Parasitic Diseases to the control of visceral leishmaniasis in China. *Adv. Parasitol.* **2020**, *110*, 185–216. [CrossRef] [PubMed]
42. Xiong, G. Current situation of kala-azar epidemic in China. *End Dis. Bull.* **1972**, *7*, 113–125. (In Chinese)
43. Wang, Z.; Xiong, G.H.; Guan, L. Achievement on the epidemiology and control of kala-azar in China. *Chin. J. Epidemiol.* **2000**, *21*, 51–54. (In Chinese)
44. Guan, L. Present situation of visceral leishmaniasis and prospect for its control in China. *Chin. J Parasitol. Parasit. Dis.* **2009**, *27*, 394–397. (In Chinese)
45. Guan, L.; Shen, W. Recent advances in visceral leishmaniasis in China. *Southeast Asian J. Trop. Med. Public Health* **1991**, *22*, 291–298. [PubMed]
46. Guan, Z.; Chen, C.; Huang, C.; Zhang, H.; Zhou, Y.; Zhou, Y.; Wu, J.; Zhou, Z.; Yang, S.; Li, L. Epidemiological features and spatial-temporal distribution of visceral leishmaniasis in mainland China: A population-based surveillance study from 2004 to 2019. *Parasit Vectors* **2021**, *14*, 517. [CrossRef]
47. Zheng, Y.; Bai, Y.; Tie, P.; Yan, C.; Wang, T.; Wang, J.; Wang, S.; Cai, J. Epidemiological characteristics of visceral leishmaniasis in Shanxi province, 2010—2019. *Chin. J. Parasitol. Parasit Dis.* **2021**, *39*, 352–358. (In Chinese)
48. Tang, Z.; Gao, P.; Zhou, K.; Liu, J.; Guo, X. A survey on sandfly in Linzhou city of Henan province. *Chin. J. Hyg Insect. Equip.* **2017**, *23*, 397–398. (In Chinese)
49. Hong, A.; Zampieri, R.A.; Shaw, J.J.; Floeter-Winter, L.M.; Laranjeira-Silva, M.F. One Health Approach to Leishmaniases: Understanding the Disease Dynamics through Diagnostic Tools. *Pathogens* **2020**, *9*, 809. [CrossRef]
50. Firouraghi, N.; Mohammadi, A.; Hamer, D.H.; Bergquist, R.; Mostafavi, S.M.; Shamsoddini, A.; Raouf-Rahmati, A.; Fakhar, M.; Moghaddas, E.; Kiani, B. Spatio-temporal visualisation of cutaneous leishmaniasis in an endemic, urban area in Iran. *Acta Trop.* **2022**, *225*, 106181. [CrossRef] [PubMed]
51. Ocampo, C.B.; Guzmán-Rodríguez, L.; Moreno, M.; Castro, M.D.M.; Valderrama-Ardila, C.; Alexander, N. Integration of phlebotomine ecological niche modelling, and mapping of cutaneous leishmaniasis surveillance data, to identify areas at risk of under-estimation. *Acta Trop.* **2021**, *224*, 106122. [CrossRef] [PubMed]
52. Townsend, J. Mapping disease transmission risk: Enriching models using biogeography and ecology. *Emerg. Infect. Dis.* **2015**, *21*, 1489. [CrossRef]
53. Hu, Z.; Chen, D.; Chen, X.; Zhou, Q.; Peng, Y.; Li, J. CCHZ-DISO: A Timely New Assessment System for data quality or model performance from Da Dao Zhi Jian. *Nature* **2022**.
54. Zhou, Q.; Chen, D.; Hu, Z.; Chen, X. Decompositions of Taylor diagram and DISO performance criteria. *Int. J. Climatol.* **2021**, *41*, 5726–5732. [CrossRef]
55. Hu, Z.; Chen, X.; Zhou, Q.; Chen, D.; Li, J. DISO: A rethink of Taylor diagram. *Int. J. Climatol.* **2019**, *39*, 2825–2832. [CrossRef]
56. Hao, Y.; Hu, X.; Gong, Y.; Xue, J.; Zhou, Z.; Li, Y.; Wang, Q.; Zhang, Y.; Li, S. Spatio-temporal clustering of Mountain-type Zoonotic Visceral Leishmaniasis in China between 2015 and 2019. *PLoS Negl. Trop. Dis.* **2021**, *15*, e0009152. [CrossRef]
57. Chen, H.; Gao, J.; Li, K.; Yang, Z.; Peng, H.; Ma, Y. Ecological niches of sandfly (Diptera: Psychodidae) in the extension region of Loess Plateau, China: An endemic focus of visceral leishmaniasis. *Chin. J. Vector Biol. Control* **2019**, *30*, 597–602. (In Chinese)
58. Zheng, C.; Xue, C.; Wu, W.; Zhou, X. Epidemiological characteristics of Kala-azar disease in China, during 2005–2015. *Chin. J. Epidemiol.* **2017**, *38*, 431–434. (In Chinese)
59. Cross, E.R.; Hyams, K.C. The potential effect of global warming on the geographic and seasonal distribution of Phlebotomus papatasi in southwest Asia. *Environ. Health Perspect.* **1996**, *104*, 724–727. [CrossRef]

atmosphere

MDPI

Article

Analysis of Correlation between Quality of Life and Subjective Evaluation of Air Quality—Empirical Research Based on CHARLS 2018 Data

Yuanfang Du [1,2,*], Shibing You [1], Mengyu Zhang [1], Ze Song [1], Weisheng Liu [3] and Dongju Li [4]

1 School of Economics and Management, Wuhan University, Wuhan 430072, China; 00001839@whu.edu.cn (S.Y.); 2019201050157@whu.edu.cn (M.Z.); songze@whu.edu.cn (Z.S.)
2 Mathematical Department, Tibet University, Lhasa 850000, China
3 School of Economics, Jiangxi University of Finance and Economics, Nanchang 330013, China; liu.wilson.edu@gmail.com
4 School of Statistics and Big Data, Henan University of Economics and Law, Zhengzhou 450046, China; 20100452@huel.edu.cn
* Correspondence: ruyier521@whu.edu.cn

Abstract: This paper mainly focuses on the relationship between the subjective evaluation of air quality and the quality of life (QOL) of middle-aged and elderly residents in China. The 2018 China Health and Retirement Longitudinal Study (CHARLS) project database is the key sources of data, from which 16,736 valid samples were used in our research. Multivariate linear regression analysis and binomial logistic regression model were applied to detect the impact of the subjective evaluation of air quality on QOL, which was evaluated in two dimensions, which are health utility and experienced utility, using the health utility EQ-5D score and the experienced utility of life satisfaction score. Our results show that there is a significant positive correlation between the subjective evaluation of air quality and the two dimensions of QOL. Age, education, marital status and sleep status also have a relatively great impact on the QOL of residents. This worked studied the overall QOL of middle-aged and elderly residents in China, while policy suggestions regarding high-quality air public goods are also given in the paper.

Keywords: air quality satisfaction; quality of life; binomial logistic regression; health utility value; experienced utility

Citation: Du, Y.; You, S.; Zhang, M.; Song, Z.; Liu, W.; Li, D. Analysis of Correlation between Quality of Life and Subjective Evaluation of Air Quality—Empirical Research Based on CHARLS 2018 Data. *Atmosphere* **2021**, *12*, 1551. https://doi.org/10.3390/atmos12121551

Academic Editors: Zengyun Hu, Xuguang Tang and Qinchuan Xin

Received: 14 October 2021
Accepted: 22 November 2021
Published: 24 November 2021

1. Introduction

Quality of life (QOL) is a multidimensional concept that not only perceives and evaluates people's physical, psychological, social belonging and comprehensive conditions, but also involves people's living environment. Compared with the health-related QOL in the medical field, the research on QOL in the field of social science is more extensive in content, focusing on other non-medical indicators reflecting QOL, such as education, employment, income, social security, living environment, etc. Therefore, the significance of QOL research goes far beyond health itself, which largely reflects the collection of the impact of macro-social factors on individual life quality or the QOL.

In recent years, environmental problems have become increasingly serious, and environmental pollution has greatly threatened human physical and mental health, life and work. People pay much more attention to public goods, such as water and air. The relevant literature shows that there is consensus at home and abroad that air quality plays an important role in measuring people's QOL. There are a number of studies that targeted scientifically measured air quality and its impact on QOL. It has been argued that there is a significant positive correlation between air quality and QOL [1,2]. Liao Li et al. [3] have pointed out, that the objective measurement of air quality indirectly affects residents' life satisfaction. Air quality also has varying degrees of impact on human (physical) health.

Harold J. Rickenbacker et al. revealed a significant relationship between indoor particulate matter (PM) and individual dimensions of QOL [4]; air pollution significantly reduces the life satisfaction of Chilean residents [5]; and the increase in PM 2.5 concentration may reduce the average life expectancy [6].

From the perspective of research content, this is mainly reflected in the research related to the objective indicators of air quality. There are few studies on the relationship between the subjective assessment of air quality and the QOL of residents. However, as Liao, X. et al. studied the influencing factors of respondents' perception of air quality, and found that relevant indicators of air quality, such as $PM_{2.5}$, PM_{10}, SO_2 and NO_2 concentration, would have a negative impact on respondents' perception of air quality [7]. Shi, X. et al. found that there is a high correlation between the objective air quality index and subjective air quality perception [8]. Therefore, the air quality status of the place of residence directly affects their subjective evaluation of the air quality. Thus, in the selection of air quality evaluation indicators, compared with previous studies, which mainly used the objective indicators published by the meteorological department, this study focuses on residents' subjective perception of air quality.

Since previous studies on QOL rarely involved the subjective evaluation of air quality, it is of great practical significance to explore the relationship between the subjective evaluation of air quality and residents' QOL. Therefore, this study attempts to provide some meaningful supplements and discussions in this field.

This study creatively puts forward the "two-dimensional" research perspective of QOL, which divides the QOL into two different dimensions—the health utility of the QOL and the experienced utility of the QOL—and performed beneficial exploration and research into these two dimensions to investigate the correlation between the subjective evaluation of air quality and the utility value of QOL. The correlation between the EuroQol five-dimensional questionnaire (EQ-5D) [9] score of health utility of QOL and the subjective evaluation and individual characteristics of air quality was detected using the multi factor linear regression analysis model. The impact of air quality on life satisfaction according to the experienced utility is analyzed using logistic regression analysis. The two models produced consistent results regarding a significant positive relationship between air quality satisfaction and QOL. Additionally, other explanatory variables and their related variables are significantly correlated with QOL.

2. Data Description and Variable Selection

2.1. Data Source

This article was based on the China Health and Retirement Longitudinal Study (CHARLS) 2018 data [10], which cover 459 village-level units within 150 county-level units in 28 provinces and municipalities (Tibet, Ningxia, and Hainan are excluded) in Mainland China. Using python (version 3.8.8 Wilmington, DE, USA) and Jupyter notebook software (version 5.7.4 New York, NY, USA) to clean the missing values and outliers of the sample, 16,736 middle-aged and elderly people aged or above 45 years old were finally included as samples.

2.2. Descriptive Statistics of Data

Through the literature review, it was found that the QOL of residents is affected by many factors, such as individual characteristics, personal life perception, income level, daily behavior patterns and so on. This paper took the utility score of residents' QOL as the explained variable. The explanatory variables included air quality satisfaction, health satisfaction, marriage satisfaction, children satisfaction, age, sex, residence, marital status, drinking, smoking, sleeping status, education background and yearly individual income.

In this paper, the resident health utility score was obtained from EQ-5D to measure the health-related quality of life of middle-aged and elderly Chinese residents. From the descriptive statistics (Table 1) of the full sample, it was found that the sociodemographic characteristics of the residents are quite different. The mean value of health utility score of residents' QOL (EQ-5D) was 0.7417 ± 0.2262, the minimum value was −0.149 and the maximum value was 1. Health state index scores generally ranged from less than 0 (where 0 is a health state equivalent to death; negative values are valued as worse than death) to 1 (perfect health), with higher scores indicating higher health utility, though health state preferences can differ between countries [9]. The standard deviation of health utility score shows that the overall fluctuation of health utility level of residents' QOL was relatively small. The mean value of life satisfaction of residents' experienced utility of life quality was 2.7519 ± 0.7963. The standard deviation of experienced utility score shows that the overall level of residents' life satisfaction had little fluctuation and 89% residents were satisfied with life. This shows that the overall QOL of the middle-aged and elderly Chinese residents interviewed was relatively good. The mean value of respondents' satisfaction with air quality was 2.8405 ± 0.8309. Overall, 799 residents were extremely satisfied with the air quality of the year, 4404 respondents were very satisfied, and 8761 residents were relatively satisfied with air quality—that is, the proportion of air quality satisfaction was 83%.

Table 1. Descriptive statistics section.

Type	Variable	Min	Max	Mean	Standard Deviation
Y1	Utility Value [a]	−0.149	1	0.74	0.23
Y2	Life Satisfaction [b]	1	5	2.75	0.80
X1	Air Quality Satisfaction [b]	1	5	2.84	0.83
X2	Sex	0	1	0.48	0.50
X3	Age (years)	45	108	61.64	9.43
X4	Residence Areas	0	1	0.74	0.44
X5	Education Background [c]	1	11	3.52	1.91
X6	Marital Status	1	5	1.40	1.01
X7	Smoking Status	0	1	0.04	0.19
X8	Drinking	0	1	0.35	0.48
X9	Sleeping Status (h)	0	15	6.20	1.94
X10	Individual yearly Income ($)	0	86,870.78	2420.70	1705.87
X11	Health Satisfaction [b]	1	5	3.06	0.92
X12	Marital Satisfaction [b]	1	5	2.93	1.29
X13	Children Satisfaction [b]	1	5	2.42	0.81

[a] Scale range = −0.149 (worse than death)–1 (perfect health), where 0 is a health state equivalent to death. [b] Scale range = 1 (completely satisfied)–5 (not at all satisfied.), where 1 = completely satisfied, 2 = very satisfied, 3 = somewhat satisfied, 4 = not very satisfied, 5 = not at all satisfied. [c] Scale range = 1 (no formal education (illiterate))–11 (11 = doctoral degree/Ph.D.), where 2 = did not finish primary school, 3 = sishu/home school, 4 = elementary school, 5 = middle school, 6 = high school, 7 = vocational school, 8 = two-/three-year college/associate degree, 9 = four-year college/Bachelor's degree, 10 = Master's degree.

According to China's legal retirement age, 7925 respondents were between 45 and 60 years old, indicating that among the 16,736 samples, nearly half were middle-aged and elderly people who were on-the-job or capable of working. From the perspective of the gender of respondents, the proportion of males (52%) and females (48%) was relatively balanced. There was a big difference in data distribution between urban and rural areas, with 12,529 respondents living in rural areas. It may be that the CHARLS questionnaire collection is more focused on rural areas; in terms of the annual income level of the interviewees, there were 4976 interviewees with an annual income of USD 0 and 7145 individuals with an annual income of less than USD 144.64, while the highest annual income was 86,870.78$. The sample mean was 16,736, and the standard deviation was 11,793, indicating that there was a large gap in the income level of the interviewees. The average educational background of respondents was 0.3546 ± 0.4784, and the number of respondents with junior middle school education or below was 14,580 (87%), indicating that the overall educational level of middle-aged and elderly groups in the surveyed sample was not high. The proportion of abnormal sleep (less than 5 h or more than 9 h) accounted for 27% of those interviewed, indicating that nearly one-third of the middle-aged and elderly residents surveyed had sleep problems. The number of smokers and drinkers were 5818 and 654, respectively, indicating that the vast majority of respondents had a relatively healthy daily lifestyle. In other subjective perceptions, the mean value of health satisfaction was 3.0572 ± 0.9234, the mean value of marriage satisfaction was 2.9259 ± 1.2855, and the mean value of children satisfaction was 2.4155 ± 0.8060, indicating that middle-aged and elderly residents in China generally had high levels of satisfaction with their health, marriage and children.

2.3. *Variable Selection*

2.3.1. Explained Variables

In this paper, the QOL health utility score EQ-5D (Y1) and experienced utility score for residents' life satisfaction (Y2) were taken as the explained variables, respectively, and the multivariate linear regression model and binary logistic regression model were established, respectively.

The resident health utility score was obtained from the EuroQol Group's three-level EuroQol five-dimensional questionnaire (EQ-5D−3L) to measure the health-related quality of life of middle-aged and elderly Chinese residents. The EQ-5D−3L descriptive system comprises the following five dimensions, each describing a different aspect of health: mobility (MO), self-care (SC), usual activities (UA), pain/discomfort (PD), and anxiety/depression (AD). Each dimension is divided into three levels: no problems, some problems, extreme problems (labelled 1–3). By convention, the EQ-5D−5L health states are presented in a short form using five-digit numbers in which the digits represent the levels of functioning for the dimensions in order of presentation (MO, SC, UA, PD, and AD). For example, state 11,223 indicates no problems with mobility and self-care, some problems with performing usual activities, moderate pain or discomfort and extreme anxiety or depression, while state 11,111 indicates no problems regarding any of the five dimensions [9]. The health utility score (EQ-5D) of this paper is selected from CHARLS questionnaire, using "DB006: Do you have difficulty with stooping, kneeling, or crouching?" in the consideration of the mobility (MO) of the elderly; using "DB017: Because of health and memory problems, do you have any difficulties with preparing hot meals?" in the consideration of the self-care ability (SC) of the elderly; using "DB016: Because of health and memory problems, do you have any difficulties with doing house-hold chores?" in the consideration of the usual activities level (UA) of the elderly; using "DA041: Are you often troubled with body pains?" in the pain/discomfort (PD) evaluation of the elderly; and using "DC011: The degree of feeling depressed" in the consideration of anxiety/depression (AD) in the elderly [11].

In this study, the EQ-5D health utility value was calculated by using the health-related QOL utility value integration system of Chinese residents constructed by Liu et al. 2014 [12]. The calculation formula is:

Health utility value = 1 (a constant term)—the standard coefficient corresponding to different levels of each dimension—N3 (an additional term which should be subtracted if extreme difficulty occurs in any dimension), i.e.,

$$\begin{aligned} U = 1 - (0.039 + 0.099|_{MO2} + 0.105|_{SC2} + 0.074|_{UA2} + 0.092|_{PD2} + 0.086|_{AD2} + 0.246|_{MO3} \\ + 0.208|_{SC3} + 0.193|_{UA3} + 0.236|_{PD3} + 0.205|_{AD3}) - 0.022|_{N3} \end{aligned} \quad (1)$$

Here, MO2, SC2, UA2, PD2 and AD2 indicate that mobility, self-care ability, daily activity ability, pain/discomfort and anxiety/depression are at level 2. MO3, SC3, UA3, PD3 and AD3 indicate that the above dimensions are at level 3. N3 means that at least one of the five dimensions is at level 3. For instance, the value for "33233" was 1 − (0.039 + 0.246 + 0.208 + 0.074 + 0.236 + 0.205) − 0.022= − 0.03.

The QOL experienced utility score for life satisfaction is a sequential classification variable, with the value ranging from 1 to 5 according to the degree of satisfaction, where 1 = completely satisfied, 2 = somewhat satisfied, 4 = not very satisfied, and 5 = not at all satisfied. Among the 16,736 surveyed residents, the proportion of extremely satisfied, very satisfied and relatively satisfied was 88.69%, indicating that the overall life satisfaction of Chinese residents is relatively high. Since this study adopted the discrete dependent variable binary logic model, the explained variables were represented by 0 and 1. During modeling, we merged "extremely satisfied, very satisfied, relatively satisfied" into 1, and "not very satisfied, not satisfied at all" into 0.

2.3.2. Core Explanatory Variables

This paper took residents' subjective evaluation of air quality satisfaction (X1) as the core variable that affects residents' QOL in 2018. The subjective evaluation question regarding air quality was expressed as "How satisfied are you with the air quality this year (2018)?" The value for this was 1–5, where 1 = completely satisfied, 2 = very satisfied, 3 = somewhat satisfied, 4 = not very satisfied, 5 = not at all satisfied. The discrete distribution of air quality satisfaction, QOL health utility EQ-5D and experienced utility for life satisfaction is shown in Figure 1. For samples with a standard normal distribution, only a few values were outliers. There are only six outliers in the right figure of Figure 1, indicating that the distribution has almost no tail and a large degree of freedom. The outliers in the left figure are concentrated on the side with the lower health utility value, indicating that the distribution of air quality satisfaction and the health utility value shows a slight leftward skew.

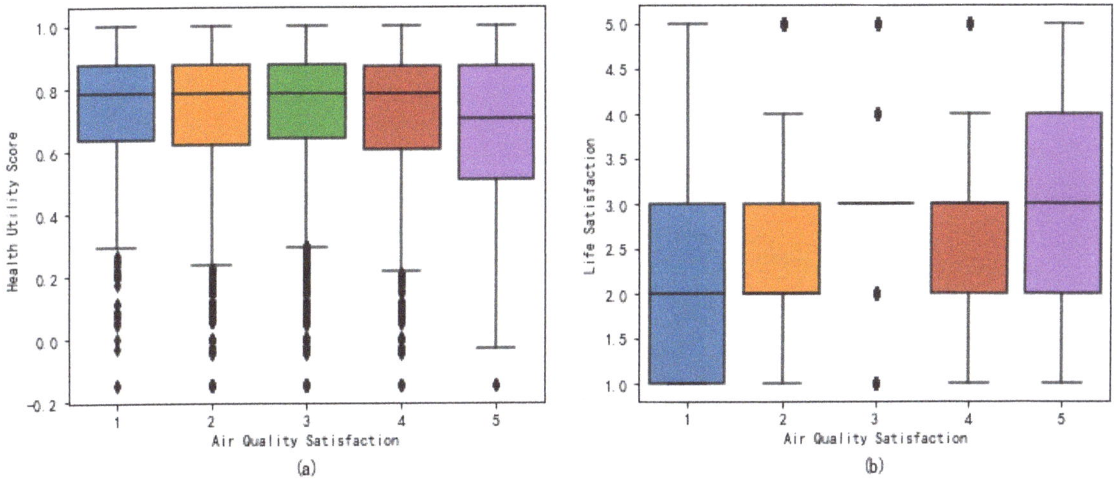

Figure 1. Distribution of air quality satisfaction and QOL utility data. (**a**) The comprehensive distribution characteristics of air quality satisfaction and Health Utility Score (EQ-5D). (**b**) The comprehensive distribution characteristics of air quality satisfaction and Life Satisfaction. (The five colors represent air quality satisfaction, respectively: completely satisfied, very satisfied, somewhat satisfied, not very satisfied, not at all satisfied. Black dots present the outlier values).

2.3.3. Control Variables

In order to eliminate the estimation error caused by missing variables as much as possible, other individual characteristics that may affect the health utility of residents' QOL were introduced as control variables—sex, residence areas, education background, marital status, smoking status, drinking, sleeping status and individual yearly income—to control the impact of personal characteristics and lifestyle on residents' QOL regarding the health utility. When taking the QOL regarding the experienced utility for life satisfaction as the explanatory variable for logical progression, in addition to the above control variables, three satisfaction indicators of health satisfaction, marital satisfaction and children satisfaction were added as supplementary control variables to control the impact of residents' subjective evaluation indicators on residents' life satisfaction. The value was 1–5, where 1 = completely satisfied, 2 = very satisfied, 3 = somewhat satisfied, 4 = not very satisfied, 5 = Not at all satisfied. The specific assignment of variables is shown in Table 2.

Table 2. Dummy variables assignment.

Variables	Assignment and Meaning
Sex (X2)	Male = 0; Female = 1
Residence Areas (X4)	Other = 0; Village = 1
Education Background (X5)	Primary school or below = 0; Secondary school or above = 1
Marital Status (X6)	Widowed/Divorced/Separated/Unmarried = 0; Married = 1
Smoking Status (X7)	Non-smokers = 0; Smoking history = 1
Drinking (X8)	No drinking = 0; Drinking = 1
Sleeping Status (h) (X9)	Abnormal (<5 h or >9 h) = 0; Normal ([5 h,9 h]) = 1
Health Satisfaction (X11)	1 = Completely satisfied, 2 = Very satisfied, 3 = Somewhat satisfied, 4 = Not very satisfied, 5 = Not at all satisfied
Marital Satisfaction (X12)	1 = Completely satisfied, 2 = Very satisfied, 3 = Somewhat satisfied, 4 = Not very satisfied, 5 = Not at all satisfied
Children Satisfaction (X13)	1 = Completely satisfied, 2 = Very satisfied, 3 = Somewhat satisfied, 4 = Not very satisfied, 5 = Not at all satisfied

For education background, the score spanned from 1 to 11, where 1 = no formal education (illiterate), 2 = did not finish primary school, 3 = sishu/home school, 4 = elementary school, 5 = middle school, 6 = high school, 7 = vocational school, 8 = two-/three-year college/associate degree, 9 = four-year college/Bachelor's degree, 10 = Master's degree, 11 = doctoral degree/Ph.D. In Table 2, education background (X5) ranges from 0 to 1. As a control variable, education background did not significantly affect the size of the regression coefficient of the explained variables in the model, which made the problem description more concise. Therefore, it was used as a dummy variable reflecting education level for modeling.

The frequency chart of the subjective perception of middle-aged and elderly Chinese residents is shown in Figure 2, from which it can be seen that the subjective perception of residents is generally normally distributed.

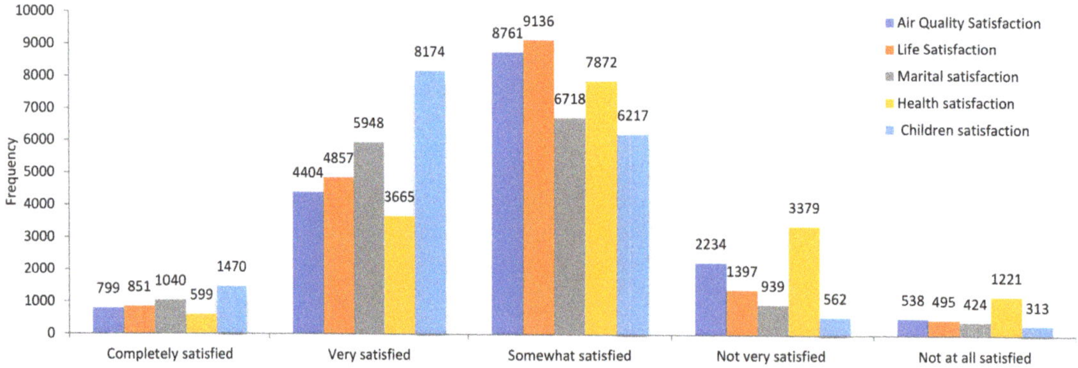

Figure 2. Satisfaction of various indicators of middle-aged and elderly residents in China.

Considering the actual significance of the data, the variables age (years) and yearly individual income used the original data to participate in the regression.

3. Models and Methods

3.1. T Test and Pearson Correlation Analysis

In order to ensure the accuracy of the sample and the correctness of the model, the *t*-test and the Pearson correlation coefficient test were carried out on the influencing factors of QOL. The factors that were significant at the significance level of 5% in both tests were included in the regression model.

3.2. Multicollinearity Test

In order to avoid the possible multiple collinearities between explanatory variables, this paper used the Variance Inflation Factor (VIF) method to test the possible collinearity between variables before establishing the model. This method mainly judges whether there is multicollinearity between variables through the size of variance inflation factor. We took the explanatory variable Xi (i = 1, 2, . . . , 13) as the dependent variable, and the other explanatory variables other than Xi as the independent variables and established a linear regression model to obtain the decisive factor. The calculation formula of variance inflation factor is as follows:

$$\mathrm{VIF} = \frac{1}{1 - \mathrm{R}_{\mathrm{i}}^{2}} \tag{2}$$

where R^2 is the determination coefficient of regression to other explanatory variables when the explanatory variable Xi (i = 1, 2, . . . , 13) is the dependent variable.

The multicollinearity test showed that the value of variance expansion factor VIF between explanatory variables was between 1.02 and 1.45, indicating that there was no multicollinearity between explanatory variables.

3.3. Multivariate Linear Regression Model

In order to test the impact of air quality satisfaction on the EQ-5D score of the health utility of residents' QOL, based on the existing research experienced and available data, a multi factor linear regression model was set:

$$\mathrm{QOL}_{\mathrm{i}} = u_i + \alpha \mathrm{AQS}_{\mathrm{i}} + \beta Control_{\mathrm{i}} + \varepsilon_{\mathrm{i}} \tag{3}$$

Here, the core explanatory variable is air quality satisfaction (AQS), and the control variables include variables related to QOL: age, residence areas, sex, marital status, drinking, smoking status, sleeping status, education background and individual yearly income. u_i represents a fixed effect, ε_{i} represents a random disturbance item, and i represents the individual respondent.

3.4. Binomial Logistic Regression Model

The binomial logistic regression model was established to carry out comprehensive evaluation among various influencing factors, to analyze the influencing factors of life satisfaction. This method can better solve the problem of interdependence among influencing factors.

The binomial logistic regression model is as follows:

$$\begin{cases} \mathrm{Z}_{\mathrm{i}} = Ln\frac{\mathrm{P}_{\mathrm{i}}}{1-\mathrm{P}_{\mathrm{i}}} = u_{\mathrm{i}} + \alpha \mathrm{AQS}_{\mathrm{i}} + \beta Control_i + \varepsilon_{\mathrm{i}} \\ g(z) = \frac{1}{1+e^{-Z}} \\ P(Y_2 = 1|w, x) = g(z) \\ P(Y_2 = 0|w, x) = 1 - g(z) \\ \mathrm{odds} = \frac{\mathrm{P}}{1-\mathrm{P}} \end{cases} \tag{4}$$

where Pi is the probability of life satisfaction, the probability of life dissatisfaction is (1-pi), Pi $\in$ (0,1), and odds is the ratio of the probability of life satisfaction to the probability of life dissatisfaction. Zi represents the explained variable residents' life satisfaction, and the core explanatory variable is air quality satisfaction. The control variables include variables related to QOL: health satisfaction, marriage satisfaction, children satisfaction, age, residence areas, sex, marital status, drinking, smoking status, sleeping status, education background and individual yearly income. u_i is a fixed effect, ε_{i} is a random disturbance item, and i is the individual respondent.

As the subjective perception factors entering the logistic regression equation were air quality satisfaction, health satisfaction, marriage satisfaction and child satisfaction, the assignment of the classification level was arranged from small to large according to its logical meaning, which is opposite to the logical order of the dependent variable (life satisfaction) (0 = dissatisfaction, 1 = satisfaction). Therefore, when the estimated coefficient value of these subjective perception variables was negative in the regression model, it indicated that the smaller the value of this variable, the greater the possibility of life satisfaction. Among other control variables, education level, individual yearly income and sleeping status had the same logical order with life satisfaction, indicating that the greater the value of these variables, the greater the probability of life satisfaction.

4. Analysis of Influencing Factors of QOL

4.1. Single Factor Analysis of QOL

The results of the univariate analysis of the QOL are shown in Table 3. The results of correlation analysis show that air quality satisfaction, sex, age (years), education background, marital status, drinking, sleeping status and individual year income were the significant influencing factors of health utility value (EQ-5D) ($p < 0.05$). There was no difference between residence and health utility value in the t-test ($p > 0.05$), and the residence area factor was not included in the linear regression model.

Table 3. *T*-test and Pearson's correlation of influencing factors.

Variables	Explained Variable: EQ-5D Score		Explained Variable: Life Satisfaction	
Explanatory Variables	***t* Test t(*p*)**	**Pearson's Correlation r(*p*)**	***t* Test t(*p*)**	**Pearson's Correlation r(*p*)**
Air Quality Satisfaction (X1)	−315.101 *** (0.0)	−0.065 *** (3.99×10^{-17})	−284.030 *** (0.0)	−0.155 *** (1.87×10^{-90})
Sex (X2)	52.367 *** (0.0)	−0.186 *** (1.37×10^{-130})	80.315 *** (0.0)	−0.065 *** (4.57×10^{-17})
Age (years) (X3)	−835.15 *** (0.0)	−0.188 *** (1.35×10^{-133})	−832.93 *** (0.0)	0.04 *** (2.34×10^{-7})
Residence Areas (X4)	−0.275 (0.784)	−0.104 *** (3.76×10^{-41})	34.556 *** (1.041×10^{-256})	−0.053 *** (7.10×10^{-12})
Education Background (X5)	94.647 *** (0.0)	0.234 *** (7.76×10^{-207})	120.050 *** (0.0)	0.058 *** (6.45×10^{-14})
Marital Status (X6)	−39.619 *** (0.0)	−0.128 *** (4.21×10^{-62})	5.625 *** (1.87×10^{-8})	0.058 *** (4.67×10^{-14})
Smoking Status (X7)	305.164 *** (0.0)	0.022 ** (0.004)	295.453 *** (0.0)	−0.005 (0.523)
Drinking (X8)	96.699 *** (0.0)	0.170 *** (3.09×10^{-108})	121.996 *** (0.0)	0.051 *** (2.86×10^{-11})
Sleeping Status (X9)	−6.678 *** (2.46×10^{-11})	0.245 *** (1.02×10^{-226})	29.489 *** (1.63×10^{-188})	0.107 *** (1.45×10^{-43})
Individual yearly Income (X10)	−68.389 *** (0.0)	0.177 *** (3.16×10^{-118})	−68.388 *** (0.0)	0.078 *** (1.01×10^{-23})
Health Satisfaction (X11)			−287.616 *** (0.0)	−0.323 *** (0.0)
Marital Satisfaction (X12)			−199.239 *** (0.0)	−0.201 *** (2.09×10^{-151})
Children Satisfaction (X13)			−228.362 *** (0.0)	−0.199 *** (2.52×10^{-148})

* $p < 0.05$, ** $p < 0.01$, *** $p < 0.001$. Notes: p values are in parentheses. To save space, only parameters of key interest are presented, but results of the full models are available upon request.

The fourth and fifth columns of Table 3 show the results of the *t*-test and Pearson correlation analysis of life satisfaction with the 13 influencing factors. The results of the analysis showed that air quality satisfaction was correlated with life satisfaction at a significance level of 0.1%. Sex, age (years), residence areas, educational background, marital status, drinking, sleeping status, individual yearly income, health satisfaction, marital satisfaction and children satisfaction were influential factors in experiencing utility value for life satisfaction ($p < 0.05$). Smoking status was not significant in Pearson's correlation test ($p > 0.05$), and smoking status was not included in the logistic regression model. This is inconsistent with the research results of Wang, H. et al. [13].

4.2. Multi-Factor Linear Regression Analysis of Factors Influencing Health Utility

The linear regression model was constructed by taking the significant variables of univariate analysis, air quality satisfaction, sex, age (years), education background, marital status, smoking status, drinking, sleeping status and individual yearly income as explanatory variables and the health utility value of QOL (EQ-5D) as the explained variable to test the effect of air quality satisfaction on the health utility of residents' QOL, and the regression results are shown in Table 4, where Model 1 is the estimation result without introducing control variables, while Model 2 and Model 3 are the estimation results after introducing control variables and correcting for heteroskedasticity.

Table 4. Multiple linear regression analysis of factors influencing health utility values (n = 16,736).

Variables	Health Utility EQ-5D Score		
Explanatory Variables	**Model 1**	**Model 2**	**Model 3**
Air Quality Satisfaction (X1)	−0.018 *** (0.002)	−0.0235 *** (0.002)	−0.0235 *** (0.002)
Sex (X2)		−0.0464 *** (0.004)	−0.0472 *** (0.004)
Age (years) (X3)		−0.0032 *** (0.000)	−0.0032 *** (0.000)
Education Background (X5)		0.0403 *** (0.004)	0.0402 *** (0.004)
Marital Status (X6)		0.0262 *** (0.006)	0.0260 *** (0.006)
Smoking Status (X7)			−0.0139 (0.008)
Drinking (X8)		0.0330 *** (0.004)	0.0332 *** (0.004)
Sleeping Status (X9)		0.0978 *** (0.004)	0.0977 *** (0.004)
Individual Income (X10)		9.876×10^{-7} *** (9.85×10^{-8})	9.94×10^{-7} *** (9.88×10^{-8})
Intercept	0.7919 *** (0.006)	0.8932 *** (0.017)	0.8944 *** (0.017)
Adj.R^2	0.004	0.145	0.145
AIC	−2321	−4862	−4863
BIC	−2305	−4793	−4786

* $p < 0.05$, ** $p < 0.01$ and *** $p < 0.001$. Notes: Standard errors are in parentheses. To save space, only parameters of key interest are presented, but results of the full models are available upon request.

The multi-factor linear regression model shows the significant effect of air quality satisfaction on the health utility score of residents' QOL at the significance level of 0.001. From Model 2, on average, for each increase in air quality satisfaction level (the larger the level, the lower the air quality satisfaction), the health utility value of residents' QOL decreases by 2.35 percentage points, when all other indicators are equal. This indicates that poorer air quality satisfaction reduces the level of health utility of residents' QOL, and that there is a significant positive relationship between air quality satisfaction and residents' QOL.

The regression results of other explanatory variables also provide some enlightenment. The regression coefficient of sex is significantly negative in the estimation, indicating that the QOL health utility value of female respondents is significantly lower than that of males, and the fluctuation of QOL health utility scores is somewhat greater than that of males, which may be due to the fact that most Chinese women bear the dual pressures of work and family life, and lack the ways and environment to relax. Secondly, there is a significant negative correlation between age and the health utility value of QOL, which decreases with age. The estimated value of the regression coefficient of education level is significantly positive, indicating that education can improve the ability, cognitive level and psychological resilience of residents in China, and bring about the improvement of material living standards, thus indirectly improving the QOL. The marriage factor is significant at the level of 0.001, indicating that there are significant differences in the QOL between different marital status groups. From the marital status variables, it is shown that the QOL health utility of married groups is higher, while the health utility of divorced, widowed or single groups is lower, which may be because married people obtain more utility in the form of family support. The estimated linear regression coefficient of alcohol consumption is significantly positive—a possible reason for this is that in contrast with smoking (not significant) addiction, most middle-aged and elderly people tend to drink alcohol in moderation, and only on important festivals or occasions, and that a small amount of alcohol is beneficial to their health to some extent. At the same time, the improvement of national health awareness, the increase in healthy life publicity, the diversification of publicity forms and the development of the Internet have all made middle-aged and elderly people pay more attention to the healthy lifestyle of avoiding smoking and drinking. The effect of sleep status on the health effect of the QOL is significantly positive, indicating that residents with good sleep quality are healthier than those with abnormal sleep conditions. The estimated coefficient of personal annual income is significant in the regression, but the regression coefficient is very small. In general, the higher the income of residents, the more and better medical care and services they can access, and the better their QOL. However, due to the possible "happiness paradox" phenomenon, an increase in an individual's annual income does not necessarily lead to an improvement in the QOL.

According to the absolute value of the estimated value of the regression coefficient, the magnitude of the effect of each of the variables entering the regression equation on the utility value of QOL can be ranked, in descending order, as sleep status, gender, education, alcohol consumption, marital status, air quality satisfaction, age and annual personal income.

4.3. *Multi-Factor Logistic Regression Analysis of Influencing Factors of Experienced Utility*

The significant influencing factors which were significant in the univariate analysis include air quality satisfaction, sex, age (years), residence areas, education background, marital status, drinking, sleeping status, individual yearly income, health satisfaction, marital satisfaction and child satisfaction. Life satisfaction was taken as the dependent variable to construct a multi-factor binary logistic regression model.

The regression results are shown in Table 5, where Model 3 is the estimation result without introducing control variables, and Model 4, Model 5 and Model 6 are the estimation results after introducing the control variables and gradually removing insignificant factors using the backward elimination method. Exp(β) represents the estimated value of the change multiple of the ratio of the probability of life satisfaction to the probability of life dissatisfaction (odds) caused by the increase in one unit of the i-th explanatory variable and reflects the magnitude of the effect of each explanatory variable on the explained variable. The value of Exp(β) indicates that the lower the classification level of the independent variable, the greater the probability that the resident is satisfied with his or her life.

4.3.1. Results

At the 0.1% significance level, the estimated value of the regression coefficient of air quality satisfaction is negative—that is, it is significant. In Model 4, its Exp(β) value is 0.7225—that is, when all other indicators are equal, the air quality satisfaction increases by one level (the higher the level, the worse the air quality satisfaction), and the ratio of the probability of dissatisfaction life to the probability of life satisfaction (odds) is 0.7225 times the original value. This indicates that poorer air quality satisfaction can reduce life satisfaction, which means that there is a significant positive correlation between air quality satisfaction and the experienced utility level of residents' QOL. This is consistent with the research result that the objective measurement of air quality will indirectly affect residents' life satisfaction [3]. The estimated value of the logistic regression coefficient of the control variable age is significantly positive. According to the actual situation in China, it is stipulated that 45–59 years old represents early old age, 60–79 years old represents old age, and 80 years old or older is the longevity period. (https://baike.sogou.com/, accessed on 7 October 2021) The respondents' life satisfaction is low in early old age, which is consistent with the fact that they are at a specific age and need to face the greatest pressures and responsibilities in their life, such as heavy work, purchasing a house and raising children.

Table 5. Binomial logistic regression analysis of factors influencing experienced utility (n = 16736).

	Life Satisfaction				
Explanatory Variables	**Model 3**	**Model 4**	**Model4 Exp (β)**	**Model 5**	**Model 6**
Air Quality Satisfaction (X1)	0.6568 *** (0.0084)	−0.3250 *** (0.0343)	0.7225	−0.3258 *** (0.0343)	−0.3259 *** (0.0343)
Sex (X2)				0.0485 (0.0593)	0.0585 (0.0644)
Age (years) (X3)		0.0301 *** (0.0032)	1.03055	0.0304 *** (0.0032)	0.0305 *** (0.0032)
Residence Areas (X4)		−0.1635 * (0.0740)	0.8492	−0.1563 * (0.0745)	−0.1562 * (0.0745)
Education Background (X5)		0.1958 ** (0.0651)	1.2163	0.2050 ** (0.0660)	0.2043 ** (0.0661)
Marital Status (X6)		−0.8260 *** (0.1159)	0.4378	−0.8311 *** (0.1162)	−0.8317 *** (0.1162)
Drinking (X8)					0.0267 (0.0667)
Sleeping Status (X9)		0.3691 *** (0.0604)	1.4464	0.3729 *** (0.0605)	0.3729 *** (0.0605)
Individual yearly Income (X10)		0.0000 *** (0.0000)	1.0	0.0000 *** (0.0000)	0.0000 *** (0.0000)
Health Satisfaction (X11)		−0.9954 *** (0.0330)	0.3696	−0.9972 *** (0.0331)	−0.9962 *** (0.0332)
Marital Satisfaction (X12)		−0.4975 *** (0.0328)	0.6080	−0.5019 *** (0.0333)	−0.5021 *** (0.0333)
Children Satisfaction (X13)		−0.3426 *** (0.0326)	0.7099	−0.3392 *** (0.0329)	−0.3392 *** (0.0329)
Intercept		7.5124 *** (0.3326)		7.4694 *** (0.3366)	7.4480 *** (0.3407)
Pseudo R^2	−0.131	0.226		0.226	0.226
AIC	13,357.42	9168.41		9169.75	9171.59
BIC	13,365.14	9253.39		9262.93	9272.01

* $p < 0.05$, ** $p < 0.01$, *** $p < 0.001$. Notes: Standard errors are in parentheses. To save space, only parameters of key interest are presented, but results of the full models are available upon request.

Secondly, the regression coefficient of residence is significantly negative in the estimation ($p < 0.05$), indicating that the possibility of life satisfaction of rural respondents is greater than that of non-rural respondents. The higher the education level and the better the sleep status, the greater the probability of life satisfaction of middle-aged and elderly residents. The logistic regression coefficients of gender, drinking and smoking are not significant in this estimation.

The regression coefficients of health satisfaction, marital satisfaction and child satisfaction were significantly negative ($p < 0.001$), and the Exp(β) values were 0.3696, 0.6080 and 0.7099, respectively, indicating that the higher the classification level of subjective perception factors, the more the marginal effect of life satisfaction of the residents showed a considerable positive effect. That is, health satisfaction, marital satisfaction and children satisfaction show significant positive correlations with life satisfaction.

According to the magnitude of Exp(β) values in Model 4, the magnitude of the effect of each variable on life satisfaction that finally entered the logistic regression equation can be ranked, in descending order, as sleeping status, education background, age (years), individual yearly income, areas. air quality satisfaction, children satisfaction, marital satisfaction, marital status, and health satisfaction.

If the probability of life satisfaction is set to p (the probability of life dissatisfaction is (1-p), $p \in (0,1)$), and the logit transformation of p is the dependent variable, a logistic regression model is constructed according to the logistic regression formula using direct logistic regression coefficients:

$$\log(p) = 7.5124 - 0.3250 \times X_1 + 0.0301 \times X_3 - 0.1635 \times X_4 + 0.1958 \times X_5 - 0.8260 \times X_6 + 0.3691 \times X_9 - 0.9954 \times X_{11} - 0.4975 \times X_{12} - 0.3426 \times X_{13} \quad (5)$$

4.3.2. Model Evaluation

To evaluate the accuracy of the model, all 16,736 samples were divided into a training set and a test set in logistic regression—11,715 samples in the training set and 5021 samples in the test set were evaluated using logistic regression analysis. The overall prediction fit accuracy of the regression equation classification prediction was 0.89 (classification truncation was taken as 0.50), and 14,895 samples out of 16,736 samples were correctly classified. The accuracy of life satisfaction classification was 0.89.

From the Receiver Operating Characteristic curve (ROC) is shown in Figure 3, it can be seen that the Area Under Curve (AUC) value of the optimal critical point of the area covered by the ROC curve is 0.83, because a larger value represents a better effect of logistic regression analysis, which means the model has a good prediction effect. The recall rate of the number of truly positive cases predicted is 0.99, and the recall rate of the number of truly negative cases predicted is 0.17. The overall discriminant accuracy of the estimated samples is 88.99%—that is, the overall accuracy of the model prediction is good.

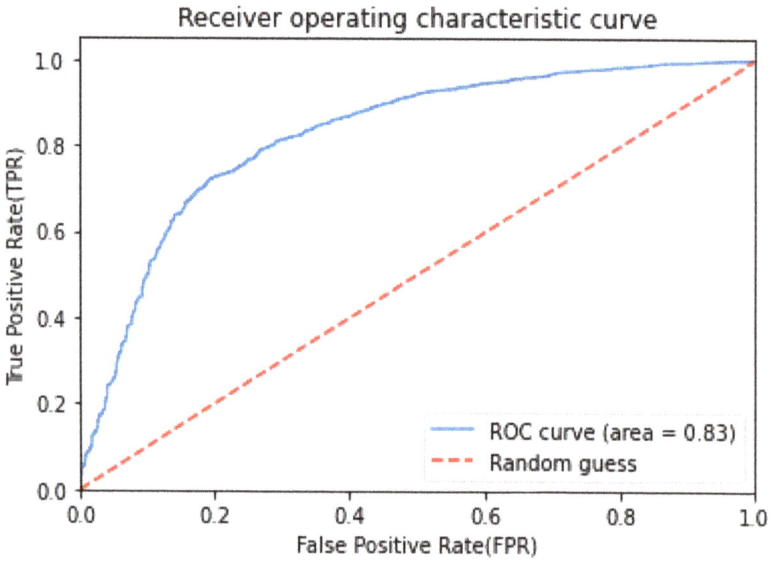

Figure 3. ROC curve of logistic regression. The red line (TPR = FPR) corresponds to the "random guess" model, and the point on the upper left of the red line (TPR > FPR) indicates that the judgment is generally correct.

4.4. Robustness Analysis

The regression results remain robust in the following robustness tests:

(1). The replacement of the dependent variable. The "experienced utility" life satisfaction of QOL is used to replace the EQ-5D score of health utility for regression analysis. After the replacement, the explanatory variable air quality satisfaction is still positive at the significant level of 1%. The better the air quality satisfaction, the greater the possibility of life satisfaction, which is consistent with the results of the article findings [1,2]. This shows that the subjective evaluation of air quality is indeed positively correlated with residents' QOL, and the improvement of air quality helps to improve residents' QOL. Additionally, other explanatory variables and their related variables are significantly correlated with QOL.

(2). Add control variables. Considering that life satisfaction indicators of QOL experienced utility are subjective perception data, which are influenced by residents' cognitive level and other aspects, variables such as health satisfaction, marital satisfaction and child satisfaction were added as control variables. After adding the control variables, the regression also shows that the estimated coefficients of air quality satisfaction are significantly positive, and the estimated values of the regression coefficients of the subjective perception category control variables are all significant, and the conclusion maintains that the higher the air quality satisfaction, the better the QOL of the residents.

(3). The replacement model test. A multi-factor linear regression model was established for the analysis of the correlation between air quality satisfaction and health utility values. The multi-factor logistic regression model was used to test the relationship between air quality satisfaction and life satisfaction, and the positive effect of air quality satisfaction on residents' life satisfaction passed the significance test at the 1% significance level, which was consistent with the statistical results of the multiple linear regression model. The worse the air quality satisfaction, the lower the residents' life satisfaction and the worse the QOL, and the regression results remain robust.

5. Conclusions and Discussion

This paper uses the 2018 China Health and Retirement Longitudinal Study (CHARLS) database, using multi-factor linear regression and binary logistic regression methods to examine the relationship between subjective air quality assessment and the QOL of Chinese middle-aged and elderly residents. The two models produced consistent results regarding a significant positive relationship between air quality satisfaction and QOL. This is consistent with the finding that there is a significant positive correlation between air quality (outdoor air-PM_{10}) and QOL [2]. Exploring the relationship between subjective evaluation of air quality and the QOL of residents has important practical significance. From the government's point of view, the local air quality reflects the government's environmental governance performance to a certain extent. With regard to the subjective feelings of residents, the air quality of the place of residence will directly affect the quality of their daily life, thereby affecting their subjective evaluation of the air quality status.

This study analyzed the impact of air quality satisfaction on the QOL from the two dimensions of the health utility and experienced utility of the QOL and concludes that there is a significant positive correlation between air quality satisfaction and QOL. The impact of gender and income on the QOL of middle-aged and elderly residents is statistically significant ($p < 0.05$), but this impact is indirect and limited. The influence of lifestyle factors on the QOL cannot be ignored. The significance of the impact of the interviewee's residence and smoking status on the QOL and health utility needs to be further verified.

In this study, we used the CHARLS 2018 database to analyze the impact of the subjective assessment of air quality on the QOL. We put forward the "two-dimensional" research perspective of QOL, which divides the QOL into two different dimensions: the health utility of the QOL and the experienced utility of the QOL and perform beneficial exploration and research to investigate the correlation between subjective evaluation of air quality and QOL. One of the main shortcomings of this study is that the subjective assessment of air quality is not comprehensive enough, and there is no specific index assessment involving air quality. Despite this shortcoming, this research better reflects the problems of environment-related QOL, and draws several conclusions:

Firstly, China should reform the current performance assessment system and increase the weight of environmental assessment indicators. The government should thus carry out the air quality satisfaction assessment of urban and rural residents based on questionnaire surveys, and take the public satisfaction assessment as one of its main data sources in the assessment system of atmospheric environment governance, in order to improve the assessment method of comprehensive improvement of atmospheric environment, and thus to improve the living standards of residents by enabling the local government to provide more high-quality air public goods.

Secondly, since the improvement of air quality can significantly enhance residents' air quality and life satisfaction, the government needs to improve not only the local air quality, but also residents' subjective perception of air quality. The government should establish a government-led model of air governance with the participation of stakeholders, encourage and guide the public to participate in air quality protection actions, and actively promote the participation of individuals and civil environmental groups in air governance actions.

Thirdly, by establishing an atmospheric environmental education system to increase the publicity of atmospheric environmental knowledge improve the public's attention to air quality, the government will increase the public's environmental awareness and enrich environmental knowledge, as well as strengthening the publicity regarding the significance of individual environmental behaviors. This will provide residents with the belief that personal behavior can have a profound impact on atmospheric environmental protection.

Due to the limitations of the complexity of the factors affecting the QOL and the availability of data, this article still has the following drawbacks: on the one hand, this article mainly considers the influence of the subjective assessment of the overall air quality of residents on the QOL and does not involve the evaluation of specific indicators of air quality. On the other hand, this article lacks analysis of the correlation between other evaluation indicators and residents' QOL, such as the quality of the ecological environment, drinking water, the green environment, the pollution of rivers and lakes, the degree of soil pollution, and noise pollution. The interaction between the subjective evaluation of ecological environment quality and the QOL also needs to be studied in the future.

Author Contributions: Y.D. conducted the analysis of conditions, numerical simulation and drafted the manuscript; S.Y. initiated the study; M.Z. and Z.S. coordinated the data collection; Y.D. and M.Z. were involved in the data interpretation. W.L. and D.L. contributed to checking and proofreading. All authors have read and agreed to the published version of the manuscript.

Funding: This research received no external funding.

Institutional Review Board Statement: Not applicable.

Informed Consent Statement: Not applicable.

Data Availability Statement: The data presented in this study are openly available in China Health and Retirement Longitudinal Study (CHARLS) project database.

Acknowledgments: The authors would like to thank the editor and anonymous reviewers for their valuable comments and suggestions to this paper.

Conflicts of Interest: The authors declare no conflict of interest.

References

1. Suwandee, S.; Anupunpisit, V.; Ratanamaneichat, C.; Sukcharoen, N.; Boonpen, P. Quality of Life and Environment of Communities Along Saen Saeb Canal: A Survey Foundation of the Physical and the Current Situation (Phase I). *Procedia Soc. Behav. Sci.* **2013**, *88*, 205–211. [CrossRef]
2. Darçın, M. Association between air quality and Quality of Life. *Environ. Sci. Pollut. Res.* **2014**, *21*, 1954–1959. [CrossRef] [PubMed]
3. Liao, L.; Liao, H. Role of the Radiative Effect of Black Carbon in Simulated PM (2.5) Concentrations during a Haze Event in China. *Atmos. Oceanic Sci. Lett.* **2014**, *7*, 434–440. [CrossRef]
4. Rickenbacker, H.J.; Vaden, J.M.; Bilec, M.M. Engaging Citizens in Air Pollution Research: Investigating the Built Environment and Indoor Air Quality and Its Impact on Quality of Life. *J. Archit. Eng.* **2020**, *26*, 04020041. [CrossRef]
5. Mendoza, Y.; Loyola, R.; Aguilar, A.; Escalante, R. Valuation of Air Quality in Chile: The Life Satisfaction Approach. *Soc. Indic Res.* **2019**, *1*, 367–387. [CrossRef]
6. Chen, Y.; Ebenstein, A.; Greenstone, M.; Li, H. Evidence on the impact of sustained exposure to air pollution on life expectancy from China's Huai River policy. *Proc. Natl. Acad. Sci. USA* **2013**, *110*, 12936–12941. [CrossRef] [PubMed]
7. Liao, X.; Tu, H.; Maddock, J.E.; Fan, S.; Lan, G.; Wu, Y.; Yuan, Z.K.; Lu, Y. Residents' perception of air quality, pollution sources, and air pollution control in Nanchang, China. *Atmos. Pollut. Res.* **2015**, *6*, 835–841. [CrossRef]

8. Shi, X.; Li, X.; Chen, X.; Zhang, L. Objective air quality index versus subjective perception: Which has a greater impact on life satisfaction? *Environ. Dev. Sustain.* **2021**. [CrossRef]
9. EuroQol Research Foundation. EQ-5D-3L User Guide. 2018. Available online: https://euroqol.org/publications/user-guides (accessed on 12 April 2021).
10. China Health and Retirement Longitudinal Study (CHARLS). Wave 4. 2018. Available online: http://charls.pku.edu.cn/pages/data/111/zh-cn.html (accessed on 15 May 2021).
11. Tingting, W.U.; Yun, L.U.; Dandan, A.I. Analyzing the Health-related Quality among the Elderly with Multiple Chronic Conditions in Jiangsu: Estimation Based on the EQ-5D Value Sets. *Chin. Gen. Pract.* **2020**, *23*, 47–51.
12. Liu, G.G.; Wu, H.; Li, M.; Gao, C.; Luo, N. Chinese time trade-off values for EQ-5D health states. *Value Health* **2014**, *17*, 597–604. [CrossRef] [PubMed]
13. Wang, H.M.; Bell, J.F.; Edwards, T.C.; Patrick, D.L. Weight status, quality of lifeand cigarette smoking among adolescents in Washington State. *Qual. Lifesearch* **2013**, *22*, 1577–1587. [CrossRef]

atmosphere

MDPI

Article

Study on the Mechanism of Haze Pollution Affected by Urban Population Agglomeration

Xuesong Li [1,*], Min Zhou [1], Wenyu Zhang [1], Kewei Yu [2] and Xin Meng [3,*]

1 School of Economics and Management, Wuhan University, Wuhan 430072, China; zhoumin1998@whu.edu.cn (M.Z.); zhangwenyu0525@whu.edu.cn (W.Z.)
2 Institute of Industrial Economics of CASS, Beijing 100044, China; yukewei@sz.icbc.com.cn
3 Department of Business Management, Hubei Water Resources Technical College, Wuhan 430072, China
* Correspondence: xsli@whu.edu.cn (X.L.); Mengxinlxs@163.com (X.M.); Tel.: +86-1388-6199-990 (X.L.); +86-1387-1299-388 (X.M.)

Abstract: Population agglomeration and haze pollution are two major problems that urban development will inevitably face in the future. Population agglomeration has a spatial impact on smog pollution through scale and intensive effects. This paper uses panel data from 236 prefecture-level cities in China from 2001 to 2012 to verify the impact of urban population agglomeration on haze pollution and its mechanism based on a spatial lag model. The research shows that: (1) China's urban haze pollution has a significant positive spatial spillover effect, and presents a spatial distribution state of high-high and low-low agglomeration. (2) There is a significant "N-type" nonlinear relationship between urban population agglomeration and haze pollution. (3) At present, the scale effect of urban population agglomeration in China is greater than the intensification effect, and the scale effect as well as intensification effect have opposite effects on haze pollution. This shows that urban layout should be scientifically planned, urban population should be reasonably controlled, production efficiency should be improved, and green development should be promoted to deal with haze pollution. (4) The spillover effect of urban population agglomeration on haze pollution is significantly greater than the direct effect, indicating that local haze pollution is more likely to be affected by spatially related regions, indicating that strengthening regional coordination and cooperation and joint prevention and control are necessary to control haze pollution.

Keywords: collective effect; haze pollution; scale effect; special spillover effect; urban population agglomeration

Citation: Li, X.; Zhou, M.; Zhang, W.; Yu, K.; Meng, X. Study on the Mechanism of Haze Pollution Affected by Urban Population Agglomeration. *Atmosphere* **2022**, *13*, 278. https://doi.org/10.3390/atmos13020278

Academic Editors: Zengyun Hu, Xuguang Tang and Qinchuan Xin

Received: 23 December 2021
Accepted: 2 February 2022
Published: 7 February 2022

1. Introduction

Haze pollution has recently emerged as a significant atmospheric environmental hazard harming China's economic and social development. According to data from China's Ministry of Environmental Protection's (MEP) National Air Quality Report released at the end of 2016, only 84 of the country's 338 cities at the prefecture level and above met the annual average air quality standards, with the majority of cities suffering from haze pollution. As a result, China has taken steps to improve air quality and pollution monitoring, raise the required standards for pollutants such as inhalable particulate matter and nitrogen dioxide, control the demand for motor vehicles and industrial emissions, adjust industrial structures, promote clean energy, and establish a medium- and long-term pollution prevention and control mechanism. However, the primary sources of haze pollution are pollutant concentrations, atmospheric conditions, and air humidity. Pollutant concentration is both a required material basis for creating haze pollution and intimately tied to human life. The demand for motor vehicles and industrial emissions is based on population agglomeration and human activities. According to the United Nations Human Settlements Programme's World Cities Report (2016), published on May 18, 2016, the top 600 big cities currently house one-fifth of the world's gross population and generate up to

60% of global GDP. Population agglomeration has a significant impact on cities' economic, social, and environmental development. On 1 April 2017, The Central Committee of the Communist Party of China (CPC) and the State Council decided to establish the Xiong'an New Area at the national level, aiming at relieving Beijing of functions nonessential to its role as the capital and exploring a new model of optimal development in densely populated areas. Therefore, how to view the spatial spillover effect of urban population agglomeration on haze pollution and reasonably plan and control the scale of urban development is a topic worth delving into in-depth.

With rising environmental awareness, attention to urban haze pollution in China has gradually increased, but due to data availability issues, more literature on the research of haze pollution in China has just surfaced in the last two years. In terms of the study topic, most haze studies in the available literature examine the socioeconomic factors that influence the creation of haze pollution. The majority of the literature acknowledges the importance of population size, economic expansion, energy or industrial structures, and environmental regulation in the formation of haze pollution and focuses on one of these. In analyzing the link between haze pollution and economic growth, Guan et al. [1] used the environmental input–output model to suggest that economic considerations mainly influence the increase in pollutant emissions in Chinese cities. Furthermore, in order to investigate the existence of a non-linear relationship between economic development and haze pollution, Ma Limei et al. [2] found that China is still in a stage where haze pollution concentrations continue to increase with regional per capita GDP levels, based on the results of a spatial lag model and a spatial error model. Shao Shuai et al. [3] came to a different conclusion. They looked into it further with a dynamic spatial lag model and discovered that local haze pollution in China has a sizeable spatial agglomeration effect. Some scholars [4] have jointly reached consistent conclusions from two research perspectives: the degree of energy price distortion and the proportion of high coal-consuming industries' output value in regional GDP, respectively. They jointly determined that a high-energy-consumption energy structure would considerably contribute to regional haze pollution. Scholars have also paid close attention to the role of environmental regulation in haze pollution mitigation. According to Huang Shoufeng [5], the shadow economy is an essential aspect in investigating the impact of environmental policy on haze reduction. Furthermore, Quan Shiwen and Huang Bo [6] use Beijing's environmental policies as examples, emphasizing that the embedding effect between environmental policies is an essential influencing aspect when developing and evaluating environmental policies. According to the study, pollutant concentrations were shown to be closely associated with stringent pollution control measures. More specifically, there are studies that have also focused on haze pollution in Beijing but only choose the period during the Olympics held in 2008. Zhang, Wang, et al. [7] developed a system to quantify haze pollution and showed the data they observed for the 2008 Beijing Olympics based on the outcomes of various emission control measures quantized by ground-based and satellite monitoring.

Many articles have used the difference-in-differences (DID) and propensity score matching (PSM) models as research methodologies. Shi Qingling et al. [8], for example, began with the extraordinary event of the "Two Sessions" held at the local level, whereas Zhang Shengling et al. [9] investigated changes in haze pollution management in China before and after the event, based on the outbreak of social opinion on haze. Furthermore, some authors [10,11] employed simultaneous equations and structural equations to link haze pollution causes and economic growth factors using industrial coal consumption as a loop, forming closed equations. We investigate how to balance environmental conservation and economic growth in this article. Furthermore, because of the spatial correlation of haze pollution, the spatial econometric method has progressively received the acceptance of many academics in recent years, but the relevant literature is still relatively small. Ma Limei et al., for example, quantified local energy by the total output value of eight high-coal-consumption industries as a fraction of regional GDP using a spatial Durbin model (SDM). The relationship between local haze pollution, energy structure, and traffic patterns was investigated by assessing transport parameters such as vehicle pressure and congestion. The study found that the spatial spillover effect of haze pollution is more significant at the national level but the causes driving haze pollution vary among areas. On the other hand, Xiang Kun and Song Deyong [12] employed the SDM to investigate the factors impacting haze pollution at the provincial level in China and reached conclusions that are broadly consistent with those of Ma Limei and other scholars.

2. Method

The methods section mainly explains the mechanism of the impact of population agglomeration on haze pollution and presents the design of the spatial error model for empirical verification. Firstly, there are two research hypotheses put forward through literature reading and model construction: one is that there is a nonlinear relationship between urban population agglomeration and haze pollution, and the other is that the impact of urban population agglomeration on haze pollution in China is mainly divided into scale effects and intensive effects. Secondly, in order to verify the above assumptions, this paper adopts the spatial error model and refers to the STIRPAT model to select control variables for empirical verification. In addition, we explain the data sources and present the results of descriptive statistical analysis. Finally, the applicability of the spatial error model is verified from the perspective of statistics and econometrics.

2.1. *Analysis of Theoretical Mechanism*

The action mechanism of urban population agglomeration on haze pollution is presently divided into two main views. The first is that population agglomeration will increase pollutant emissions in the atmosphere via the scale effect, making hazy weather more likely. According to some scholars [3,13], urban population agglomeration will cause traffic congestion, housing tensions, and high demand for heating and gas, which will generate a scale effect and increase pollutants in the atmosphere, resulting in haze pollution. On the contrary, another school of thought holds that increasing human density will lower pollution emissions in the atmosphere via the collective effect, aiding in the regulation of hazy weather. Although urban population agglomeration can cause a slew of issues, cities can serve the purpose of increasing resource efficiency. The agglomeration of a large number of people in cities allows them to take advantage of the city's public transportation and pollution abatement infrastructure more fully, resulting in the collective effect and, as a result, a reduction in the emission of air pollutants, ultimately leading to a reduction in the haze pollution problem [14]. The reason why the two viewpoints indicated above reach different and opposing results is found in their development of two mesomeric effect models of urban population agglomeration on haze pollution: the scale effect and the collective effect, the routes of which are depicted in Figure 1.

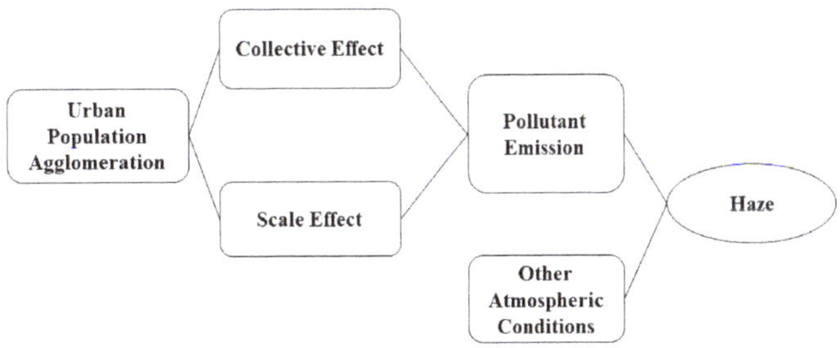

Figure 1. Schematic Diagram of Two Action Mechanisms of Urban Population Agglomeration on Haze Pollution.

2.1.1. Scale Effect

From the economic standpoint, the scale effect of urban population agglomeration on haze pollution is caused by externalities. The final consequence of combining the effects of each individual's behavioral decisions is the influence of urban population agglomeration on haze pollution. According to a study conducted by Zeng Xiangang et al. [15], even though more than half of Beijing residents were aware of the severe health risks that haze would pose to themselves and their families, a sizable number of people were still unwilling to reduce the social health risks of haze in exchange for a lower quality of life and a higher cost of living. This study also shows that personal interests tend to take precedence over social interests in individual behavioral choices. As a result, as a city's population proliferates, so do human activities and demands. At the same time, if people continue to think in terms of their interests and do not change their lifestyle habits, for example, in terms of travel, they will continue to drive motor vehicles, resulting in a large number of vehicle emissions due to an excess of motor vehicles and traffic congestion. This will have a scale effect on the haze pollution problem, with significant negative externalities for the urban environment.

2.1.2. Collective Effect

The formation of cities comprises population urbanization and two crucial features: non-agriculture industries and social modernization [16]. When a high population agglomeration leads to a city being formed, the collective effect becomes more significant. First, the spatial agglomeration of production activities in the urban industrial sector will lead production elements of various forms, accompanied by related industrial activities, to the city. As a result of these, the city's allocable resources will be more plentiful. Second, as the urban industrial sector gradually improves its productivity, its public infrastructure will improve. At the same time, the city's resource allocation will become more efficient. Finally, during urban development, distinct functional zones will be developed inside the city, depending on the location of the land, and there will be a concentration of similar activities within the same functional zone. Land intensification will increase, while social services will be improved. As a result, the urban collective effect to address the problem of air pollution prevention and control can efficiently deploy all types of plentiful resources and increase pollution prevention and control efficiency. For example, establishing a comprehensive public transportation and sewage system can strengthen pollutant treatment and prevention, establish a regional coordination management system, coordinate air pollution prevention and treatment, and ultimately reduce pollutant per capita emission to reduce haze pollution.

2.1.3. Reaction Mechanisms of the Two Effects

The two effects stated above result in two reaction mechanisms that influence haze pollution caused by urban population concentration. As indicated in Figure 2, w denotes haze concentration, and p denotes urban population density. Point A represents the equilibrium state attained with a haze concentration of w* and a population density of p* under the condition of urban environment carrying capacity U2. As population density rises, two mesomeric effects emerge—the scale effect and the collective effect. (1) If only the scale effect is taken into account, the increase in urban population density means more discharge of production and living pollutants. As shown in Figure 2, the combination line L of population density and haze concentration will be rotated to the right around its intersection with the vertical axis, giving rise to the line L1. If only the scale effect is taken into account, the environmental carrying capacity at U2 stays unaltered. Hence, the line L1 can be switched to L2 and intersect the environmental carrying capacity curve U2 at point B. At this time, the equilibrium point moves from the initial position A to B, and the haze concentration in the equilibrium state is w_1*. w_1* − w*, which is a positive value, represents the scale effect of population density increase, and it means that the scale effect will make the increase in urban population density lead to an increase in haze pollution concentration.

(2) Consider the collective effect, which occurs as urban population density rises and people rush to the city. While people thoroughly enjoy the city's socialized services, the city's comprehensive public transportation system and pollution control and emission reduction system can reduce pollution generation. This is conducive to further strengthening the regulation and unified management of pollution caused by human activities. As a result, the city's environment carrying capacity grows, and the curve U2 shifts to U1, intersecting with the combination line L1 of haze pollution and population density at point C. At this point, the equilibrium state of haze pollution concentration is w_0*. The haze concentration has decreased, based on w_1*. At this stage, the total effect of urban population density on haze pollution is w_0* − w*, indicating that an increase in urban population density eventually increases haze pollution concentration. (3) The scale effect and the intensification effect work together to produce a total effect, so when only the intensification effect is considered, the impact of the increase in population density on haze pollution can be obtained by subtracting the two, the result of which is w_0* − w_1*. It is a negative value, which suggests the scale effect makes the increase in urban population density reduce the concentration of haze pollution. Of course, Figure 2 only depicts one of the scenarios in which urban population agglomeration influences haze pollution due to the scale and collective combined effects. The direction and breadth of the influence of urban population agglomeration on haze pollution would also vary depending on the scale and collective effects. As a result, we suggest the following hypothesis:

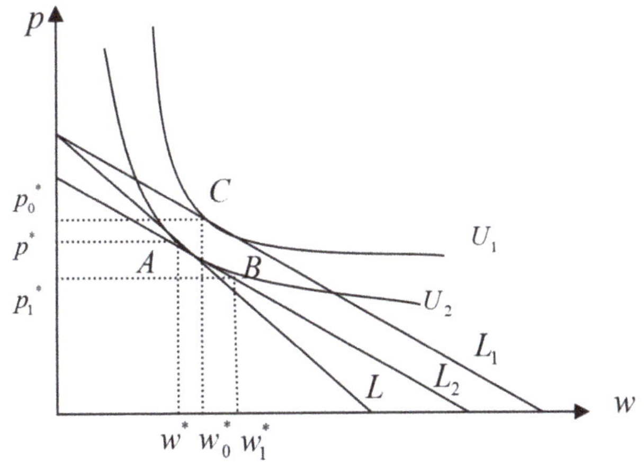

Figure 2. Schematic Diagram of the Reaction Mechanisms of Urban Population Agglomeration Affecting Haze Pollution. w denotes haze concentration, and p denotes urban population density. Line L is a combination of population density and haze concentration. Curve U represents environmental carrying capacity. Point A represents the equilibrium state attained with a haze concentration of w* and a population density of p* under the condition of urban environment carrying capacity U2.

H1: The correlation between urban population agglomeration and haze pollution may be non-linear.

H2: At this point, the impact of urban population agglomeration on haze pollution in China can be split into two mesomeric effects: the scale and the collective. The ultimate impact of human agglomeration on haze pollution is closely related to the dominant mesomeric effect.

2.2. Model Specification and Variable Description

2.2.1. Model Specification

Haze pollution is a meteorological issue caused by a combination of natural and human factors. Both its creation and dispersion are strongly impacted by atmospheric motion. The concentration of haze pollution in a region is equal to the haze produced in this region, minus the haze that spreads to other regions, plus the haze that spreads to this region from other regions. As a result, haze pollution seems to have a spatial correlation. If this spatial correlation is overlooked, it is impossible to adequately evaluate the relationship between other parameters such as urban population agglomeration and haze pollution, and the spatial spillover effect of haze itself cannot be accurately measured. As a result, a spatial econometric model is used for the analysis to make the estimation of the influence of urban population agglomeration on haze pollution more effective and accurate.

Due to the relative ease of the measuring course and the prominent economic connotation, the spatial lag model (SLM) and the spatial error model (SEM) have become two of the more basic and often employed models in spatial econometrics.

The following is the specific form of the spatial lag model:

$$y_{it} = \delta \sum_{j=1}^{N} w_{ij} y_{jt} + x_{it}\beta + u_i + \varphi_t + \varepsilon_{it} \tag{1}$$

In Equation (1), w_{ij} is the element of the spatial weight matrix W, u_i represents spatial fixed effects, φ_t represents time fixed effects, and ε_{it} represents the random error vector. y_{it} represents the explained variable, and x_{it} represents the primary explanatory variable and other control variables.

The following is the specific form of the spatial error model:

$$y_{it} = x_{it}\beta + u_i + \varphi_t + v_{it} \tag{2}$$

$$v_{it} = \lambda \sum_{j=1}^{N} w_{ij} u_{it} + \varepsilon_{it} \tag{3}$$

The connotations in Equations (2) and (3) are nearly identical to those in Equation (1). On the other hand, the spatial error model acknowledges that other elements in the spatially related areas will have a spillover effect on the haze in the region.

The inverse distance principle is employed in this research to build the spatial weight matrix, which means that the reciprocal of the distance between two locations is used as the weight. The nearer the distance between the two places, the greater the weight; the further the distance, the lesser the weight. Furthermore, for simplicity of later measurement, the spatial weight matrix is frequently normalized by rows to equalize the influence of other spatial regions on the local region. However, to ensure that the mutual ratio between each element of the inverse distance spatial weight matrix remains unchanged and to maintain the weight matrix's economic interpretation, this paper adopts an alternative weight matrix treatment proposed by Elhorst [17] and Kelejian et al. [18], i.e., each element of the inverse distance spatial weight matrix is divided by its largest eigenroot to obtain the normalized matrix.

2.2.2. Description of Primary Variables

First, in order to better control the influence of other factors on haze pollution so that the effect of population agglomeration on haze pollution will be measured more accurately, this study is based on the STIRPAT model widely used in the field of environmental economy. Select variables from four factors including environment, population, wealth, and technology are used to construct an empirical model of urban population agglomeration and smog pollution. The specific expression form of the STIRPAT model is shown in Equation (4):

$$I_{it} = aP_{it}^{b} A_{it}^{c} T_{it}^{d} e \tag{4}$$

In Equation (4), I is the environmental quality, P is the population, A is the per capita wealth, T is the technology, e is the error term, a is the model coefficient, and b, c, and d are the solve-for parameters. Equation (5) is produced by taking the logarithm of both sides of Equation (4):

$$\ln I_{it} = \ln a + b \ln P_{it} + c \ln A_{it} + d \ln T_{it} + \ln e \tag{5}$$

In this work, the relevant proxy variable will be used by the four key environmental factors in the STIRPAT model. Furthermore, in order to investigate the action mechanism of urban population agglomeration on haze pollution further—the scale effect and the collective effect—and to integrate various aspects of the relationship between urban population agglomeration and haze pollution, such as influence, representativeness, and data availability, we finally use the urban transportation sector as an example to assess and compare the influence of urban agglomeration on haze pollution. The bus ownership per 10,000 people was chosen as a proxy variable for the collective effect, and the occupying amount of civilian vehicles was chosen as a proxy variable for the scale effect. Table 1 displays the relevant variable descriptions:

Table 1. Description of Primary Variables.

Variables	Abbreviation	Connotation of the Variable	Average Value	Minimum Value	Maximum Value
Haze Concentration	Ln*mean*	Every 3 years' moving average of $PM_{2.5}$ concentration, and take the logarithm	3.7905	2.1031	4.6946
Population Density	Ln*denpop*	The average annual population of the municipal district divided by the built-up area, and take the logarithm	6.6746	2.5751	9.5453
Opening-Up Level	Ln*fdi*	The proportion of actual utilized foreign investment in local GDP in the municipal district, and take the logarithm	−4.1187	−8.8040	−0.8981
Economic Development Level	Ln*rpgdp*	After constant price treatment (base year–2001), the per capita GDP of the municipal district, and take the logarithm	9.9910	7.7630	12.1075
Industrial Structure	Ln*indgdp*	The proportion of secondary industry added value in local GDP in the municipal district, and take the logarithm	−0.6914	−1.8163	0.9517
Fixed-Asset Investment	Ln*fasset*	The total fixed-assets investment of the whole city (excluding farmers), and take the logarithm	14.0532	9.7746	17.9906
Scientific Research-Capability	Ln*sci*	The proportion of scientific research employment in the total urban employment in the municipal district, and take the logarithm	−4.1422	−6.3652	−2.1148
Scientific Research-Investment	Ln*exp_sci*	The proportion of scientific research investment in local financial expenditure in the municipal district, and take the logarithm	−5.0789	−8.5620	−2.3067
Public Transportation	Ln*busp*	The bus ownership per 10,000 people in the municipal district, and take the logarithm	1.6412	−1.1394	4.7074
Civilian Vehicles	Ln*vehicle*	The occupying amount of civilian vehicles of the whole city, and take the logarithm	11.6146	8.6995	15.3479

2.3. Data Sources

The explanatory variable in this paper, namely $PM_{2.5}$ concentration values, is obtained from the SEDAC. In 2012, China amended the Ambient Air Quality Standard (AAQS) (GB3095-2012) for the third time, revising the concentration limited values standard for some detecting indicators and increasing monitoring and reporting of $PM_{2.5}$ concentration limited values. The new AAQS was trialed in 74 prefecture-level cities across China in the same year. On 1 January 2016, the new AAQS was extended and applied across the country. Furthermore, the $PM_{2.5}$ statistics issued by China's MEP are generated from ground detection stations, which are more accurate but cannot accurately quantify the average $PM_{2.5}$ concentration in a specific area. It is not possible to make a long-term assessment of the haze pollution condition in various regions of China based on $PM_{2.5}$ statistics published by China's MEP in recent years. As a result, the $PM_{2.5}$ concentration data in this study are derived from the global annual average $PM_{2.5}$ surface concentration data given by SEDAC. Following the notion of Donkelaar et al. [19], the Battelle Memorial Institute (BMI) and Columbia University transformed the annual average $PM_{2.5}$ concentration data from aerosol concentration calculations using the relevant chemical model. Furthermore, the aerosol concentrations utilized for measurements are provided by satellite-mounted equipment, which improves the accuracy of the recorded data.

In terms of data processing, this research differs from other studies in that the variable of population density does not use the "population density" indicator published in the China Urban Statistical Yearbook for each prefecture-level city. The reason for this is that the "population density" indicator in the China Urban Statistical Yearbook is primarily calculated by dividing the gross population of the prefecture-level city at the end of the year by the administrative division area (including suburbs and some rural areas). This technique does not adequately reflect each prefecture-level city's actual "urban population density." As a result, the population density data in this study are calculated by dividing the annual average population of each prefecture-level city by the built-up area.

Other explanatory variables from CEInet Statistics Database and China City Statistics Yearbook (1999–2013) include the actual amount of foreign investment, per capita GDP, the proportion of secondary industry added value in the GDP, fixed-asset investment (excluding farmers), the bus ownership per 10,000 people, development land area, land area, scientific research employment, total urban employment, local financial expenditure, and local financial expenditure on scientific research. The civilian vehicles occupying amount is from China Statistical Yearbook for Regional Economic (2000–2013). The interpolation method was used to fill in the missing data that the above methods could not fill.

In summary, due to data availability constraints, the sample of 287 prefecture-level cities in China that underwent land withdrawal and city transformation after 1998 was deleted. After interpolation, the sample with an interpolation value less than 0 was removed, leaving a final sample of 2360 (236 $\times$ 10). Because the haze data were collected from a foreign-language website, the recording method was every three years' moving average of the annual average $PM_{2.5}$ concentration values. Therefore, the data for the other variables were likewise taken as a three-year moving average. The time span of the overall sample includes 10 time units over the period 2001–2012.

2.4. *Descriptive Analysis*

2.4.1. Analysis of the Current State of Urban Haze Pollution

The latest annual average $PM_{2.5}$ concentrations from 2010 to 2012 were plotted on a spatial distribution map (Figure 3) based on the annual average $PM_{2.5}$ data provided by Columbia University's Socioeconomic Data and Applications Center (SEDAC), which are the data used for the empirical analysis in the following section as well.

As shown in Figure 3, areas that tend to be black represent the increasing severity of the haze pollution, and these areas are primarily located in the North China Plain and near the eastern coast of China. Furthermore, the annual average haze concentrations in the Sichuan Basin and Qinling Region area are higher, owing to the region's topography, which is dominated by mountains and basins, and pollutants are not easily dispersed.

2.4.2. An Analysis of the Current State of Urban Population Density

The term "population density" refers to the population density data in China City Statistical Yearbook (2016). Because we focuse on urban population agglomeration, the term "population density" refers to the population density (per capita/km^2) of municipal districts as reported in the China City Statistical Yearbook 2016. Figure 4 depicts China's population density distribution map in conjunction with the country's administrative map.

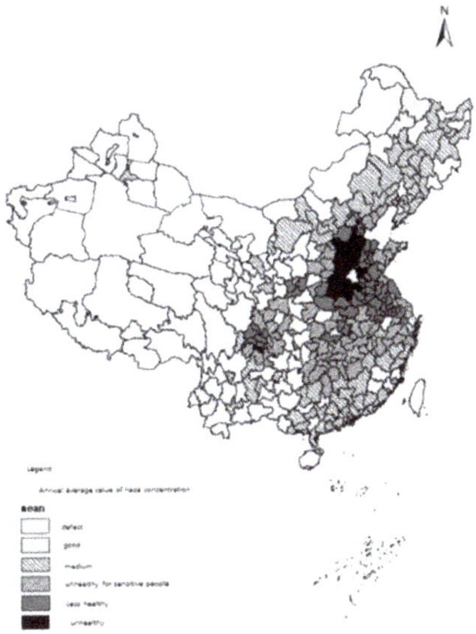

Figure 3. Annual Average Concentration of Haze from 2010 to 2012 in Chinese Prefecture-Level Cities. Source: Map of Global Annual Average Surface $PM_{2.5}$ Concentrations from SEDAC. https://sedac.ciesin.columbia.edu/ (accessed on 20 December 2021).

Figure 4. Spatial Distribution Map of Point Density in Chinese Population Density in 2015.Data Source: China City Statistical Yearbook 2016. https://www.tongjinianjian.com/111090.html (accessed on 20 December 2021).

Figure 4 depicts China's population distribution, marked by a dense population in the southeast and a sparse population in the northwest, with two unique distribution patterns along the Heihe–Tengchong line. China's most densely populated places are the Beijing–Tianjin–Hebei region, the eastern coastal region, and the southeastern coastal areas. These regions contain three main Chinese city clusters: the Beijing–Tianjin–Hebei region, the Yangtze River Delta, and the Pearl River Delta. These locations are also primarily consistent with China's higher levels of economic growth. Furthermore, Henan and Sichuan provinces have a high population density. These locations are relatively flat and resource-rich, making them ideal for human production and living activities.

To better compare with the annual average haze concentration in Figure 3, the population density data of each prefecture-level city from 2010 to 2012 were also taken as annual averages to generate the results shown in Figure 5. When comparing Figure 5 to Figure 3, it is clear that the regions with higher population density are also the regions with higher haze concentrations, which is especially evident in the Beijing–Tianjin–Hebei region, which is both an exceptionally densely populated region of China and a region with higher haze pollution levels. Furthermore, while some cities in the south are more densely populated than others, the number of populated cities in the north is higher overall. Figure 3 depicts the annual average haze concentration map, which shows that haze pollution is more severe in the north than in the south.

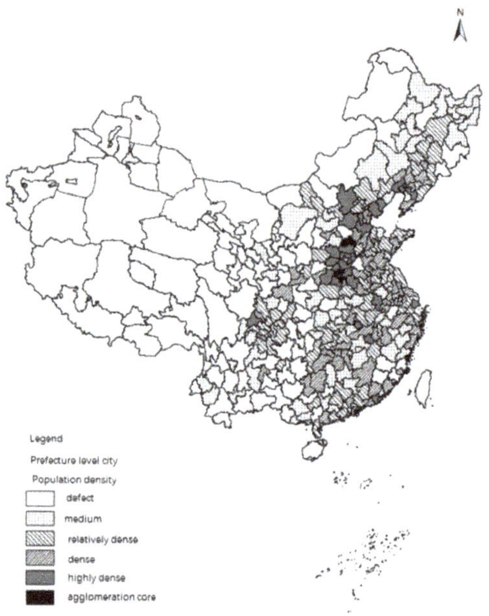

Figure 5. Annual Average Population Density Map of Chinese Prefecture-Level Cities during 2010–2012. Data Source: China City Statistical Yearbook (2012–2013). https://www.tongjinianjian.com/111090.html (accessed on 20 December 2021).

The following conclusions can be inferred from the descriptive statistics and empirical analyses shown above: (1) In China, the distribution of haze pollution is relatively concentrated. The North China Plain, the eastern coastal areas, and Xinjiang Uygur Autonomous Region are China's most polluted hazy areas today. (2) China's population distribution is characterized by a high density in the southeast and a low density in the northwest. The greater the region's level of economic development, the greater the urban population agglomeration. Finally, a comparison of the regional distribution maps (Figures 3 and 5) reveals that densely populated areas are essentially identical to areas with severe haze pollution. There is undoubtedly a spatial correlation between urban population density and haze concentration.

2.5. Testing of Spatial Panel Models

A necessary condition for the use of spatial econometric models is that the single variable or variables are spatially related and affect the validity of the estimated coefficients. We estimate a spatial correlation between haze pollution and population density based on the following description of the regional distribution of annual average $PM_{2.5}$ concentration values and urban population density in each prefecture-level city in China. However, the spatial relation should not be determined solely based on assumptions. Moran's I index is used in this article to analyze the spatial relation for the core variable in this spatial panel data—the annual average $PM_{2.5}$ concentration values. $PM_{2.5}$ concentrations are moving averages taken for every three years due to data availability limits. This section chooses $PM_{2.5}$ concentration data for three sub-samples (2001–2003, 2005–2007, and 2010–2012) and depicts Moran's I scatter plots, as seen in Figure 6:

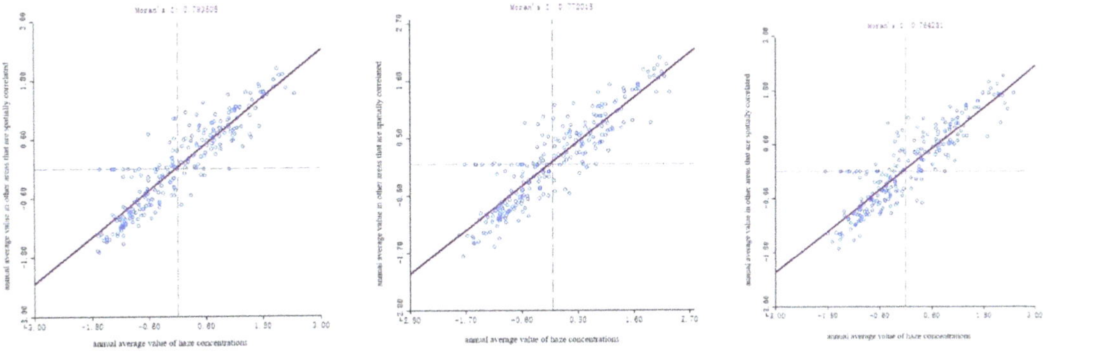

(a) (b) (c)

Figure 6. Scatter Plots of $PM_{2.5}$ in Prefecture-Level Cities in Some Years under the Geographical Proximity Weight Matrix. Source: Satellite Map of Global Annual Average Surface $PM_{2.5}$ Concentrations From SEDAC. The diagonal lines mean trend line of these scatter points. (**a**) Moran's I scatter plot of $PM_{2.5}$ concentration values from 2001 to 2003. (**b**) Moran's I scatter plot of $PM_{2.5}$ concentration values from 2005 to 2007. (**c**) Moran's I scatter plot of $PM_{2.5}$ concentration values from 2010 to 2012.

Under the "chariot" contiguity principle, if two areas share a common boundary, they are considered proximate, and the index is set to 1. In contrast, if two areas do not have a common boundary, they are not proximate, and the index is set to 0. Based on the "chariot" contiguity principle, a spatial weight matrix of geographical proximity is built, and the global Moran's I index of haze concentration is calculated. In academics, the global Moran's I statistic is a widely used metric for assessing global spatial auto-correlation. According to the Moran's I scatter plot of haze concentrations in Figure 6, the abscissa is the annual average value of haze concentrations during the representative year period. The ordinate is the annual average value of haze concentrations in other areas that are spatially correlated during the representative year period. Moran's I indexes for the annual average value of haze concentrations in 2001, 2005, and 2010 are all bigger than 1. Furthermore, most of the scatter plot points are clustered in the first and third quadrants. These phenomena exhibit a spatial distribution of high-high and low-low clustering, reflecting the positive spatial correlation of haze pollution.

If a spatial autocorrelation exists, a suitable spatial econometric model must be chosen. This study uses the Lagrange multiplier test and a robust Lagrange multiplier test to pick a spatial lag model and a spatial error model based on the findings of the general panel model estimation. In addition, to determine the category of fixed effect in the spatial econometric model, a likelihood ratio (LR) test will be performed on the entity fixed effect and the temporal fixed effect. Table 2 displays the outcomes of the testing:

Table 2. Moran's I Index and Lagrange Multiplier Test Results.

Test Index	Mixed Effect	Spatial Fixed Effect	Temporal Fixed Effect	Two-Way Fixed Effects
LM-LAG	5465.49 ***	6412.10 ***	4380.32 ***	2008.70 ***
robust LM-LAG	811.51 ***	192.81 ***	654.34 ***	71.41 ***
LM-ERR	5662.97 ***	15,496.40 ***	4843.55 ***	1938.22 ***
robust LM-ERR	1008.98 ***	9277.11 ***	1117.57 ***	0.93
Moran I	0.25 ***	0.42 ***	0.23 ***	0.15 ***
LR Space		9337.32 ***		
LR Time		1110.81 ***		

Note: *** in the table represent the 1% significance levels, while LM-LAG and LM-ERR here refer to the chi-square value and p-value derived from the Lagrangian test for the absence of spatial lag and the LM test for the absence of spatial error, respectively.

The test results for Moran's I index are all greater than 0 for all four effects, as shown in Table 2, and pass the test at the 1% level of significance. These test results suggest that the data within the spatial scope examined in this research have a substantial positive spatial correlation. They also increase the persuasiveness of the spatial econometric model employed for the estimate in this paper. Regardless of whether the model employs fixed effects and what fixed effects are used, the combined likelihood ratio test results show that the spatial and temporal fixed effects pass the test at the 1% significance level. As a result, a two-way fixed-effects model with spatial and temporal fixed effects is preferable. The Lagrangian multiplier lag test, the Lagrangian multiplier error test, and the robust Lagrangian multiplier lag test all pass the significance test at the 1% level. However, the robust Lagrangian error test does not. Furthermore, when the Lagrangian multiplier lag test's χ^2 is compared to the Lagrangian multiplier error test's χ^2, the former is greater than the latter. When the robust Lagrangian multiplier lag test's χ^2 is compared to the robust Lagrangian multiplier error test's χ^2, the result is the same. As a result, the spatial lag model is preferable to the spatial error model for analyzing this spatial panel data.

3. Results and Discussions

Next, we will assess whether there is a non-linear relationship between urban population agglomeration and haze pollution as a whole. The spatial spillover effect of urban population agglomeration on haze pollution will also be assessed. We will then examine which of the two mechanisms of urban population agglomeration on haze pollution is dominant. Namely, the scale and collective effects provide some empirical reference for future policy formulation on urban population agglomeration and haze pollution control.

3.1. The Influence Effect of Urban Population Density on Haze Pollution

Based on the results of the Lagrange multiplier test and the likelihood ratio test described above, the spatial error model [1] and the spatial lag model results for two-way fixed effects were measured. However, the spatial lag model with two-way fixed effects is still the primary basis for future study. The results of the three spatial lag models estimated using spatial fixed effects, temporal fixed effects, and two-way fixed effects in space and time are shown in Table 3.

Table 3 indicates that the spatial spillover effects of haze pollution are all significantly positive and that urban population agglomeration has a considerable positive effect on haze pollution. δ^2 epresents the spillover effect of haze pollution, that is, the impact of local haze pollution on other spatially related areas. The two-way fixed effects spatial lag model, in particular, indicates that this spatial spillover effect leads to 0.9826. At the 1% significance level, it also passes the estimated coefficient test. This implies that haze pollution in the region will have a spillover effect on other spatially related areas and that the more severe the haze pollution in the region, the more severe the haze pollution in its neighboring areas.

Table 3. Spatial Lag Model Regression Results Under Three Fixed Effects. The first column is the variable name, and the last three rows represent the estimation results of the three spatial lag models under the variable space fixed effect, time fixed effect, and space and time double fixed effect.

Variable	Spatial Fixed Effect	Temporal Fixed Effect	Two-way Fixed Effect
Ln*idenpop*	1.3117 *** (2.6217)	−0.7101 *** (−2.8983)	1.3180 *** (2.6064)
Ln*idenpop2*	−0.1918 *** (−2.7357)	0.1584 *** (3.7865)	−0.1928 *** (−2.7215)
Ln*idenpop3*	0.0093 *** (2.8657)	−0.0088 *** (−3.7784)	0.0093 *** (2.8529)
Ln*fdi*	0.0002 (0.1203)	0.0141 ** (2.2996)	0.0000 (0.0009)
Ln*rpgdp*	−0.0013 (−0.1630)	−0.2492 *** (−14.9650)	0.0018 (0.2100)
Ln*indgdp*	0.0741 *** (5.6274)	0.2853 *** (8.8712)	0.0696 *** (5.1666)
Ln*fasset*	0.0084 ** (2.0860)	0.0915 *** (10.3035)	0.0116 ** (2.2569)
Ln*sci*	−0.0027 (−0.6482)	−0.0338 *** (−3.1389)	−0.0059 (−1.2067)
Ln*exp_sci*	−0.0060 *** (−2.6021)	0.0285 *** (2.7951)	−0.0076 *** (−2.6213)
δ^2	0.9650 *** (137.7361)	0.9870 *** (575.3820)	0.9826 *** (258.5835)
R^2	0.9907	0.6698	0.9907
σ^2	0.0029	0.0914	0.0028

Note: *** and ** in the table represent 1% and 5% significance levels, respectively.

As shown in Table 3, the results of the spatial lag model based on the two-way effect of spatial fixed and temporal fixed effects show that the population density's absolute term, quadratic term, and cubic term all pass the estimated coefficient test at the 1% significance level. The absolute term, quadratic term, and cubic term coefficients are positive, negative, and positive, respectively. This concludes a significant non-linear relationship with an "N-shaped" curve between urban population agglomeration and haze pollution. According to current estimates, the inflection point values (persons/km^2) are between 490 and 2053, and the impact of urban population agglomeration on haze pollution may be separated into three stages: First stage: The population density is relatively low to the left of the first inflection point value, and it essentially belongs to small and medium-sized cities in China. These cities have not invested sufficiently in constructing various public services and supporting facilities. The majority of small and medium-sized cities are experiencing accelerated urban economic development, with increased investment attraction and new factories. As a result, contaminants are difficult to treat efficiently, pollution sources are expected to persist unabated, and urbanization will likely worsen haze pollution. Second stage: Urban population agglomeration and haze pollution have an inverse connection between the two inflection points. This could be because the urban population has been reduced. Public infrastructure has been improved, so that the amount of pollution controlled and treated exceeds the amount of pollution emitted during urban development, resulting in a virtuous cycle between population agglomeration and the urban environment. Third stage: To the right of the second inflection point, the urban population agglomeration and haze pollution once again have a positive relation, but it differs from the first stage in that the city's public facilities and services have been improved, but the city's continued growth in population density has caused several problems. For example, in first-tier cities such as Beijing, Shanghai, and Shenzhen, the permanent population is significant, and the migrant population is massive, with core urban districts growing overcrowded and a high number of white collars commuting every day between the city center and the suburbs. The enormous permanent population has created a significant demand for housing and appliances in megacities, while longer commuting times have exacerbated traffic congestion and motor vehicle emissions. The total number of pollutants emitted is so large that the city's self-regulatory function is "overloaded," finally exceeding the carrying capacity of the urban environment and increasing the haze pollution problem.

In conclusion, there is an "N-shaped" association between urban agglomeration and haze pollution, and Hypothesis 1 is tested. So, are there two action mechanisms in this non-linear relationship between urbanization and haze pollution, and what are the magnitudes of the specific effects?

3.2. Tests of Two Effects of Urban Population Agglomeration on Haze Pollution

According to the above-mentioned theoretical mechanism analysis, the scale and collective effects are essential in the influence of urban population agglomeration on haze pollution. The empirical results of other socioeconomic parameters other than population density in Table 3 can validate this tentatively.

First, the city scale effect has a significant positive impact on haze pollution. (1) Both the secondary industry added value as a proportion of GDP and the amount of fixed-asset investment, which reflect the industrial structure of the city, have a significant positive effect on haze pollution, indicating that the city scale effect concentrates and increases a large number of industrial activities, contributing to the aggravation of the haze pollution problem. The most significant pollutants that contribute to haze pollution are toxic pollutants produced during industrial production. This analysis also reveals that some regions of China's historic "three highs and one low" extensive growth model have contributed to haze pollution. (2) The majority of urban fixed-asset investment (excluding farmers) is spent on infrastructure and building investment, which reflects the degree of construction investment in the city to some extent. However, dust from construction sites is undeniably one of the major pollutants contributing to haze pollution. Too many construction projects, widespread renovations, and blind expansions can have a scale effect, aggravating haze pollution and potentially exacerbating the situation.

Second, the urban collective effect will exacerbate haze pollution. The urban agglomeration is usually accompanied by a cluster of additional talents, technology, and other elements that will help reduce haze pollution through knowledge spillover and technological innovation. As demonstrated in Table 3, the proportion of scientific research employment hurts haze pollution. However, it does not pass the significance level test. The amount of local fiscal investment in science and technology has a considerable detrimental impact on haze pollution. This finding suggests that, in combatting haze, the significance of the urban intensification effect should be fully recognized and that adequate government involvement and policy direction are also required. The proportion of science and technology expenditure in local financial expenditure has a considerable negative effect on haze pollution. This finding suggests that, in the process of combatting haze, the significance of the urban collective effect should be fully recognized, along with adequate government involvement and policy guidance. Increasing the percentage of science and technology expenditure in local financial expenditure will give more incentives for researchers to speed the transformation of research findings and develop strategies to eliminate haze. However, both of the above indicators are fairly tiny, which may be because technology's contribution is mainly reflected in two areas: enhancing production efficiency and green emission reduction technology. The former focuses primarily on reducing industrial emissions through technological means; however, the lower the energy consumption, the lower the costs; thus, demand for industrial products increases rather than decreases, and total pollution emissions do not necessarily decrease. The latter focuses primarily on reducing industrial emissions and the treatment of pollutants emitted from industrial production, which plays a more direct role in pollution reduction. At the same time, this outcome is consistent with the reality that, at this level, technological factors play a minimal role in eliminating haze.

The preceding is preliminary proof of the opposite effect of scale and collective effects on haze pollution during the urban population agglomeration process. A more in-depth examination will follow. Which effect dominates in urban population agglomeration due to haze pollution? The final effect of urban agglomeration on haze pollution is connected to the outcome. It is impossible to describe all of the components involved in the scale and collective effects of urban agglomeration due to data availability constraints. As a result, this article seeks to analyze one of the main viewpoints—urban traffic—while also comparing the degrees of the scale and collective effects caused by urban population agglomeration. Based on actual monitoring data and component analysis, the literature [20,21] discovered that motor exhaust emissions, dust from construction sites, burning of crops such as straw, coal firing, and other secondary products are the primary sources of pollutants (or secondary aerosols) at this stage. This shows that the urban transportation sector was representative and significant for the study. The number of civilian vehicles owned by each prefecture-level city is chosen as a proxy variable for the scale effect in terms of variables' setting. The bus ownership per 10,000 people in each prefecture-level city is chosen as a proxy variable for the collective effect. Both are mediators in the Sobel test for the effect of urban population agglomeration on haze pollution. This study assesses if the scale effect or the collective impact is substantial and whether the effect dominates at this stage. Table 4 displays the outcomes of the testing:

Table 4. Mesomeric Effect of Urban Population Agglomeration on Haze. The first line is the name of the measured index, and the last two lines represent the results of the Sobel test when Bus Ownership per 10,000 people or Occupying Amount of Civil Vehicles are used as proxy or mediator variables, respectively.

Index	Bus Ownership per 10,000 People	Occupying Amount of Civil Vehicles
Sobel	−0.0063 *** (0.0018)	0.0123 *** (0.0029)
Goodman-1(Aroian)	−0.0063 *** (0.0018)	0.0123 *** (0.0029)
Goodman-2	−0.0063 *** (0.0018)	0.0123 *** (0.0029)
a coefficient	0.0926 *** (0.0139)	0.1859 *** (0.0167)
b coefficient	−0.0677 *** (0.0170)	0.0664 *** (0.0142)
Indirect effect	−0.0063 *** (0.0018)	0.0123 *** (0.0029)
Direct effect	0.2900 *** (0.0115)	0.2772 *** (0.0118)
Total effect	0.2837 *** (0.0115)	0.2896 *** (0.0115)
Mediation/total effct	−0.0221	0.0426
Indirect effect/direct effect	−0.0216	0.0445
Total effect/direct effect	0.9784	1.0445

Note: *** in the table denote 1% significance levels. Standard deviations are shown in parentheses.

In general, the mediator's influence is composed of two parts: the effect of the explanatory variable on the mediator (Path a) and the effect of the mediator on the explanatory variable (Path b). In addition, Path c_1 refers to the overall effect of the primary explanatory variable on the explanatory variable without taking the mediator into account. When the mediator is taken into account, Path c_2 refers to the effect of the primary explanatory variable on the explanatory variable. According to Table 5, reviewing the specific paths in the mesomeric model reveals that the effect of Path c_2 is greater than Path c_1 when the bus ownership per 10,000 people is used as the mediator, indicating that encouraging the public to use public transportation and improving public infrastructure services are integral parts of the fight against haze pollution. However, when the occupying account of civilian vehicles is utilized as a mediator, the effect of Path c_2 is substantially smaller. The effect of the primary explanatory variable on the explanatory variable is much smaller than the total effect once the mediator is taken into account. The increased number of civilian vehicles will surely raise the emission of toxic gas in vehicle exhaust, "reinforcing" the beneficial effect of urban population agglomeration on haze pollution. As a result, Hypothesis 2's prediction that two mesomeric mechanisms negatively impact haze pollution is substantially validated.

Table 5. Direct, Indirect, and Total Effects of Different Factors on Haze Pollution.

Variable \ Effect	Direct Effect	Indirect Effect	Total Effect
Lndenpop	1.5923 ***	62.4951 **	64.0874 **
Lndenpop2	−0.2330 ***	−9.1437 **	−9.3767 **
Lndenpop3	0.0113 ***	0.4432 ***	0.4545 ***
Lnfdi	0.0000	−0.0007	−0.0007
Lnrpgdp	0.0022	0.0880	0.0903
Lnindgdp	0.0832 ***	3.2550 ***	3.3382 ***
Lnfasset	0.0141 **	0.5511 **	0.5652 **
Lnsci	−0.0070	−0.2727	−0.2797
Lnexp_sci	−0.0091 ***	−0.3554 **	−0.3644 **

Note: *** and ** in the table denote 1% and 5% significance levels, respectively.

So, which of the two mesomeric effects of urban population agglomeration on haze pollution reigns supreme at this point? As shown in Table 5, the bus ownership per 10,000 people and the occupying amount of civilian vehicles in the city pass the mesomeric effect test at the 1% significance level. It also demonstrates that two action mechanisms significantly exist in the effect of urban agglomeration on haze pollution. When the bus ownership per 10,000 persons in the city is used as the mediator, the indirect effect is −0.0063, and the proportion of this mesomeric effect in the total effect is −0.0221. When the occupying amount of civilian vehicles is utilized as the mediator, the indirect effect is 0.0123. The proportion of this mesomeric effect in the total effect is 0.0426. A comparison demonstrates that the mesomeric effect in the first situation contributes much less to the total effect than it does in the second situation. The "collective effect," indicated by the bus ownership per 10,000 persons in the city, is far less than the "scale effect," expressed by the occupying amount of civilian vehicles. This demonstrates that, at this time, the scale effect outweighs the effect of population agglomeration on haze pollution. The findings of this empirical study are also typically congruent with reality. With a significant number of people congregating in China's first- and second-tier cities, the demand for transportation, housing, and life has skyrocketed. However, many creative technology achievements have not been rapidly transformed and adapted to urban life, and people's consuming habits and attitudes have not significantly altered. Although first-tier cities such as Beijing and Shanghai have long used traffic restrictions and a lottery for license plates to govern urban traffic, the number of individuals buying vehicles and playing the lottery is increasing; despite repeated national attempts to regulate and control housing prices, real estate developers around the country are developing new construction in significant numbers. On the contrary, due to technical progress not being fully mature, relative discomfort, and consumer attitudes, new energy cars and electric vehicles have not been widely marketed and used. As a result, to tackle the problem of haze pollution, we might try to consider and design policies to suppress the scale effect caused by urban population agglomeration and improve the collective effect.

3.3. *Subdivision of the Effect of Urban Population Agglomeration on Haze Pollution*

The degree of various socioeconomic factors' effects on haze pollution is direct and indirect. A difference will be drawn here between direct and indirect effects. Direct effects refer to the direct action of changes in regional socioeconomic variables, while indirect effects are the spatial spillover effects of changes in regional socioeconomic variables. The estimated direct effects, indirect effects, and total effects of the explanatory variables on the core variables are shown in Table 5. The results suggest that population agglomeration has a significant positive effect on haze pollution, both directly and indirectly, which is consistent with our prior findings. Other factors' effects on haze pollution's direct and indirect effects remain essentially consistent. Population density, the proportion of secondary industry added value in GDP, fixed-asset investment (excluding farmers), and the proportion of scientific research expenditure in local governmental expenditure pass the coefficient tests with varying degrees of significance. It is also worth mentioning that the indirect effect of all socioeconomic factors on haze pollution is substantially more significant than the direct effect, demonstrating that haze pollution is vulnerable to the interaction of socioeconomic factors across regions.

4. Conclusions

4.1. Conclusions

Based on the preceding analysis, the paper's conclusions are: (1) Positive spatial correlation is a characteristic of haze pollution. Haze pollution in the region will be influenced by natural and artificial variables such as atmospheric movements and industrial transfers, affecting locations geographically proximate to the region. Similarly, air pollution in other places will impact urban air quality in the region. (2) There is a significant "N-shaped" non-linear relationship between urban population agglomeration and haze pollution. This suggests that, allowing for other socioeconomic factors, as urban population density increases, haze pollution tends to increase, then drop, then increase again. The reason for this is that at different levels of urban population agglomeration, the intensity and efficiency (the collective effect) of investment in public infrastructure and services and demand (the scale effect) for housing, home appliances, and motor vehicles fluctuate. (3) In China, the scale effect of urban population agglomeration on haze pollution is currently more significant than the collective effect. Most Chinese cities are still in the "accelerated development stage," with the rate of population urbanization significantly outpacing the rate of social service urbanization. Because urban infrastructure such as public transportation, health care, and education are still in the early stages of investment and construction, cities cannot exploit the benefits of efficient resource allocation fully, nor can they "regulate" and "alleviate" urban environmental issues. (4) The scale and collective effects negatively impact haze pollution, as evidenced by disparities in the impact of associated socioeconomic factors. The proportion of secondary industry added value in GDP and the urban fixed-asset investment have a significant positive effect on haze pollution in the region and a significant positive spatial spillover effect on spatially related areas. Meanwhile, local government investment in scientific research has a considerable impact on haze pollution prevention and control and a negative spatial spillover effect on haze pollution. (5) The indirect effect (or spatial spillover effect) of urban population agglomeration on haze pollution is much more significant than the direct effect. The same can be said about the effects of other socioeconomic factors on haze pollution. This demonstrates that other socioeconomic factors in spatially related areas are more likely to influence local haze pollution than local socioeconomic factors.

4.2. Policy Implications

(1) Effective haze pollution prevention and management necessitates completely using the collective effect induced by urban population agglomeration and successfully controlling the scale effect caused by it. Specifically:

(2) City layouts should be scientifically designed, and the size of the urban population should be reasonably regulated. The aims of city regulation should include transferring some of the urban functions of super-large and large cities, improving the social and economic development of small and medium-sized cities, attracting people to move to

them, and so on. Simultaneously, a city should increase its investment in the development in urban pollution control and emission reduction systems as well as improvements to public infrastructure. Governments at all levels should tighten their control over the approval criteria for urban fixed-asset investment projects while streamlining the approval process and severely controlling duplication of urban construction projects. Governments should also manage to increase active public engagement, stimulate the use of public transportation, and improve resource use efficiency. They should create a sound urban management system to assist city inhabitants in developing green travel habits by providing subsidies, fines, and traffic restrictions.

(3) Regional cooperation should be strengthened to coordinate policies on haze prevention and control. Joint prevention and control is a necessary way to combat haze. Each region could not avoid haze pollution itself but chose to strengthen communication and cooperation in order to complete a systematic regional coordination mechanism for atmospheric pollution prevention and control sooner. The keys to controlling haze pollution include modifying urban industrial structures, restricting high energy-consuming enterprises, and improving emission reduction and pollution control technologies.

(4) It is important to increase production efficiency, encourage green development and increase source and process management. We will raise awareness of "green production" and "green consumption" on both the supply and demand through policy advice and political assistance. We will strengthen the power of innovation and research in energy conservation and emission reduction and the ability to transform scientific and technological achievements, raise environmental awareness among manufacturing enterprises and the general public, and bring the market's role in regulating urban production and lifestyle into play.

Author Contributions: Conceptualization, X.L. and X.M.; methodology, X.L.; software, M.Z.; validation, X.L., M.Z. and W.Z.; data curation, K.Y.; writing—original draft preparation, X.L.; writing—review and editing, M.Z.; supervision, W.Z.; project administration, X.M.; funding acquisition, X.L. All authors have read and agreed to the published version of the manuscript.

Funding: This study was funded by the General Project funded by the National Social Science Foundation: Research on the impact mechanism and policy optimization of agricultural subsidies on environmental efficiency, grant number 20BJL038, the General Project funded by National Social Science Foundation: Research on the same group effect of technological innovation under the spatial economic social connection, grant number 19BJL083, and Key Project of Hubei Social Science Foundation: Research on major strategic issues of scientific planning in the 14th five year plan of Hubei Province, grant number HBSKZD2020033.

Institutional Review Board Statement: Not applicable.

Informed Consent Statement: Not applicable.

Data Availability Statement: The data in this study are publicly available in the Map of Global Annual Average Surface $PM_{2.5}$ Concentrations from SEDAC and China City Statistical Yearbook.

Acknowledgments: The author would like to thank the editors and anonymous reviewers for their valuable comments and suggestions on this paper.

Conflicts of Interest: The authors declare no conflict of interest.

References

1. Guan, D.; Su, X.; Zhang, Q.; Peters, G.P.; Liu, Z.; Lei, Y.; He, K. The Socioeconomic Drivers of China's Primary $PM_{2.5}$ Emissions. *Environ. Res. Lett.* **2014**, *9*, 1–9. [CrossRef]
2. Ma, L.M.; Zhang, X. The Spatial Effect of China's Haze Pollution and the Impact from Economic Change and Energy Structure. *China Ind. Econ.* **2014**, *4*, 19–30.
3. Shuai, S.; Xin, L.; Cao, J.; Yang, L. China's Economic Policy Options for Governing Fog Pollution—Based on the Perspective of Spatial Spillover Effect. *Econ. Res. J.* **2016**, *9*, 73–88.
4. Leng, Y.; Du, S.Z. Energy price distortion and haze pollution: The Evidence from China. *Ind. Econ. Res.* **2016**, *1*, 71–79.
5. Shoufeng, H. Environmental regulation, shadow economy and smog pollution. *Econ. Perspect.* **2016**, *11*, 33–44.

6. Quan, S.W.; Huang, B. Embedding Effects in Evaluation of Multiple Environmental Policies—Evidences from Beijing's Haze and Sand Control Policies. *China Ind. Econ.* **2016**, *8*, 23–39.
7. Zhang, X.Y.; Wang, Y.Q.; Lin, W.L.; Zhang, Y.M.; Zhang, X.C.; Gong, S.; Zhao, P.; Yang, Y.Q.; Wang, J.Z.; Hou, Q.; et al. Changes of atmospheric composition and optical properties over Beijing—2008 Olympics monitoring campaign. *Bull. Am. Meteorol. Soc.* **2009**, *90*, 1633–1651. [CrossRef]
8. Shi, Q.L.; Guo, F.; Chen, S.Y. "Political Blue Sky" in Fog and Haze Governance—Evidence from the Local Annual "Two Sessions" in China. *China Ind. Econ.* **2016**, *5*, 40–56.
9. Zhang, S.L.; Li, Y. Different Government Smog Governance Strategies in Response to Public Opinion. *Comp. Econ. Soc. Syst.* **2016**, *3*, 52–60.
10. Chen, S.Y.; Chen, D.K. Energy Structure, Haze Governance and Sustainable Growth. *J. Environ. Econ.* **2016**, *1*, 59–75.
11. Ma, L.M.; Liu, S.L.; Zhang, X. Study on Haze Pollution Induced by Energy Structure and Transportation: Based on Spatial Econometric Model Analysis. *Finance Trade Econ.* **2016**, *1*, 147–160.
12. Xiang, K.; Song, D.Y. Spatial Analysis of China's $PM_{2.5}$ Pollution at the Provincial Level. *China Popul. Resour. Environ.* **2015**, *25*, 153–159.
13. Qin, M.; Liu, X.Y.; Tong, Y.T. Does Urban Sprawl Exacerbate Haze Pollution—An Empirical Study of Fine Particles($PM_{2.5}$) in Chinese Cities. *Finance Trade Econ.* **2016**, *11*, 146–160.
14. Glaeser, E. *Triumph of the City: How Our Greatest Invention Makes Us Richer, Smarter, Greener, Healthier, and Happier*; Runquan, L., Translator; Shanghai Social Sciences Press: Shanghai, China, 2012; pp. 113–131.
15. Zeng, X.G.; Xie, F.; Zong, Q. Behavior Selection and Willingness to Pay of Reducing $PM_{2.5}$ Health Risk: Taking Residents in Beijing as an Example. *China Popul. Resour. Environ.* **2015**, *25*, 127–133.
16. Chun, L. On China's Urbanization: Knowledge Rebuilding and Practice Rethinking. *Popul. Res.* **2013**, *37*, 3–15.
17. Elhorst, J.P. Dynamic Models in Space and Time. *Geogr. Anal.* **2001**, *33*, 119–140. [CrossRef]
18. Kelejian, H.H.; Robinson, D.P. *Spatial Correlation: A Suggested Alternative to the Autoregressive Model—New Directions in Spatial Econometrics*; Springer: Berlin/Heidelberg, Germany, 1995; pp. 75–95.
19. Donkelaar, A.; Martin, R.V.; Brauer, M.; Kahn, R.; Levy, R.; Verduzco, C.; Villeneuve, P.J. Global Estimates of Exposure to Fine Particulate Matter Concentrations from Satellite-based Aerosol Optical Depth. *Environ. Health Perspect.* **2010**, *118*, 847–855. [CrossRef] [PubMed]
20. Lv, B.L.; Zhang, B.; Bai, Y.Q. A systematic analysis of $PM_{2.5}$ in Beijing and its sources from 2000 to 2012. *Atmos. Environ.* **2016**, *124*, 98–108. [CrossRef]
21. Li, X.Y. Empirical Analysis of the Smog Factors in Beijing-Tianjin-Hebei Region. *Ecol. Econ.* **2016**, *32*, 144–150.

atmosphere

Article

The Hot Topics, Frontiers and Trends about Research on the Relationship between Air Pollution and Public Health—Visual Analysis Based on Knowledge Map

Yu Gao [1], Shibing You [1,*], Yiping Xu [2] and Na Wang [3,*]

[1] Economics and Management School, Wuhan University, Wuhan 430072, China; gao.yu@whu.edu.cn
[2] Economic College, Wuhan Business University, Wuhan 430056, China; 20160116@wbu.edu.cn
[3] School of Business, Hubei University, Wuhan 430062, China
* Correspondence: sbyou@whu.edu.cn (S.Y.); 20110034@hubu.edu.cn (N.W.)

Abstract: It is of great practical significance to analyze the hot issues, research frontiers, and trends concerning the relationship between air pollution and public health and to adopt reasonable strategies to control air pollution and prevent health hazards for follow-up research in this field. Unlike traditional literature reviews, this paper adopts a visual, flexible, and scientifically systematic approach to the analysis, which makes these analysis results more intuitive and comprehensive. Based on the core collection of the Web of Science and CNKI databases, this paper uses CiteSpace software to draw and comment on the maps of Chinese and English keywords, publishing time, author, country, and research institutions in this field. The results show the following: (1) The number of studies on the relationship between air pollution and health has increased year by year; researchers have formed sub cooperation networks, and the trend of cooperation and exchange has become more and more obvious in recent years; the impact of air pollution on health is a hot topic in the world. (2) Research hot topics mainly focus on the selection of air pollutants, health economic consequences of air pollution and the global burden of disease it causes, health indicators, research samples, which are gradually being refined, the synergistic governance of air pollution, and climate change. (3) The analysis of research frontiers and trends reveals that, first, the study of air pollutants in the English literature has undergone a refinement from nitrogen dioxide to fine particulate matter, and the sources of air pollutants in the Chinese literature have undergone changes in the petrochemical industry, indoor formaldehyde pollution, and haze. Second, atmospheric pollution has a significant negative impact on health, increasing the incidence of respiratory and cardiovascular diseases, and even causing death. Third, sustained exposure to pollution then causes greater damage to health and will be a key direction for future research. Fourth, the literature not only studies the correlation but also emphasizes the causal inference between air pollution and health and measures the economic costs associated with health. Finally, air pollution and climate change need to be governed synergistically. The article points out that the three areas of sustained pollution exposure, indirect consequences of negative health effects of air pollution, and air pollution and climate change may be the future focus of the field.

Keywords: air pollutants; sustained exposure to pollution; respiratory and cardiovascular diseases; CiteSpace; co-occurrence keywords; burst words

Citation: Gao, Y.; You, S.; Xu, Y.; Wang, N. The Hot Topics, Frontiers and Trends about Research on the Relationship between Air Pollution and Public Health—Visual Analysis Based on Knowledge Map. *Atmosphere* **2022**, *13*, 892. https://doi.org/10.3390/atmos13060892

Academic Editors: Zengyun Hu, Xuguang Tang and Qinchuan Xin

Received: 19 May 2022
Accepted: 26 May 2022
Published: 31 May 2022

1. Introduction

Air pollution is a growing hazard to human health. It causes respiratory and other diseases, and is an important cause of morbidity and mortality. The London smog incident in December 1952, an air pollution disaster, caused more than 12,000 people to die of respiratory-related diseases [1]. The environmental disaster and a subsequent series of air pollution events have gradually attracted the attention of academia, world leaders, and the public on the relationship between air pollution and health.

Since the 1980s, the rapid urbanization and industrialization of developed countries have led to an increase in air pollutant emissions, which has intensified the urgency of studying and solving this problem. Based on a large random sample of data from developed countries, studies found that people living in cities with a high degree of urbanization and industrialization had significantly higher rates of respiratory cancer, atherosclerosis, and ischemic heart disease than in other areas, and that these risks increased over time [2–5]. In addition, air pollution also significantly increases infant mortality [6,7]. The above conclusions are drawn from the research based on the samples of developed countries such as the United States and Europe.

Developing countries, especially those that rely heavily on manufacturing and fossil energy for their development, are facing serious pollution challenges as their level of industrialization and urbanization increases. Some scholars have conducted a series of studies based on data from developing countries, considering that the health effects of air pollution in developing countries may differ in many ways from those in developed countries, such as differences in regulatory environment, public health services, education level, and initial level of pollution and health [8,9]. The OECD estimates that nearly 1.5 million people die each year from exposure to particulate matter, more than from malaria or water pollution. As pollution levels rise, the organization claims that this figure may exceed 3.5 million people per year by 2050, with most of deaths occurring in rapidly industrialized countries such as India and China.

China is typical in terms of air pollution research. There are three reasons for this. First, in recent years, China has experienced a significant decline in environmental quality with its high economic growth, which provides a direct case study for the field. At the same time, the Chinese government has taken various measures to increase its efforts to combat air pollution, and pollution concentrations have declined significantly, which provides scholars with good conditions for quasi-experimental research. In addition, the accuracy, availability and accessibility of pollution data in China have been qualitatively improved, which provides scholars with a good research basis. Considering that pollution levels and environmental policies in China vary greatly in space and time, as well as that the quality of pollution data has increased significantly, scholars have continued to pay attention to this issue, producing a large number of relevant representative studies. In view of this, this paper summarizes relevant research on CNKI in order to find the different impacts of air pollution on health in different periods and the research progress of the relationship between air pollution and health. At the same time, in view of the comprehensiveness and universality of the global research on the relationship between air pollution and health, this paper summarizes and compares the research results of air pollution and health in the web of science database and makes a more comprehensive analysis and supplement to the research on the relationship between air pollution and health. However, faced with a large number of studies, it is difficult to systematically and scientifically sort out and summarize the research results by simply reading and summarizing the literature. Based on this, this paper uses CiteSpace software to conduct atlas quantitative research on the literature related to the relationship between the two so as to master the research hot topics, frontiers and overall trends in this field. Unlike traditional literature reviews, the contribution of this paper is to identify the main areas of research activity on the relationship between air pollution and health, and the stages of evolution of the research topics involved, in a visual, flexible and scientifically systematic way, making the results of the analysis more intuitive and comprehensive.

The structure of the rest of this paper is arranged as follows: the second part shows the data and methods, the third part discusses the general overview of air pollution and health research, the fourth part includes the analysis of research progress in the field of air pollution and health, the fifth part is the conclusions draw from our research, and the sixth part represents a discussion of our findings.

2. Data, Methods and Descriptions of Indicators

2.1. Data Sources

This paper takes the Chinese core journals in CNKI and the core collection of Science Citation retrieval Web of Science as the data source to conduct general measurement and statistics on the research on the relationship between air pollution and public health; we visualize and analyze the dataset with CiteSpace. Since there are few relevant documents in each year before 1991, which has little statistical significance, the retrieval time area of Chinese and English studies are uniformly set from 1991 to 2021. In the CNKI core journal database, the search strategy is set as the theme "atmosphere pollution" or "air pollution" and "health". The search condition of the Web of Science core collection is TS = (air pollution and health). A total of 1111 Chinese documents and 29,422 English documents meet the search strategy.

2.2. Research Methods and Descriptions of Indicators

This paper uses CiteSpace visualization software for bibliometric analysis and literature knowledge mapping. The CiteSpace software was developed by Dr. Chaomei Chen, which combines social network analysis, association rule analysis and other methods to analyze the development dynamics of research fields through knowledge mapping, and explore the evolution trends, research hot topics, and research frontiers of research fields.

Based on the bibliometric method, 1111 Chinese studies and 29,422 English studies were retrieved as the research database of the relationship between air pollution and health, and the literature database was imported into CiteSpace software for the analysis of publication time, authors, issuing institutions, issuing countries, research hot topics, and research frontiers. Among them, the selection of evaluation indexes about research hot topics and research frontiers are explained as follows.

First, there is the issue of research hot topics. Keywords are a high condensation of the research content of the literature, and high-frequency keyword co-occurrence analysis, keyword clustering analysis, keyword centrality, and research content of highly cited literature can be used as the analysis indicators of research hot topics. Keyword co-occurrence analysis is a bibliometric method to identify research hot topics in the field by revealing or expressing how frequently keywords at the core of the literature appear in that research field. Based on the co-occurrence keyword spectrum, it can be found that the keyword co-occurrence network clusters into irregular regions, each of which corresponds to a label. The order is from zero to large, and the smaller the number, the more keywords are contained in the clusters; each cluster is composed of multiple closely related words. Centrality indicates the amount of information flow between keywords and the degree of control over the literature network, which can reflect the importance and connectivity of the relative position of the keyword nodes in a research field and is used to test whether the high-frequency keywords are in the core position and reflect the position of the nodes in the overall network. In addition, this paper also selects the highly cited literature in this field in both the English and Chinese literature as an auxiliary analysis, and the highly cited literature can also reflect the research hot topics in this field.

Second, there is the issue of research frontiers. The identification and tracking of research frontiers can provide the evolutionary dynamics of the research field, identify the research progress in each stage of the field and predict the development trend, mainly through the analysis results of the burst keywords for judgment. Among them, burst keywords are keywords with a sudden increase in frequency in a certain time period, emphasizing sudden change, which can reflect the research hot topics and trends in a specific time period. By identifying and tracking research frontiers, we can detect the sudden growth of research interest in a certain subject area and can predict the development trend of the field and identify the direction of further development.

3. General Situation of Air Pollution and Health Research

3.1. Chronological Distribution

The publication of the research on the relationship between air pollution and health published in Chinese core journals is shown in Figure 1. The overall number of published papers shows an upward trend. The number of papers published before 2002 was less than 20, and the number began to increase significantly after 2002. The number of papers published in 2018 reached 99. The impact of air pollution on health has attracted extensive attention in the academic community.

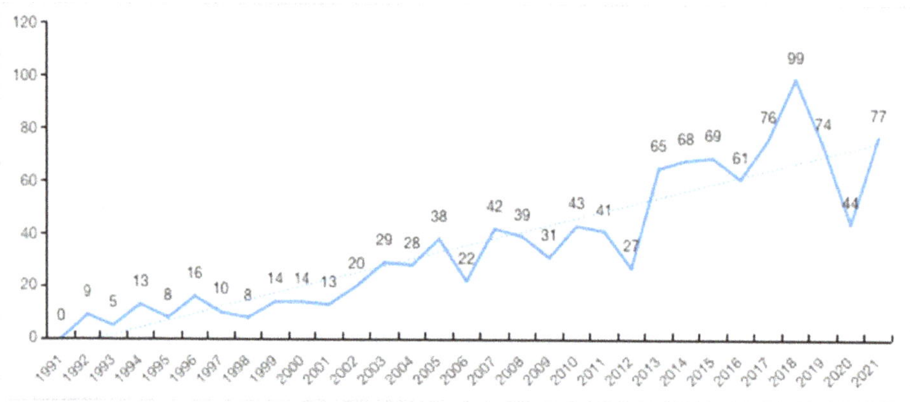

Figure 1. The number of Chinese studies with time.

According to the statistical analysis results of 29,422 English documents received in the core collection of Web of Science, the number of documents on the relationship between air pollution and public health is on the rise, especially after 1999, from 212 in 1999 to 4061 in 2021, as shown in Figure 2.

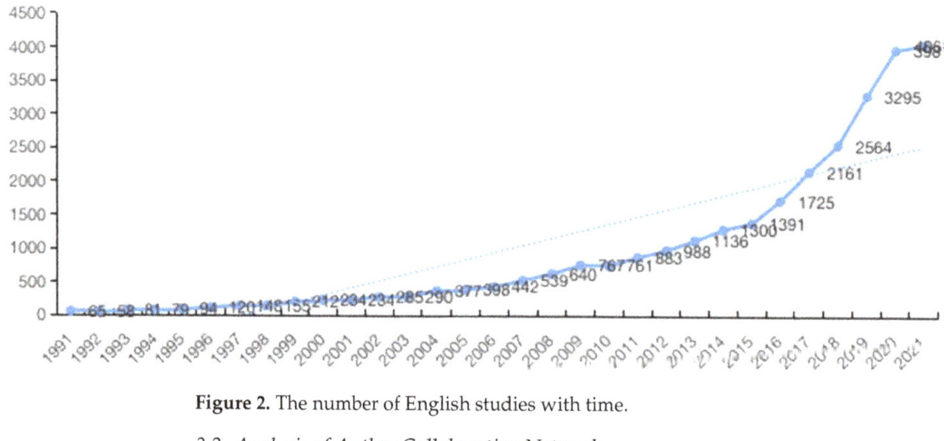

Figure 2. The number of English studies with time.

3.2. Analysis of Author Collaboration Network

Figures 3 and 4 show the author collaboration network of the Chinese and English literature. The number of nodes, the number of links and the network density of the network can be seen by the parameters in the upper left corner. The number of nodes is the number of authors in the graph, and the number of links is the number of author collaborations; as long as the authors have appeared in the same document, there will be a link between them. The network density measures whether the authors collaborate closely with each other. In addition, the size of the circles of the nodes in the graph represents the number of articles published by the authors, and the more articles published, the larger the circles. The color of the line corresponds to the color bar at the top of the graph. If the color of the line corresponds to the brown color on the left side of the color bar, it means that the authors collaborated a long time ago, and if the color of the line corresponds to the green color on the right side, it means that the authors collaborated recently.

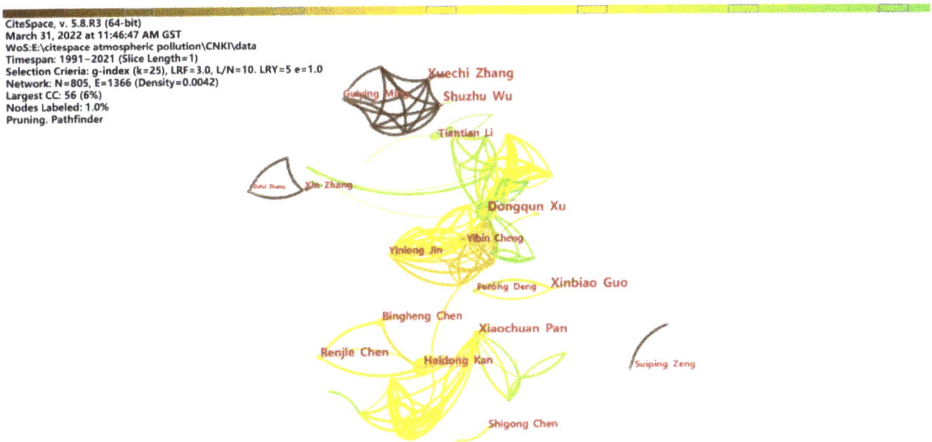

Figure 3. Chinese literature author cooperative network.

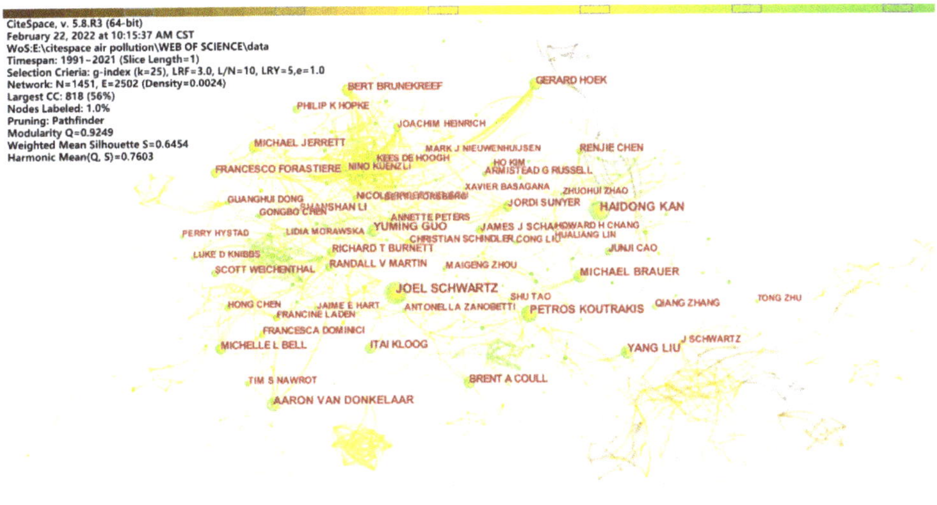

Figure 4. English literature author cooperative network.

The author network mapping shows some of the major authors generated based on the Chinese literature, where the number of nodes is 805, the number of connections is 1366, and the density is 0.0042. This small density value indicates that the network of the whole mapping is more fragmented. As can be seen from Figure 3, the scholars with a larger number of nodes are Haidong Kan, Dongqun Xu, Bingheng Chen, Xiaochuan Pan, Xinbiao Guo, Tiantian Li, Renjie Chen, Shuzhu Wu and Xin Zhang, and larger nodes indicate more publications. In addition, the mapping shows several author sub-networks, implying that scholars communicate and cooperate with each other, and the sub-collaboration networks centered on teachers such as Xiaochuan Pan, Dongqun Xu, and Yinlong Jin are more obvious. The color of the connecting line corresponds to the color bar at the top of the figure. If the color of the connecting line corresponds to the brown color on the left side of the color bar, it means that the authors collaborated many years ago, and the green color on the right side means that the authors collaborated in recent years. It can be seen that the color of the collaboration network line of Xiaochuan Pan and Dongqun Xu is yellow and green, which indicates that the teachers have maintained their research in this field and contacted and cooperated with many other scholars for a long time, while several teachers, such as Haidong Kan, Bingheng Chen and Renjie Chen, who have also published more articles, have formed a mutual collaboration network, but have less communication with other scholars in comparison. The color of Shuzhu Wu's collaborative network line shows that the time of publication and collaboration has been very early.

The number of nodes in the network map of English literature authors is 1451, the number of connections is 2502, and the density is 0.0024. This small density value indicates that the network of the whole mapping is more fragmented. As can be seen from Figure 4, the scholars with larger nodes are Joel Schwartz, Haidong Kan, Yuming Guo, Yang Liu, Petros Koutrakis, Aaron van Donkelaar and Michael Brauer. From the point of view of node color, these scholars have been researching for a long time and continue to plow deeply in this field. It is worth mentioning that Prof. Haidong Kan's English posting volume is very high, but his cooperation network is not large compared with other scholars with a large amount of writing, combined with the previous Chinese literature analysis. Prof. Haidong Kan is a world leader in the number of Chinese and English publications in this field. In addition, the cooperation networks are mostly green and yellow, indicating that with the development of time, scholars pay more and more attention to mutual communication and cooperation; the cooperation has increased in recent years; and that there are many cooperative papers among various scholars.

3.3. Analysis of Research Institutions

The research institutions that published Chinese research in this field of the relationship between air pollution and health are shown in Figure 5. Three institutions—Peking University, CDC, and Fudan University—lead other institutions in China in terms of the number of Chinese publications, with Peking University having the highest number of publications, and all of the above institutions have more than 60 publications. Figure 6 shows the number of publications in English by research institution, where the top five institutions in terms of the number of publications are the University of California System, Harvard University, Chinese Academy of Sciences, Harvard T H Chan School of Public Health, and Peking University. As we can see, Peking University is still on the list, with a large number of publications in both English and Chinese, and has accomplished much in this field.

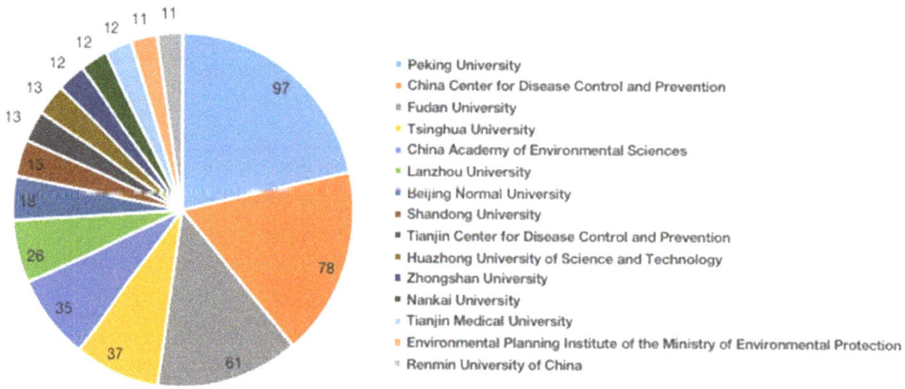

Figure 5. Chinese publication institutions of air pollution and health research.

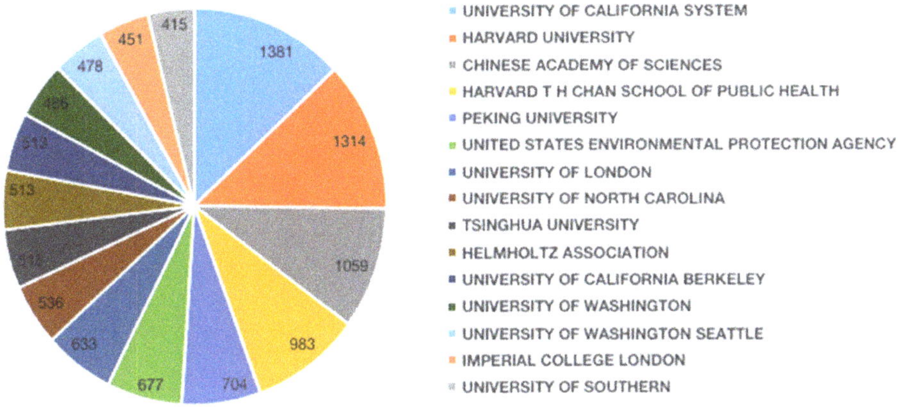

Figure 6. English publication institutions of air pollution and health research.

3.4. Country Analysis

According to the map of national co-occurrence knowledge drawn by the Web of Science database (Figure 7), the USA ranked first in node size and number of connections, indicating that the United States not only published the most papers in this field but also had the most extensive foreign exchanges and cooperation. The United States, China, the United Kingdom, Canada, Italy, Germany, India, and Spain are the countries with the highest number of publications, indicating that these countries have strong scientific research capacity in the field of the relationship between air pollution and health. On the one hand, it can be seen from the analysis results of the above research institutions that many universities in these countries have strong scientific research and academic strength. On the other hand, most of these countries are mature industrialized developed countries, which have accumulated rich experience in the impact of air pollution on health. It can also be seen that the study of the impact of air pollution on health has become a hot issue worldwide.

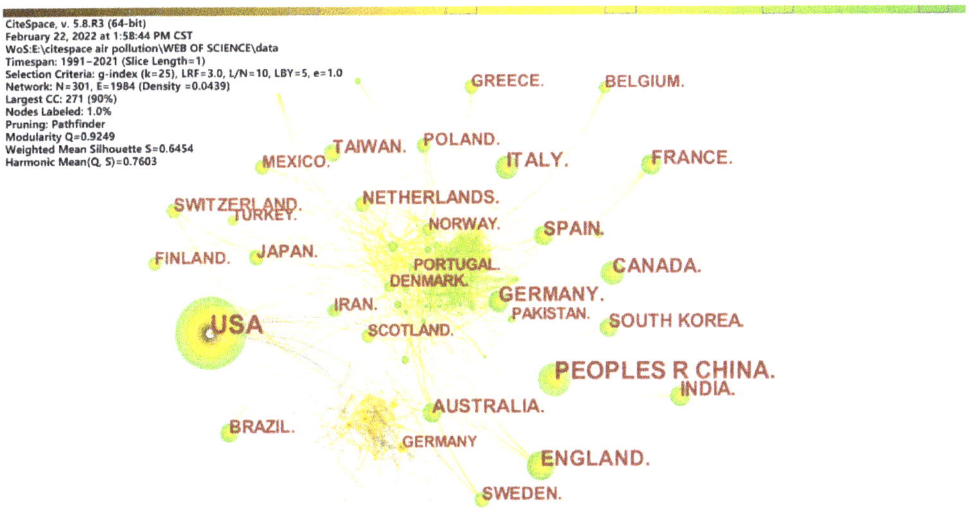

Figure 7. Co-occurrence knowledge mapping of the issuing countries.

4. Research Progress

4.1. Analysis of Research Hot Topics

The article analyzes research hot topics through high-frequency keyword co-occurrence analysis, keyword clustering analysis, keyword centrality, and highly cited literature, of which the keyword co-occurrence graph, keyword clustering graph, centrality, and highly cited literature tables are shown in Figures 8 and 9, Tables 1–3.

Among them, keyword co-occurrence analysis is a bibliometric method to determine the research hot topics in the field by judging the frequency of keywords appearing in that research field. Keyword clustering is based on the co-occurrence keyword spectrum for analysis, which can be found in the keyword co-occurrence network clustered into irregular regions, where each region corresponds to a label. The order is from small to large, and the smaller the number, the more keywords are contained in the clusters; each cluster is composed of multiple closely related words. In addition, centrality indicates the amount of information flow between keywords and the degree of control over the literature network, which can reflect the importance and connectivity of the relative position of the subject term nodes in a research field and is used to test whether high-frequency keywords are in a central position and reflect the position of the nodes in the overall network.

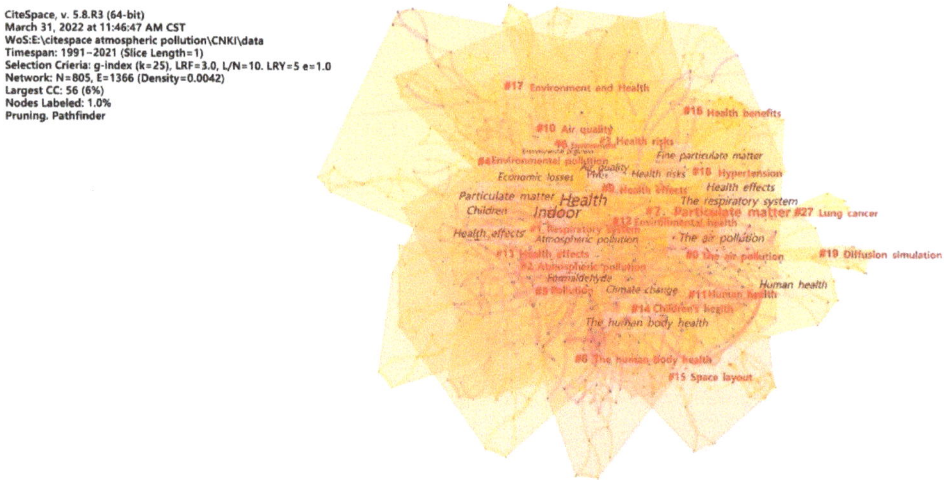

Figure 8. Chinese keyword clustering view.

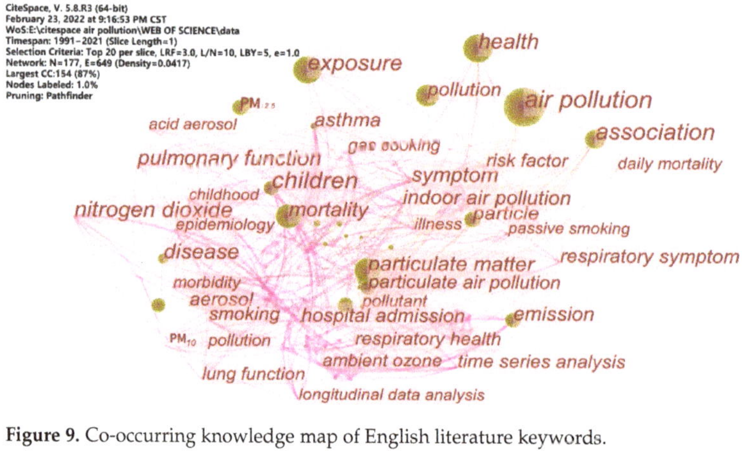

Figure 9. Co-occurring knowledge map of English literature keywords.

Table 1. Highly cited Chinese literature.

Subject	Literature	Main Content	Cited Frequency
Health effects, health economic losses	[10]	Exposure response relationships for health effects of particulate matter	275
	[11]	PM_{10} and economic losses to health	306
	[12]	$PM_{2.5}$ and premature death, respiratory disease	402
	[13]	PM_{10}, $PM_{2.5}$ and daily mortality	243
	[14]	A comprehensive review of the health effects of haze	338
	[15]	A review of the health effects of TSP, PM_{10}, and $PM_{2.5}$	200
	[16]	Haze pollution causes, solutions	243
Economic losses from environmental pollution	[17]	Estimation of economic losses from environmental pollution	230
	[18]	Estimation of economic losses from environmental pollution	207
Climate change	[19]	Air pollution and climate change	196

Table 2. Main research hot topics in this field in English literature.

Number	Keyword	Frequency	Year	Centrality
1	Air Pollution	92,062	1991	0.3
2	Health	4072	1991	0.07
3	Exposure	3505	1991	0.11
4	Particulate Matter	3350	1994	0.15
5	Mortality	2580	1991	0.05
6	Pollution	2252	1991	0.07
7	Association	1803	1992	0.13
8	Impact	1533	2006	0
9	$PM_{2.5}$	1392	2007	0.03
10	Particle	1348	1994	0.04
11	Risk	1319	2001	0
12	Emission	1301	1992	0.15
13	Disease	1011	1991	0.08
14	Children	829	1991	0.21
15	Quality	825	2009	0
16	Particulate Air Pollution	823	1994	0.04
17	Air Quality	770	2016	0

Table 3. Highly cited English literature.

Literature	Published Journals and Years	Main Content	Cited Frequency
[20]	LANCET 2017	Global burden of disease due to ambient air pollution	2432
[21]	LANCET 2012	Global burden of disease	6961
[22]	LANCET 2015	Global burden of disease	1690
[23]	LANCET 2018	Global burden of disease	1666
[24]	LANCET 2020	Global burden of disease	1414
[25]	NATURE 2015	Outdoor Air Pollution and Premature Mortality	2564
[26]	NATURE 2014	Secondary Aerosol and Particulate Pollution	2581

CiteSpace software provides two metrics, Modularity Q (Q value) and Mean Silhouette (S value), as a basis for judging the effectiveness of mapping based on network structure and the clarity of clustering. In general, the Q value is generally in the interval [0,1), the structure of the delineated associations is significant for Q value > 0.3, the clustering is reasonable for S value > 0.5, and the clustering is efficient and convincing for S value > 0.7.

4.1.1. Research Hot Topics in Chinese Literature

Analysis of Chinese Literature Keyword

The clustering view of Chinese literature keywords is shown in Figure 8. The parameters in the upper left corner show that the Modularity Q value is 0.6846 and the Mean Silhouette value is 0.8782, indicating that both parameter values are within the ideal range, indicating that the mapping effect is good and the clustering effect is significant. The keyword co-occurrence network in the figure is clustered into many irregular regions, where each region has its own clustering label; the smaller the clustering number, the more keywords contained in the clusters. Figure 8 shows part of the clustering labels.

As shown in Figure 8, in addition to the subject terms of air pollution and health studied in this paper, the hot topics of the Chinese literature related to the study of the relationship between the two are mainly respiratory system, health risks and benefits, particulate matter, health effects and effects, air quality, human health, children's health, spatial layout, hypertension, diffusion simulation, and lung cancer incidence. These high-frequency and high-centrality keywords represent, to some extent, the hot issues of research concerns on the relationship between air pollution and health in the Chinese literature during 1991–2021. Based on these words, the hot topics are summarized as follows. First, the types of air pollutants with health effects; second, the health effects of air pollution, which are found by the clustering results that the health effects of air pollution are mainly focused on making the public suffer from respiratory and cardiovascular diseases; third, the study groups are more all-age samples, as many articles focus on their effects on children's health; fourth, what the spatial layout of air pollution is and the simulation of pollutant dispersion; and fifth, the causal effects of air pollution on health, and the assessment of health risks and benefits.

Analysis of Chinese Highly Cited Literature

The Chinese highly cited literature covers three main topics: first, the health effects of air pollution, including health risks and health economic losses; second, the economic losses due to environmental pollution; and third, the synergistic management of air pollution and meteorological changes.

Kan et al. (2002) [10] established an exposure–response relationship for particulate matter–health effects in China based on Chinese air pollution and health data, establishing the relative risk of occurrence of adverse health effects in the population for each unit increase in atmospheric particulate matter concentration at each health effect endpoint from morbidity to mortality. This literature has been cited many times and has been used by many scholars to evaluate the health risks of atmospheric particulate matter pollution and the health economic losses in China.

Chen et al. (2010) [11] roughly estimated the related health economic losses based on the annual average PM_{10} concentration and health data in China in 2006, and the article found that the health losses caused by PM_{10} pollution were mainly premature death, respiratory diseases, and cardiovascular diseases, and based on the outpatient and hospitalization costs incurred in dealing with these health problems, the health economic losses were calculated to be CNY 341.403 billion. Among them, the loss caused by premature death accounts for the largest percentage. High $PM_{2.5}$ exposure resulted in a significant increase in acute population health risk in Beijing during the study period, with a significant increase in sudden deaths, hospitalizations for respiratory diseases, hospitalizations for cardiovascular diseases, pediatric outpatient visits, and internal medicine outpatient visits. The article measured the economic loss of health of the population based on the population risk evaluation [12]. Based on the data from Shanghai, the same conclusion was obtained that PM_{10} and $PM_{2.5}$ pollution significantly cause health risks and have potential health hazards for the acute population [13]. Bai (2004), Wang (2005), and Wang (2014) [14–16], on the other hand, conducted a review of the literature on the health effects of haze and particulate matter.

In the discussion of economic losses of environmental pollution, the losses involved mainly include economic losses caused by air pollution, water pollution, solid waste, and other pollution, among which economic losses caused by air pollution are further divided into economic losses of human health, industrial production, household cleaning and corrosion of construction materials, so economic losses of health caused by air pollution are only a part of the environmental pollution losses [17,18].

Another highly cited paper on climate change and air pollution points out that global climate change is mainly caused by increasing greenhouse gas emissions, while air pollution is mainly caused by aerosol particles, and the causes of both are largely common, mainly caused by emissions from fossil fuel combustion. Since the two problems (air pollution and climate change), not only coexist but also have the same causes to a large extent, it is necessary to formulate a synergistic or coupled management strategy to address both problems from the same source [19].

4.1.2. Research Hot Topics in English Literature

Analysis of English Literature Keyword

Figure 9 shows the results of co-occurrence keyword analysis in the English literature. In this figure, each node represents a keyword, the size of the node indicates the frequency of the keyword, and the connection between the nodes is the co-occurrence relationship between the keywords. Table 2 shows the keyword centrality, which indicates the amount of information flow between keywords and the degree of control over the literature network. The larger the centrality, the more important the node is in the network. The larger the circle of nodes, the greater the centrality of the keyword, and the more it can represent the research hot topic in the research field.

In Figure 9 and Table 2, excluding the article's subject word "air pollution" and "health", the key words that appear more frequently in the field of air pollution-health relationship research are mainly "exposure", "association", "mortality", "particulate matter ", "particulate air pollution", "pollutant", "disease", "emission", "children", "asthma", "$PM_{2.5}$", "risk". From Table 2, the keywords with high centrality are mainly "air pollution", "children", "particulate matter", "emission", "association", "exposure". These high-frequency and high-centered keywords represent, to some extent, the hot issues in the field of air pollution and health that generally concerned the international academic community during 1991–2021. Based on these words, this paper classifies the hot topics into four categories: types of air pollutants that have health effects, ways of air pollutants affecting health, health consequences caused by air pollution, and groups affected by air pollution.

Combining the relevant literature and the analysis of the above results, firstly, from some high-frequency keywords—"particulate matter", "particulate air pollution", "pollutant", and "$PM_{2.5}$"—we can see that the field focuses on the exploration of the types of air pollutants that affect health, and particulate matter is the main source of pollution affecting health. The WHO reported that atmospheric particulate matter can have harmful effects on public health in both developed and developing countries [27]. Second, keywords such as "exposure" and "emission" reflect the pathways through which air pollutants affect health. Much of the research studies the dispersion and emission of pollutants. The typical sources of air pollution such as outdoor motor vehicle emissions, coal combustion emissions and heavy industrial pollutant emissions, and indoor emissions of pollutant gases caused by smoking, cooking, and formaldehyde emissions from interior decoration, all lead to higher levels of human exposure to pollutants. One of the research components of the health effects of air pollution is to determine the exposure–response relationship of pollutants, where either short-term or long-term exposure to pollutants can have adverse health effects on humans [28], which also involves the third research hot topic attributed above—the health consequences caused by air pollution. Third, the adverse health effects of air pollution are mainly reflected in the positive correlation between the concentration levels of air pollutants and the incidence of respiratory diseases as well as other diseases and human mortality. Fourth, atmospheric pollution can cause damage to the health of people at different ages, and the academic community has focused more on choosing children as a group as a research sample [29,30].

Analysis of English Highly Cited Literature

Exposure to ambient air pollution increases morbidity and mortality and is one of the major contributors to the global burden of disease [20], so it is necessary to cite articles that study the topic of disease burden when conducting research on the relationship between air pollution and health. Most of the highly cited research in the English-language literature focuses on studies of the global burden of disease, where the contribution of different risk factors to the burden of disease often changes significantly and where the health effects of the burden of disease are both general and more geographically specific, so that new studies are regularly updated with improved methods, new combinations of risks and risk effects, and new data on the association between risk exposure levels and risk outcomes assessments to report on the details and underlying causes of population health, thereby helping policymakers identify successes in disease control that can be emulated and opportunities for improvement [21–24].

As research progresses, we gain an increasingly detailed understanding of trends in exposure and each risk, and also gain insight into the magnitude of health losses attributable to the risk and how changes in exposure have contributed to health trends.

In terms of research on air pollution as one of the risk factors in the global burden of disease, one study quantified the global burden of disease due to air pollution based on the global range of pollution exposure and estimated the relative risk of death from cardiovascular and respiratory diseases using a combined exposure–response function for each cause of death, which found that air pollution contributed significantly to the global burden of disease in 2015, with $PM_{2.5}$ as the fifth-ranked risk factor for mortality, and exposure resulting in 4.2 million deaths and 103.1 million disability-adjusted life years (DALYs) in 2015, accounting for 7.6% of total global deaths and 4.2% of global DALYs, while reducing exposure has the potential to provide significant health benefits [20]. Lelieveld J et al. (2015) [25] studied the global contribution of outdoor air pollution sources to premature mortality, and the article used a global atmospheric chemistry model to study the association between premature mortality and seven air pollution emission source categories, calculating that outdoor air pollution, mainly $PM_{2.5}$, contributes to 3.3 million premature deaths per year worldwide. In addition, one of the highly cited articles is a study on air pollution in China, which found that secondary aerosols contribute significantly to particulate pollution in Chinese haze weather, and that severe haze pollution events are largely secondary aerosol driven, with contributions of 30–77% and 44–71% to $PM_{2.5}$ and organic aerosols, respectively. Therefore, reducing emissions of secondary aerosol precursors from fossil fuel combustion and biomass burning, in addition to mitigating primary particulate emissions, may be important for controlling $PM_{2.5}$ levels and reducing the environmental, economic, and health impacts of particulate pollution in China [26].

4.2. Research Frontier and Trend Analysis

Research frontiers are theoretical trends and new topics that are emerging in a given time period. CiteSpace software uses a burst detection algorithm to extract burst keywords from articles in order to detect sudden growth in research interest in a subject area and to identify and track the research frontiers and trends in the subject area. The burst keywords strength can reflect the influential research topics over a period of time. In Figures 10 and 11, the red line segment indicates the time period of the keyword burst, which is a visual representation of the begin and end years. In this paper, the analysis of research frontiers is based on the analysis of burst keywords, and, combined with the analysis of the literature where the burst words are located, we make a comprehensive judgment, generalization, and discussion.

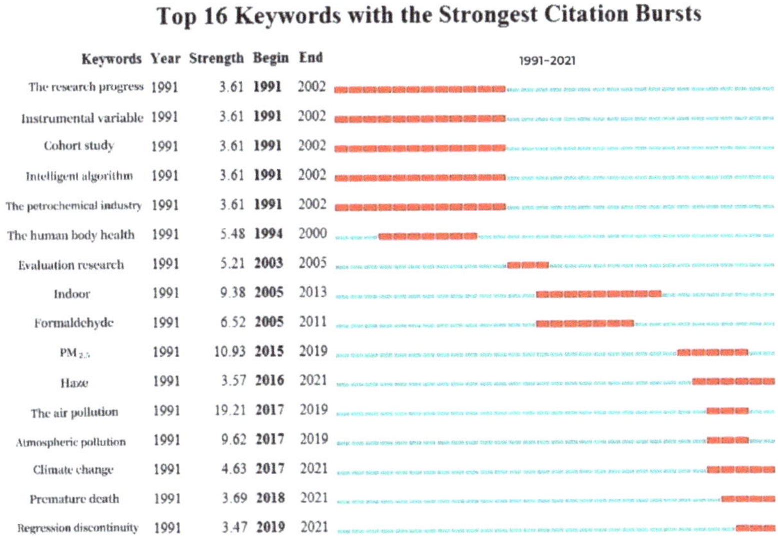

Figure 10. Chinese literature keyword bursts.

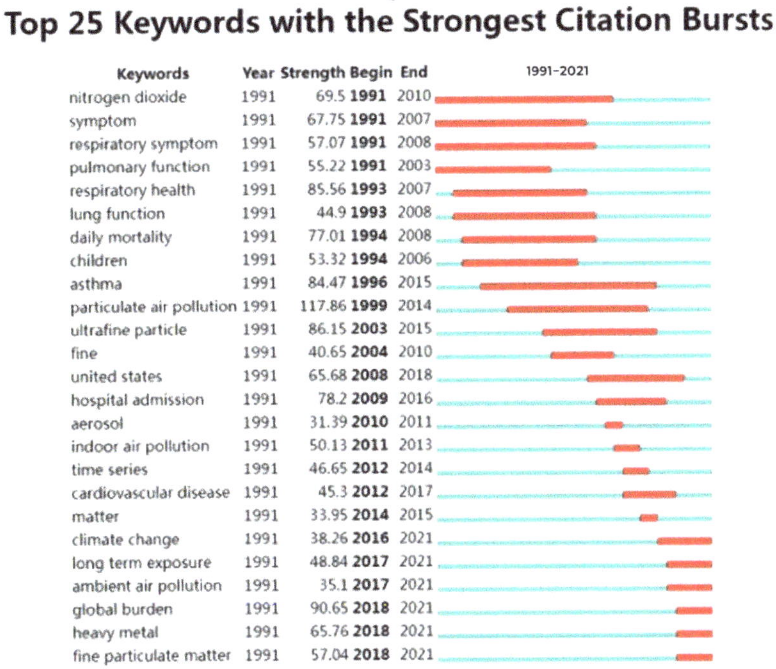

Figure 11. English literature keyword bursts.

4.2.1. Research Trends in Chinese Literature

The burst keywords map of Chinese literature related to this paper is shown in Figure 10, and a total of 16 burst keywords were obtained. For example, it can be seen that the burst keywords during 1991–2002 are "research progress", "instrumental variables", "definite group research", "intelligent algorithm", "petrochemical industry". Most of the keywords with large burst strength are atmospheric pollutants. It is worth noting that phrases such as "climate change", "premature death", and "regression discontinuity" have burst in recent years and continue to be used today, so they can be regarded as the main research trends in this field in China at present. The Chinese research frontier can be broadly summarized into three stages of evolution: early research frontier, middle research frontier, and latest research frontier.

The early research frontier (1991–2005) reflected the research themes of "instrumental variables", "definite group research", "intelligent algorithms", "petrochemical industry", "human health" and "evaluation research", which were mostly focused on the selection of research methods at this time, while it can be seen that most of the air pollutant emissions from crude heavy industries during this period came, and the petrochemical industry was a key concern at that time. For example, PAHs released from petrochemical areas are carcinogenic to humans, and Xia (2014) [31] explored the health risks caused by PAHs through the respiratory route and studied their effects on residents' exposure.

The research themes in the mid-term research frontier (2005–2013) mainly focused on indoor pollution and human health hazards, with the burst keywords "indoor" and "formaldehyde". At that time, due to the lack of supervision, inadequate laws and regulations, and low quality of products, the industry failed to develop with high quality in parallel with economic development, resulting in serious threats to human health caused by indoor air pollution, which became a research hot topic in the field, and poor-quality interior decoration led to formaldehyde pollution, which in turn affected the health of the public's eyes and respiratory system [32].

The latest research frontiers (2013–2021) are "$PM_{2.5}$", "haze", "air pollution", "climate change", "premature death" and "regression discontinuity", which can be summarized into three aspects. Firstly, air pollution was focused on the public view in that period, and the research of major pollutants mostly focused on haze, $PM_{2.5}$ and other particulate matter, which is due to the fact that, at that stage, haze events were frequent in China, and $PM_{2.5}$ pollution was its root cause, and such air pollution was prone to causing human respiratory, cardiovascular, neurological and other diseases, and even lead to cancer [33]. The serious health consequences have caused great concern in society.

Secondly, premature human deaths caused by air pollution are also another key direction of academic concern. Air pollutant emissions lead to premature human deaths, which in turn cause health economic losses. At this stage, the country has implemented several environmental policies to control air pollution to avoid premature death of the population and save health economic losses, thereby increasing health benefits. Therefore, many scholars in this field also evaluate the causal effects of environmental policies on air quality and health [34,35].

Thirdly, studies have found that the causes of climate change are largely the same as those caused by air pollution; climate change is due to the increase in greenhouse gas concentrations, which are mostly caused by the combustion emissions of fossil fuels. This is consistent with the causes of air pollution, so the actions and strategies taken in the two works on air pollution control and greenhouse gas emission control are also consistent [19]. In addition, some studies have estimated the effects of air pollution by exploring climate change, such as wind and thermal inversion as IV [8,36], and wind can disperse pollutants and have a spillover effect on health. Thermal inversion occurs when warm air settles on the cold air and, in turn, atmospheric pollution becomes trapped and cannot be dispersed, greatly increasing air pollution concentrations.

Fourthly, Figure 10 shows that compared with the research methods in the early research frontier period, the research methods of the latest studies mostly use causal inference methods, such as regression discontinuity. Thistlethwaite first proposed regression discontinuity in 1960, which is a quasi-natural experiment, and Angrist used the method to study the effect of class size on achievement [37]. In recent years, the method has become one of the most widely used methods in causal inference and policy evaluation mostly applied in the fields of education economics, labor economics, environmental health economics and regional economics. The use of regression discontinuity for research on the effect of atmospheric pollution on health mostly focused on mental health, population well-being, population migration, and health insurance needs [38–41].

4.2.2. Research Trends in English Literature

The keyword burst graph of the English literature shows 25 burst keywords (Figure 11). Unlike the visualization results of burst keywords in the Chinese literature, the overall results of the English literature do not seem to present obvious phase characteristics, so this section will analyze the contents of the figure from a broader perspective and summarize the characteristics of research frontiers in the English literature. From different evaluation indicators, first, the keywords with high burst strength are "particulate air pollution", "global burden", "respiratory health", "asthma", "hospital admission", "daily mortality", "nitrogen dioxide", "symptom", "heavy metal", and "united states". This represents that these words are influential research topics during their burst cycle. Second, many of the hot words have a longer burst time, and the keywords with a longer burst cycle include: "respiratory symptom", "symptom", "lung function", "daily mortality", "children", "ultrafine particle", "pulmonary function", "united states", "nitrogen dioxide", and "asthma". Third, keywords with burst end dates between 2010 and 2021 represent the more frontier research themes of recent years, mainly including: "nitrogen dioxide", "asthma", "particulate air pollution", "ultrafine particle", "fine", "united states", "hospital admission", "aerosol", "indoor air pollution", "time series", "cardiovascular disease", "matter", "climate change", "long term exposure", "ambient air pollution", "global burden", "heavy metal", and "fine particulate matter". Fourth, the keywords that have suddenly broken out in recent years and have been hot until now represent the latest frontier research themes in the field, which mainly include "climate change", "long term exposure", "ambient air pollution", "global burden", "heavy metal", and "fine particulate matter". According to the above analysis, this paper will discuss the key research frontiers and the development stages of research frontiers combined with relevant literature based on two perspectives: the strength of keyword burst and the evolution of burst keywords over time.

Key Research Frontiers in English Literature

If the keywords have a high burst strength and a long burst period, they can represent the core research hot topics of the field in a certain period of time, and the keywords that meet this characteristic include "asthma", "daily mortality", "symptom" and "united states". Among them, the impact cycle of "asthma" is from 1996 to 2015, the impact cycle of "daily mortality" is from 1994 to 2008. The impact cycle of "asthma" is from 1991 to 2007, and the impact cycle of "united states" is from 2008 to 2018. The keywords "asthma", "daily mortality" and "symptom" represent the health consequences caused by air pollution, and air pollutants can cause health damage to the human respiratory system, triggering respiratory diseases and presenting symptoms of asthma, cough, bronchitis, and even leading to an increase in mortality among the population. It is worth noting that air pollution in the United States has been a key concern for a long time, since the United States is a highly developed industrial country that has experienced severe pollution during its rapid industrial development. Based on this background, the U.S. decided to establish the Clean Air Act, which not only helped the U.S. to improve air quality effectively, but also thus became a model for many countries in the world to learn from in the construction of the Clean Air Act. Many scholars have studied the health hazards of environmental pollution and evaluated the effects of environmental policies based on the U.S. sample.

Research Development Stage in English Literature

From an overall perspective, the research frontiers embodied by the burst keywords can be mainly categorized into three major categories, namely, types of air pollutants and pollution modes, health effects of air pollution, and synergistic management of air pollution. With the passage of time, there are stage characteristics for each topic category, as shown in Table 4.

Table 4. Evolution of research frontiers in English literature.

Topic	Stage	Burst Keyword	Begin-End	Burst Period	Strength
Air pollution	Early stage	Nitrogen dioxide	1991–2010	20	69.5
		Particulate air pollution	1999–2014	16	117.86
		Ultrafine particle	2003–2015	13	86.15
		Fine	2004–2010	7	40.65
	Mid-term stage	Indoor air pollution	2011–2013	3	50.13
		Matter	2014–2015	2	33.95
	Latest stage	Long term exposure	2017–2021	5	48.84
		Heavy metal	2018–2021	4	65.76
		Fine particulate matter	2018–2021	4	57.04
Health effect	Early stage	Symptom	1997–2007	11	67.75
		Respiratory symptom	1991–2008	8	57.07
		Pulmonary function	1991–2003	3	55.22
		Respiratory health	1993–2007	5	85.56
		Lung function	1993–2008	6	44.9
		Daily mortality	1994–2008	5	77.01
		Asthma	1996–2015	20	84.47
	Mid-term stage	Hospital admission	2009–2016	8	78.2
		Cardiovascular disease	2012–2017	6	45.3
	Latest stage	Global burden	2018–2021	4	90.65
Collaborative Governance		Aerosol	2010–2011	2	31.39
		Climate change	2016–2021	6	38.26

(1) Types of atmospheric pollutants and pollution modes. Scholars have mostly researched atmospheric pollutants such as NO_2, atmospheric particulate matter, and heavy metal pollution, and the research hot topics about the types of atmospheric pollutants have gradually evolved and refined. First, the early air pollutants that have been paid attention to include nitrogen dioxide, particulate air pollution, ultrafine particles, and fine particles, and these air pollutants have been paid attention to for a long time. Second, after 2010, indoor air pollution and matter began to be widely studied, but the impact cycle of these two key words is only two years. Third, the latest research frontiers of atmospheric pollutants are long-term exposure, heavy metal, and fine particulate matter. In recent years, haze pollution has received great attention, and fine particulate matter ($PM_{2.5}$) is the primary pollutant causing haze pollution, which has the characteristics of fine particles and large specific surface area, and it easily adsorbs more pollutants, especially for heavy metals with strong adsorption, resulting in secondary aerosol, which makes the heavy metal pollutants in fine particulate matter enter the human body through the respiratory exposure pathway and then cause health hazards [42,43]. Due to the more serious health hazards caused by heavy metal and fine particulate matter, they have gradually become a research hot topic in recent years. Of course, long-term exposure is a way for air pollution to affect health, and compared to short-term pollution exposure, long-term pollution exposure will be considered to cause greater damage to human health [44,45], and therefore is also the focus of regulation.

(2) The impact of atmospheric pollution on human health is mainly reflected in causing respiratory disease, cardiovascular disease, and death, so health indicators are mostly used for respiratory disease, cardiovascular disease, and hospitalization rate and daily mortality rate, and the health hazards caused by atmospheric pollution have created a serious global disease burden. In addition, Figure 11 shows that the burst interval of "children" is 1994–2006, and the early studies mostly focus on the effects of air pollution on children's health, while the later samples are no longer limited to children.

Table 4 shows that, first, in the 1991–2008 period, scholars mostly explored the health consequences of mortality and respiratory disease; second, in the 2009–2017 period, hospital admission and cardiovascular disease began to be used as indicators of human health; third, in the 2018–2021 period, the burst keyword is global burden.

Numerous epidemiological studies have shown that air pollutants can produce long-term [46,47] or short-term [48,49] health damage to the human respiratory system, triggering respiratory diseases and presenting symptoms of asthma, cough, and bronchitis; moreover, exposure to air pollution is significantly associated with increased mortality in the population, and the effects on mortality from cardiovascular and respiratory diseases are even more significant [50–52]. In addition, as mentioned in the previous analysis of research hot topics, the health loss caused by air pollution contributes significantly to the global burden of disease.

(3) Synergistic governance. As with the Chinese literature, the English literature also focuses on the relationship between air pollution and climate change, and aerosols are one of the topics of interest in this field since atmospheric aerosols are the main cause of air pollution, greenhouse gas emissions cause climate change, and the formation of aerosols is consistent with the increase in greenhouse gas emissions (both caused by fossil fuel combustion emissions). Therefore, atmospheric pollution and climate change need to be managed synergistically.

In addition, scholars usually use changes in meteorological conditions to address the endogeneity of air quality, such as wind and thermal inversion [8,36]. Wind can disperse pollutants and thus may produce spillover effects of air pollution, and the increase in pollution levels can have significant negative health effects downwind in the short term; thermal inversion occurs when warm air settles over cold air, and when it occurs, atmospheric pollutants are trapped and cannot be dispersed, thus greatly increasing ground-level air pollution concentrations. Therefore, thermal inversion is another commonly used air pollution research instrumental variable for air pollution studies.

5. Conclusions

In this paper, we use Chinese core journals in the CNKI and Web of Science core collection as data sources and use the CiteSpace software to visualize and analyze the basic situation, research hot topics, and research frontiers of the relationship between air pollution and health, and objectively explain the basic development trend of research results in this field in recent years, and we believe that the research results will become richer and more in-depth in the future.

(1) Through the overall study of the health effects of air pollution, it is found that: firstly, the number of relevant Chinese and English studies has gradually increased with time, indicating that this field has attracted widespread attention from the academic community. Secondly, through the analysis of the author co-occurrence knowledge map, it can be seen that Prof. Haidong Kan is highly accomplished in this field, and the number of articles published in both Chinese and English is far ahead. In addition, several sub-collaborative networks of authors have been formed throughout the network map, and English-language publishers have increasingly focused on collaborative exchanges in recent years. Thirdly, the research institution with the highest number of Chinese publications is Peking University, which has published more articles in both English and Chinese in this field, and the institution with the highest number of English publications is University of California System. In addition, the United States not only has the largest number of articles but is also the country with the most extensive foreign exchange and cooperation.

(2) Keyword co-occurrence analysis, clustering analysis, and analysis of highly cited literature were used to explore the research hot topics in this field. First, the results of keyword co-occurrence and clustering show that the research hot topics in the Chinese and English literature are generally similar. They mainly focus on four aspects: the types of air pollutants that affect health, the pathways through which air pollutants affect health, the health and economic consequences caused by air pollution, and the groups affected by air pollution. Second, the analysis of the highly cited literature

showed that scientists often explored the themes of quantifying the global burden of disease due to air pollution, health risks and health economic losses from air pollution, and synergistic management of air pollution and climate changes.

(3) The frontiers and trends of the Chinese and English literature are summarized through the analysis of burst keywords. The Chinese literature has obvious stage research characteristics, and this paper divides its evolution process into three stages: the early research frontier (1991–2005) stage has more burst keywords focusing on the choice of research methods, and other burst keywords show that the industries causing air pollution are also mainly focused on the petrochemical industry. The medium-term research frontier (2005–2013) focuses on the impact of indoor pollution on human health. The research themes of the latest research frontier (2013–2021) are mainly focused on air pollution caused by haze, the relationship between climate change and air pollution, and premature death caused by air pollution. The research frontiers in the English literature can be summarized into three main categories: types of atmospheric pollutants and pollution patterns, health economic effects of atmospheric pollution, and synergistic governance of atmospheric pollution and climate change. Here, each category has a significant stage evolution.

6. Discussion

This article provides a broad overview of existing research results studying the relationship between air pollution and health, which provide strong evidence that air pollution around the world has significant health effects, deepen our understanding of environmental issues, and serve as a cautionary tale for newly industrializing countries in the prevention of air pollution. Some of the findings have already begun to have policy implications, and we anticipate that research in this area is likely to make an increasing contribution to environmental policymaking in the relevant countries covered in the article.

Combining the above analysis in this paper and the related literature, we believe that several possible important directions for future research on the relationship between air pollution and health are still worth further deepening and expanding. First, studying the health consequences of sustained pollution exposure is one of the most important research topics for the future. How to collect credible evidence of the long-term effects of air pollution, how to break through the difficulty of identifying exogenous changes in long-term pollution exposure, and breakthroughs in research methods may have to be continued in depth in future studies.

Second, the negative impacts of air pollution are not only an issue in terms of direct physical health damage, but air pollution affects long-term human capital acquisition, human cognition, and productivity by affecting physical health in turn. Given this, the economic loss of health based on direct health consequence measures is greatly underestimated, and therefore the negative health impacts of air pollution should be further explored fully in future studies.

Third, through this paper, we can clearly find that $PM_{2.5}$, PM_{10}, and NO_2 air pollutants still seriously affect public health, and we have to continue to make efforts for air pollution control. For example, we can continue to research the sources and control of air pollutants, air pollution policy formulation and policy evaluation, and air pollution synergistic governance. Among them, the synergistic management of air pollution and climate change is one of the latest research frontiers in this field. The common emission sources imply that there are many links between air pollution and climate change, and the implementation of climate policies can also reduce air pollution and thus benefit health, so the question concerning whether there is an optimal combination of air pollution and climate change policies is also worth further exploration by scientists.

Author Contributions: Y.G.: conceptualization, methodology, software, validation, formal analysis, investigation, resources, data curation, writing—original draft preparation, writing—review and editing, visualization, supervision, project administration. S.Y.: conceptualization, methodology, software, validation, resources, writing—original draft, writing—review and editing, supervision, project administration, funding acquisition. Y.X.: writing—review and editing, supervision, project administration, software. N.W.: validation, supervision, writing—review and editing, methodology, software. All authors have read and agreed to the published version of the manuscript.

Funding: This research received no external funding.

Institutional Review Board Statement: Not applicable.

Informed Consent Statement: Not applicable.

Data Availability Statement: This paper takes the Chinese core journals in CNKI and the core collection of Science Citation retrieval Web of Science as the data source to make general measurement and statistics on the research on the relationship between air pollution and public health.

Conflicts of Interest: The authors declare no conflict of interest.

References

1. Bell, M.L.; Davis, D.L. Reassessment of the lethal London fog of 1952: Novel indicators of acute and chronic consequences of acute exposure to air pollution. *Environ. Health Perspect.* **2001**, *109* (Suppl. S3), 389–394. [PubMed]
2. Brunekreef, B.; Holgate, S.T. Air pollution and health. *Lancet* **2002**, *360*, 1233–1242. [CrossRef]
3. Pope, C.A., III; Arden, C. Lung cancer, cardiopulmonary mortality, and long-term exposure to fine particulate air pollution. *JAMA.* **2002**, *287*, 1132–1141. [CrossRef] [PubMed]
4. Nyberg, F.; Gustavsson, P.; Järup, L.; Bellander, T.; Berglind, N.; Jakobsson, R.; Pershagen, G. Urban air pollution and lung cancer in stockholm. *Epidemiology* **2000**, *11*, 487–495. [CrossRef] [PubMed]
5. Hoek, G.; Brunekreef, B.; Goldbohm, S.; Fischer, P.; van den Brandt, P.A. Association between mortality and indicators of traffic-related air pollution in the Netherlands: A cohort study. *Lancet.* **2002**, *360*, 1203–1209. [CrossRef]
6. Currie, J.; Neidell, M. Air Pollution and Infant Health: What Can We Learn from California's Recent Experience? *Q. J. Econ.* **2005**, *120*, 1003–1030.
7. Luechinger, S. Air Pollution and Infant Mortality: A Natural Experiment from Power Plant Desulfurization. *J. Health Econ.* **2014**, *37*, 219–231. [CrossRef]
8. Arceo-Gomez, E.O.; Hanna, R.; Oliva, P. Does the effect of pollution on infant mortality differ between developing and developed countries? Evidence from Mexico City. *Econ. J.* **2016**, *126*, 257–280. [CrossRef]
9. Greenstone, M.; Jack, B.K. Envirodevonomics: A research agenda for an emerging field. *J. Econ. Lit.* **2015**, *53*, 5–42. [CrossRef]
10. Kan, H.; Chen, B. Relationship between atmospheric particulate matter exposure and population health effects in China. *J. Environ. Health* **2002**, *19*, 422–424.
11. Chen, R.; Chen, B.; Kan, H. Health economics evaluation of atmospheric particulate matter pollution in 113 Cities in China. *Chin. Environ. Sci.* **2010**, *30*, 410–415.
12. Xie, Y.; Chen, J.; Li, W. An assessment of PM2.5 related health risks and impaired values of Beijing residents in a consecutive high-level exposure during heavy haze days. *Environ. Sci.* **2014**, *35*, 1–8.
13. Dai, H.; Song, W.; Gao, X.; Chen, L.; Hu, M. Study on relationship between ambient PM10, PM2. 5 pollution and daily mortality in a district in Shanghai. *Health Res.* **2004**, *33*, 293–297.
14. Bai, Z.; Cai, B.; Dong, H.; Bian, H. Health effects of dust-haze. *Environ. Pollut. Prev.* **2006**, *28*, 198–201.
15. Wang, P.; Dai, C.; Zhang, C. Research status and health effects of particulate matter in urban atmosphere. *China Environ. Monit.* **2005**, *21*, 83–87.
16. Wang, Y.; Zhang, J.; Wang, L.; Hu, B.; Tang, G.; Liu, Z.; Sun, Y.; Ji, D. Research significance, status and prospect of air haze pollution in Beijing-Tianjin-Hebei region. *Prog. Earth Sci.* **2014**, *29*, 388–396.
17. Zheng, Y.; Yan, L.; Qian, Y. Economic loss estimation of environmental pollution in China in the mid-1990s. *Manag. World* **1999**, *2*, 189–197.
18. Xia, G.; Zhao, Y. Economic measurement and research on environmental pollution losses in China. *Manag. World* **1995**, *6*, 198–205.
19. Ding, Y.; Li, Q.; Liu, Y.; Zhang, L.; Song, Y.; Zhang, J. Air pollution and climate change. *Meteorology* **2009**, *35*, 13.
20. Cohen, A.J.; Brauer, M.; Burnett, R.; Anderson, H.R.; Frostad, J.; Estep, K.; Balakrishnan, K.; Brunekreef, B.; Dandona, L.; Dandona, R.; et al. Estimates and 25-year trends of the global burden of disease attributable to ambient air pollution: An analysis of data from the Global Burden of Diseases Study 2015. *Lancet* **2017**, *389*, 1907–1918. [CrossRef]
21. Lim, S.S.; Vos, T.; Flaxman, A.D.; Danaei, G.; Shibuya, K.; Adair-Rohani, H.; AlMazroa, M.A.; Amann, M.; Anderson, H.R.; Andrews, K.G.; et al. A comparative risk assessment of burden of disease and injury attributable to 67 risk factors and risk factor clusters in 21 regions, 1990–2010: A systematic analysis for the Global Burden of Disease Study 2010. *Lancet* **2012**, *380*, 2224–2260. [CrossRef]

22. Forouzanfar, M.H.; Alexander, L.; Anderson, H.R.; Bachman, V.F.; Biryukov, S.; Brauer, M.; Burnett, R.; Casey, D.; Coates, M.M.; Cohen, A.; et al. Global, regional, and national comparative risk assessment of 79 behavioural, environmental and occupational, and metabolic risks or clusters of risks in 188 countries, 1990–2013: A systematic analysis for the Global Burden of Disease Study 2013. *Lancet* **2015**, *386*, 2287–2323. [CrossRef]
23. Murray, C.J.L. Global, Regional, and National Comparative Risk Assessment of 84 Behavioural, Environmental and Occupational, and Metabolic Risks or Clusters of Risks for 195 Countries and Territories, 1990–2017: A Systematic Analysis for the Global Burden of Disease Study 2017. *Lancet* **2018**, *392*, 1923–1994. [CrossRef]
24. Vos, T.; Lim, S.S.; Abbafati, C.; Abbas, K.M.; Abbasi, M.; Abbasifard, M.; Abbasi-Kangevari, M.; Abbastabar, H.; Abd-Allah, F.; Abdelalim, A.; et al. Global burden of 369 diseases and injuries in 204 countries and territories, 1990–2019: A systematic analysis for the Global Burden of Disease Study 2019. *Lancet* **2020**, *396*, 1204–1222. [CrossRef]
25. Lelieveld, J.; Evans, J.S.; Fnais, M.; Giannadaki, D.; Pozzer, A. The contribution of outdoor air pollution sources to premature mortality on a global scale. *Nature* **2015**, *525*, 367–371. [CrossRef]
26. Huang, R.-J.; Zhang, Y.; Bozzetti, C.; Ho, K.-F.; Cao, J.-J.; Han, Y.; Daellenbach, K.R.; Slowik, J.G.; Platt, S.M.; Canonaco, F.; et al. High secondary aerosol contribution to particulate pollution during haze events in China. *Nature* **2014**, *514*, 218–222. [CrossRef]
27. World Health Organization (WHO). *Air Quality Guidelines for Particulate Matter, Ozone, Nitrogen Dioxide and Sulfur Dioxide-Global Update 2005-Summary of Risk Assessment*; WHO: Geneva, Switzerland, 2005.
28. You, Y.; Bai, Z. Advances in research on exposure and health effects of atmospheric particulate matter. *J. Ecotoxicol.* **2012**, *7*, 123–132.
29. Chay, K.Y.; Greenstone, M. The Impact of Air Pollution on Infant Mortality: Evidence from Geographic Variation in Pollution Shocks Induced by A Recession. *Q. J. Econ.* **2003**, *118*, 1121–1167. [CrossRef]
30. Janke, K. Air Pollution, Avoidance Behavior and Children's Respiratory Health: Evidence from England. *J. Health Econ.* **2014**, *38*, 23–42. [CrossRef]
31. Xia, L. Health Risk Assessment of Polycyclic Aromatic Hydrocarbons in the Ambient Air of Petrochemical Zone and Its Impact on the Internal Exposure of Surrounding Residents. Master's Thesis, Jinan University, Guangzhou, China, 2014.
32. Hu, H.; Zhang, X.; Huang, W.; Ma, J.; Luo, H.; Huang, L.; Zhu, Y. Air pollution and health effects of post-decoration living rooms. *J. Environ. Health* **2004**, *21*, 47–49.
33. Chen, Y.; Ebenstein, A.; Greenstone, M.; Li, H. Evidence on the Impact of Sustained Exposure to Air Pollution on Life Expectancy from China's Huai River Policy. *Proc. Natl. Acad. Sci. USA* **2013**, *110*, 12936–12941. [CrossRef] [PubMed]
34. Ma, G.; Yu, F.; Zhang, Y.; Peng, F. Effects of the implementation of the Action Plan for the Prevention and Control of Air Pollution. Evaluation of the implementation effect of the Air Pollution Prevention and Control Action Plan and its impact on life expectancy per capita in China. *Environ. Sci. Res.* **2019**, *32*, 1966–1972.
35. Wu, W.; Xue, W.; Wang, Y.; Lei, Y.; Feng, T.; Cai, Z. Assessment of the environmental health effects of the implementation of the Air Pollution Control Action Plan. *Environ. Sci.* **2019**, *40*, 2961–2966.
36. Chen, S.; Chen, T. Air quality and public health: An example of sulfur dioxide emissions from thermal power plants. *Econ. Res.* **2014**, *49*, 13.
37. Angrist, J.D. *Mostly Harmless Econometrics: An Empiricist Companion*; Princeton University Press: Princeton, NJ, USA, 2010.
38. Li, W.; Zou, P. Air pollution and residents' mental health–Estimation based on breakpoint regression. *J. Beijing Univ. Technol.* **2019**, *21*, 10–21.
39. Chu, D.; He, P.; Liang, R.B. Subjective air pollution and residents' happiness—A microdata validation based on a breakpoint regression design. *Dyn. Econ.* **2017**, *2*, 88–101.
40. Zhao, Q.; Sun, J. The impact of air pollution on the demand for commercial health insurance—An empirical study based on the Qinling and Huaihe regions. *Lanzhou Acad. J.* **2017**, *2*, 88–101.
41. Zhao, Q.; Sun, J. A study on the impact of air pollution on the health expenditure of Chinese residents. *Econ. Surv.* **2021**, *38*, 151–160.
42. Bari, M.A.; Kindzierski, W.B. Characterization of air quality and sources of fine particulate matter ($PM_{2.5}$) in the City of Calgary, Canada. *Atmos. Pollut. Res.* **2018**, *9*, 534–543. [CrossRef]
43. Hu, X.; Zhang, Y.; Ding, Z.; Wang, T.; Lian, H.; Sun, Y.; Wu, J. Bio accessibility and health risk of arsenic and heavy metals (Cd, Co, Cr, Cu, Ni, Pb, Zn and Mn) in TSP and $PM_{2.5}$ in Nanjing, China. *Atmos. Environ.* **2012**, *57*, 146–152. [CrossRef]
44. Beelen, R.; Hoek, G.; Brandt, P.A.V.D.; Goldbohm, R.A.; Fischer, P.; Schouten, L.J.; Armstrong, B.; Brunekreef, B. Long-term exposure to traffic-related air pollution and lung cancer risk. *Epidemiology* **2008**, *19*, 702–710. [CrossRef] [PubMed]
45. Ebenstein, A.; Fan, M.; Greenstone, M.; He, G.; Zhou, M. New evidence on the impact of sustained exposure to air pollution on life expectancy from China's Huai River Policy. *Proc. Natl. Acad. Sci. USA* **2017**, *114*, 10384–10389. [CrossRef] [PubMed]
46. Künzli, N.; Bridevaux, P.O.; Liu, L.S.; Garcia-Esteban, R.; Schindler, C.; Gerbase, M.W.; Sunyer, J.; Keidel, D.; Rochat, T. Traffic-related air pollution correlates with adult-onset asthma among never-smokers. *Thorax* **2009**, *64*, 664–670. [CrossRef] [PubMed]
47. Zanobetti, A.; Bind, M.; Schwartz, J. Particulate air pollution and survival in a COPD cohort. *Environ. Health* **2008**, *7*, 48–56. [CrossRef] [PubMed]
48. Ko, F.W.; Tam, W.; Wong, T.W.; Chan, D.P.; Tung, A.H.; Lai, C.K.; Hui, D.S. Temporal relationship between air pollutants and hospital admissions for chronic obstructive pulmonary disease in Hong Kong. *Thorax* **2007**, *62*, 780–785. [CrossRef] [PubMed]

49. Segala, C.; Poizeau, D.; Neukirch, F.; Aubier, M.; Samson, J.; Gehanno, P. Air pollution, passive smoking, and respiratory symptoms in adults. *Arch. Environ. Health Int. J.* **2004**, *59*, 669–676. [CrossRef] [PubMed]
50. Atkinson, R.W.; Ross Anderson, H.; Sunyer, J.; Ayres, J.O.; Baccini, M.; Vonk, J.M.; Boumghar, A.; Forastiere, F.; Forsberg, B.; Touloumi, G.; et al. Acute effects of particulate air pollution on respiratory admissions—Results from APHEA 2 project. *Am. J. Respir. Crit. Care Med.* **2001**, *164*, 1860–1866. [CrossRef]
51. Burnett, R.T.; Smith-Doiron, M.; Stieb, D.; Cakmak, S.; Brook, J.R. Effects of particulate and gaseous air pollution on cardiorespiratory hospitalizations. *Arch. Environ. Health Int. J.* **1999**, *54*, 130–139. [CrossRef]
52. Brook, R.D.; Rajagopalan, S.; Pope, C.A., III; Brook, J.R.; Bhatnagar, A.; Diez-Roux, A.V.; Holguin, F.; Hong, Y.; Luepker, R.V.; Mittleman, M.A.; et al. Particulate matter airpollution and cardiovascular disease: An update to the scientific statement from the American heart association. *Circulation* **2010**, *121*, 2331–2378. [CrossRef]

MDPI
St. Alban-Anlage 66
4052 Basel
Switzerland
www.mdpi.com

Atmosphere Editorial Office
E-mail: atmosphere@mdpi.com
www.mdpi.com/journal/atmosphere

www.ingramcontent.com/pod-product-compliance
Lightning Source LLC
Chambersburg PA
CBHW061545220926
51850CB00005B/253
9783036591384